Mathematics

for Caribbean Schools

Students' Book 2

Althea A Foster Terry Tomlinson Third Edition

Orders: please contact Hachette UK Distribution, Hely Hutchinson Centre, Milton Road, Didcot, Oxfordshire, OX11 7HH.
Telephone: +44 (0)1235 827827. Email education@hachette.co.uk Lines are open from 9 a.m. to 5 p.m., Monday to Friday.
You can also order through our website: www.hoddereducation.com

First published in 1988 by Longman Group Ltd
Second edition published 2000
This edition published 2007
Published from 2015 by Hodder Education,
An Hachette UK Company
Carmelite House
50 Victoria Embankment
London EC4Y 0DZ
www.hoddereducation.com

2023
Imp 7

ISBN: 978–1–4058–4778–0

Set in 9.5/12 pt Stone Serif

Printed and bound by CPI Group (UK) Ltd, Croydon, CR0 4YY

Acknowledgements
The Publishers wish to acknowledge the work of J B Channon, A McLeish Smith, H C Head and M F Macrae which laid the foundation for this series.

Preface

This series of four volumes, of which this is the second, is intended for use primarily by students who are preparing to sit the certificate examinations held by the Caribbean Examinations Council and by the individual countries in the Caribbean.

Each volume represents material which may be covered by the average student in one year approximately (although some students may need more time) so that there is ample time for the series to be completed over a four- to five-year period. Emphasis has been placed on detailed explanation of concepts, principles and methods of working out problems. In addition, many problems have been included, both as worked examples to illustrate particular approaches to solving problems in the teaching text, and also as exercises for practice and reinforcement of the concepts. This has been done in a deliberate attempt to provide a stimulus to teachers in developing their strategies for teaching different topics; but, especially, to provide guidance to students as they work or revise on their own.

At the end of each chapter a summary of the main points developed in the chapter has also been included. Key reference words are printed in bold type throughout the text. The text has been arranged sequentially so that each chapter may use material covered in previous chapters. However, in order to use relevant examples at some points in the text, ideas/concepts, which may need a short explanation or reminder by the teacher, may be introduced. A few chapters are independent of the previous chapters and so may be omitted without loss of continuity at a first working. However, it is believed that the text will be used most efficiently by working through the chapters in the given order.

We are indebted to both the teachers and the students whose questions and responses over the years have undoubtedly influenced our thinking and our approach to the teaching of the subject as exemplified in this series.

This revised edition seeks to incorporate changes in keeping with amendments to the syllabuses in the various Caribbean countries and that of the Caribbean Examinations Council. Attention has also been given to the suggestions of Caribbean teachers whose positive reaction and responses to the series have been most encouraging.

We also wish to acknowledge the work of J B Channon, A McLeish Smith, H C Head and M F Macrae which laid the foundation for this series.

To the teacher

This edition of the series of four texts, revised with respect to content and its sequencing and to pedagogy (to a lesser extent), will be found useful in providing help and guidance in how the topics are taught and the order in which they are taken. The texts do not attempt to prescribe specific approaches to the teaching of any topic. Teachers are free to adapt or modify the suggested approaches. It is the teacher who must decide on the methodology to be used to create the most suitable learning conditions in the classroom and to provide challenging activities which motivate the students to think and yet give them a chance to succeed in finding solutions.

It is vitally important that teachers use the *'Oral'* exercises to initiate class discussion in the careful development of concepts. Whenever the opportunity arises, teachers are urged to use, and thus reinforce, concepts taught earlier, so that, for example, the use of estimation and approximation in the calculation of numerical values is practised throughout the course.

In addition, it is widely accepted that learning is aided by doing. Thus concrete/practical examples and real-life applications must be provided, whenever possible, as well as the use of pictures, flow charts and other diagrammatic representations to deepen the understanding of abstract/theoretical ideas. Some problems in the exercises require the use of diagrams which are tedious and/or time-consuming to produce. In order to keep down costs to the student/school and reduce the tedium and time wastage, it is suggested that teachers should use a copying machine for producing the necessary material.

Group work

In some instances, students may improve their performance by working in a group where the insecurity and stigma of not knowing *'where to start'* in solving a problem is not as manifest. Throughout the texts, therefore, in addition to the *'Oral'* exercises, we have also identified relevant exercises in which the open-ended questions are applicable to *'Group Work'*, namely, in Chapters 1, 4, 6, 11, 14, 17, 18, 19 and 23. After alternative strategies for tackling a problem have been discussed by the group, the actual solution(s) may be carried out in the initial group, in smaller groups, or by students working individually. It is important that adequate *time and thought* be given to the process of trying different approaches, and to considering the reasonableness of the 'answer'.

'Good' social behaviour, such as listening to and respecting another person's suggestions and working co-operatively in teams, is also a worth-while long-term benefit of such group activities.

In an attempt to assist the teacher and the student in quickly identifying and reviewing necessary background knowledge, we have included information to be referenced under the heading *'Pre-requisites'* at the beginning of each chapter. It must be remembered that the main new ideas of each chapter are highlighted in the 'Summary' at the end of the chapter. In this edition additional questions have been included in a series of Practice Exercises at the end of each chapter. The point at which the *'Revision Exercises and Tests'* are used is a matter for the individual teacher's choice. If the series of related chapters has been taught as sequenced in the printed text, then the revision material may be used at that time, or may be omitted until the entire book has been completed. This material may also be used as further supplemental problems for the quicker students. In order to comply with the requests of teachers, a *'Practice Examination (Papers 1 and 2)'* has also been included at the end of the book.

Use of the electronic calculator

The widespread use of the electronic calculator in today's world demands that suitable attention to its potential and usefulness be given in the early years of secondary schooling. However, it is also very important to ensure that the calculator, by its premature introduction, is not misused and seen only as a 'number cruncher'.

In this series the use of the calculator is formally introduced in this volume.

Finally, it is unfortunate that Mathematics is perceived by a large majority of students as a 'necessary evil', a subject that they have to 'get at CXC' in order to become employable. Teachers have a significant responsibility in helping to change this attitude, and in having students appreciate that, by acquiring the skills and techniques to solve problems in mathematics, they also acquire the tools and the ability to solve problems in the real world.

To the student

Before attempting the problems in the exercises, study and discuss the worked examples until you understand the concept. The oral exercises are intended to encourage discussion. This will help to clarify lingering misunderstandings. In particular, in solving word problems you first have to get thoroughly familiar with the problem. The next step involves translating the problem into mathematical symbols and language, for example, into an equation or an inequality, or into a graph. The next steps are applying the required mathematical operations, and finally, checking the original word problem. Remember to check that the variables are in the same units. Another useful hint is to look for patterns in similar problems and in the methods of solution. Concrete materials such as cans, coins, balls, stones and boxes are very useful aids for clearing up doubts – not for wasting time!

A calculator is an excellent machine when used wisely, but you must bear in mind that it needs to be used by a clear-thinking human who fully comprehends the mathematical concepts. Dividing 4.0 by 3 and giving your answer as 1.333 333 3 indicates, among other things, a lack of appreciation of the idea of accuracy.

Nothing can replace the neat appearance of an answer that is well set out – the date, the page and exercise from which the problem has been taken, a statement of the facts given, the necessary calculations performed and the conclusions drawn which result in a final answer. This whole process helps you to think clearly.

Finally, the more a concept is applied, the clearer it becomes. Thus, PRACTICE and more practice in working out examples is an essential ingredient for success.

Althea A. Foster
E. M. Tomlinson

Contents

Number systems (3)

Sets of numbers

Pre-requisites

- sets of whole numbers

Sets of numbers

Numbers are used for different purposes. We use them for counting and for measuring. The numbers used for counting are the set of natural numbers, N.

N = {1, 2, 3, 4, ...}

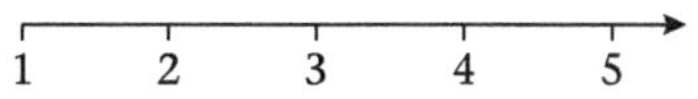

Fig. 1.1

The set of whole numbers W, includes the numeral 0.

W = {0, 1, 2, 3, 4, ...}

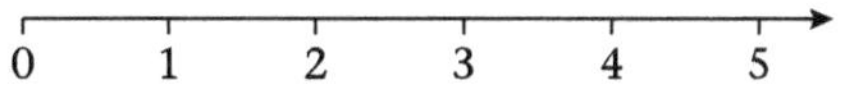

Fig. 1.2

The set of natural numbers is a subset of the set of whole numbers. $N \subset W$

These are all positive numbers. Negative numbers were used in Book 1 to measure temperatures below 0 °C. The set of positive and negative numbers together is called the set of integers, Z.

Z = {... −2, −1, 0, 1, 2, 3, ...}
$N \subset W \subset Z$

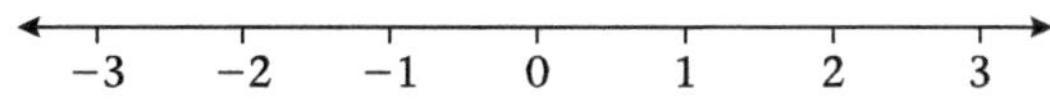

Fig. 1.3

When measuring length, time and other quantities, there is need for smaller quantities. Units are divided into smaller quantities, fractions. The numbers are extended to include quantities such as $3\frac{1}{2}$, $4\frac{3}{4}$, $2\frac{1}{2}$ km, 5.2 seconds, 3.75 kg. All these numbers together are called the set of **rational numbers**, Q.
Q = {... $-2\frac{1}{2}$, −2, $-1\frac{1}{2}$, −1, 0, 1, $1\frac{1}{2}$, ...}

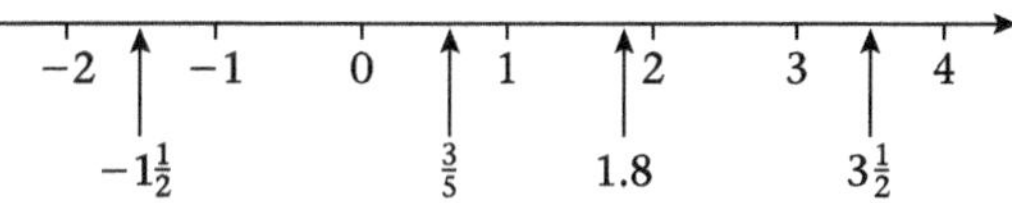

Fig. 1.4

The set of integers Z is a subset of the set of rational numbers. $N \subset W \subset Z \subset Q$. All integers can be written as rational numbers.

For example, $7 = \frac{7}{1}$, $84 = \frac{84}{1}$

Rational numbers are described as numbers written in the form p/q where p and q are integers ($q \neq 0$).

There are some numbers that cannot be expressed as a rational number. For these numbers an approximate value is used. π is one of these numbers and the approximate value used is sometimes 3.14, or 3.142, depending on the degree of accuracy.

Square roots of numbers that are not perfect squares, cannot be written in the form p/q. Examples are $\sqrt{2}$, $\sqrt{5}$, $\sqrt{7}$. These numbers are called **irrational numbers**.

All the sets of numbers are called real numbers, R, and all these numbers can be shown on the number line. The sets of numbers are all related to each other.

$N \subset W \subset Z \subset Q \subset R$

Exercise 1a

1. List any three members of each of the following sets of numbers.
   (a) Z = {integers}
   (b) W = {whole numbers}
   (c) Q = {rational numbers}
   (d) N = {natural/counting numbers}

2 Give
   (a) an element of W that is not an element of N
   (b) an element of Z that is not an element of N.

3 Replace the * in each of the statements with $\in$ or $\notin$.
   (a) $-2 * Z$
   (b) $-2 * W$
   (c) $-2 * N$
   (d) $-2 * Q$
   (e) $-2 * R$

4 Mark on a number line, the position of any member of the following subsets of numbers: W, Q, Z, N.

Use of the calculator

The calculator, like any other machine, can be excellent when used wisely. The calculator helps in doing complicated calculations and it does some calculations faster.

Sometimes we need to choose the activity for which the calculator is used in the same way we would choose between

- the sewing needle and the sewing machine
- the machette and the lawn mower
- the dishwasher and washing up by hand

Exercise 1b (Oral)

For which of the following calculations would you choose to use the calculator?

1 $2 + 4 + 8$

2 $3 \times 4 \times 5$

3 $31 \times 42 \times 52$

4 $25 - 15$

5 $2520 - 197$

6 $84 \div 7$

7 $385 \div 64 \div 12$

8 $\sqrt{36}$

9 $\sqrt{16\,641}$

10 $\sqrt{324} \times 17 \div 22$

Exercise 1c (Oral)

Using the calculator:

1 Find the value of $59 \div 252$. Give the answer to 2 decimal places.

2 Find the value of $9 \div 11$. What do you notice? To which subset of numbers does this number belong?

3 Find at least one other number that behaves like $9 \div 11$.

4 Find the square root of 61. How does this number differ from the result of $9 \div 11$? To which subset of numbers does this belong?

5 Find at least one other number that behaves like $\sqrt{61}$.

Laws in arithmetic

Commutative law

What do you notice about these? (You may use the calculator.)

(a) $56 + 45$ and $45 + 56$
(b) 62×7 and 7×62
(c) $22 \div 8$ and $8 \div 22$

The order in which numbers are written for addition and multiplication is not important. This is not the same for subtraction and division.

Addition and multiplication are **commutative**. Subtraction and division are not commutative.

Associative law

What can you say about

(a) $(42 + 25) + 63$ and $42 + (25 + 63)$
(b) $(19 \times 3) \times 4$ and $19 \times (3 \times 4)$?

Although the brackets tell the order in which the operations must be done, the addition and multiplication can be done in any order. Addition and multiplication are **associative**.

Would division and subtraction be associative? Check the following:

$9 \div (6 \div 3)$ and $(9 \div 6) \div 3$
$15 - (12 - 6)$ and $(15 - 12) - 6$

Division and subtraction are not associative.

Distributive law

63×4 may be written as

$$(60 + 3) \times 4 = 4 \times (60 + 3)$$

which is the same as

$$\begin{aligned} & 4\,(60 + 3) \\ &= 4 \times 60 + 4 \times 3 \\ &= 240 + 12 \\ &= 252 \end{aligned}$$

Similarly,

$$\begin{aligned} 89 \times 5 &= (90 - 1) \times 5 \\ &= 90 \times 5 - 1 \times 5 \\ &= 450 - 5 \\ &= 445 \end{aligned}$$

Multiplication is said to be **distributive** over addition and subtraction.

The three laws may be stated as:

$$\left.\begin{aligned} a + b &= b + a \\ a \times b &= b \times a \end{aligned}\right\} \text{Commutative law}$$

$$\left.\begin{aligned} (a + b) + c &= a + (b + c) \\ (a \times b) \times c &= a \times (b \times c) \end{aligned}\right\} \text{Associative law}$$

$$\left.\begin{aligned} a\,(b + c) &= ab + ac \\ a\,(b - c) &= ab - ac \end{aligned}\right\} \text{Distributive law}$$

Exercise 1d

State which law is illustrated by each of the following.

1. $23 + 68 = 68 + 23$
2. $(105 + 82) + 64 = 105 + (82 + 64)$
3. $(p \times q) \times r = p \times (q \times r)$
4. $p\,(q + r) = p \times q + p \times r$
5. $a * (b - c) = a * b - a * c$
6. Using any one example, show that subtraction is not commutative.
7. (a) Expand $p(q + r)$.
   (b) Using numerical values for p, q, r show that multiplication is distributive over addition.
8. Using numerical examples, show that the order of operation is not important for multiplication.

Number patterns

The multiples of 3 can be given in a row, or sequence:

3, 6, 9, 12, 15, 18, 21, ...

They can also be shown by shading on a 1–100 number square as in Fig 1.5.

1	2	3	4	5	6	7	8	9	10
11	12	13	14	15	16	17	18	19	20
21	22	23	24	25	26	27	28	29	30
31	32	33	34	35	36	37	38	39	40
41	42	43	44	45	46	47	48	49	50
51	52	53	54	55	56	57	58	59	60
61	62	63	64	65	66	67	68	69	70
71	72	73	74	75	76	77	78	79	80
81	82	83	84	85	86	87	88	89	90
91	92	93	94	95	96	97	98	99	100

Fig. 1.5

These are both examples of **number patterns**.

Extending number patterns

Example 1

Find the next four terms in the sequence 1, 2, 4, 7, 11, 16, ...

method: Find the differences between one number and the next.

sequence: 1 , 2 , 4 , 7 , 11 , 16

differences: 1 2 3 4 5

Notice the pattern in the differences. The differences increase by 1 each time. The next term in the sequence is found by adding 6 to 16. This gives 22. The next term is found by adding 7 to 22, and so on. The next four terms are: ..., 22, 29, 37, 46.

Exercise 1e

1. Complete the gaps in the following sequences. (You may use your calculator)
   (a) Multiples of 4: 4, 8, 12, 16, ..., 100
   (b) Multiples of 6: 6, 12, 18, 24, ..., 96
   (c) Multiples of 8: 8, 16, 24, 32, ..., 96
   (d) Multiples of 9: 9, 18, 27, 36, ..., 99

2 (Group work)
Work in groups of four.
Make four 1–100 number squares, one for each member of the group.
On one number square, shade all the multiples of 4 which you found in question 1. Repeat on the other number squares for the multiples of 6, 8 and 9.
(a) Which number squares are not shaded by any member of the group?
(b) Which number squares are shaded by all members of the group?
(c) Which number squares are shaded by only one member of the group?

3 Find the next four terms of the following sequences. You may use your calculator.
(a) 2, 5, 8, 11, 14, ...
(b) 1, 6, 11, 16, 21, ...
(c) 1, 12, 23, 34, 45, ...
(d) 10, 9, 8, 7, 6, ...
(e) 0, 1, 3, 6, 10, ...
(f) 1, 2, 4, 8, 16, ...
(g) 1, 3, 7, 13, 21, ...
(h) 1, 2, 5, 10, 17, ...
(i) 1, 4, 9, 16, 25, ...
(j) 1, 1, 2, 3, 5, 8, 13, 21, ...

4 A trader stacks some tins in triangles as shown in Fig. 1.6 below.

Fig. 1.6

(a) Copy and complete Table 1.1 for the number of tins in Fig 1.6.

Table 1.1

Number of tins in bottom row	1	2	3	4
Total number of tins in stack	1	3		

(b) Extend and complete the table for 5, 6, 7, 8 tins.

Table 1.2

Index form	1^2	2^2	3^2	4^2	...	10^2
Number	1	4	9	16	...	100

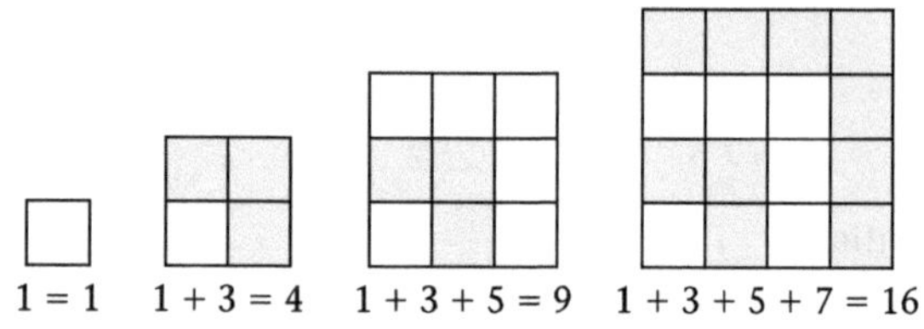

Fig. 1.7

5 (a) Copy and complete the sequence of square numbers shown in Table 1.2.
(b) Copy the pattern in Fig. 1.7 on to squared paper. Extend the pattern by drawing 5×5, 6×6 and 7×7 squares. Is it true that 7^2 = sum of the first seven odd numbers?
(c) Copy the pattern in Table 1.3 and complete it for the numbers 5, 6 and 7. Write down the sequence formed by the Total column. What do you notice?

Table 1.3

Number	Pattern	Total
1	1	1
2	1 + 2 + 1	4
3	1 + 2 + 3 + 2 + 1	9
4	1 + 2 + 3 + 4 + 3 + 2 + 1	16
5		
6		
7		

6 Copy and complete the pattern below. Use your calculator. ($a^3 = a \times a \times a$) What do you notice?

$1^3 = 1 = 1^2$
$1^3 + 2^3 = 9 = 3^2 = (1 + 2)^2$
$1^3 + 2^3 + 3^3 = 36 = 6^2 = (1 + 2 + 3)^2$
$1^3 + 2^3 + 3^3 + 4^3$
$1^3 + 2^3 + 3^3 + 4^3 + 5^3$

7 The numbers 1, 3, 6, 10, 15, 21 are known as **triangle numbers** (see Exercise le, question 4). Write the next three numbers in the sequence. What is the pattern?

8 The numbers 1, 4, 9, 16, 25 ... are known as **square numbers** (see Exercise 1e, question 5). Write the next five numbers in the sequence. What is the pattern?

9 The sequence 1, 1, 2, 3, 5, 8, ... is known as the **Fibonacci sequence**. Each term is the sum of the two previous terms. Write down the first ten terms of the Fibonacci sequence.

10 **Group work**
Work in groups of four.
(a) What patterns do you notice in question 9?
(b) Discuss **three** different ways of working out the sum of the cubes.
(c) Using any **one** method work out the sum of the cubes of numbers 1–7, 1–8, 1–9, 1–10.
Compare answers. Did you all get the same answers?

Summary

Real numbers are used in everyday life. The set of real numbers, R, includes all other numbers: Natural counting numbers, N; Whole numbers, W; Integers, Z; Rational numbers, Q.
Thus $N \subset W \subset Z \subset Q \subset R$

Operations on numbers obey the following laws:

Commutative law: $a + b = b + a$; $a \times b = b \times a$

Associative law: $(a + b) + c = a + (b + c)$;
$(a \times b) \times c = a \times (b \times c)$

Distributive law: $a\ (b + c) = ab + ac$;
$a\ (b - c) = ab - ac$.

Number patterns are sequences of numbers which follow a specific pattern. Examples of number patterns are

1, 3, 6, 10, 15, 21, ... known as **triangle numbers**

1, 4, 9, 16, 25, ... known as **square numbers**

1, 1, 2, 3, 5, 8 ... known as the **Fibonacci sequence**

Practice Exercise P1.1

1 For each of the following, list the first ten members of each set:
(a) even numbers
(b) odd numbers
(c) prime numbers
(d) composite numbers
(e) square numbers

2 List any five members of each of the following sets of numbers;
(a) rational numbers
(b) integers
(c) whole numbers
(d) irrational numbers
(e) counting numbers

3 Find the next four terms in each of the following sets of numbers:
(a) {1, 2, 4, 8, 16, ...}
(b) {1, 3, 7, 13, 21, ...}
(c) {6, 13, 20, 27, 34, ...}
(d) {1, 1, 2, 3, 5, 8, ...}
(e) {1, 3, 6, 10, 15, 21, ...}
(f) {0, 3, 6, 9, ...}
(g) {2, 5, 8, 11, 14, ...}
(h) {0, 1, 3, 6, 10, ...}
(i) {1, 4, 9, 16, 25, ...}
(j) {1, 2, 5, 10, 17, ...}

Practice Exercise P1.2

1 Write down the name of each number pattern. Some of the patterns do not start at the beginning.

(a)	2	4	6	8	10
(b)	1	4	9	16	25
(c)	3	5	7	9	11
(d)	6	9	12	15	18
(e)	2	4	8	16	32
(f)	8	12	16	20	24

2. Shawn is practicing to write the letters T, H and M. The numbers show how many straight lines she draws in writing the Ts .
   (a) Complete the numbers underneath the Ts.

T	TT	TTT	TTTT	TTTTT
2	4	....	....	....

   (b) What is the number pattern called?
   (c) Work out the numbers for the letter H.
   H HH HHH HHHH HHHHH
   (d) Work out the numbers for the letter M.
   M MM MMM MMMM MMMMM

3. Write down the first five numbers of these patterns, as shown in the example.

   Example 5 $\boxed{+2}$
   Working 5, 5 + 2 = 7, 7 + 2 = 9,
   9 + 2 = 11, 11 +2 = 13
   Answer 5 7 9 11 13

   (a) 8 $\boxed{+1}$ (b) 20 $\boxed{+5}$
   (c) 18 $\boxed{-3}$ (d) 4 $\boxed{+0.1}$
   (e) 5 $\boxed{\times 2}$ (f) 4 $\boxed{\div 2}$

4. Write down the rule for each of the following number patterns.

(a)	4	6	8	10	12
(b)	7	11	15	19	23
(c)	60	50	40	30	20
(d)	1	8	15	22	29
(e)	6	5.9	5.8	5.7	5.6
(f)	2	20	200	2000	

5. (a) Calculate the differences between the numbers of this pattern.

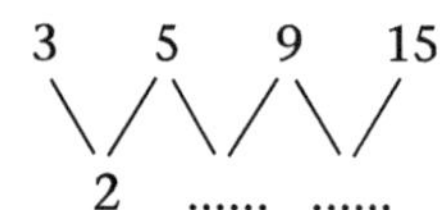

   difference 2 ....

   (b) Find the next two numbers.
   (c) Use differences to find the next two numbers in the following pattern:
   1 4 8 13 19

Practice Exercise P1.3

1. Find the value of each of the following, using an electronic calculator if necessary:
   (a) 29 + 43 × 7
   (b) 5 × (71 – 17)
   (c) (3 + 47) ÷ 25
   (d) 5 × 71 – 17
   (e) 3 + 47 ÷ 25
   (f) (24 × 5) ÷ (18 ÷ 6)

2. For each of the calculations in question 1, state the law(s) that the calculation illustrates.

3. For each of the following
   (a) work out the calculation
   (b) state which set or sets of numbers (W, Z, Q, or R) that the answer belongs to:
   (i) 27 ÷ 7
   (ii) $\sqrt{24 \div 8}$
   (iii) 3949 × 11 × 0
   (iv) 59 – 25 × 3

Practice Exercise P1.4

Work out the following. Write down the stages in your calculations as shown in question 1.

1. 5 × 12 – 15 = 60 – 15 = 45
2. 16 + 5 × 11
3. 13 × 4 + 16
4. 45 – 16 × 2
5. 60 ÷ 4 + 45
6. 73 + 66 ÷ 11
7. 75 – 60 ÷ 3
8. 56 ÷ 14 × 5
9. 8 × 39 ÷ 13
10. 15 × 42 ÷ 6

Practice Exercise P1.5

For each of the following calculations
(a) state whether it is true or false
(b) give a reason for your answer

1. 42 – 26 = 26 – 42
2. 35 + (7 × 4) = 35 + 7 × 4
3. 34 + 23 + 0 = 34 + 23
4. 6 × (14 + 3) = 6 × 14 + 6 × 3
5. (20 ÷ 5) ÷ 2 = 20 ÷ (5 ÷ 2)
6. 33 × 1 = 33
7. 20 × 5 × 2 = (20 × 5) × 2

Practice Exercise P1.6

1. Calculate the following.
   (a) $64 \div 8 \div 2$
   (b) $50 - (8 - 3) \times 6$
   (c) $(16 - 3)(18 \div 6)$
   (d) $2[16 - (9 - 5)]$
   (e) $[2 + 2(3 + 6)]^2$
   (f) $60 \div [24 \div (96 \div 12)]$
   (g) $7\{[20 - (5 + 6)] \div 3\}$
   (h) $30 - \{5 + 2[(9 + 12) \div 3]\}$
   (i) $\{12 - [15 - (20 - 13)]\}$
   (j) $-5(7 - 13)$
   (k) $(-6 + 9)(5 - 9)$
   (l) $7 - [20 \div (11 - 16)]$

2. Calculate the following, correct to 2 significant figures.
   (a) $12.8(14.2 + 29.5)$
   (b) $18.2 - [0.16 - 2(0.03 + 1.42)]$
   (c) $130 \div \{16 + [(32 + 17) \div 2 + 1]\}$
   (d) $0.7 + [11.2 - 10(5.3 - 4.9)]$
   (e) $0.2[(9 - 4) \div 2 + 1]$
   (f) $\{8^2 - [6^2 \div (3 + 3^2)]\} \div 2$

Practice Exercise P1.7

For each question, write a calculation involving brackets. Solve the problem.

1. \$Bl buys \$E1.42
   Bert spent \$B1300, \$B2800 and \$B500 on buying \$E during three trips.
   How many \$E did he buy altogether?

2. Calculate the total area of five of these square metal plates (Fig. 1.8). The hole in each plate is also square.

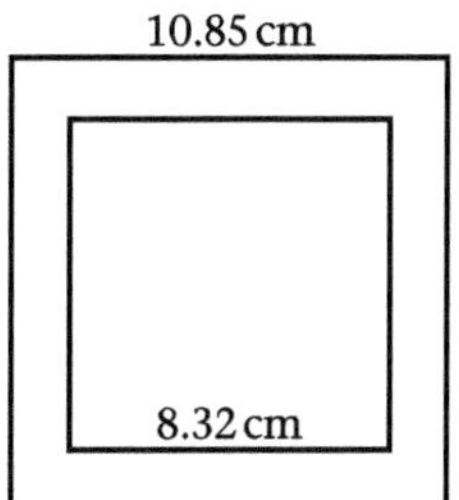

Fig. 1.8

3. Racetracks cost \$96, cars cost \$22, controllers cost \$16.50 Four children share the cost of a racetrack and three car kits. Each kit contains two cars and two controllers.
   (a) How much does each person pay?
   (b) What change do they receive if they are paying with two \$200 notes?

Practice Exercises P1.8

By using two or more of the four basic arithmetic operations at the same time, we can make a new binary operation and combine two quantities to get a result.

1. Work out the calculations for the following binary operations: * and #.
   (a) $p * q$ means $\frac{2}{3}p \times \frac{1}{4}q$
   (i) $18 * 12$ (ii) $24 * (7 * 6)$
   (b) $m \# n$ means $m^2 + 3n$
   (i) $3 \# 2$ (ii) $(3 \# 2) \# 1$

2. Show whether the binary operations given in question 1 obey
   (a) the commutative law
   (b) the associative law

Chapter 2

Geometrical constructions (1)

Triangles, parallel and perpendicular lines

Pre-requisites

- basic units of measurement; measurement of angles; properties of plane shapes

In geometry, to **construct** means to draw accurately. Accurate construction depends on using measuring instruments properly. Make sure that you have a pencil, a ruler, a pair of compasses, a protractor and a set square before beginning work on this chapter.

Drawing accurately

Note the following points:

1 All constructions should be made with a pencil. Use a hard pencil with a sharp point. This will give thin lines which will be more accurate.

2 Check that your ruler has an undamaged straight edge. A damaged ruler is useless for construction work.

3 Check that your compasses are working properly and are not too loose. Loose compasses can be tightened with a small screwdriver.

4 All construction lines must be seen. Do not rub out anything which leads to the final result.

5 Always take great care, especially when drawing a line through a point.

Constructing triangles

Freehand sketching

Before starting any construction you should make a rough sketch of what you are going to draw. This does not need to be accurate. Do not use a ruler. Make a **freehand sketch** and show the dimensions (i.e. measurements) on the sketch.

Constructing a triangle, given the lengths of all three sides

Work through the following example.

Example 1

Construct triangle ABC so that AB = 6 cm, AC = 5 cm *and* BC = 4 cm.

(a) Make a freehand sketch (Fig. 2.1). This shows what you have to do.

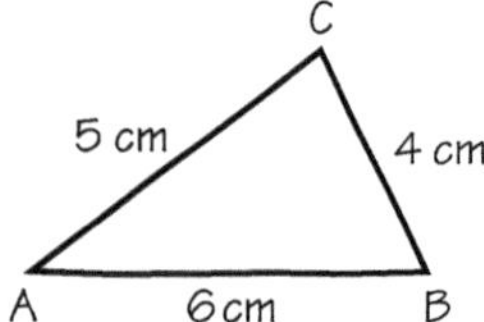

Fig. 2.1

(b) Draw a straight line a little longer than 6 cm (Fig. 2.2). Mark A with a thin line. Measure AB = 6 cm and mark off B.

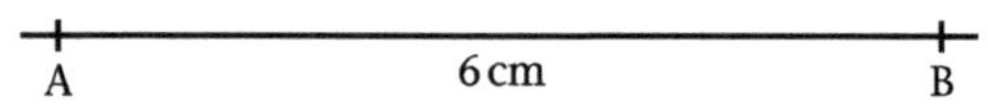

Fig. 2.2

(c) Open a pair of compasses to 5 cm. Put the point of the compasses on A. Draw an arc above AB. Every point on this arc is 5 cm from A (Fig. 2.3).

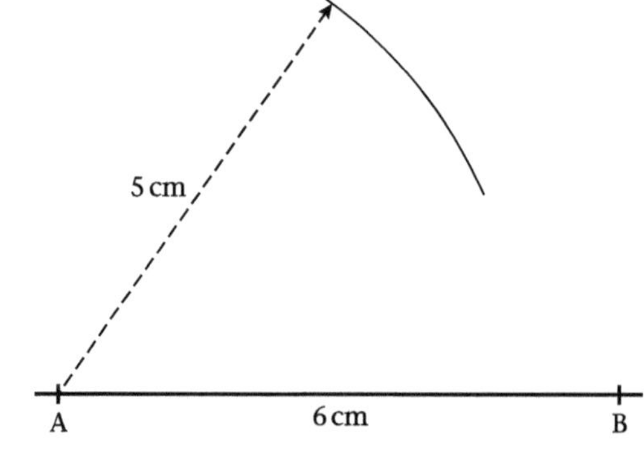

Fig. 2.3

(d) Now open the compasses to 4 cm. Put the point of the compasses on B. Draw a second arc to cut the first arc (Fig. 2.4).

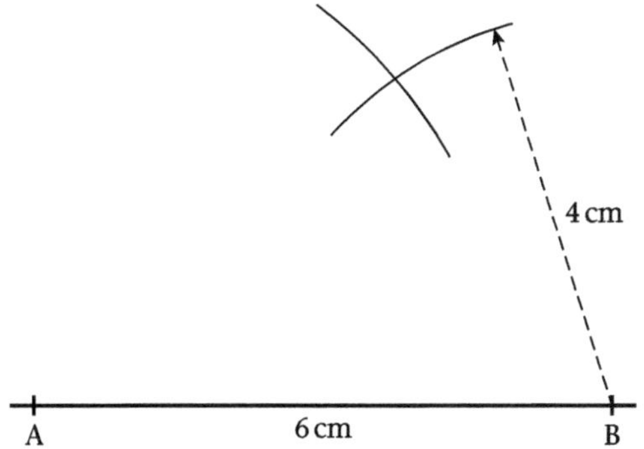

Fig. 2.4

(e) The point where the arcs intersect is 5 cm from A *and* 4 cm from B. This is the point C. Join AC and BC to complete the triangle ABC (Fig. 2.5).

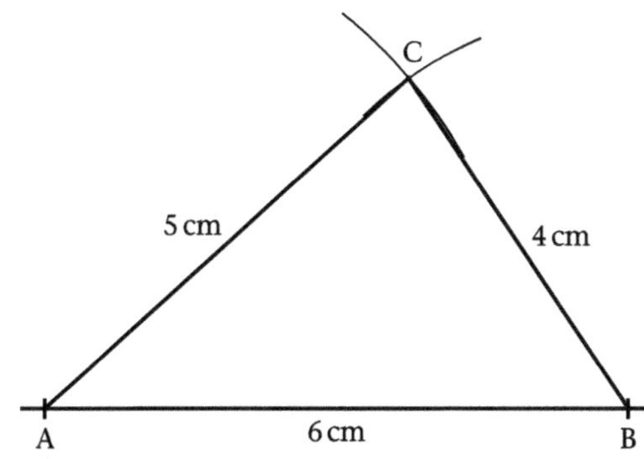

Fig 2.5

Note: Always leave construction lines, such as arcs, on your drawing.

Exercise 2a

Use ruler and compasses in this exercise.

1. Construct triangles with sides of the lengths given in the sketches in Fig. 2.6.

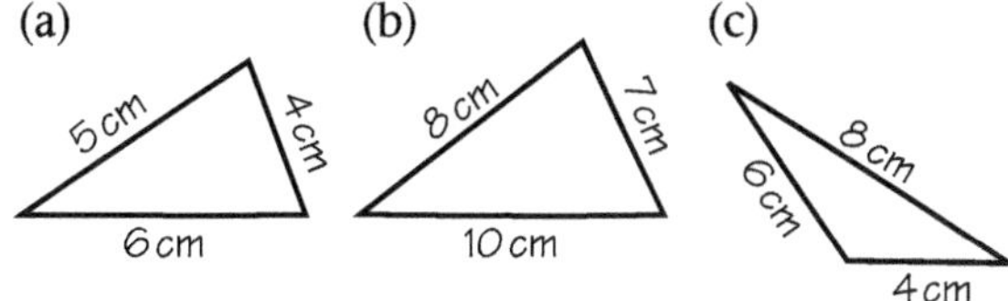

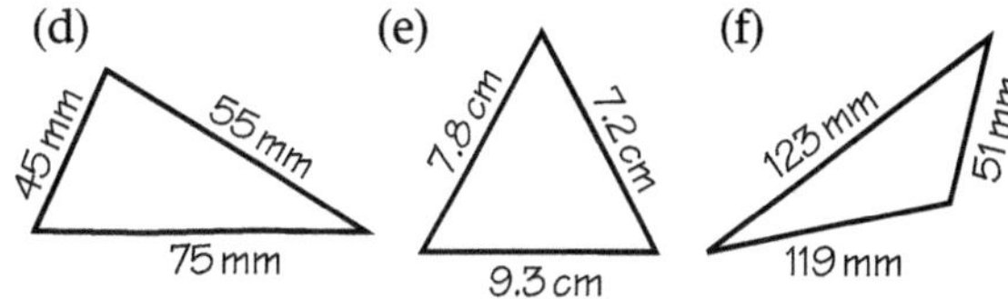

Fig. 2.6

2. Construct triangles with sides of the lengths given below. Make rough sketches first.
   (a) 8 cm, 7 cm, 6 cm
   (b) 9 cm, 5 cm, 8 cm
   (c) 10.5 cm, 7.5 cm, 7.5 cm
   (d) 105 mm, 95 mm, 45 mm

3. Try to construct a triangle with sides 10 cm, 3 cm and 6 cm. What happens?
   Try to construct another triangle with sides 8 cm, 3 cm and 4 cm. What happens?
   Complete this sentence: The sum of the lengths of any two sides of a triangle must be ... than the length of the third side.

4. Make an accurate construction of the diagram in Fig. 2.7. *Hint*: Start by drawing BD.

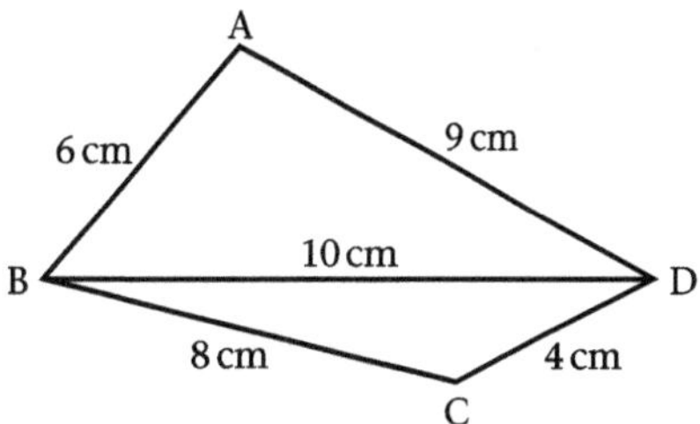

Fig 2.7

When your drawing is complete, draw the line AC. Measure AC.

5. Draw triangle ABC in which AB = 10 cm, BC = 6 cm and CA = 11 cm. On AB mark a point D such that AD = 4.5 cm. Draw CD. Measure the length of CD.

6. Draw triangle PQR in which PQ = 7.9 cm, QR = 7.5 cm and PR = 12.6 cm. With centre R and radius 10.5 cm, draw an arc to cut PQ at X. Measure the length of PX.
   Note: Draw a freehand sketch first.

Constructing a triangle, given two angles and one side

Work through the following example.

Example 2

Construct $\triangle ABC$ such that $BC = 5\,\text{cm}$, $A\hat{B}C = 30°$ *and* $A\hat{C}B = 70°$.

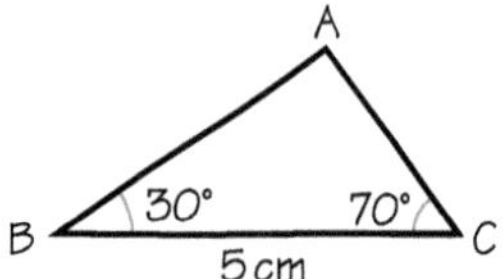

Fig. 2.8

(a) Make a freehand sketch (Fig. 2.8).

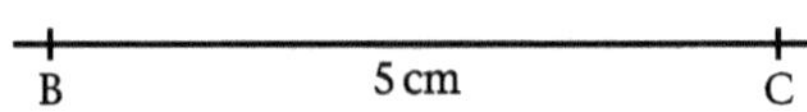

Fig. 2.9

(b) Draw a line a little longer than 5 cm. Mark off BC = 5 cm (Fig. 2.9).

(c) Place the centre of a protractor over B. Mark a point P and draw a line from B at an angle of 30° to BC (Fig. 2.10).

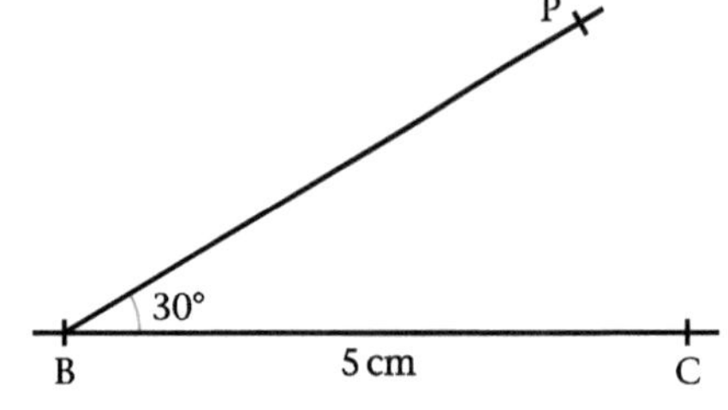

Fig. 2.10

(d) Now place the centre of the protractor over C. Construct a line from C at 70° to CB. A is the point where the lines intersect above BC (Fig. 2.11).

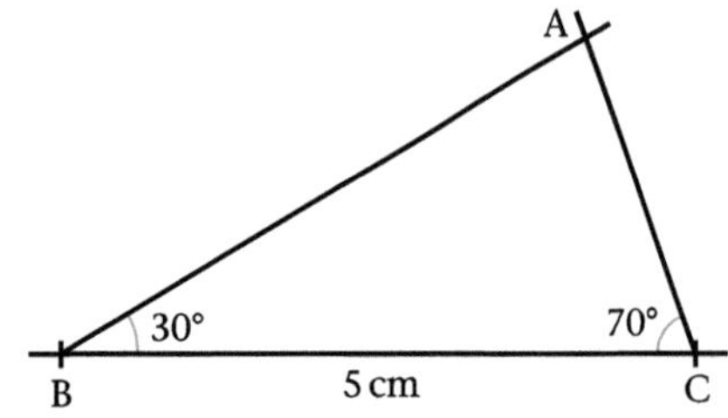

Fig. 2.11

Notice that you need to know the two angles at the end of the given line. If two angles of a triangle are known, it is always possible to find the third angle.

Example 3

Sketch $\triangle PQR$ in which $PQ = 10\,\text{cm}$, $P\hat{Q}R = 40°$ and $P\hat{R}Q = 75°$.

Calculate the third angle of the triangle.

Rough sketch (Fig. 2.12).

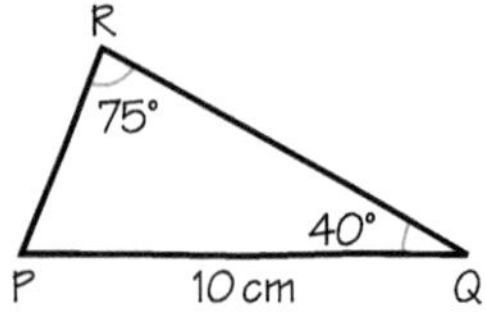

Fig. 2.12

$$R\hat{P}Q = 180° - (75° + 40°) = 180° - 115°$$
$$= 65°$$

It would now be possible to construct triangle PQR accurately. This shows the value of making a rough sketch.

Exercise 2b

1 Construct the triangles in the sketches in Fig. 2.13.

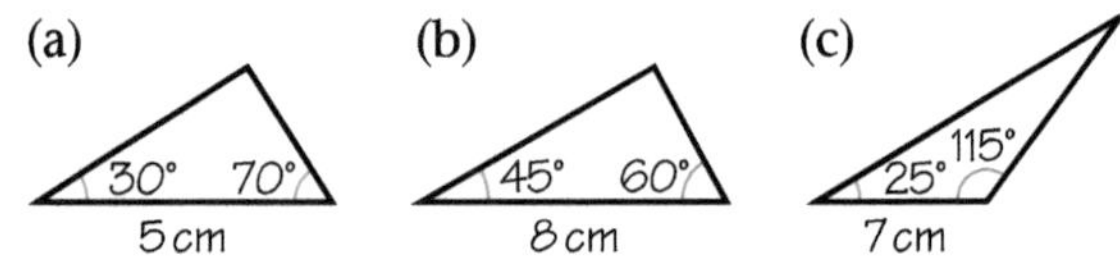

Fig. 2.13

2 Calculate the third angle in each of the triangles sketched in Fig. 2.14. Make a new sketch and construct each triangle accurately.

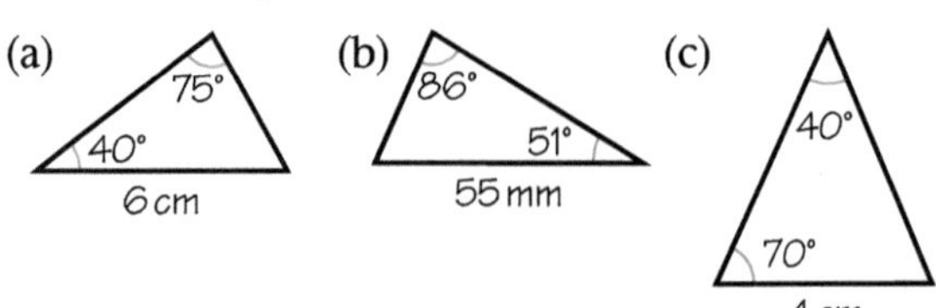

Fig. 2.14

3. Construct the following triangles accurately. In each case make a freehand sketch before drawing accurately.
   (a) $\triangle ABC$ such that AB = 5 cm, $\widehat{CAB} = 50°$, $\widehat{ABC} = 70°$
   (b) $\triangle PQR$ such that QR = 8 cm, $\widehat{PQR} = 62°$, $\widehat{QRP} = 33°$
   (c) $\triangle DEF$ such that DF = 6.5 cm, $\widehat{EDF} = 55°$, $\widehat{EFD} = 48°$
   (d) $\triangle XYZ$ such that XY = 6 cm, $\widehat{XYZ} = 40°$, $\widehat{XZY} = 60°$

4. Construct $\triangle ABC$ such that $\widehat{ABC} = 28°$, BC = 4 cm and $\widehat{BCA} = 125°$. M is the mid-point of BC. Measure AM.

Constructing a triangle, given two sides and the angle between them

Example 4

Construct $\triangle ABC$ in which AB = 4 cm, BC = 5 cm and $\widehat{ABC} = 50°$.

(a) Draw a freehand sketch (Fig. 2.15).

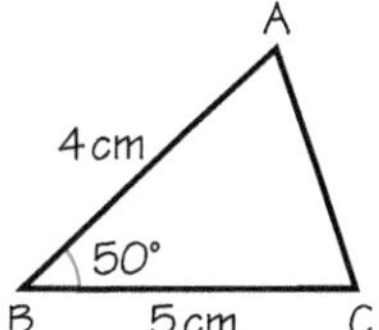

Fig. 2.15

(b) Draw a straight line BC of length 5 cm (Fig. 2.16).

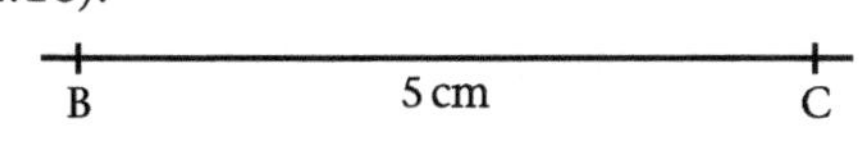

Fig. 2.16

(c) Place the centre of a protractor over B. Draw a line from B at an angle of 50° to BC (Fig. 2.17).

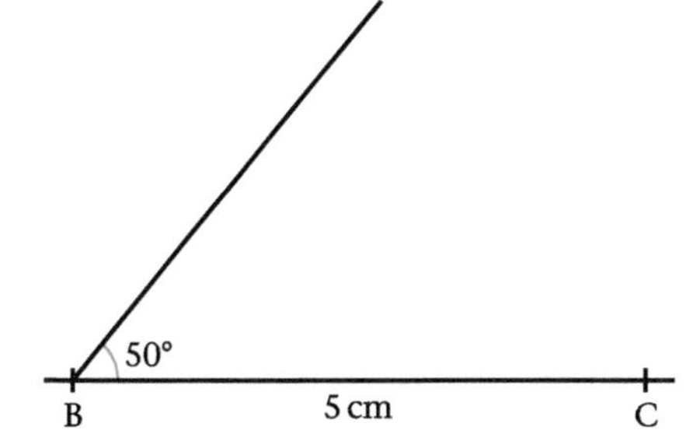

Fig. 2.17

(d) Mark a point A on this line such that BA = 4 cm. Join AC to complete $\triangle ABC$ (Fig. 2.18).

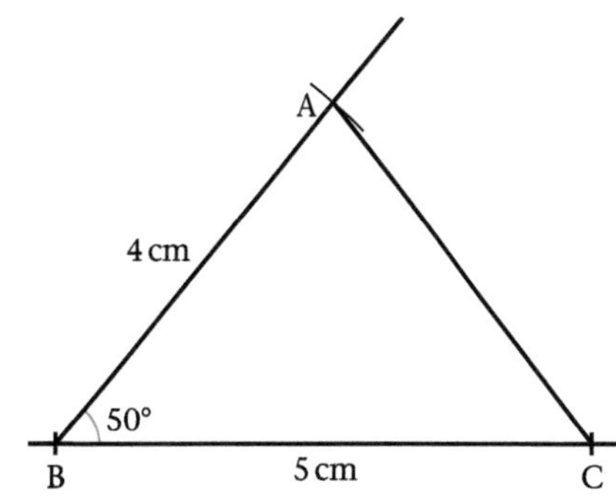

Fig. 2.18

Exercise 2c

1. Construct the triangles sketched in Fig. 2.19. Remember to complete the triangles.

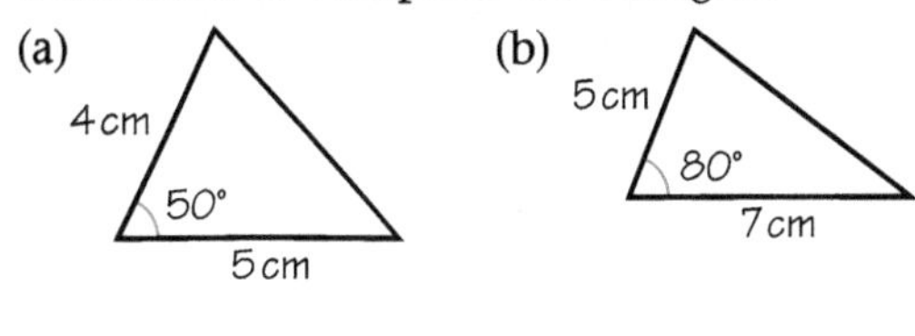

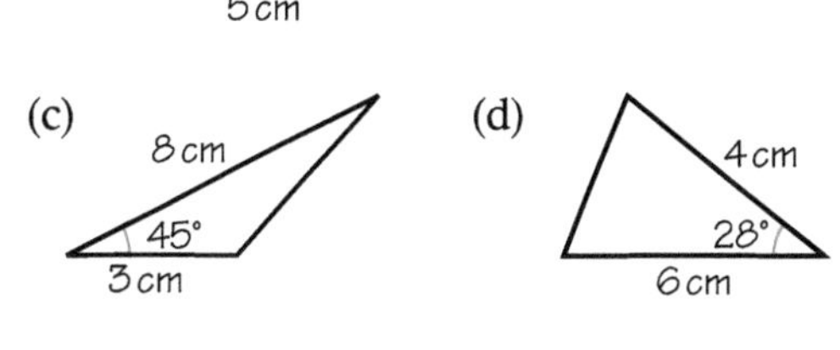

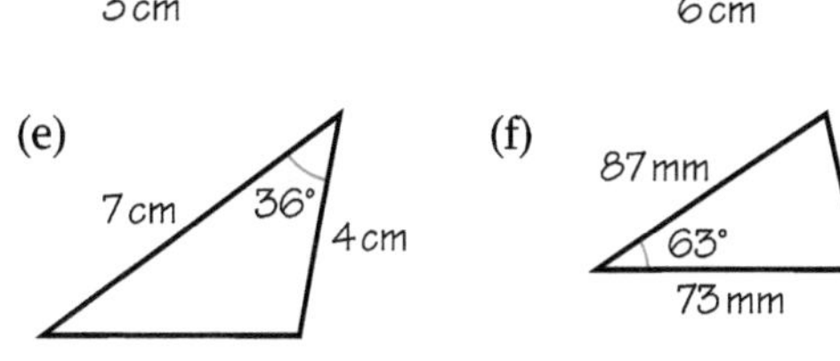

Fig. 2.19

2. In each of the following, two sides of a triangle and the angle between them are given. Construct each triangle and measure the length of the third side.
   (a) 6 cm, 90°, 8 cm
   (b) 8 cm, 60°, 5 cm
   (c) 6.2 cm, 42°, 7.9 cm
   (d) 47 mm, 56°, 74 mm
   (e) 56 mm, 60°, 35 mm
   (f) 5.4 cm, 43°, 4.8 cm

3. Construct the figure shown in Fig. 2.20.
Hint: start by constructing $\triangle ABC$.

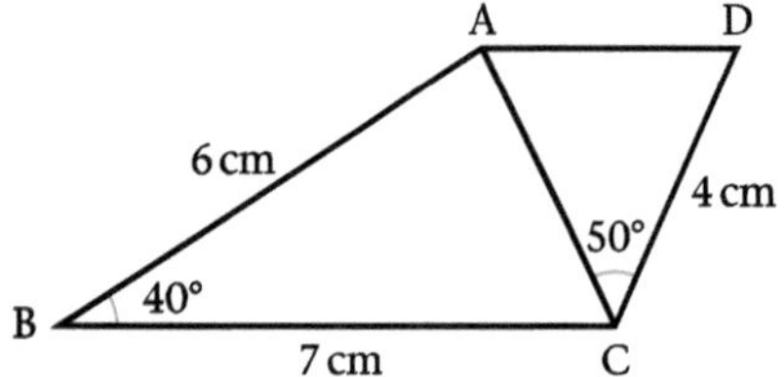

Fig. 2.20

When the drawing is complete, measure the length of AD.

Parallel lines

Constructing parallel lines using ruler and set square only

Follow Example 5 by copying the work on to a sheet of blank paper.

Example 5

Construct a line through P so that it is parallel to AB (Fig. 2.21).

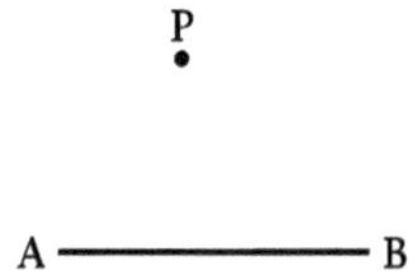

P

Fig. 2.21 A —— B

(a) Place a set square so that one edge is accurately along AB (Fig. 2.22).

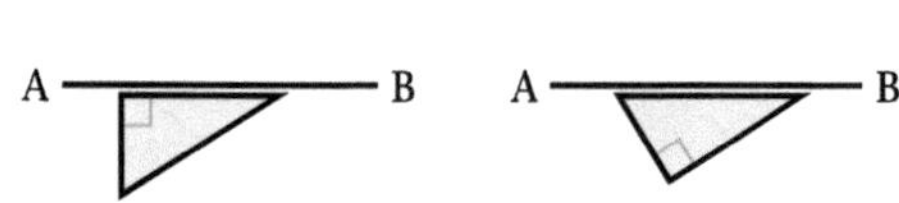

Fig. 2.22

(b) Place a ruler along one of the other edges of the set square (Fig. 2.23). (Use the left-hand edge if you are right handed.)

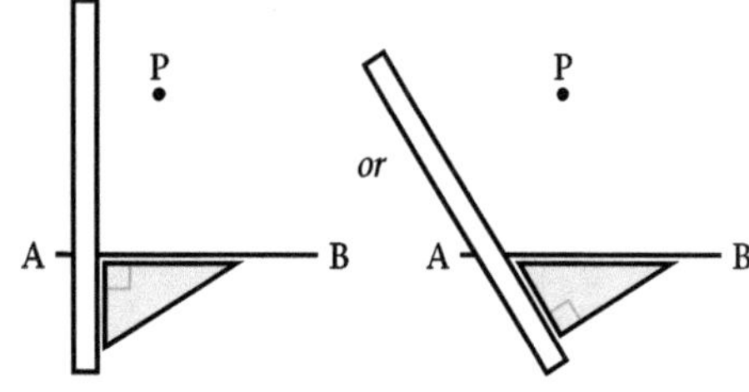

Fig. 2.23

(c) Hold the ruler firmly. Slide the set square along the ruler towards P. Stop when the edge that was on AB reaches P. Draw a line along this edge of the set square through P (Fig. 2.24).

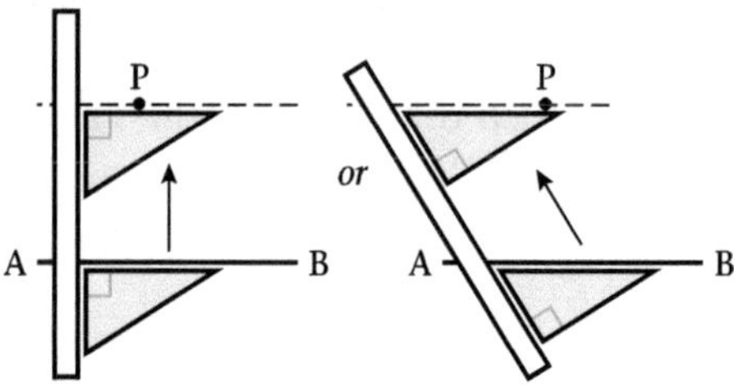

Fig. 2.24

Notice, in Example 5, that since corresponding angles are equal, the line through P and the line AB are parallel.

Exercise 2d

Work on blank (i.e. unruled) paper.

1. Draw a straight line AB on a piece of unruled paper. Mark a point P which is not on the line. Use a ruler and set square to draw a line through P which is parallel to AB.

2. Use a ruler and set square to draw four lines which are parallel to each other.

3. Draw any triangle near the centre of a sheet of paper. Use a ruler and set square to draw another triangle with sides parallel to those of the first triangle.

4. Draw $A\hat{B}C = 70°$ so that the arms BA and BC are 3 cm and 5 cm long. See the sketch in Fig. 2.25. Draw a line through A parallel to BC. Draw a line through C parallel to BA to make parallelogram ABCD.

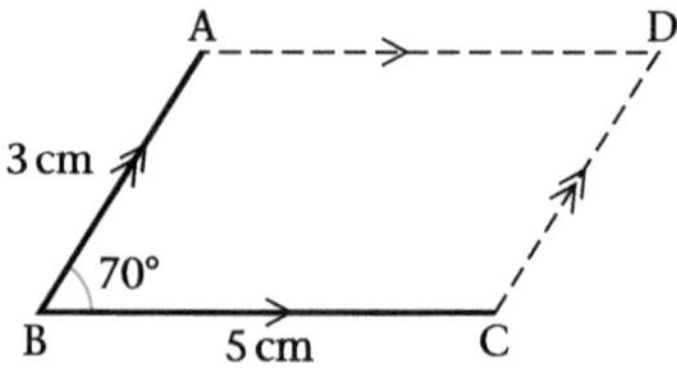

Fig. 2.25

Measure the diagonals AC and BD.

5 Construct a parallelogram ABCD with AB = 3 cm, BC = 4 cm and $\widehat{B} = 75°$. Draw the diagonals AC and BD to intersect at point X. Measure XA and XC. Measure XB and XD.

6 Construct a rhombus so that one of its acute angles is 65° and each side is 4 cm long. Measure the lengths of its diagonals.

7 Construct the trapezium ABCD shown in Fig. 2.26.

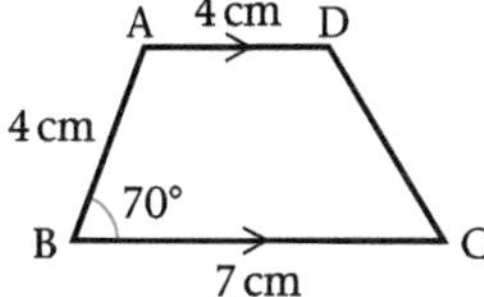

Fig. 2.26

Measure DC and the three unknown angles.

8 Construct trapezium ABCD so that BC = 6 cm, AB = 3 cm, $\widehat{B} = 80°$ and $\widehat{C} = 70°$. Measure BD.

Perpendicular lines

When two lines meet at right angles we say that they are **perpendicular** to each other.

In Fig. 2.27, XY is perpendicular to AB *and* AB is perpendicular to XY.

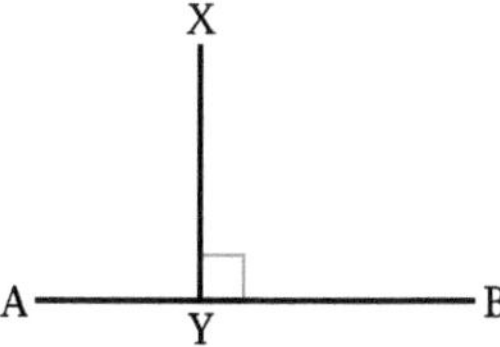

Fig. 2.27

Lines which meet perpendicularly are very common. Adjacent edges of a door frame are perpendicular to each other. The margin in an exercise book is perpendicular to the ruled lines. There are many other examples.

The perpendicular distance of a point from a line

In Fig. 2.28 P is a point and *l* is any line.

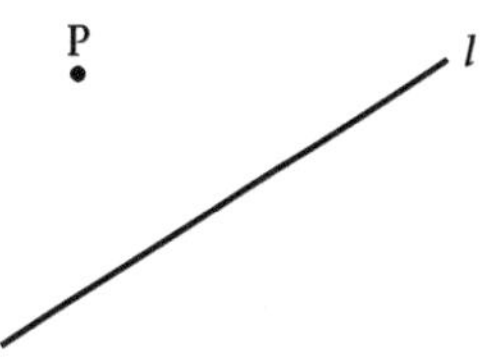

Fig. 2.28

P is different distances from different points D_1, D_2, D_3, D_4, ... on the line (Fig. 2.29).

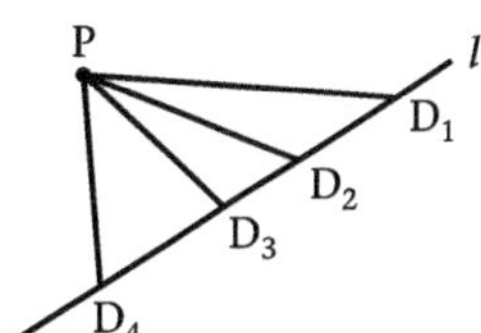

Fig. 2.29

Measure the distance from P to each point marked D_1, D_2, D_3, ... on the line *l*. What do you notice?

One point on the line is nearest to P. This is the point D such that PD is perpendicular to *l* (Fig. 2.30).

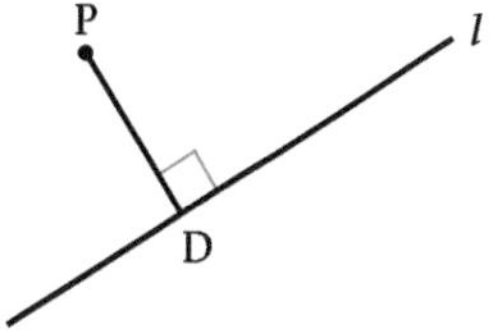

Fig, 2.30

The length PD is the **perpendicular distance** of P from line *l*.

Constructing a perpendicular from a point to a line using ruler and set square

(a) Place a ruler along the given line (Fig. 2.31).

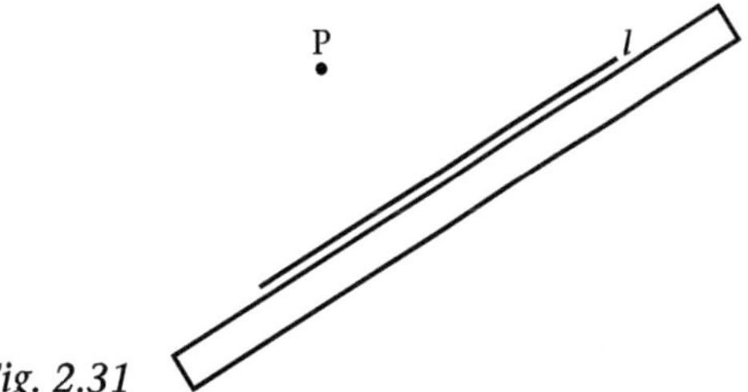

Fig. 2.31

(b) Use the two edges of a set square which are the arms of its right angle. Place one of these edges along the ruler. Slide the set square along the ruler until the other edge reaches P (Fig. 2.32).

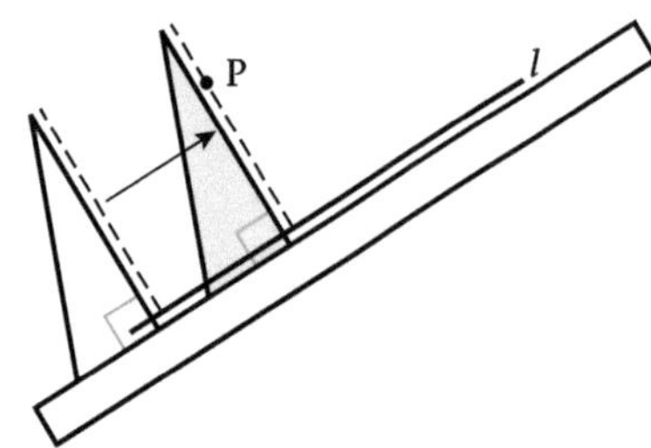

Fig. 2.32

(c) Hold the set square firmly. Draw the line through P to meet the line perpendicularly (Fig. 2.33).

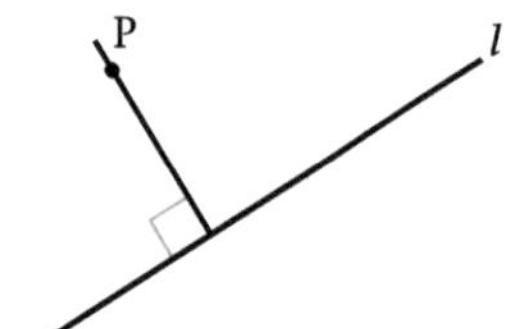

Fig. 2.33

Exercise 2e

1. Mark a point P near the middle of a page in your exercise book. Mark P between two of the ruled lines in your exercise book. Use ruler and set square to draw and measure the perpendicular distance of P from
   (a) the line which is four lines above P,
   (b) the line which is six lines below P,
   (c) the left-hand margin.

2. Trace the diagram, Fig. 2.34, into your exercise book. Draw and measure the perpendicular distances of P from the five sides of the pentagon.

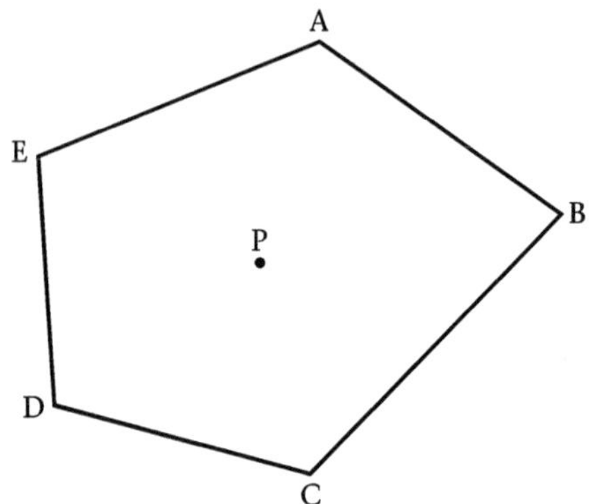

Fig. 2.34

3. Use ruler and set square to construct the following:
   (a) a square of side 3 cm
   (b) a square of side 5 cm
   (c) a rectangle 3 cm by 6 cm
   (d) a rectangle 4 cm by 5 cm

4. Use ruler and set square to construct a pair of parallel lines which are
   (a) 3 cm apart,
   (b) 5 cm apart,
   (c) 4.2 cm apart,
   (d) 57 mm apart.

5. Turn back to the parallelograms you constructed in Exercise 2d, questions 4 and 5. In each case, measure the perpendicular distances between (a) parallel lines AD and BC, (b) parallel lines AB and DC. Hence calculate the area of each parallelogram in two ways.

Summary

To **construct** means to draw accurately.

Triangles can be constructed using ruler, protractor and compasses. **Parallel lines** and **perpendicular lines** can be constructed using ruler and set square.

Practice Exercise P2.1

Remember to leave all construction lines on your drawing and always close a triangle or other shape.

1. Using ruler and protractor, construct triangle PQR, with PR = 8.5 cm, $P\hat{Q}R = 60°$, RQ = 8 cm.
   Draw the side PR.
   Measure and state the length of PR.

2. Using ruler and protractor,
   (a) construct a triangle ABC with AB = 7 cm, $C\hat{A}B = 60°$ $A\hat{B}C = 45°$.
   (b) Measure and state the length of BC.

3. Using ruler and protractor only,
   (a) construct a triangle LMN with LM = 7.5 cm, $L\hat{M}N = 30°$, $M\hat{L}N = 45°$
   (b) Using a set square and ruler draw the line NP through N parallel to LM.

4. Using ruler, protractor and a set square
   (a) draw a line PQ = 7.2 cm
   (b) construct an angle $Q\hat{P}R$ of 65° and draw PR = 5.6 cm.
   (c) construct a line QS||PR and QS = 5.6 cm.
   (d) Join RS to form the parallelogram PQRS.
   (e) Measure and state the size of $P\hat{Q}S$.

5. Using ruler, protractor and set square, construct a rhombus so that one of its acute angles is 55°, and each side is 5 cm long. Measure the lengths of the diagonals of the rhombus.

6. (a) Construct a triangle ABC with AB = 7.4 cm, AC = 6.5 and BC = 6 cm.
   (b) Using ruler and compasses only construct the perpendicular from C to AB.

7. Using ruler, set square and compasses
   (a) construct a triangle ABC with AB = 6.5 cm, BC =7 cm, AC = 5.5 cm.
   (b) Construct the perpendicular from A to BC, B to AC and from C to AB.
   What do you notice?

Chapter 3

Directed numbers (2)

Subtraction and division

Pre-requisites

- factors and multiples; addition and multiplication of directed numbers; simplifying algebraic expressions

Subtraction

The subtraction $4 - 3 = 1$ can be written in directed numbers as: $(+4) - (+3) = +1$. We show this subtraction on the number line in Fig. 3.1. Follow the arrows from START to FINISH.

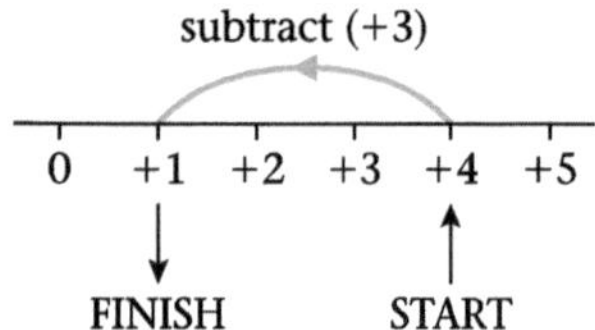

Fig. 3.1

Similarly, we can find the value of $3 - 7$. In directed numbers this is: $(+ 3) - (+ 7)$. Follow the arrows on the number line in Fig. 3.2 from START to FINISH.

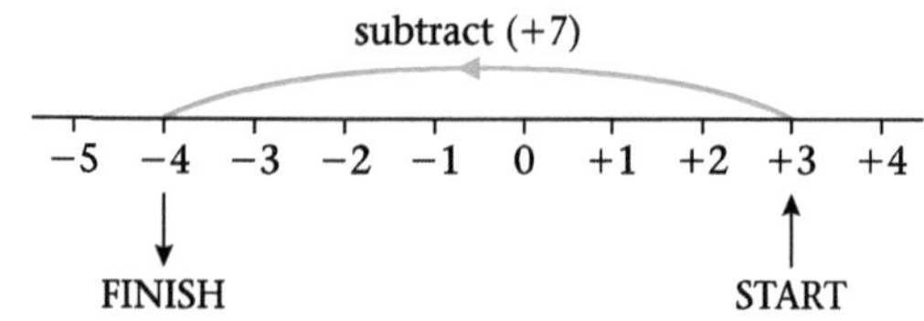

Fig. 3.2

Thus: $3 - 7 = -4$

To subtract a positive number: move that number of places to the left on the number line.

Consider the subtraction $5 - 3$. One way of doing this is to say, 'What must be added to 3 to get 5?' Since subtraction is the inverse of addition, if $3 + 2 = 5$, then $5 - 3 = 2$.

Now consider the subtraction $5 - (-3)$. 'What must be added to -3 to get 5?' From earlier work on the number line we know that $-3 + 8 = 5$. Thus: $5 - (-3) = 8$.

This can be shown on the number line as in Fig. 3.3. START at +5 and FINISH at +8. The only way this can happen is if we move 3 places to the *right.*

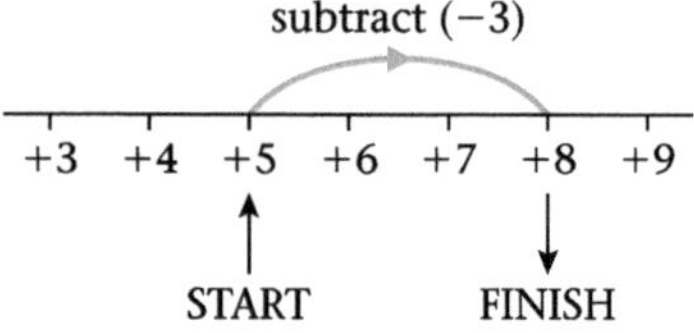

Fig 3.3

To subtract a negative number: move that number of places to the right on the number line.

Notice that to subtract a negative number is equivalent to adding a positive number of the same value; for example see Fig. 3.4.

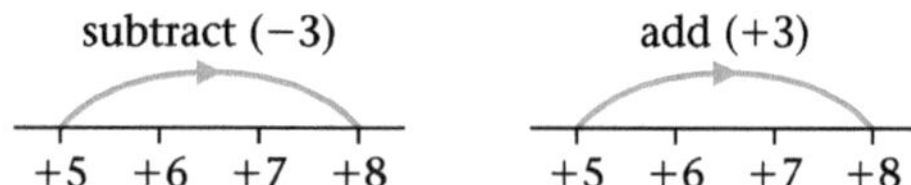

Fig 3.4

Exercise 3a (Oral)

1. Complete the following number patterns.

(a) $5 - (+2) = +3$
$5 - (+1) =$
$5 - \quad 0 =$
$5 - (-1) =$
$5 - (-2) =$

(b) $(-3) - (+3) = -6$
$(-3) - (+2) =$
$(-3) - (+1) =$
$(-3) - \quad 0 =$
$(-3) - (-1) =$

(c) $-2 - (+1) = -3$
$-2 - \quad 0 =$
$-2 - (-1) =$
$-2 - (-2) =$
$-2 - (-3) =$

2. Simplify the following.
   (a) $8 - (-3)$ (b) $(-8) - (-3)$
   (c) $(+3) - (-8)$ (d) $(-3) - (-8)$
   (e) $(+9) - (-9)$ (f) $(-4) - (-4)$
   (g) $(+6) - (-1)$ (h) $(-1) - (-2)$
   (i) $-7 - (-6)$ (j) $-30 - (-80)$
   (k) $20 - (-10)$ (l) $-40 - (-60)$
   (m) $-41 - (-26)$ (n) $-59 - (-37)$
   (o) $76 - (-74)$ (p) $7 - (-8) + (-5)$
   (q) $-3 - (-21) -18$

Exercise 3b

1. Which is greater:
   (a) -3 or 4, (b) -4 or -7,
   (c) -5 or 3, (d) 0 or -5,
   (e) 2 or -15, (f) -7 or -10?

2. The temperature during the day in a cold country is 9 °C. At night the temperature falls by 13 °C. What is the temperature at night?

3. Elias and Steve have no money, but Elias owes $12 to Steve. When Friday comes they both get the same wages. Elias repays his debt to Steve. Steve now has more money than Elias. How much more?

4. What must be added to:
   (a) 3 to make 8,
   (b) -1 to make 2,
   (c) 16 to make 4,
   (d) 3 to make -8,
   (e) -35 to make -27,
   (f) 26 to make -4?

5. What must be subtracted from:
   (a) 12 to make 8,
   (b) 6 to make -10,
   (c) -2 to make -7,
   (d) 8 to make 12,
   (e) -3 to make 4,
   (f) -10 to make -3?

6. A man has $23.46 in his bank account. He writes a cheque for $39.50. How much will he be overdrawn?

7. In the year AD 45 a man was 63 years old. In what year was he 5 years old?

8. Simplify the following.
   (a) $-7 - (-16) -3$
   (b) $1 + (-4) - (-3)$
   (c) $800 - (+500) - (-150)$
   (d) $-50 + (-25) - (+45)$
   (e) $6x - 9x - (-5x)$
   (f) $-4y + 12y - (-10y)$

9. Simplify the following.
   (a) $1\frac{3}{4} - 2\frac{1}{4}$
   (b) $-2.8 + 6.3$
   (c) $4.8 - (-3.9)$
   (d) $-1\frac{1}{2} - (-3\frac{2}{3})$
   (e) 7.2 °C − 9.6 °C
   (f) − 5.4 °C + 8.6 °C

Division

When directed numbers are multiplied together,

two like signs give a positive result;
two unlike signs give a negative result.

For example,

$(+3) \times (+8) = +24$
$(-3) \times (-8) = +24$
$(+3) \times (-8) = -24$
$(-3) \times (+8) = -24$

The same rule is true for division. For example,

$(+24) \div (+3) = (+8)$
$(-24) \div (-3) = (+8)$
$(+24) \div (-3) = (-8)$
$(-24) \div (+3) = (-8)$

Since division is the inverse of multiplication, we may ask, for example, 'what must be multiplied by $+3$ to give $+24$?'
Since $(+3) \times (+8) = (+24)$, the answer is $+8$.

Or 'what must be multiplied by -3 to give $+24$?' Similarly, since $(-3) \times (-8) = (+24)$ the result must be -8.

Example 1

Divide (a) -36 by 9, (b) -4 by -12.

(a) $-36 \div 9 = -\frac{36}{+9} = -\left(\frac{36}{9}\right) = -4$

(b) $-4 \div (-12) = \frac{-4}{-12} = +\left(\frac{4}{12}\right) = +\frac{1}{3}$

Example 2

Simplify $\frac{(-6) \times (-5)}{-10}$

$\frac{(-6) \times (-5)}{-10} = \frac{+30}{-10} = -\left(\frac{30}{10}\right) = -3$

Exercise 3c

1. Divide:
   (a) -18 by 3, (b) -18 by -3,
   (c) 18 by -3, (d) 36 by $+4$,
   (e) -20 by -4, (f) -8 by -1,
   (g) -22 by -11, (h) 24 by -8,
   (i) 33 by -3, (j) -3 by -18,
   (k) -6 by 12, (l) -5 by -15.

2. Simplify the following.
   (a) $\frac{(-2) \times (+12)}{-6}$ (b) $\frac{(-6) \times (-10)}{-4}$
   (c) $\frac{36}{(-2) \times (-9)}$ (d) $\frac{(-3) \times (-15)}{9}$
   (e) $\frac{9 \times 20}{-3}$ (f) $\frac{-28 \times (-3)}{21}$
   (g) $\frac{30}{(-5) \times (-4)}$ (h) $\frac{(-1) \times (-5)}{-10}$

Directed algebraic terms

(a) Just as $5a$ is short for $5 \times a$, so $-5a$ is short for $(-5) \times a$.

(b) Just as m is short for $1 \times m$, so $-m$ is short for $(-1) \times m$.

(c) Algebraic terms and numbers can be multiplied together. For example,

$$\begin{aligned} 4 \times (-3x) &= (+4) \times (-3) \times x \\ &= -(4 \times 3) \times x \\ &= -12 \times x \\ &= -12x \end{aligned}$$

$$\begin{aligned} (-2y) \times (-8y) &= (-2) \times y \times (-8) \times y \\ &= (-2) \times (-8) \times y \times y \\ &= +(2 \times 8) \times y^2 \\ &= +16y^2 \text{ or } 16y^2 \end{aligned}$$

d) Division with directed numbers is also possible. For example,

$$\begin{aligned} 18a \div (-6) &= \frac{(+18) \times a}{(-6)} \\ &= -\left(\frac{18}{6}\right) \times a \\ &= (-3) \times a = -3a \end{aligned}$$

$$\begin{aligned} \frac{-33x^2}{-3x} &= \frac{(-33) \times x \times x}{(-3) \times x} \\ &= +\left(\frac{33}{3}\right) \times x = 11x \end{aligned}$$

Read Example 3 carefully.

Example 3

	simplify	working	result
(a)	$-5 \times 2y$	$= (-5) \times (+2) \times y$ $= -(5 \times 2) \times y$	$= -10y$
(b)	$-6 \times -4x$	$= (-6) \times (-4) \times x$ $= +(6 \times 4) \times x$	$= 24x$
(c)	$-3a \times 7$	$= (-3) \times a \times (+7)$ $= (-3) \times (+7) \times a$	$= -21a$
(d)	$4y \times -5$	$= (+4) \times y \times (-5)$ $= (+4) \times (-5) \times y$	$= -20y$
(e)	$-3a \times -6b$	$= (-3) \times a \times (-6) \times b$ $= (-3) \times (-6) \times a \times b$	$= 18ab$
(f)	$\frac{-14a}{7}$	$= \frac{(-14) \times a}{(+7)}$ $= -\left(\frac{14}{7}\right) \times a$ $= -2 \times a$	$= -2a$
(g)	$-\frac{1}{3}$ of $36x^2$	$= \frac{(+36) \times x^2}{(-3)}$ $= -\left(\frac{36}{3}\right) \times x^2$	$= -12x^2$

Exercise 3d

Simplify the following.

1. $(-5) \times a$
2. $x \times (-4)$
3. $(-x) \times 3$
4. $3 \times (-c)$
5. $(-6) \times (-x)$
6. $(-y) \times (-9)$
7. $3 \times (-2a)$
8. $(-3) \times (-2a)$
9. $(-3) \times 2a$
10. $(-9) \times 4x$
11. $(-5x) \times -8$
12. $(-5x) \times 8$
13. $(-8x) \times (-3x)$
14. $6d \times (-3d)$
15. $21a \div (-7)$
16. $(-6x) \div (-2)$
17. $20y \div (-10)$
18. $18z \div (-3)$
19. $\frac{-16x}{8}$
20. $\frac{28x}{-4}$
21. $\frac{17x}{-17}$
22. $\frac{-49x}{-7}$
23. $\frac{1}{5}$ of $(-45y)$
24. $\left(-\frac{1}{10}\right)$ of $100z$

Exercise 3e

1. Write down the next four terms of the following patterns.
   (a) $+6, +4, +2, 0, -2, \ldots, \ldots, \ldots, \ldots$
   (b) $2 \times (+3), 2 \times (+2), 2 \times (+1), 2 \times 0,$
   $2 \times (-1), \ldots \times \ldots, \ldots \times \ldots,$
   $\ldots \times \ldots, \ldots \times \ldots$
   (c) $2 \times (+3) = +6$
   $2 \times (+2) = +4$
   $2 \times (+1) = +2$
   $2 \times 0 = 0$
   $2 \times (-1) = -2$
   $\ldots \times \ldots = \ldots$
   $\ldots \times \ldots = \ldots$
   $\ldots \times \ldots = \ldots$
   $\ldots \times \ldots = \ldots$
   (d) $+16, +12, +8, +4, 0, \ldots, \ldots, \ldots, \ldots$
   (e) $(+4) \times 4, (+3) \times 4, (+2) \times 4,$
   $(+1) \times 4, 0 \times 4, \ldots \times \ldots,$
   $\ldots \times \ldots, \ldots \times \ldots, \ldots \times \ldots$
   (f) $(+4) \times 4 = +16$
   $(+3) \times 4 = +12$
   $(+2) \times 4 = +8$
   $(+1) \times 4 = +4$
   $0 \times 4 = 0$
   $\ldots \times \ldots = \ldots$
   $\ldots \times \ldots = \ldots$
   $\ldots \times \ldots = \ldots$
   $\ldots \times \ldots = \ldots$
   (g) $-15, -10, -5, 0, +5, \ldots, \ldots, \ldots, \ldots$
   (h) $(+3) \times (-5), (+2) \times (-5),$
   $(+1) \times (-5), 0 \times (-5), (-1) \times (-5),$
   $\ldots \times \ldots, \ldots \times \ldots, \ldots \times \ldots, \ldots \times \ldots$
   (i) $(+3) \times (-5) = -15$
   $(+2) \times (-5) = -10$
   $(+1) \times (-5) = -5$
   $0 \times (-5) = 0$
   $(-1) \times (-5) = +5$
   $\ldots \times \ldots = \ldots$
   $\ldots \times \ldots = \ldots$
   $\ldots \times \ldots = \ldots$
   $\ldots \times \ldots = \ldots$

2. Simplify the following.
   (a) $(-3) \times 4$ (b) $(-4) \times (-7)$
   (c) $5 \times (-3)$ (d) $(-6) \times (-1)$
   (e) $24 \times (-2)$ (f) $(-8) \times (-7)$
   (g) $(-1) \times (-1)$ (h) $0 \times (-3)$
   (i) $(-1) \times 1$

3. The bottom of a water well is -6 metres from ground level. An oil well is 15 times deeper. Find the distance of the bottom of the oil well from ground level.

4. What must be multiplied by:
   (a) 5 to make 40, (b) 1 to make -7,
   (c) -2 to make 12, (d) -3 to make -30,
   (e) 4 to make -36, (f) -9 to make 27?

5. Simplify the following.
   (a) $(+20) \div (-10)$ (b) $(-26) \div (+2)$
   (c) $(-6) \div (-6)$ (d) $(+18) \div (-3)$
   (e) $(-27) \div (+3)$ (f) $(+30) \div (+15)$
   (g) $\frac{-14}{-7}$ (h) $\frac{+50}{-25}$
   (i) $\frac{-60}{+12}$

6. What must be divided by:
   (a) 6 to make 2, (b) 6 to make -2,
   (c) -3 to make 5, (d) 8 to make -3,
   (e) -2 to make 11, (f) -3 to make 30?

7. Simplify the following.
   (a) $(-7) \times (-3) \times (-2)$
   (b) $(-1) \times (-1) \times (-1)$
   (c) $\frac{(-8) \times (+5)}{10}$ (d) $\frac{+60}{(-3) \times (-5)}$
   (e) $(-3) \times (-2)^2$ (f) $\frac{(-2) \times (-12)}{(+6)^2}$

8. Simplify the following.
   (a) $(-1\frac{1}{2}) \times 3\frac{2}{3}$ (b) $(-2.8) \times (-0.2)$
   (b) $\frac{4}{9}$ of $(-2\frac{4}{7})$ (d) $7 \times (-6.2)$

9. Simplify the following.
   (a) $1\frac{3}{4} \div (-2\frac{1}{4})$ (b) $(-4.8) \div (-6)$
   (c) $-7\frac{1}{5} \div (-9\frac{3}{5})$ (d) $-8.4 \div 7$

10. Simplify the following.
    (a) $7a \times (-6)$ (b) $2 \times (-11y)$
    (c) $4 \times (-4c)$ (d) $(-10) \times (-3y)$
    (e) $(-2x) \times (-9y)$ (f) $3a \times (-7b)$

11. Simplify the following.
    (a) $12x \div (-4)$ (b) $(-15b) \div 3$
    (c) $(-42y) \div (-7)$ (d) $(-3ac) \div a$
    (e) $(-8bd) \div (-4d)$ (f) $14x^2 \div (-2x)$

12 Simplify the following.

(a) $(-\frac{1}{8})$ of $(-48t)$ (b) $\frac{1}{4}$ of $(-36n)$

(c) $\dfrac{-36x^2}{-12x}$ (d) $\dfrac{22ab}{-2a}$

(e) $\dfrac{-39xy}{3y}$ (f) $\dfrac{-63m^2}{-21m}$

Summary

(a) To subtract a positive number, move to the left on the number line, that is, the same as to add a negative number.
(b) To subtract a negative number, move to the right on the number line, that is, the same as to add a positive number.
(c) When two numbers of the same sign are divided, one by the other, the result is positive.
(d) When two numbers of different sign are divided, one by the other, the result is negative.

Practice Exercise P3.1

1 Arrange the numbers from the largest to the smallest in each of the following sets :
(a) 14, 5, 12, 7, −17, 18
(b) 12, −15, 18, 21, −6, −9
(c) −17, −34, 23, −14, −18, 36
(d) 30, −43, −18, −11, −32, −12

2 How much bigger is the first number than the second? (You can use a number line to help you.)
(a) 9, 4 (b) 4, −2 (c) 3, −3
(d) 2, −4 (e) −2, −4 (f) −4, −9

Practice Exercise P3.2

1 Subtract the second number from the first
(a) −6, −2 (b) −3, −4 (c) −5 , −13
(d) −2, 3 (e) −3, 2 (f) −4 , 3
(g) −6, 8 (h) −3, 5 (i) 3, −5
(j) −8, 9 (k) −12, 3 (l) −18, 4
(m) −21, −17 (n) 21, 28 (o) −23, −56

2 Divide the first number by the second
(a) −3, 6 (b) 3, −6 (c) −21, −7
(d) 21, 28 (e) 24, −6 (f) −30, −15

Practice Exercise P3.3

Simplify and evaluate the following calculations:

1 $-5 - (-7)$ 2 $-7 - (-3)$
3 $-7 + (-3) - 1$ 4 $(-5) - 9 - 6$
5 $9 - (-6) + (-8)$ 6 $9 \div (-3) \div (-6)$
7 $-16 - (-7)$ 8 $15 \div (-3)$
9 $4 \times (-4) \div (-8)$ 10 $-12 \div (-4) \div (-3)$
11 $19 - (-6) + (-18)$ 12 $-20 \times (-4) \div (-2)$

Practice Exercise P3.4

Find the missing number

1 $-5 + * = (-7)$ 2 $-5 - * = (-7)$
3 $* - (-3) = 1$ 4 $* + (-3) = 1$
5 $9 - * = (-3)$ 6 $-13 \times * = -39$
7 $-18 - * = 6$ 8 $15 \div (-3) = *$
9 $* \div (-3) = (-6)$ 10 $-12 \times (-2) = *$
11 $-27 \div * = 9$ 12 $-16 - * = -9$
13 $19 - * + (-18) = 6$ 14 $3 \times (-4) \times * = 24$

Practice Exercise P3.5

1 Find the difference between these temperatures (using a number line may help you)
(a) −2 °C, 4 °C (b) 1 °C, 8 °C
(c) 4 °C, 9 °C (d) −3 °C, 2 °C
(e) −1 °C, 1 °C (f) −7 °C, 3 °C
(g) −8 °C, 0 °C (h) 5 °C, −3 °C
(i) 7 °C, −7 °C

2 Write the temperature that is
(a) 3 °C hotter than −2 °C
(b) 2 °C colder than −3 °C
(c) 9 °C hotter than −7 °C
(d) 10 °C colder than 1 °C
(e) 11 °C hotter than −5 °C

3 Find the difference between these temperatures.
(a) −3 °C and −5 °C (b) −7 °C and 2 °C
(c) −3 °C and −9 °C (d) 4 °C and −9 °C

4 Write down the new temperature if these changes occur.
(a) −9 °C increases by 3 °C
(b) −2 °C increases by 4 °C
(c) 6 °C decreases by 4 °C
(d) 2 °C decreases by 5 °C
(e) −2 °C decreases by 5 °C
(f) −9 °C decreases by 3 °C

Chapter 4
Mappings, ordered pairs

Pre-requisites
- relations; arrow diagrams; flow charts

An arrow diagram is used to show a relationship between the members of two sets. The relationship between the members of the Jones family (Book 1) is shown in Fig 4.1.

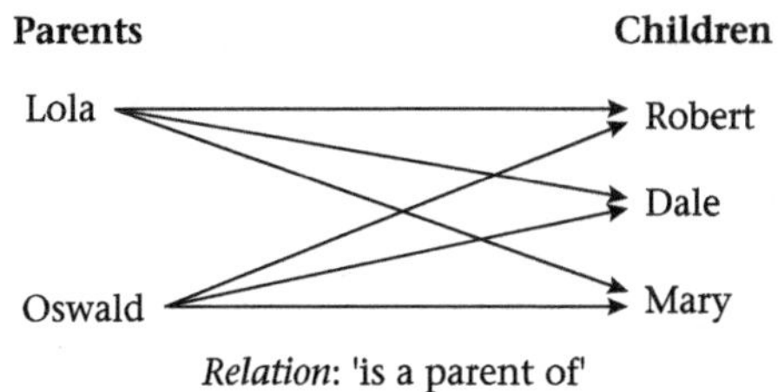

Relation: 'is a parent of'

Fig. 4.1

Mappings

Let us now look at the children of the Jones family and their favourite subjects in school.

Robert likes Mathematics best
Dale likes History best
Mary likes Mathematics best

The relationship between the children and their favourite subjects can be shown by writing it instead like this:

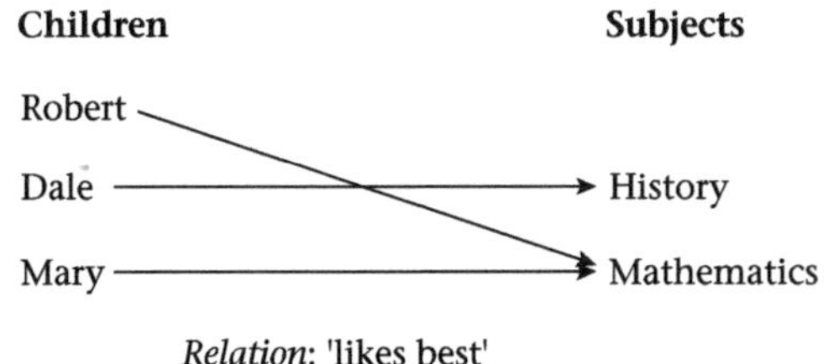

Relation: 'likes best'

Fig. 4.2

Notice the differences between Fig. 4.1 and Fig. 4.2.

(a) In Fig. 4.1, there is more than one arrow from the members of the first set to members of the second set.

(b) In Fig. 4.2, there is only one arrow from each of the members of the first set to the members of the second set. Fig. 4.2 indicates a special relation.

Example 1

Draw an arrow diagram mapping the members of the Jones family to their sex (male or female).

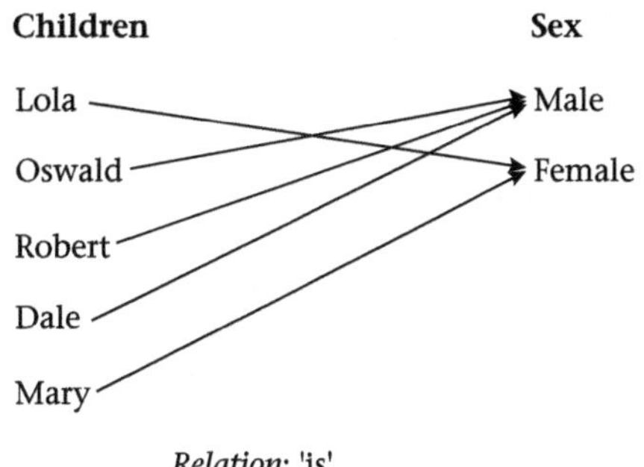

Relation: 'is'

Fig. 4.3

Notice that, in Fig. 4.3, one and *only* one arrow goes from *each* member of the first set to the second set. This diagram is an example of a **mapping**.

A mapping is a relation between two sets of objects in which each object in the first set 'goes to' one (and only one) object in the second set.

Note that there must be an arrow from *all* the objects of the first set to an object in the second set.

The relationship is sometimes called the **rule** of the mapping.

Exercise 4a

1. Consider the set of students in the same row of the class as yourself. Find out the favourite colour of each person in the row. Draw an arrow diagram showing the relationship 'has the favourite colour'. Write the name of each colour in the second set once only.

2. Using the set of students in your class and the months of the year, draw a mapping to illustrate the relation 'was born in'.

   student $\xrightarrow{\text{was born in}}$ month

3. Draw a diagram showing the relation between the set of students in your class and the set of shoe sizes, using the rule:

   student → size of student's shoes.

4. Draw a diagram to show the following mapping:

   First set: {the date of the first 7 days of the present month}
   Second set: {days of the week}
   Rule of mapping: date of the month → days of the week.

5. Draw diagrams to show the following mappings:

   (a) First set: {Grenada, Jamaica, Trinidad, Barbados, Antigua}
   Second set: {Kingston, St Johns, St Georges, Bridgetown, Port of Spain}
   Rule: country → capital of that country.

   (b) First set: {Montego Bay, Arima, Roseau, Castries, Georgetown, Chaguaramas, Ocho Rios, St Christopher}
   Second set: {St Kitts, Dominica, Guyana, Jamaica, Trinidad, St Lucia}
   Rule: place → country in which it is.

   What is the difference between mappings (a) and (b)?

6. Draw a diagram to show the following mapping:

   First set: {school subjects}
   Second set: {less than 4, 4, 5, 6, 7, 8, more than 8}
   Rule: subject → numbers of lessons per week which you have in that subject

7. (Group work)

   (a) Each member of the group should find out the number of legs that the ant, centipede, crab, spider have.

   (b) Draw a diagram to show the following relations:
   First set: {ant, cat, cow, dog, man, centipede, crab, spider}
   Second set: {1, 2, 3, 4, 5, 6, 7, 8, 10, more than 10}
   Rule: animal → number of legs.

   (c) Is the relation in (b) a mapping?

One-to-one mapping

In Exercise 4a question 5(a), notice that each member of the second set has one and only one arrow leading to it. That means St Johns is the capital of Antigua *only*. A mapping like this is called a **one-to-one** mapping. Exactly one object in one set is related to exactly one object in the other set. This shows that there are exactly as many countries as there are capitals.

Exercise 4b

1. Which of the mappings in Exercise 4a are one-to-one?

2. Which of the following mappings are one-to-one for the set of students in your class?

   (a) Student → student's Mathematics textbook
   (b) Student → home address
   (c) Student → subject disliked most
   (d) Student → number on class register
   (e) Student → nationality

3. Which of these mappings are one-to-one?

   (a) A dollar bill → number on dollar bill
   (b) Coin → year on coin
   (c) Date in the month → day of the week
   (d) The time → position of the hour hand

Example 2

Make a table showing the square of numbers from 2 up to 9.

Table 4.1

Number	2	3	4	5	6	7	8	9
Square of number	4	9	16	25	36	49	64	81

Example 3

The length of a child's pace is 30 cm. Make a table showing the distance the child covered in taking 7 paces.

Table 4.2

No. of paces	1	2	3	4	5	6	7
Distance covered (cm)	30	60	90	120	150	180	210

In Table 4.1 each number has its own square. In Table 4.2 a particular distance covered is given for each number of paces.

Both tables are examples of a relation which can be written as a mapping as shown in Tables 4.3 and 4.4.

Table 4.3

Number	2	3	4	5	6	7	8	9
↓	↓	↓	↓	↓	↓	↓	↓	↓
Square of number	4	9	16	25	36	49	64	81

Table 4.4

No. of paces	1	2	3	4	5	6	7
↓	↓	↓	↓	↓	↓	↓	↓
Distance covered (cm)	30	60	90	120	150	180	210

These are examples of a mapping table.

Exercise 4c

1. A student spends $2.50 each day of the school week on bus fares and lunch money. Make a table of the mapping:
number of days → amount of pocket-money spent. Complete the table for 8 days.
2. Make a table of the mapping:
number → square of the number.
Complete the table for numbers 10 to 18.
3. What angle does the hour hand of a clock turn through in one hour? Make a table of the mapping:
number of hours → angle turned through by the hour hand in degrees. Complete the table for one revolution of the hour hand.
4. A car is travelling at 40 km/h. Make a table of the mapping:
number of hours → distance travelled in kilometres. Complete the table for hourly intervals up to 5 hours.
5. A bus will take 32 seated passengers when fully loaded. Make a table of the mapping:
number of buses → number of passengers, when up to 5 fully loaded buses leave the bus station.
6. A household uses 12 kilowatt hours of electricity each day. Make a table of the mapping:
number of days → amount of electricity (kWh). Complete the table for 7 days.
7. The slope of a road is such that for every metre of the horizontal distance the vertical rise is 5 cm. Make a table of the mapping:
horizontal distance → vertical rise.
Complete the table for 10 metres horizontal distance at 1 metre intervals.

Ordered pairs

In Table 4.1 where each number has its own square, this information could be written:
(2, 4), (3, 9), (4, 16), (5, 25), (6, 36), (7, 49), (8, 64), (9, 81) where the first number is a member of the **domain** (the first set) and the second number is its match from the **range** (the second set). The order in which these numbers are written is important. These are called **ordered pairs**.

Would the numbers written in a different order be showing the same relationship?

Example 4

Tables 4.5 and 4.6 are two mapping tables.

Table 4.5

2	3	4	5	6	7
4	5	6	7	8	9

Table 4.6

4	5	6	7	8	9
2	3	4	5	6	7

Write the information from each table as ordered pairs and state the relation of each mapping. What do you notice?

(2, 4), (3, 5), (4, 6), (5, 7), (6, 8), (7, 9)
relation: 'is 2 less than'

(4, 2), (5, 3), (6, 4), (7, 5), (8, 6), (9, 7)
relation: 'is 2 more than'

The order in Table 4.6 is the opposite of the order in Table 4.5. The relation is reversed.

Example 5

State the rule for the following mapping. Write the information for the first seven terms as ordered pairs.

{1, 2, 3, 4, ...}
↓ ↓ ↓ ↓
{4, 8, 12, 16, ...}

Rule: number → four times the number.
(1, 4), (2, 8) (3, 12), (4, 16), (5, 20), (6, 24), (7, 28)

Example 6

(a) Copy and complete the flow chart (Fig. 4.4) for each member of the set {0, 1, 2, 3, 4}.
(b) Write the information as a set of ordered pairs.

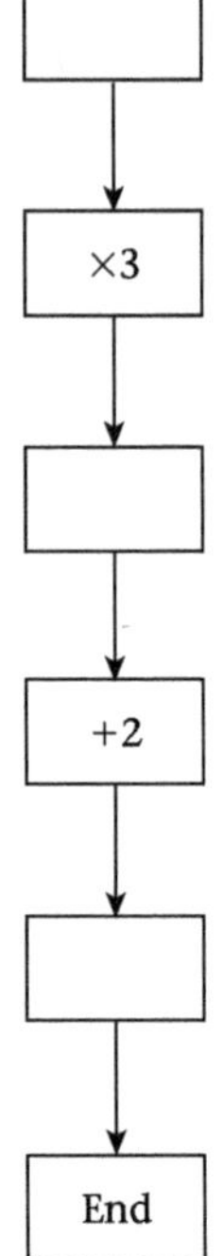

Fig. 4.4

(a)

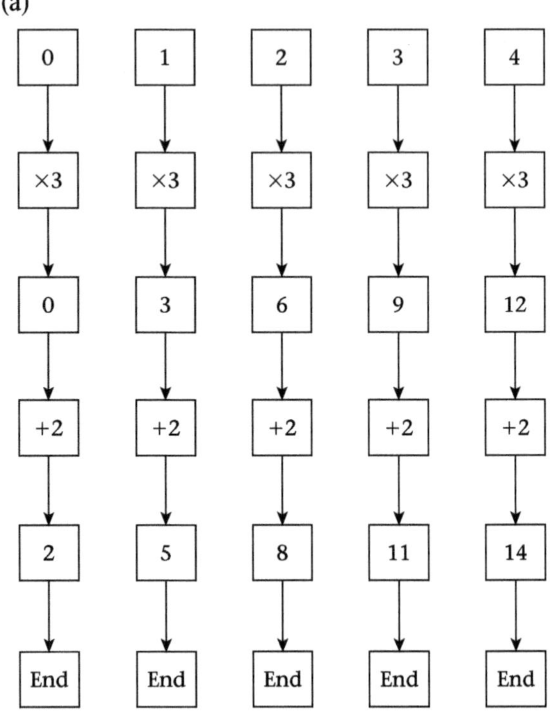

Fig. 4.5

(b) {(0, 2), (1, 5), (2, 8), (3, 11), (4, 14)}

Exercise 4d

1. Write the information from the mapping table for Exercise 4c question 4 as a set of ordered pairs.

2. An hourly paid worker is paid at the rate of \$50 per hour.
   (a) Make a mapping table: number of hours → wages. Complete the table for 8 hours.
   (b) Write the information as a set of ordered pairs.

3. If the hourly rate paid to the worker in question 2 is increased by 4, write the new information as a set of ordered pairs.

4. Water drips from a tap at the rate of 16 drops per minute. Complete a mapping table to show: number of minutes → number of drops of water. Show this information for a total of 10 minutes.

5 Using the flow chart with the members of the set {0, 1, 2, 3, 4}

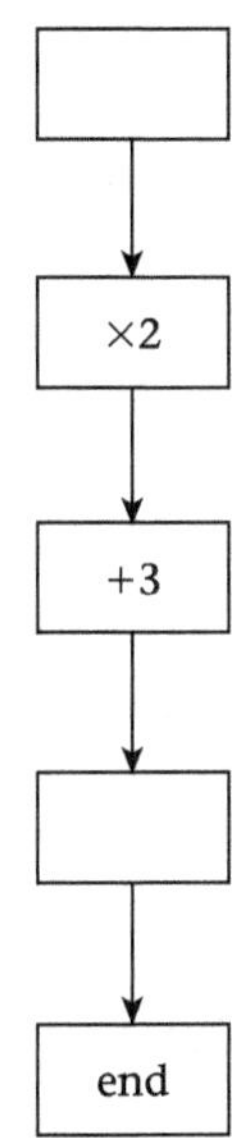

Fig. 4.6

(a) make a table of the mapping
(b) write the mapping as a set of ordered pairs.

6 Complete the flow chart below for each member of the set {2, 4, 6, 8}.

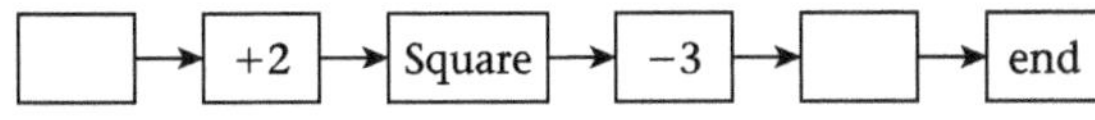

Fig. 4.7

7 Write a set of ordered pairs showing the relation: 'is 3 less than'.

8 Pocket money is shared between two brothers Jim and Joe, so that Joe gets one-half the amount Jim gets. Complete a mapping table to show how much each boy gets when Jim gets \$20, \$32, \$38, \$44, \$50 at different times.

9 Complete a flow chart for each member of the set {0, 1, 2, 3, 4, 5} to show the relation: 'multiply by 3 and then add 2'.

10 State the relation shown by the set of ordered pairs:
{(0, 2), (1, 5), (2, 8), (3, 11), (4, 14)}

Summary

A **mapping** is a relation between two sets of objects in which each object of the first set is related to one and only one object in the second set. Mappings may be many-to-one or one-to-one. Mappings may be shown as arrow diagrams, tables or ordered pairs. The order in which the members of the pairs are written is very important.

Practice Exercise P4.1

1 In each of the following examples
(a) draw arrow diagrams to show the relation between the members of the sets
(b) list the ordered pairs
(i) {3, 4, 5, 6, 8, 10, 12, 20, 21, 25} *is a multiple of*
(ii) {fly, ant, dove, mosquito, pigeon, crow} and {bird, insect} *belongs to the species*
(iii) {factors of 12} and {factors of 30} *is less than or equal to*
(iv) {1, 8, 27, 64} and {1, 2, 3, 4} *is the cube of*

2 If the domain D = {1, 2, 3, 4} and the co-domain R = {1, 2, 4, 8, 16}, for each of the relations
(i) is greater than or equal to
(ii) is the square root of
(iii) is less than
(a) draw an arrow diagram
(b) list the ordered pairs
(c) state the type of relation.

Practice Exercise P4.2

For each of the following sets of ordered pairs
(a) write the relation as $x \rightarrow$
(b) if it is a mapping, state the rule
(c) indicate whether it is a *one-to-one* or a *many-to-one* type

1 {(−3, 4), (−2, 4), (−1, 4), (0, 4)}

2 {(1, 5), (1, 6), (2, 7), (2, 8)}

3 {(6, 15), (5, 10), (4, 5), (3, 0), (2, −5)}

4 $\{(-1, 1), (0, 3), (2, 7), (5, 13)\}$

5 $\{(3, 0), (4, -1), (4, 1), (7, -2), (7, 2)\}$

Practice Exercise P4.3

For each of the following arrow diagrams
(a) list the members of
 (i) the domain (ii) the range
(b) write the relation as $x \rightarrow$
(c) state whether the relation is a mapping or not, clearly stating the reason
(d) if it is, indicate whether it is a *one-to-one* or a *many-to-one* type

1 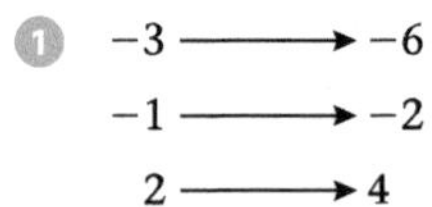

2 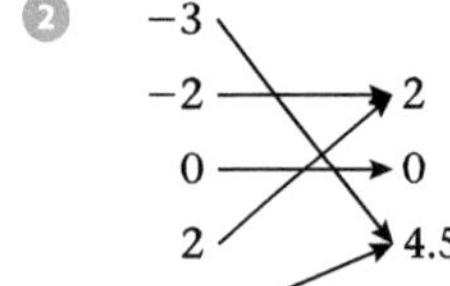

3 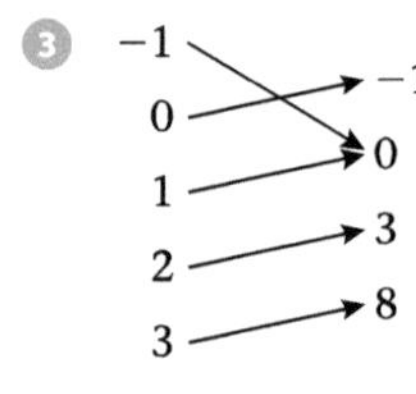

4 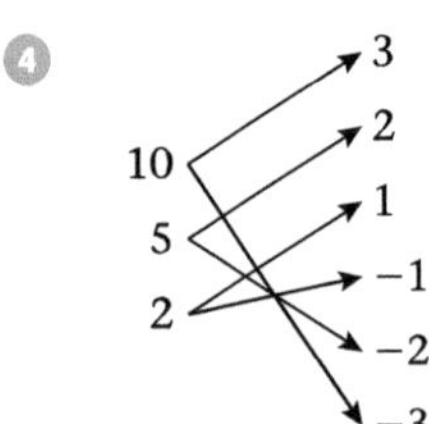

5 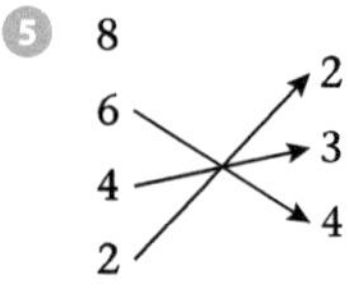

Fig. 4.8

Practice Exercise P4.4

1 (a) Find the values of
 (i) $f(-2)$, $f(-1)$, $f(0)$, $f(3)$, where $f: x \rightarrow 2x - 5$
 (ii) $F(-3)$, $F(-1)$, $F(1)$, $F(3)$, where $F: x \rightarrow 2x^3 + 1$
(b) Write each of the relations as a set of ordered pairs.

2 (a) Draw arrow diagrams for the following mappings.
(b) Write the relations as ordered pairs.
(c) State if each of the relations is a mapping and if so, whether *one-to-one* or *many-to-one*.
 (i) $x \rightarrow x^2$, for $\{-3, -2, -1, 0, 1, 2\}$
 (ii) number of hours → distance travelled, for the first 5 hours if the car is moving at 40 km/h

3 Given the relation $f(x) = 2x - 3$ for the domain $\{-2, -1, 0, 1, 2\}$,
(a) draw an arrow diagram to represent the relation
(b) find x when $f(x) = 3$.

4 (a) If $F: x \rightarrow 3x + 1$, and $F(b) = 2$, work out the value of b.
(b) If $G: x \rightarrow x^2 - 3$, and $G(c) = 1$, work out the values of c.

Practice Exercise P4.5

For the relations given, write the corresponding images for the objects in each set of ordered pairs.

1 $x \rightarrow x^2 - 5$
$\{(-2,\ ); (-1,\ ); (0,\ ); (1,\ ); (3,\ )\}$

2 $x \rightarrow 2 - 3x$
$\{(-1,\ ); (0,\ ); (2,\ ); (3,\ ); (5,\ )\}$

3 $x \rightarrow (x - 4)/5$
$\{(-1,\ ); (1,\ ); (4,\ ); (10,\ )\}$

Practice Exercise P4.6

In a universal set $\mathbf{U} = \{1, 2, 3, 4, \ldots, 10\}$, and the domain $\mathbf{D} = \{4, 5, 6, 7, 8, 9\}$.

For each of the following relations,
(a) list the members of the range using the letter indicated
(b) write the set of ordered pairs
(c) identify the relations that are mappings, clearly giving the reason
(d) write each of the mappings in the forms
 (i) $y = f(x)$, and (ii) $f: x \rightarrow$
(e) What do you notice about the relation and $f(x)$?

1 is five more than [F]

2 is a factor of [M]

3 is the square of [S]

4 is four times [T]

Chapter 5

Simplifying algebraic expressions (2)

Brackets, indices, factors, fractions

Pre-requisites

- Grouping, multiplying and dividing algebraic terms

Removing brackets

$3 \times (7 + 5)$ means first add 7 and 5, then multiply the result by 3. Suppose a pencil costs 7 cents and a rubber costs 5 cents.

Cost of a pencil and rubber
= 7 cents + 5 cents
= (7 + 5) cents = 12 cents

If 3 students each buy a pencil and rubber, then,
total cost = $3 \times (7 + 5)$ cents
= 3×12 cents
= 36 cents

There is another way to find the total cost.
3 pencils cost 3×7 cents. 3 rubbers cost 3×5 cents. Altogether,
total cost = 3×7 cents + 3×5 cents
= 21 cents + 15 cents
= 36 cents
Thus, $3 \times (7 + 5) = 3 \times 7 + 3 \times 5$.

This shows that brackets can be removed by multiplying the 3 into *both* the 7 and the 5.

Usually we do not write the multiplication sign. We just write 3(7 + 5). 3(7 + 5) is short for $3 \times (7 + 5)$. Say 3(7 + 5) as '3 multiplied into (7 + 5)' or just '3 into (7 + 5)'.

Thus, $3(7 + 5) = 3 \times 7 + 3 \times 5$

In general terms, using letters for numbers,

$$a(x + y) = ax + ay$$

Notice also that

$$3(7 - 5) = 3 \times 2 = 6$$

and $3 \times 7 - 3 \times 5 = 21 - 15 = 6$
Thus, $3(7 - 5) = 3 \times 7 - 3 \times 5$

Again, using letters for numbers,

$$a(x - y) = ax - ay$$

Remember that in algebra, letters stand for numbers. The numbers can be whole or fractional, positive or negative.

Exercise 5a

Simplify each of the following in two ways. Question 1 shows you how to do this.

1. $9(7 - 3)$ (a) $9(7 - 3) = 9 \times 4 = 36$
   (b) $9(7 - 3) = 9 \times 7 - 9 \times 3 = 63 - 27 = 36$
2. $3(7 + 2)$
3. $10(6 - 4)$
4. $9(4 + 3)$
5. $6(4 - 2)$
6. $7(1 + 9)$
7. $3(8 - 2)$
8. $3(8 + 2)$
9. $4(11 - 3)$
10. $5(10 + 2)$

Example 1

Remove brackets from the following.
(a) $8(2c + 3d)$
(b) $4y(3x - 5)$
(c) $(7a - 2b)3a$

(a) $8(2c + 3d) = 8 \times 2c + 8 \times 3d$
$= 16c + 24d$
(b) $4y(3x - 5) = 4y \times 3x - 4y \times 5$
$= 12xy - 20y$
(c) $(7a - 2b)3a$

Notice that we can multiply into the bracket from the right.
$(7a - 2b)3a = 7a \times 3a - 2b \times 3a$
$= 21a^2 - 6ab$

Exercise 5b (Oral)

Remove brackets from the following.

1. $3(x + y)$
2. $4(9 - n)$
3. $2(a - 7)$
4. $5(2p - q)$
5. $6(3a + 4b)$
6. $7(7x + 2y)$
7. $2a(4 + 7c)$
8. $4y(6x - 8)$
9. $3x(7 - 8p)$
10. $3q(- p + q)$

11. $2a(-a-b)$
12. $5d(-b-2c)$
13. $(a+2b)2$
14. $(u-v)5$
15. $(3a+b)3$
16. $(5-3d)3a$
17. $(7c-2)7c$
18. $(4s+t)6s$

The multiplier outside the bracket can be a directed number.

Example 2

Remove brackets from the following.
(a) $-7(2n+3)$
(b) $-5z(x-9y)$
(c) $(3-5a)(-2a)$

(a) $-7(2n+3)$
$= (-7)\times(+2n)+(-7)\times(+3)$
$= (-14n)+(-21)$
$= -14n-21$

Notice that $+(-21) = -(+21) = -21$.
Adding a negative number is equivalent to subtracting a positive number of the same value.

(b) $-5z(x-9y)$
$= (-5z)\times(+1)\times x-(-5z)\times(9y)$
$= (-5xz)-(-45yz)$
$= -5xz+45yz$
$= 45yz-5xz$

Notice that
$-(-45yz) = +(+45yz) = +45yz$.

(c) $(3-5a)(-2a)$
$= (+3)\times(-2a)-(+5a)\times(-2a)$
$= (-6a)-(-10a^2)$
$= -6a+10a^2$
$= 10a^2-6a$

Example 2 shows that **when the multiplier is negative, the signs inside the bracket are changed when the bracket is removed.**

Exercise 5c (Oral)

Remove brackets from the following.

1. $-3(m+n)$
2. $-2(u+v)$
3. $-4(a+b)$
4. $-5(a-b)$
5. $-8(x-y)$
6. $-9(p-q)$
7. $-4n(3m+2)$
8. $-7y(3-5y)$
9. $-a(4a+6)$
10. $-2a(-a-3b)$
11. $-5x(11x-2y)$
12. $-p(p-5q)$
13. $(2c+8d)(-2)$
14. $(-5m+2n)(-9m)$
15. $(10+3t)(-7t)$
16. $-x(2x-11y)$
17. $-3(12a-5)$
18. $-5a(-5x-7y)$

Example 3

Remove brackets and simplify the following.
(a) $3(6a+3b)+5(2a-b)$
(b) $2(3x-y)-3(2x-3y)$
(c) $x(x-7)+4(x-7)$

(a) $3(6a+3b)+5(2a-b)$
$= 18a+9b+10a-5b$
$= 18a+10a+9b-5b$
$= 28a+4b$

Notice that like terms are grouped.

(b) $2(3x-y)-3(2x-3y)$
$= (+2)\,(3x-y)+(-3)\,(2x-3y)$*
$= 6x-2y+(-6x)+(+9y)$*
$= 6x-2y-6x+9y$
$= 6x-6x-2y+9y$
$= 7y$

* You will be able to leave out the first two steps after trying a few exercises.

(c) $x(x-7)+4(x-7)$
$= x\times x-x\times 7+4x-28$
$= x^2-7x+4x-28$
$= x^2-3x-28$

Exercise 5d

Remove brackets and simplify the following.

1. $3a+2(a+2b)$
2. $6x+3(2y-x)$
3. $7p+5(p-q)$
4. $5c+3(1+2c)$
5. $8a+5(2a-b)$
6. $11x+3(3x+2y)$
7. $12x-2(4x+5)$
8. $9r-4(3+r)$
9. $14a-3(2b+5a)$
10. $4a-5(a-2)$
11. $3-7(5-4x)$
12. $10t-8(3-t)$
13. $5(a+2)+4(a+1)$
14. $2(5x+8y)+3(2x-y)$
15. $2(3x-y)+3(x+2y)$
16. $2(a-3b)+3(a-b)$
17. $6(4x+y)-7(3x+5y)$

18 $4(x - 2y) - 3(2x - y)$
19 $7(a - b) - 8(a - 2b)$
20 $6(a - 2b) - 3(2a + b)$
21 $x(x - 2) + 3(x + 2)$
22 $x(x + 4) + 5(x + 4)$
23 $x(x - 2) + 7(x - 2)$
24 $x(x - 8) + 4(x - 8)$
25 $a(a + 5) - 2(a + 5)$
26 $a(a + 2) - 9(a + 2)$
27 $y(y - 3) - 6(y - 3)$
28 $z(z - 1) - 10(z - 1)$
29 $x(x + a) + a(x + a)$
30 $a(a + b) - b(a + b)$

Expanding algebraic expressions

The expression $(a + 2)\ (b - 5)$ means

$$(a + 2) \times (b - 5)$$

The terms in the first bracket, $(a + 2)$, multiply each term in the second bracket, $(b - 5)$. Just as

$$x(b - 5) = bx - 5x$$

so, writing $(a + 2)$ instead of x,

$$(a + 2)(b - 5) = b(a + 2) - 5(a + 2)$$

The brackets on the right-hand side can now be removed.

$$\begin{aligned}(a + 2)(b - 5) &= b(a + 2) - 5(a + 2)\\ &= ab + 2b - 5a - 10\end{aligned}$$

$ab + 2b - 5a - 10$ is the product of $(a + 2)(b - 5)$. We often say that the **expansion** of $(a + 2)(b - 5)$ is $ab + 2b - 5a - 10$.

Example 4

Expand the following.
(a) $(a + b)(c + d)$
(b) $(6 - x)(3 + y)$
(c) $(2p - 3q)(5p - 4)$

(a) $\begin{aligned}(a + b)(c + d) &= c(a + b) + d(a + b)\\ &= ac + bc + ad + bd\end{aligned}$

(b) $\begin{aligned}(6 - x)(3 + y) &= 3(6 - x) + y(6 - x)\\ &= 18 - 3x + 6y - xy\end{aligned}$

(c) $\begin{aligned}(2p - 3q)(5p - 4) &= 5p(2p - 3q) - 4(2p - 3q)\\ &= 10p^2 - 15pq - 8p + 12q\end{aligned}$

When expanding brackets it is most important to be careful with the signs of the terms in the final product.

Exercise 5e

Expand the following.

1 $(p + q)(r + s)$
2 $(x + 8)(y + 3)$
3 $(4 + 5a)(3b + a)$
4 $(a - b)(c + d)$
5 $(x - 7)(y + 1)$
6 $(2 - p)(3p + q)$
7 $(w + x)(y - z)$
8 $(a + 6)(b - 9)$
9 $(2m + 5n)(p - 3q)$
10 $(a - b)(c - d)$
11 $(x - 4)(y - 5)$
12 $(5x - y)(x - 3y)$

It is often possible to simplify terms in the final product.

Example 5

Expand $(x + 8)(x + 5)$.

$$\begin{aligned}(x + 8)(x + 5) &= x(x + 8) + 5(x + 8)\\ &= x^2 + 8x + 5x + 40\end{aligned}$$

Notice that the middle two terms are both terms in x. They can be collected together.

$$(x + 8)(x + 5) = x^2 + 13x + 40$$

Example 6

Expand the following.
(a) $(6 + x)(3 - x)$
(b) $(2a - 3b)(a + 4b)$

(a) $\begin{aligned}(6 + x)(3 - x) &= 3(6 + x) - x(6 + x)\\ &= 18 + 3x - 6x - x^2\\ &= 18 - 3x - x^2\end{aligned}$

(b) $\begin{aligned}(2a - 3b)(a + 4b) &= a(2a - 3b) + 4b(2a - 3b)\\ &= 2a^2 - 3ab + 8ab - 12b^2\\ &= 2a^2 + 5ab - 12b^2\end{aligned}$

Example 7

Expand $(x - 3)^2$.

$$\begin{aligned}(x - 3)^2 &= (x - 3)(x - 3)\\ &= x(x - 3) - 3(x - 3)\\ &= x^2 - 3x - 3x + 9\\ &= x^2 - 6x + 9\end{aligned}$$

Exercise 5f

Expand the following.

1. $(a+3)(a+4)$
2. $(b+2)(b+5)$
3. $(m+3)(m-2)$
4. $(n-7)(n+2)$
5. $(x+2)^2$
6. $(y+1)(y-4)$
7. $(c-2)(c+5)$
8. $(d-3)(d-4)$
9. $(p-2)(p-5)$
10. $(x-4)^2$
11. $(y+1)(y+7)$
12. $(a-4)(a+6)$
13. $(b-3)(b-7)$
14. $(c+5)(c-1)$
15. $(3+d)(2+d)$
16. $(5-x)(2+x)$
17. $(3-y)(4-y)$
18. $(m+2n)(m+3n)$
19. $(a-3b)(a+2b)$
20. $(x-4y)(x-3y)$
21. $(p+2q)^2$
22. $(m+5n)(m-3n)$
23. $(a+5)(2a-3)$
24. $(3x+4)(x-2)$
25. $(2h-k)(3h+2k)$
26. $(5x+2y)(3x+4y)$
27. $(3a-2b)^2$
28. $(5h+k)^2$
29. $(5a-2b)(2a-3b)$
30. $(7m-5n)(5m+3n)$

Laws of indices

10^3 is short for $10 \times 10 \times 10$. Similarly, x^5 is short for $x \times x \times x \times x \times x$. x can be any number.

Example 8

Multiply (a) x^5 by x^3
(b) a^3 by a^2
(c) y by y^4.

(a) $x^5 \times x^3 = (x \times x \times x \times x \times x) \times (x \times x \times x)$
$= x \times x \times x \times x \times x \times x \times x \times x$
$= x^8$

(b) $a^3 \times a^2 = (a \times a \times a) \times (a \times a)$
$= a \times a \times a \times a \times a$
$= a^5$

(c) $y \times y^4 = y \times (y \times y \times y \times y)$
$= y \times y \times y \times y \times y$
$= y^5$

Notice that the index in the result is the sum of the given indices:

$x^5 \times x^3 = x^{5+3} = x^8$
$a^3 \times a^2 = a^{3+2} = a^5$
$y \times y^4 = y^1 \times y^4 = y^{1+4} = y^5$

In general: $\boldsymbol{x^a \times x^b = x^{a+b}}$

Example 9

Simplify the following.
(a) $10^4 \times 10^2$
(b) $4c^3 \times 7c^2$

In fully expanded form:

(a) $10^4 \times 10^2$
$= (10 \times 10 \times 10 \times 10) \times (10 \times 10)$
$= 10 \times 10 \times 10 \times 10 \times 10 \times 10$
$= 10^6$

(b) $4c^3 \times 7c^2 = 4 \times c \times c \times c \times 7 \times c \times c$
$= 4 \times 7 \times c \times c \times c \times c \times c$
$= 28c^5$

Or, more quickly, by adding indices:
(a) $10^4 \times 10^2 = 10^{4+2} = 10^6$
(b) $4c^3 \times 7c^2 = 4 \times 7 \times c^{3+2} = 28c^5$

Exercise 5g (Oral)

1. Simplify the following by fully expanding the terms.
(a) $x^2 \times x^3$ (b) $a^4 \times a^3$
(c) $n^5 \times n$ (d) $3a^2 \times 8a^4$
(e) $5x^3 \times 4x^7$ (f) $3c^2 \times 2c^5$

2. Simplify the following by adding the indices.
(a) $m^3 \times m^5$ (b) $a^6 \times a^4$
(c) $c^5 \times c^9$ (d) $10^2 \times 10^7$
(e) $b^8 \times b^7$ (f) $x^7 \times x$
(g) $2e^4 \times 5e^{10}$ (h) $3 \times 10^6 \times 5 \times 10^3$
(i) $5y^5 \times 3y^3$

Example 10

Divide (a) x^5 by x^3
(b) a^7 by a^4
(c) p^6 by p^5.

(a) $x^5 \div x^3 = \dfrac{x^5}{x^3} = \dfrac{x \times x \times x \times x \times x}{x \times x \times x}$
$= \dfrac{x \times x \times (x \times x \times x)}{(x \times x \times x)}$
$= x^2$

(b) $a^7 \div a^4 = \dfrac{a^7}{a^4} = \dfrac{a \times a \times a \times a \times a \times a \times a}{a \times a \times a \times a}$
$= \dfrac{a \times a \times a \times (a \times a \times a \times a)}{(a \times a \times a \times a)}$
$= a^3$

(c) $p^6 \div p^5 = \frac{p^6}{p^5} = \frac{p \times p \times p \times p \times p \times p}{p \times p \times p \times p \times p}$

$= \frac{p \times (p \times p \times p \times p \times p)}{(p \times p \times p \times p \times p)}$

$= p$

Notice that the index in the result is the index of the divisor subtracted from the index of the dividend:

$x^5 \div x^3 = x^{5-3} = x^2$

$a^7 \div a^4 = a^{7-4} = a^3$

$p^6 \div p^5 = p^{6-5} = p^1 = p$

In general: $\boldsymbol{x^a \div x^b = x^{a-b}}$

Example 11

Divide (a) 10^6 by 10^2
(b) $12a^7$ by $3a^2$.

In fully expanded form:

(a) $10^6 \div 10^2 = \frac{10^6}{10^2}$

$= \frac{10 \times 10 \times 10 \times 10 \times 10 \times 10}{10 \times 10}$

$= 10 \times 10 \times 10 \times 10 = 10^4$

(b) $12a^7 \div 3a^2$

$= \frac{12 \times a \times a \times a \times a \times a \times a \times a}{3 \times a \times a}$

$= \frac{4 \times (3 \times a \times a) \times a \times a \times a \times a \times a}{(3 \times a \times a)}$

$= 4 \times a \times a \times a \times a \times a = 4a^5$

Or, more quickly, by subtracting indices:

(a) $10^6 \div 10^2 = 10^{6-2} = 10^4$

(b) $12a^7 \div 3a^2 = \frac{12}{3} \times a^{7-2} = 4a^5$

Exercise 5h (Oral)

1. Simplify the following by fully expanding the terms.
   (a) $a^7 \div a^3$ (b) $c^4 \div c$
   (c) $\frac{d^6}{d^5}$ (d) $12x^7 \div 4x^3$
   (e) $10a^8 \div 5a^6$ (f) $4x^6 \div 4x$

2. Simplify the following by subtracting indices.
   (a) $x^6 \div x^4$ (b) $b^8 \div b^5$
   (c) $c^9 \div c^3$ (d) $\frac{a^{11}}{a^9}$
   (e) $10^9 \div 10^7$ (f) $\frac{x^7}{x}$
   (g) $18x^5 \div 9x^4$ (h) $\frac{24x^8}{6x^5}$
   (i) $\frac{8 \times 10^9}{4 \times 10^6}$

Example 12

Simplify $x^3 \div x^3$
(a) by fully expanding each term,
(b) by subtracting indices.

(a) $x^3 \div x^3 = \frac{x \times x \times x}{x \times x \times x} = 1$

(b) $x^3 \div x^3 = x^{3-3} = x^0$

From the results of parts (a) and (b) in Example 9, $x^0 = 1$.

In general: **any number raised to the power 0 has the value 1.**

Example 13

Simplify $x^2 \div x^5$
(a) by fully expanding each term,
(b) by subtracting indices.

(a) $x^2 \div x^5 = \frac{x \times x}{x \times x \times x \times x \times x}$

$= \frac{1}{x \times x \times x} = \frac{1}{x^3}$

(b) $x^2 \div x^5 = x^{2-5} = x^{-3}$

From the results of parts (a) and (b) in Example 10, $x^{-3} = \frac{1}{x^3}$

In general, $x^{-a} = \frac{1}{x^a}$

Example 14

Simplify the following:
(a) 10^{-3} (b) $x^0 \times x^4 \times x^{-2}$
(c) $a^{-3} \div a^{-5}$ (d) $(\frac{1}{4})^{-2}$

(a) $10^{-3} = \frac{1}{10^3} = \frac{1}{1000}$

(b) $x^0 \times x^4 \times x^{-2} = 1 \times x^4 \times \frac{1}{x^2} = \frac{x^4}{x^2}$

$= x^{4-2} = x^2$

or

$x^0 \times x^4 \times x^{-2} = x^{0+4+(-2)} = x^2$

(c) $a^{-3} \div a^{-5} = \frac{1}{a^3} \div \frac{1}{a^5} = \frac{1}{a^3} \times \frac{a^5}{1}$
$= a^{5-3} = a^2$

or

$a^{-3} \div a^{-5} = a^{(-3)-(-5)}$
$= a^{-3+5} = a^2$

(d) $(\frac{1}{4})^{-2} = \frac{1}{(\frac{1}{4})^2} = \frac{1}{\frac{1}{16}} = 16$

Example 15

Simplify $(x^2)^3$

(a) by fully expanding the expression
(b) by applying the law of indices.

(a) $(x^2)^3 = (x^2) \times (x^2) \times (x^2)$
$= (x \times x) \times (x \times x) \times (x \times x)$
$= x^6$

(b) $(x^2)^3 = (x^2) \times (x^2) \times (x^2)$
$= x^{2+2+2}$
$= x^6$

From the results of (a) and (b), we see that
$(x^2)^3 = x^{2 \times 3}$
$= x^6$

In general, $\boldsymbol{(x^a)^b = x^{a \times b}}$

Exercise 5i

Simplify the following.

1. 10^{-2}
2. 10^{-4}
3. 10^{-6}
4. $x^5 \times x^{-2}$
5. $a^{-2} \times a^{-3}$
6. $m^0 \times n^0$
7. $a^2 \div a^7$
8. $x^2 \div x^{-7}$
9. $p^{-2} \div p^{-7}$
10. $b^3 \div b^0$
11. $r^7 \div r^7$
12. $(a^3)^2$
13. $(\frac{1}{2})^{-3}$
14. $(\frac{1}{3})^{-2}$
15. $(\frac{1}{9})^{-1}$
16. $2a^{-1} \times 3a^2$
17. $(2a)^{-1} \times 3a^2$
18. $2a^{-1} \times (3a)^2$
19. $c^{-1} \times c^{-1}$
20. $(b^{-2})^3$

Factors of algebraic terms

All numbers, other than 1, have two or more factors. For example, the factors of 42 are: 1, 2, 3, 6, 7, 14, 21, 42.

In the same way, algebraic terms have two or more factors. For example, the expression $6ab$ has 16 factors: 1, 2, 3, 6, a, $2a$, $3a$, $6a$, b, $2b$, $3b$, $6b$, ab, $2ab$, $3ab$, $6ab$. Each factor divides exactly into $6ab$. For example,

$$6ab \div 3a = \frac{2 \times 3 \times a \times b}{3 \times a} = 2b$$

$$6ab \div 2ab = \frac{2 \times 3 \times a \times b}{2 \times a \times b} = 3$$

Example 16

Write down all the factors of $5a^2x$.

Expand $5a^2x$ as a product of separate terms:
$5a^2x = 5 \times 1 \times a \times a \times x$

The factors will contain:

numerical terms:	1, 5
terms in a:	a, $5a$
terms in x:	x, $5x$
terms in a^2:	a^2, $5a^2$
terms in ax:	ax, $5ax$
terms in a^2x:	a^2x, $5a^2x$

The factors of $5a^2x$ are: 1, 5, a, $5a$, x, $5x$, a^2, $5a^2$, ax, $5ax$, a^2x, $5a^2x$.

Notice that 1 and the term itself are always factors of an algebraic term.

Exercise 5j (Oral)

State the factors of the following.

1. $3x$
2. ab
3. x^2
4. a^2b
5. $6a$
6. $5ab$
7. xyz
8. $10m^2$
9. $25pq$
10. $8de$
11. $2a^2b^2$
12. $14ab^2$

Highest common factor

Algebraic expressions may have common factors.

Example 17

Find the HCF of $12ab^2$ and $30a^2b$.

$12ab^2 = 2 \times 2 \times 3 \times a \times b \times b$
$30a^2b = 2 \times 3 \times 5 \times a \times a \times b$

The highest product of factors that is contained in both expressions is $2 \times 3 \times a \times b = 6ab$.
$6ab$ is the HCF of $12ab^2$ and $30a^2b$.

With practice the HCF can be found without expanding each expression.

Exercise 5k (Oral)

Find the HCF of the following.

1. ax and ay
2. $3d$ and $3e$
3. ax and bx
4. $3m$ and $5m$
5. $2m$ and $2n$
6. gp and gq
7. $2xy$ and $7xz$
8. $4r$ and $12s$
9. abz and xyz
10. $3mx$ and $10nx$
11. $4pq$ and $4pr$
12. abc and abd
13. axy and bxy
14. $9pq$ and $11pq$
15. $4ax$ and $10bx$
16. amx and anx
17. x^2 and $5x$
18. $2x^2$ and $8x$
19. $5x^2$ and $10x$
20. $3a^2$ and $12a$
21. mn^2 and m^2n
22. $8de$ and $6de$
23. $10m^2$ and $15m^2$
24. $9ab^2$ and $12a^2b$
25. $18xy$ and $6xy$
26. $21ad$ and $14bd$
27. $5a^2$ and $20a^2b$
28. $16xy^2$ and $4xy$
29. $18a^2y$ and $27x^2y^2$
30. $17ax^2$ and $3by$

Factorisation

Example 18

Complete the brackets in the statement
$15ax + 10a = 5a(\quad)$.

$15ax + 10a = 5a(\quad)$.
The contents of the bracket multiplied by $5a$ should give $15ax + 10a$. See **Removing brackets** earlier. Thus, divide each term in $15ax + 10a$ by $5a$ to find the contents of the bracket.

$$\text{Contents of bracket} = \frac{15ax}{5a} + \frac{10a}{5a} = 3x + 2$$

Thus $15ax + 10a = 5a(3x + 2)$
Check: RHS $= 5a(3x + 2) = 5a \times 3x + 5a \times 2$
$= 15ax + 10a =$ LHS

In the above example, $5a$ is the HCF of $15ax$ and $10a$. $5a(3x + 2)$ is the factorised form of $15ax + 10a$. $5a$ and $(3x + 2)$ are factors of $15ax + 10a$. **Factorisation** means writing an expression in terms of its factors. Think of factorisation as the inverse of removing brackets.

Example 19

Factorise the following:
(a) $12y + 8z$
(b) $4n^2 - 2n$
(c) $24pq - 16p^2$

(a) $12y + 8z$
The HCF of $12y$ and $8z$ is 4.

$$12y + 8z = 4\left(\frac{12y}{4} + \frac{8z}{4}\right) * = 4(3y + 2z)$$

(b) $4n^2 - 2n$
The HCF of $4n^2$ and $2n$ is $2n$.

$$4n^2 - 2n = 2n\left(\frac{4n^2}{2n} - \frac{2n}{2n}\right) * = 2n(2n - 1)$$

(c) $24pq - 16p^2$
The HCF of $24pq$ and $16p^2$ is $8p$.

$$24pq - 16p^2 = 8p\left(\frac{24pq}{8p} - \frac{16p^2}{8p}\right) * = 8p(3q - 2p)$$

* With practice the first line of working can be left out. The results of factorisation can be checked by removing the brackets.

Exercise 5l

1. Copy the following statements and complete the brackets.
   (a) $9x + 3y = 3(\quad)$
   (b) $5a - 15b = 5(\quad)$
   (c) $ax + ay = a(\quad)$
   (d) $px + qx = x(\quad)$
   (e) $8am - 8bm = 8m(\quad)$
   (f) $a - ay = a(\quad)$
   (g) $2s - rs = s(\quad)$
   (h) $3ab + 5ax = a(\quad)$
   (i) $3abx + 5adx = ax(\quad)$
   (j) $9xy - 3xz = 3x(\quad)$
   (k) $12cm + 16dm = 4m(\quad)$
   (l) $15x^2 + 10x = 5x(\quad)$
   (m) $18ax + 9x = 9x(\quad)$
   (n) $4m^2 - 2m = 2m(\quad)$
   (o) $2pq - 6q^2 = 2q(\quad)$
   (p) $5a^2 + 2ax = a(\quad)$

2 Factorise the following.

(a) $12c + 6d$ (b) $4a - 8b$
(c) $6z - 3$ (d) $9x + 12y$
(e) $xy + xz$ (f) $bc + dc$
(g) $9x + ax$ (h) $abc + abd$
(i) $xyz - ayz$ (j) $pmq + amb$
(k) $3ab - 6ac$ (l) $12ax + 8bx$
(m) $3x^2 - x$ (n) $6m^2 - 2$
(o) $2abx + 7acx$ (p) $3d^2e + 5d^2$
(q) $4am^2 - 6am$ (r) $-5x^2 - 10$
(s) $-18fg - 12g$ (t) $-5xy + 10y$

Lowest common multiple

Example 20

Find the LCM of the following.

(a) $8a$ and $6a$ (b) $2x$ and $3y$

(a) $8a = 2 \times 2 \times 2 \times a$
$6a = 2 \times 3 \times a$
LCM $= 2 \times 2 \times 2 \times 3 \times a = 24a$

(b) $2x = 2 \times x$
$3y = 3 \times y$
LCM $= 2 \times 3 \times x \times y = 6xy$

Exercise 5m (Oral)

Find the LCM of the following.

1 a and b 2 x and 5
3 $2a$ and 3 4 $3a$ and $4b$
5 $2x$ and $5y$ 6 $9a$ and a
7 x and $3x$ 8 ab and bc
9 xy and yz 10 $3b$ and $2b$
11 x^2 and x 12 $3a$ and a^2
13 $3m$ and m^2n 14 $2a^2$ and $9ab$
15 $3x^2y$ and $2xy^2$ 16 $6ab$ and $5b^2$

Algebraic fractions

Equivalent fractions

Equivalent fractions can be made by multiplying or dividing both the numerator and the denominator of a fraction by the same quantity.

(i) by multiplication

(a) $\frac{a}{2} = \frac{a \times 3}{2 \times 3} = \frac{3a}{6}$

(b) $\frac{a}{2} = \frac{a \times a}{2 \times a} = \frac{a^2}{2a}$

(c) $\frac{3}{d} = \frac{3 \times 2b}{d \times 2b} = \frac{6b}{2bd}$

(ii) by division

(d) $\frac{4x}{6y} = \frac{4x \div 2}{6y \div 2} = \frac{2x}{3y}$

(e) $\frac{5ab}{10b} = \frac{5ab \div 5b}{10b \div 5b} = \frac{a}{2}$

Example 21

Fill the blanks in the following

(a) $\frac{3a}{2} = \frac{\quad}{10}$ (b) $\frac{5ab}{12a} = \frac{\quad}{12}$ (c) $\frac{9bc}{12b} = \frac{3c}{\quad}$

(a) Compare the two denominators.
$2 \times 5 = 10$

The denominator of the first fraction has been multiplied by 5. The numerator must also be multiplied by 5.

$\frac{3a}{2} = \frac{3a \times 5}{2 \times 5} = \frac{15a}{10}$

(b) The denominator of the first fraction has been divided by a. The numerator must also be divided by a.

$\frac{5ab}{12a} = \frac{5ab \div a}{12a \div a} = \frac{5b}{12}$

(c) Divide both numerator and denominator by $3b$.

$\frac{9bc}{12b} = \frac{9bc \div 3b}{12b \div 3b} = \frac{3c}{4}$

Exercise 5n (Oral)

Fill the blanks in the following.

1 $\frac{8b}{5} = \frac{\quad}{15}$ 2 $\frac{a}{b} = \frac{ax}{\quad}$ 3 $\frac{2x}{5} = \frac{\quad}{10}$

4 $\frac{3a}{8} = \frac{6a}{\quad}$ 5 $\frac{3}{12a} = \frac{\quad}{4a}$ 6 $\frac{9c}{21d} = \frac{\quad}{7d}$

7 $\frac{9ah}{6ak} = \frac{3h}{\quad}$ 8 $\frac{bm}{my} = \frac{\quad}{y}$ 9 $\frac{\quad}{3a} = \frac{1}{a}$

10 $\frac{\quad}{8yz} = \frac{3x}{2y}$ 11 $\frac{2c}{a} = \frac{6c^2}{\quad}$ 12 $\frac{3m}{1} = \frac{\quad}{2d}$

Adding and subtracting algebraic fractions

As with numerical fractions, algebraic fractions must have common denominators before they can be added or subtracted.

Example 22

Simplify the following.

(a) $\frac{5a}{8} - \frac{3a}{8}$ (b) $\frac{5}{2d} + \frac{7}{2d}$ (c) $\frac{1}{x} - \frac{1}{3x}$

(d) $\frac{4}{a} + b$ (e) $\frac{1}{u} + \frac{1}{v}$ (f) $\frac{5}{4c} - \frac{4}{3d}$

(a) $\frac{5a}{8} - \frac{3a}{8} = \frac{5a - 3a}{8} = \frac{2a}{8} = \frac{2a \div 2}{8 \div 2} = \frac{a}{4}$

(b) $\frac{5}{2d} + \frac{7}{2d} = \frac{5 + 7}{2d} = \frac{12}{2d} = \frac{12 \div 2}{2d \div 2} = \frac{6}{d}$

(c) $\frac{1}{x} - \frac{1}{3x}$

The LCM of x and $3x$ is $3x$.

$$\frac{1}{x} - \frac{1}{3x} = \frac{3 \times 1}{3 \times x} - \frac{1}{3x}$$
$$= \frac{3}{3x} - \frac{1}{3x}$$
$$= \frac{3 - 1}{3x} = \frac{2}{3x}$$

(d) $\frac{4}{a} + b = \frac{4}{a} + \frac{b}{1}$

The LCM of a and 1 is a.

$$\frac{4}{a} + \frac{b}{1} = \frac{4}{a} + \frac{a \times b}{a}$$
$$= \frac{4}{a} + \frac{ab}{a}$$
$$= \frac{4 + ab}{a}$$

This does not simplify further.

(e) $\frac{1}{u} + \frac{1}{v}$

The LCM of u and v is uv.

$$\frac{1}{u} + \frac{1}{v} = \frac{1 \times v}{uv} + \frac{1 \times u}{uv}$$
$$= \frac{v}{uv} + \frac{u}{uv}$$
$$= \frac{v + u}{uv}$$

This does not simplify further.

(f) $\frac{5}{4c} - \frac{4}{3d}$

The LCM of $4c$ and $3d$ is $12cd$.

$$\frac{5}{4c} - \frac{4}{3d} = \frac{5 \times 3d}{12cd} - \frac{4 \times 4c}{12cd}$$
$$= \frac{15d}{12cd} - \frac{16c}{12cd}$$
$$= \frac{15d - 16c}{12cd}$$

Exercise 5o

Simplify the following.

1. $\frac{7a}{5} - \frac{4a}{5}$
2. $\frac{9b}{2} - \frac{3b}{2}$
3. $\frac{4x}{9} + \frac{8x}{9}$
4. $\frac{x}{2} + \frac{x}{3}$
5. $\frac{a}{4} - \frac{a}{20}$
6. $\frac{6x}{5} - \frac{3x}{4}$
7. $\frac{2}{a} + \frac{5}{a}$
8. $\frac{3}{5y} - \frac{1}{5y}$
9. $\frac{5}{2x} + \frac{7}{2x}$
10. $\frac{4}{3a} - \frac{1}{3a}$
11. $\frac{7}{8y} + \frac{9}{8y}$
12. $\frac{1}{x} + \frac{1}{x}$
13. $\frac{1}{3x} + \frac{1}{x}$
14. $\frac{1}{a} - \frac{1}{5a}$
15. $\frac{1}{z} - \frac{1}{2z}$
16. $\frac{1}{4a} - \frac{1}{6a}$
17. $\frac{1}{3x} + \frac{1}{5x}$
18. $\frac{3}{2a} - \frac{2}{3a}$
19. $\frac{3}{x} + 1$
20. $2 - \frac{b}{a}$
21. $3 + \frac{1}{2d}$
22. $\frac{9}{x} + y$
23. $2b + \frac{3}{4}$
24. $p - \frac{3}{2q}$
25. $\frac{1}{a} + \frac{1}{b}$
26. $\frac{1}{x} - \frac{1}{y}$
27. $\frac{3}{4} + \frac{1}{m}$
28. $\frac{3}{c} - \frac{2}{d}$
29. $\frac{5}{4a} - \frac{2}{3b}$
30. $\frac{2}{5x} + \frac{1}{3y}$

Fractions with brackets

$\frac{x + 6}{3}$ is a short way of writing $\frac{(x + 6)}{3}$ or $\frac{1}{3}(x + 6)$.

Notice that all of the terms of the numerator are divided by 3.

$$\frac{x + 6}{3} = \frac{(x + 6)}{3} = \frac{1}{3}(x + 6) = \frac{1}{3}x + 2$$

Example 23

Simplify (a) $\frac{x + 3}{5} + \frac{4x - 2}{5}$,

(b) $\frac{7a - 3}{6} + \frac{3a + 5}{4}$.

(a) $\frac{x+3}{5} + \frac{4x-2}{5} = \frac{(x+3)+(4x-2)}{5}$

$= \frac{x+3+4x-2}{2}$

$= \frac{5x+1}{5}$

(b) The LCM of 6 and 4 is 12.

$\frac{7a-3}{6} - \frac{3a+5}{4}$

$= \frac{2(7a-3)}{2 \times 6} - \frac{3(3a+5)}{3 \times 4}$

$= \frac{2(7a-3) - 3(3a+5)}{12}$

$= \frac{14a - 6 - 9a - 15}{12}$ removing brackets

$= \frac{5a-21}{12}$ collecting like terms

The next example shows that further simplification is sometimes possible after collecting like terms.

Example 24

Simplify $\frac{4x+1}{3} - \frac{x-5}{12}$.

The LCM of 3 and 12 is 12.

$\frac{4x+1}{3} - \frac{x-5}{12} = \frac{4(4x+1)}{12} - \frac{(x-5)}{12}$

$= \frac{4(4x+1) - (x-5)}{12}$

$= \frac{16x + 4 - x + 5}{12}$ removing brackets

$= \frac{15x+9}{12}$ collecting like terms

$= \frac{3(5x+3)}{12}$ factorising the numerator

$= \frac{5x+3}{4}$ dividing numerator and denominator by 3

Exercise 5p

1. Simplify the following.
   (a) $\frac{2a-3}{2} + \frac{a+4}{2}$
   (b) $\frac{3b+4}{3} + \frac{2b-5}{3}$
   (c) $\frac{4c-3}{5} - \frac{2c+1}{5}$
   (d) $\frac{3x+4}{4} - \frac{x-3}{4}$
   (e) $\frac{z+5}{4} + \frac{3z-5}{2}$
   (f) $\frac{2n+7}{3} - \frac{5n+6}{4}$
   (g) $\frac{3a-2}{4} + 2a$
   (h) $3b - \frac{5b-1}{2}$
   (i) $\frac{5u-3}{4} + \frac{u-3}{6}$
   (j) $\frac{m-1}{6} - \frac{m+1}{8}$
   (k) $\frac{3c-2d}{10} + \frac{2c-3d}{15}$
   (l) $\frac{2a+3b}{9} + \frac{a-4b}{6}$

2. Simplify the following *as far as possible.*
   (a) $\frac{3x+1}{4} + \frac{x-5}{4}$
   (b) $\frac{7x-14}{10} - \frac{2x+1}{10}$
   (c) $\frac{6h+5}{7} - \frac{4h-6}{21}$
   (d) $\frac{a+3b}{3} - \frac{5a-3b}{6}$

Summary

When removing brackets **each** term inside the bracket is multiplied by the multiplier.

If the multiplier is negative, the sign of each term inside the bracket is changed after multiplication.

If brackets (that is, expressions inside brackets) are multiplied together, then each term inside each bracket must be multiplied by each term inside the other bracket – this is called **expansion.**

To simplify expressions with terms raised to different powers, the **laws of indices** are used:

$b^m \times b^n = b^{m+n}$

$b^m \div b^n = b^{m-n}$

$(b^m)^n = b^{mn}$

$b^{-m} = \frac{1}{b^m}$

Note: $b^0 = 1$

An algebraic expression may be written as a product of two or more simpler expressions – this is called **factorisation**.

To solve problems in which there are algebraic fractions, the same rules apply as for numerical fractions, e.g. for addition and subtraction, use the LCM of the denominators.

Practice Exercise P5.1

Simplify the following

1. $4uv - vu + 2uw - 3v^2 + 2u^2$
2. $6y^2 - 3xy - x^2 + 2xy$
3. $4u^2v - v^2u + 2uw^2 - 3vu^2 + 2w^2u$
4. $y^2 - 5x^2y - 4y^2x + 3xy^2 - yx^2$
5. $\frac{2}{3}ab - 2bc - 4bc + \frac{1}{3}ab$
6. $a^2 + 4ac + 2a^2 - 8ac$

Practice Exercise P5.2

Expand the following

1. $3(x + y)$
2. $a(x + y)$
3. $x(x + y)$
4. $2x(x + y)$
5. $ax(x + y)$
6. $xy(x + y)$
7. $4(2x + 3y)$
8. $a(2x + 3y)$
9. $3y(2x + 3y)$

Practice Exercise P5.3

Expand and when possible simplify the following

1. $3a(a - b)$
2. $-\frac{1}{2}c(2c - 1)$
3. $(2a + b)\,d - ad$
4. $2h\,(h - 2) - 3h^2$
5. $s(s - 2t + 3) - t(t + 2 + 2s)$
6. $(m - 2n)p + (p + n)m$
7. $k(h - 2k + 3) - h(k + 2h - 1)$
8. $-\frac{1}{3}(3jk + 6k - j)$
9. $f(fg^2 - f^2g)$
10. $mn(m^2 + 2mn - n^2)$
11. $4(u - 2) + 3(2u - 3)$
12. $6y - 3(y - 2) + 4$
13. $4(-2d + 1) + 3(d - 2) + 7$
14. $(a + 4) - (2 - a) - 3$
15. $x(x + 3y) - y(2x - 3y)$
16. $d(d^2 - 2d + 1) - d(d^2 - 2)$
17. $(t^2 - 3)t^2 + 2t\,(2t^3 + 2t^2 - 3)$
18. $v(3v - w) + w(2w + v)$
19. $k(h - 2k + 3) - h(k + 2h - 1)$
20. $2n(2u + 3v) - 5n(2u - 3v)$

Practice Exercise P5.4

Expand and when possible simplify the following

1. $4(2x + y) + 2(3x - 2y)$
2. $3(3r + s) - 4(2r - 5s)$
3. $-3m(3u + 4v) - 7m(2u - 4v)$
4. $x(4r + 6) - x(4r + 6)$
5. $a(2x - 2y) + a(x - 4y)$
6. $-2(3x + by) - (x + by)$
7. $2(a - 2b) + 3(2a + b)$
8. $u(u + 2v) + v(2u - v)$
9. $h(2h + k) + k(h - 3k)$
10. $2u(u + v) + 4u(u + v)$

Practice Exercise P5.5

Expand and when possible simplify the following

1. $x(3x - y)$
2. $(2x + y)(x + y)$
3. $(a - 2b)(2a + b)$

4. $(u + 2v)(2u - v)$
5. $(2h + k)(h - 3k)$
6. $(u + v)(u + v)$
7. $(x - y)(x + y)$
8. $(2a - b)(2a + b)$
9. $(m - 2n)(2n - m)$
10. $(2h + k)(2h + k)$

Practice Exercise P5.6

Simplify the following:

1. $2x \times 3x$
2. $2x \times 3x^2$
3. $x^3 \times x^2$
4. $2xy \times 3x^2$
5. $4d^2 \div 2$
6. $6d^2 \div d$
7. $6d^2e \div 3d$
8. $6d^3e^4 \div 3de^2$

Practice Exercise P5.7

Factorise each of the following

1. $a^4 + 4a^3$
2. $p^3rq^2 - 2pq^2$
3. $3d^2 - 6bd^3$
4. $3xy^2 + 12yz$
5. $9h^2k - 3h^2j$
6. $12ab^5c^5 - 4b^3c^3$
7. $(2a^2f)^3 + a^2f^3$
8. $3(hkn^2)^3 + (3hk)^2$
9. $(a^3x^2y)^3 + 2xy^2$
10. $12ab^3c^5 - 6bc^6$

Practice Exercise P5.8

Simplify the following:

1. 3^{-2}
2. 2^{-3}
3. $2^{-1} \times 2^3$
4. $(2b)^3$
5. $4(k^2)^3$
6. $(2v)^{-2} \times v^4$
7. $(x^{-1})^4$
8. $(\frac{1}{8})^{-2}$
9. $2h^{-3} \times 3h^{-2}$
10. $\frac{2}{(3)^{-2}}$
11. $2x^{-1} \times 3x^3$
12. $(2x)^{-1} \times (3x)^3$

Practice Exercise P5.9

Simplify the following:

1. $\frac{2x}{3} + \frac{x}{2}$
2. $2 - \frac{3}{x}$
3. $\frac{u}{2} + \frac{3u}{4}$
4. $\frac{f}{3} + 4f$
5. $3 - \frac{2}{k}$
6. $\frac{w}{3} - \frac{w}{4}$
7. $\frac{2}{h} + 4$
8. $\frac{2x}{3} - \frac{y}{6}$
9. $\frac{c}{2} - \frac{c}{6}$
10. $\frac{2}{h} - \frac{1}{3h}$
11. $3k - \frac{2k}{3h}$
12. $\frac{2x}{y} + \frac{y}{x}$
13. $\frac{2x}{3} + \frac{x - 1}{2}$
14. $\frac{2a + 3}{4} - \frac{a}{3}$
15. $\frac{2c - 3}{3} - \frac{c + d}{6}$
16. $\frac{3x + 1}{4} - \frac{y + 1}{2}$
17. $\frac{2a - b}{2} + \frac{a - b}{3}$
18. $\frac{2x - 1}{3} - \frac{3y - 1}{4}$
19. $2a - c + \frac{2b - c}{3}$
20. $\frac{2x - y}{3} - \frac{3y - x}{5}$

Measurement (1)

Volume of solids

Pre-requisites

- properties of solids; area of plane shapes; units of measurement; symmetry

Volume

The volume of a solid is a measure of the space it takes up. The cube is used as the shape for the basic unit of volume. A cube of edge 1 metre has a volume of **1 cubic metre** or **1 m³**. A cube of edge 1centimetre has a volume of 1 **cubic centimetre** or **1 cm³** (Fig. 6.1).

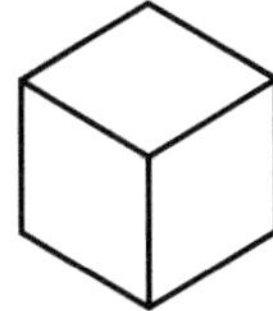

Fig. 6.1 1 cm³

It is difficult to measure volume directly. One way is to build a copy of the solid using basic units. For example, to measure the volume of the 4 cm by 3 cm by 2 cm cuboid in Fig. 6.2, a copy can be built from 1 cm³ cubes as in Figs. 6.3, 6.4 and 6.5.

The volume of the cuboid is 24 cm³.

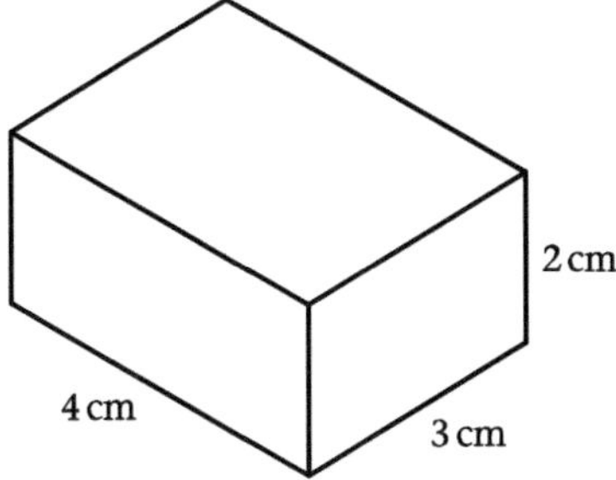

Fig. 6.2

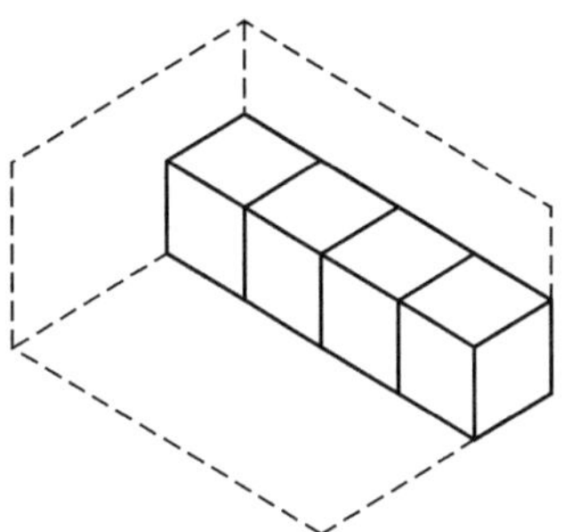

Fig. 6.3 1 row of 4 cubes

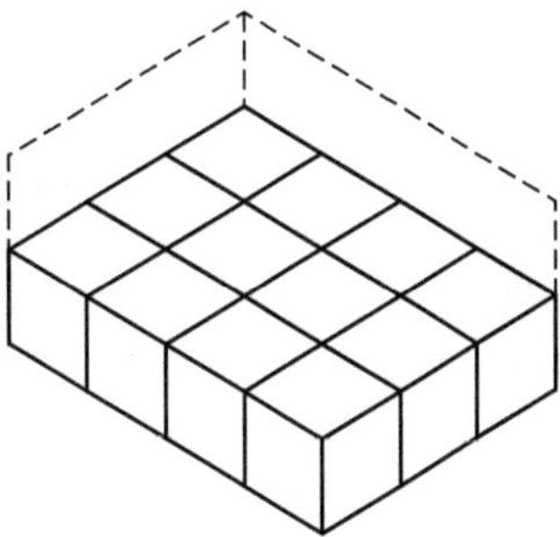

Fig. 6.4 3 rows of 4 cubes, i.e. 1 layer of 12 cubes

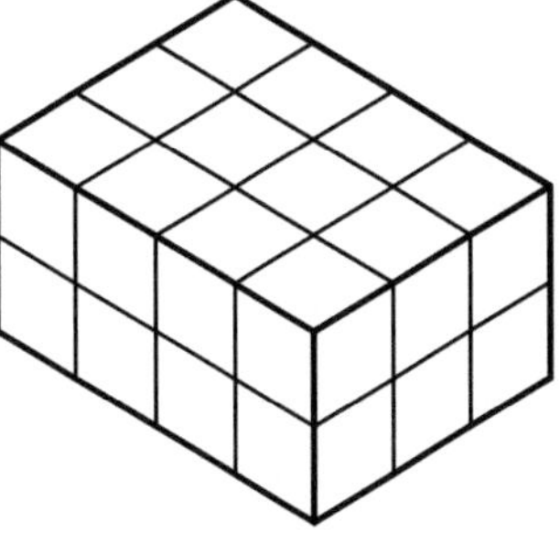

Fig. 6.5 2 layers of 12 cubes, i.e. 24 cubes altogether

Volume of a cuboid

Notice that

(a) the 4 cm by 3 cm by 2 cm cuboid in Fig. 6.2 has a volume of 24 cm³; *and*

(b) Volume $= (4 \times 3 \times 2)\,\text{cm}^3$
$= (4 \times 3)\,2\,\text{cm}^3$
$= (3 \times 2)\,4\,\text{cm}^3$
$= (4 \times 2)\,3\,\text{cm}^3$
$= 24\,\text{cm}^3$

We can find the volume of any cuboid by finding the product of its length, breadth and height:

volume of cuboid
= length × breadth × height
= area of base × height
= area of end face × length
= area of side face × breadth

Example 1

Calculate the volume of a rectangular box which measures 30 cm by 15 cm by 10 cm.

$$\text{Volume of box} = (30 \times 15 \times 10)\text{ cm}^3 = 4500\text{ cm}^3$$

Group work

Collect cuboids of different sizes such as shoe-boxes, empty cereal boxes, chocolate boxes, detergent cartons, match-boxes, toothpaste and perfume boxes, as well as a die, domino, or small solid block of wood.

Work in groups of four in two pairs, with each pair in the group responsible for a different task. Each group selects a small solid cuboid and a large box.

One pair measures the dimensions of the small solid, calculates its volume, and estimates by sight or by placing the edges of the solid along corresponding edges of the box, the number that can be fitted into the large box. In this way they get an approximate value of the volume of the space inside the box.

Meanwhile the other pair measures the dimensions of the large box and calculates its volume.

Now compare the estimated and calculated values for the solids and boxes.

Example 2

A closed room measuring 4 m long by 3 m wide has a volume of 30 m³. Find the height of the room.

$$\text{Volume of room} = 30\text{ m}^3$$
$$\text{Area of base} = (4 \times 3)\text{ m}^2 = 12\text{ m}^2$$
$$\text{Height of room} = \frac{30}{12} = 2\tfrac{1}{2}\text{ m}$$

Exercise 6a

1. Copy and complete the table of cuboids (Table 6.1).

Table 6.1

	Length	Breadth	Height	Volume
(a)	5 m	2 m	3 m	
(b)	3 cm	2 cm	8 cm	
(c)	5 m	2 m	$2\frac{1}{2}$ m	
(d)	3 cm	3 cm	3 cm	
(e)	6 cm	2 cm		24 cm³
(f)		4 m	3 m	36 m³
(g)	5 cm		4 m	120 cm³
(h)	$5\frac{1}{3}$ m	3 m	$2\frac{1}{8}$ m	

2. How many cm³ are in a cube of edge 2 cm?
3. A box has a square base of side 9 cm. Calculate the volume of the box if it is 10 cm deep.
4. A rectangular room 8 m long by 5 m wide contains 120 m³ of air. Find the height of the room.
5. Which has the greater volume, a 4 cm by 4 cm by 4 cm cube or a 3 cm by 7 cm by 3 cm cuboid?
6. A room is 3 m high and has a volume of 60 m³. Calculate the area of the floor of the room.
7. A concrete block is made by pouring 1000 cm³ of concrete into a 10 cm by 25 cm rectangular tray. How thick is the block?
8. Calculate the volume of air in a dormitory 10 m long, 5 m wide and 3 m high. If each person should have 15 m³ of air space, how many people can sleep in the dormitory?

Units of volume

The cubic metre, m³, is the basic unit of volume.

$1\text{ m} = 100\text{ cm}$
$1\text{ m}^3 = (100 \times 100 \times 100)\text{ cm}^3 = 1\,000\,000\text{ cm}^3$

Similarly

1 cm = 10 mm

$1\text{ cm}^3 = (10 \times 10 \times 10)\text{ mm}^3 = 1000\text{ mm}^3$

When calculating problems about volume, make sure that all dimensions are in the same units.

Example 3

A concrete beam is 20 m long. Its end face is a rectangle 60 cm by 40 cm. Calculate the volume of the beam.

Working in metres:

volume of beam $= (20 \times 0.6 \times 0.4)\text{ m}^3$

$= 4.8\text{ m}^3$

Exercise 6b

1. A wooden beam has a rectangular end face, 24 cm by 15 cm, and is 8 m long. Calculate the volume of the beam.
2. A rectangular tank, $1\frac{1}{2}$ m long and 1m wide, contains water to a depth of 50 cm. Calculate the volume of water in the tank.
3. A block of concrete is 1m long, 50 cm wide and 4 cm thick. Calculate the volume of the block in cm^3.
4. A block measures 22 cm by 11 cm by 7 cm. How many of these blocks will be needed to build a wall $5\frac{1}{2}$ m long, 22 cm thick and $3\frac{1}{2}$ m high?
5. How many 2 cm by 2 cm by 2 cm cubes can be packed in a box 1m long, 20 cm wide and 4 cm deep?

Volume of prisms

A cuboid can be cut into two equal right-angled triangular prisms as in Fig. 6.6.

For each prism:

volume of prism $= \frac{1}{2}lbh = (\frac{1}{2}lb) \times h$.

But, $\frac{1}{2}lb$ = area of end face because the face is a right-angled triangle.

Volume of prism = area of end face × height.

(a)

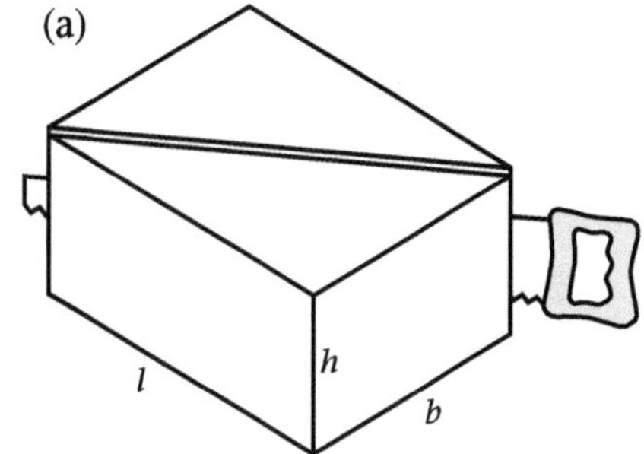

(b)

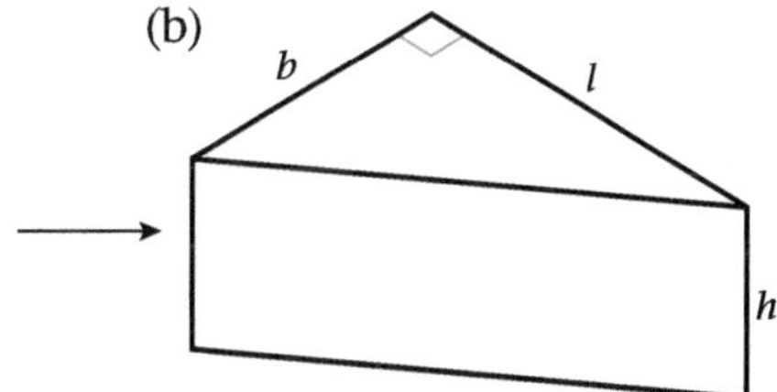

(c)

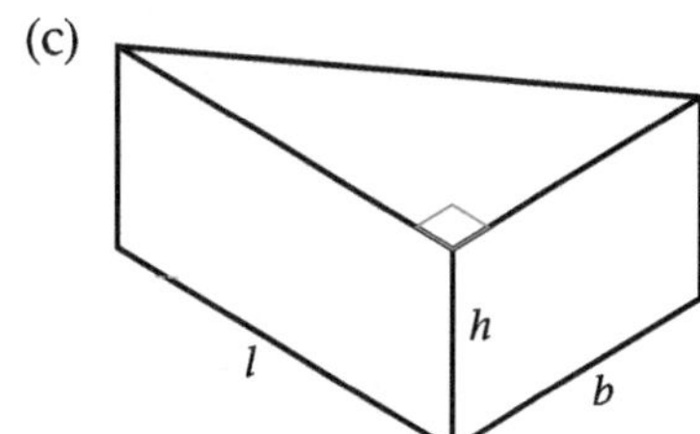

Fig. 6.6 (a) Volume $= lbh$
(b) Volume $= \frac{1}{2}lbh$
(c) Volume $= \frac{1}{2}lbh$

Example 4

Calculate the volume of the prism in Fig. 6.7.

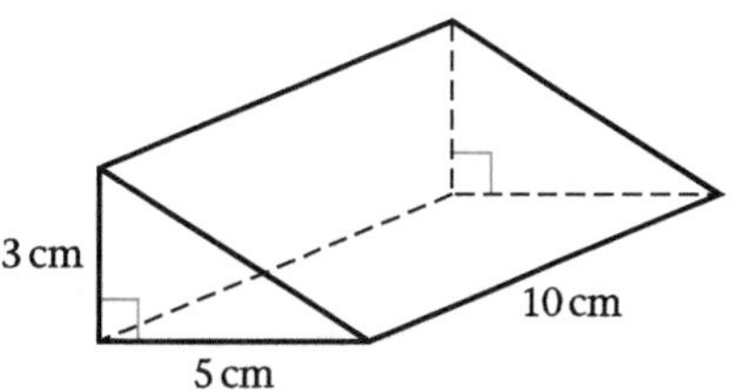

Fig. 6.7

Area of triangular face $= \frac{1}{2} \times 5 \times 3\text{ cm}^2$

Volume of prism $= \frac{1}{2} \times 5 \times 3 \times 10\text{ cm}^3$

$= 75\text{ cm}^3$

Example 5

A beam has an end face as shown in Fig. 6.8. Calculate the volume of an 8 m length of the beam.

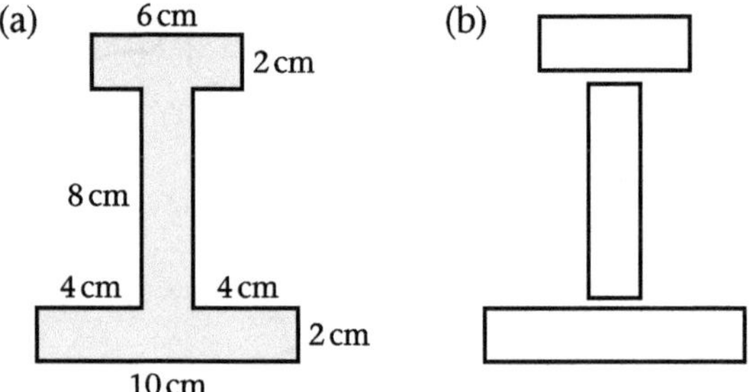

Fig. 6.8

The beam is an I-shaped prism. Its end face is made up of three rectangles, as shown in Fig. 6.8(b). This gives three cuboids.

Length of each cuboid = 8 m = 800 cm

$$\begin{aligned}\text{Volume of beam} &= (6 \times 2 \times 800 + 8 \times 2 \times 800 \\ &\quad + 10 \times 2 \times 800)\,\text{cm}^3 \\ &= 800(12 + 16 + 20)\,\text{cm}^3 \\ &= 800 \times 48\,\text{cm}^3 = 38\,400\,\text{cm}^3\end{aligned}$$

In Example 5, notice that the area of the end-face of the prism is 48 cm^2; its volume is the product of this area and the length of the prism. In general, for any prism:

volume of prism = area of end face × distance between the end faces

Exercise 6c

1. Calculate the volumes of the right-angled triangular prisms in Fig. 6.9 (a)–(f). All measurements are in cm.

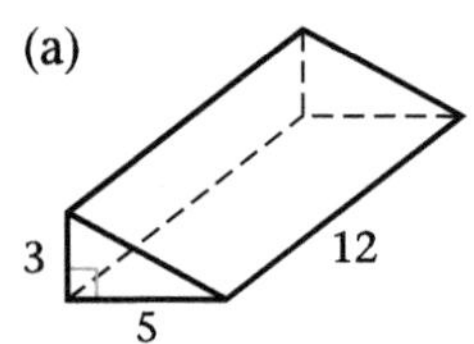

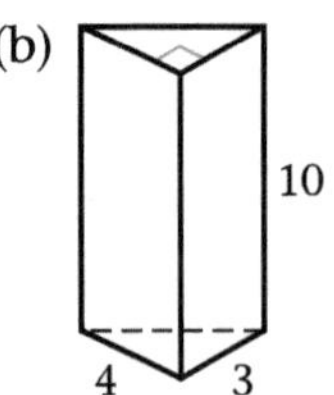

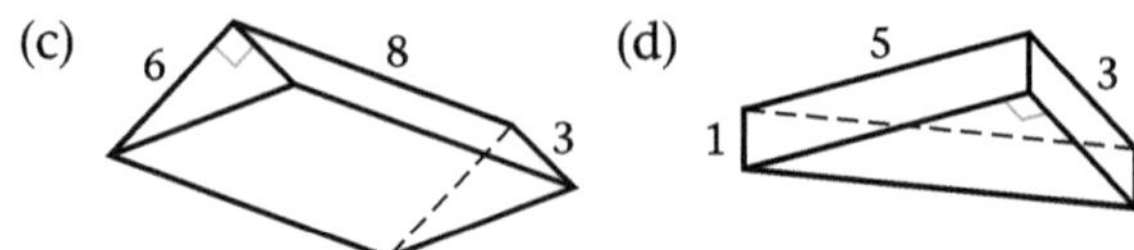

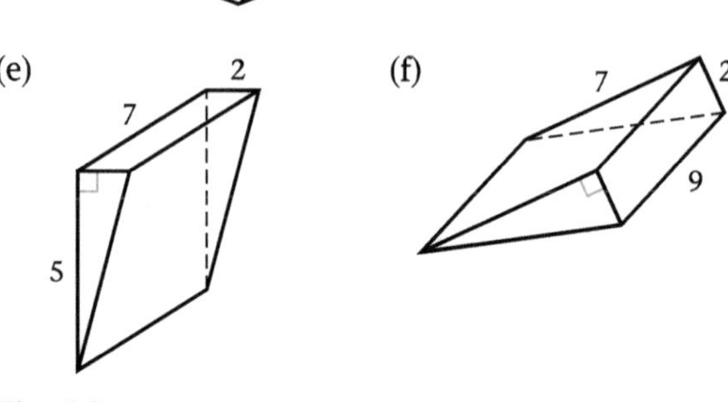

Fig. 6.9

2. The diagrams in Fig. 6.10 show the cross-sections of steel beams. All dimensions are in cm. Calculate the volumes of 5m lengths of the beams. Give your answers in cm^3.

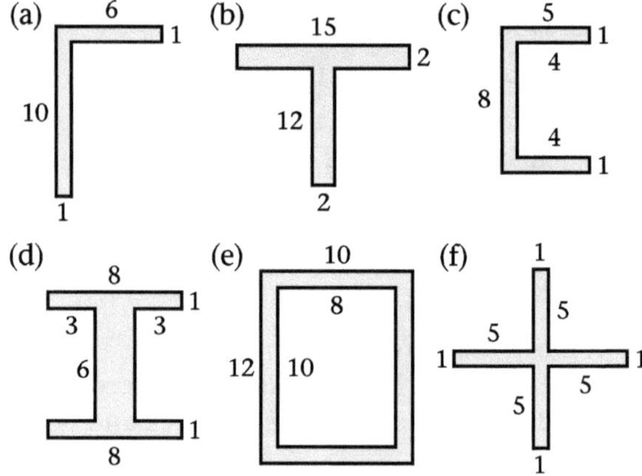

Fig. 6.10

3. The end face of the I-girder shown in Fig. 6.11 has an area of 42 cm^2. Calculate its volume in m^3.

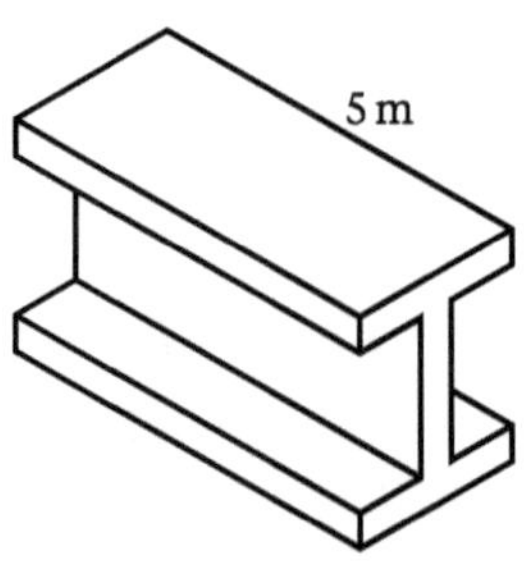

Fig. 6.11

4 Fig. 6.12 shows a special block used in building. If each hole is 15 cm by 15 cm, calculate the volume of concrete in the block in cm^3.

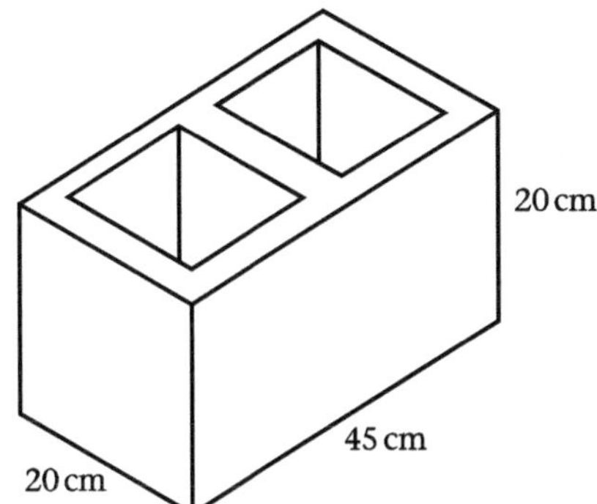

Fig. 6.12

5 Fig. 6.13 shows an open wooden box.

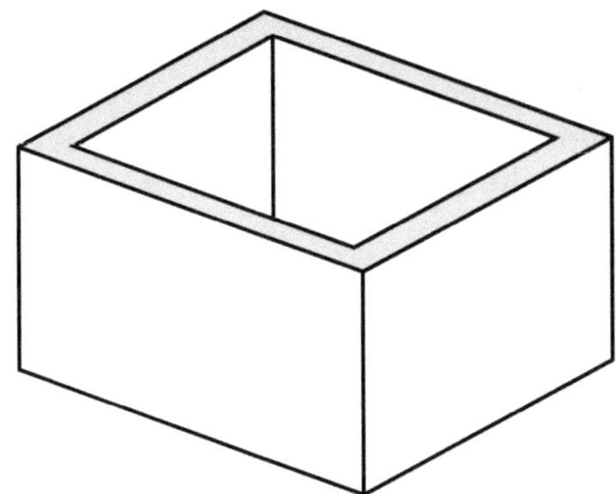

Fig. 6.13

The external dimensions of the box are 36 cm long, 24 cm wide and 12 cm deep and the wood is 1 cm thick. Calculate the volume of wood used in cm^3.

Volume of a cylinder

Volume of prism = area of end face × distance between the end faces

For example, in Fig. 6.14,

volume of cuboid $= (l \times b) \times h$
$= lbh$

volume of triangular prism $= (\frac{1}{2} \times a \times b) \times h$
$= \frac{1}{2}abh$

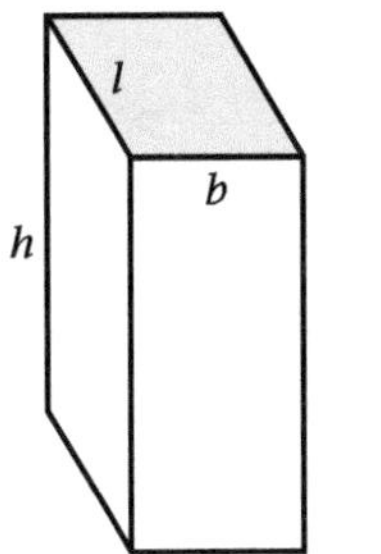

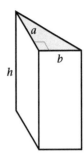

Fig. 6.14

A cylinder is a special kind of prism. Its volume can be found in the same way.

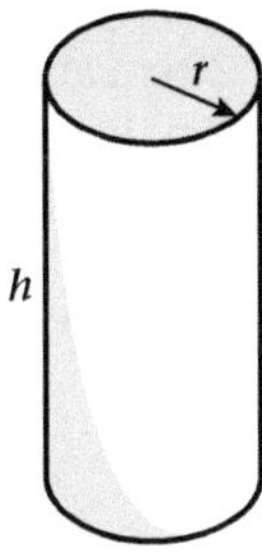

Fig. 6.15

In Fig. 6.15,

volume of cylinder = area of circular face × height of cylinder
$= \pi r^2 \times h$

Volume of cylinder = $\pi r^2 h$

Capacity

The **capacity** of a container is a measure of the space inside it.

The **litre** is the basic unit of capacity and is a measure of the space inside a container of volume 1000 cm^3 (Fig. 6.16).

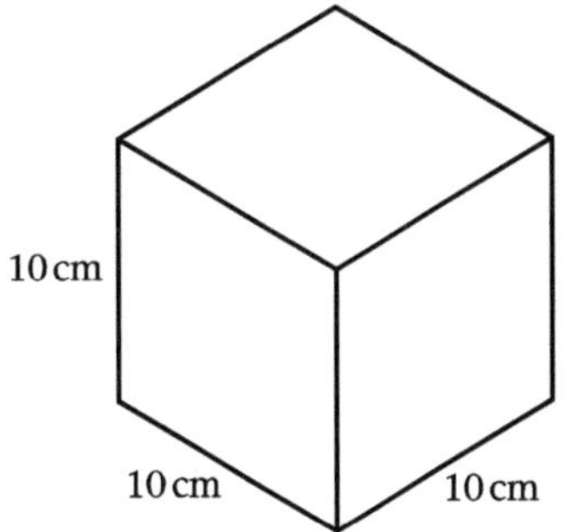

Fig. 6.16

Table 6.2

Capacity	Abbreviation	Relationship to basic unit
1 kilolitre	1 kℓ	1000 ℓ
1 litre	1 ℓ	1 ℓ
1 millilitre	1 mℓ	0.001 ℓ

Only the kilolitre, litre and millilitre are used for practical and scientific purposes.

Example 6

The height of a tin is 20 cm and the base is a square of side 14 cm. How many litres can the tin hold?

$$\begin{aligned}\text{Volume of space} &= (14 \times 14) \times 20\,\text{cm}^3\\ &= 3920\,\text{cm}^3\\ &= 3.92\ell\end{aligned}$$

Example 7

88 litres of oil are poured into a drum 40 cm in diameter. Use the value $\frac{22}{7}$ for π to find the depth of the oil in the drum.

Let the depth of the oil in the drum be d cm. Fig. 6.17 shows the information.

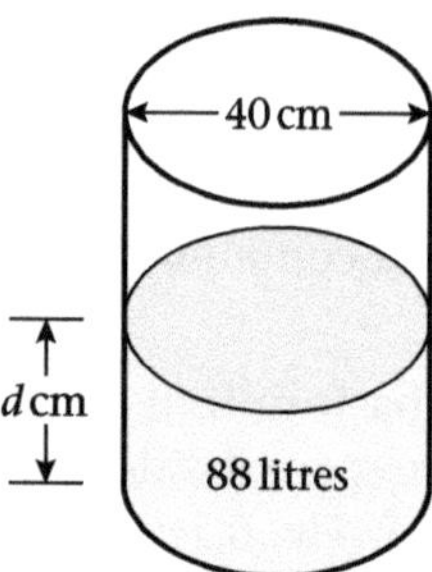

Fig. 6.17

$$\begin{aligned}\text{Volume of oil} &= 88\text{ litres}\\ &= 88 \times 1000\,\text{cm}^3\\ \text{Also, volume of oil} &= \pi r^2 d\\ &= \tfrac{22}{7} \times 20 \times 20 \times d\,\text{cm}^3\\ \text{Thus, } \tfrac{22}{7} \times 20 \times 20 \times d &= 88 \times 1000\\ d &= \frac{88 \times 1000 \times 7}{20 \times 20 \times 22}\\ &= 70\end{aligned}$$

The depth of the oil is 70 cm.

Notice that numerical simplification is left until the last step.

Exercise 6d

Use the value $\frac{22}{7}$ for π in this exercise.

1. Calculate the volume of a cylindrical steel bar which is 8 cm long and 3.5 cm in diameter.
2. The tank on a petrol lorry is a cylinder 2 m in diameter and 7 m long. Calculate its volume in m^3. Find its capacity in kilolitres.
3. The base of a storage vat is a rectangle 3 m by 5 m. The storage vat is $3\frac{1}{2}$ m deep. Calculate the capacity in kilolitres of the storage vat.
4. 99 litres of oil are poured into a cylindrical drum 60 cm in diameter. How deep is the oil in the drum?
5. How many litres of water will a cylindrical pipe hold if it is 1 m long and 7 cm in diameter?
6. A wooden roller is 1 m long and 8 cm in diameter. Find (a) its volume in cm^3, (b) its mass in grams if 1 cm^3 of the wood has a mass of 0.7 g.
7. A pudding fills a cylindrical tin 21 cm in diameter and 10 cm deep. Find its weight if 1 cm^3 of the pudding weighs 0.75 g. Give your answer to two significant figures.
8. The roof of a building is supported by four concrete pillars. Each pillar is 4 m long and 50 cm in diameter.
   (a) Calculate the total volume of the four pillars in m^3.
   (b) If 1m^3 of concrete has a mass of 2.1 tonnes, calculate the total mass of the four pillars in tonnes.

9 A cylindrical tin has a diameter of 14 cm and a height of 20 cm and is full of water. The water is poured into another tin which has a diameter of 20 cm. How deep is the water in the second tin?

Summary

The **volume** of a solid is a measure of the space the solid occupies.

Volume of a cuboid
= length × breadth × height
= area of base × height
= area of end face × length
= area of side face × breadth

Volume of a prism = area of end face × distance between end faces

The **capacity** of a container is the measure of the amount of space inside the container. The basic unit of capacity is the **litre** which is a fluid measure.

Practice Exercise P7.1

Remember unless otherwise stated, use $\pi = \frac{22}{7}$

1 A carton measures 5 m long, 2 m wide and 3 m high on the inside.
(a) Calculate the volume of the inside of the carton. Give your answer in cubic metres and cubic centimetres
(b) How many cubes of side 10 cm would be needed to pack the inside of the carton?

2 A rectangular tank measures 2 m long, 1.8 m wide and 5 m high.
(a) Calculate the volume of the tank.
(b) Calculate the volume of water in the tank when the depth of the water in the tank is 4 m.

3 A cylindrical jug has a base of diameter 14 cm and is of height 16 cm.
(a) Calculate the volume of water the jug can hold.
(b) What is the depth of water in the jug when it is half full?

4 A rectangular tank measures 5 m long, 3.5 m wide and 8 m high.
(a) Calculate the volume of water in the tank when it is full.
(b) How many cylindrical containers of radius 28 cm and height 60 cm can be filled from the tank?

5 Calculate the volume of a solid cylindrical block of radius 30 cm and height 2 m. Give your answer in cubic centimetres to 2 decimal places.

6 The length of a triangular prism is 12 cm, its triangular face has a base of 6 cm and perpendicular height of 4 cm. What is the volume of the triangular prism?

7 The total surface area of a cube is 216 cm^2.
(a) What is the area of each face of the cube?
(b) What is the length of each side of the cube?
(c) Calculate the volume of the cube.

8 Water flows through a cylindrical pipe of length 2.1 m and 8 cm in diameter.
(a) Calculate the volume of water in the pipe in cubic centimetres.
(b) If this amount of water flows into a tub every minute, calculate the amount of water that flows into the tube in 15 minutes.

9 A wooden cylinder of radius 14 cm and height 12 cm is screwed to another cylinder of radius 21 cm and height 4.8 cm to form a model.
Calculate the volume of the model.

10. A cylindrical mug is of radius 7 cm and height 6 cm.
    (a) Calculate the volume of water the cup holds when full.
    (b) If 100 mugs of water are poured into a larger container, calculate the amount of water poured into the larger container.
    (c) If the radius of the larger container is 14 cm, calculate the depth of water in this container.

11. How many jars of diameter 8 cm and height 9 cm can be filled with milk from a larger container of radius 12 cm and height 24 cm?

12. An open cylindrical tank is of radius 14 cm and height 8 m.
    (a) Calculate the volume of water the tank holds.
    (b) What is the height of the water when the tank is three-quarters full?
    (c) How much water is in the tank when it is three-quarters full?

13. A rectangular block measures 30 cm long, 15 cm wide and 3.5 cm high. Calculate the weight of the block if it is made of material that weighs 0.4 kg per 100 cm^3.

14. A jar of radius 4 cm is filled with cooking salt to a depth of 5 cm. Calculate the depth of salt when it is poured into another cylindrical container of radius 5 cm.

Chapter 7

Standard form

Large and small numbers

Pre-requisites

- decimal numbers; index form

Large numbers

There is no such thing as 'the biggest number in the world'. It is always possible to count higher. Science and economics use very large numbers. Table 7.1 gives the names and values of some large numbers.

Table 7.1

Name	Value
thousand	1000
million	1000 thousand = 1 000 000 = 1000^2
billion	1000 million = 1 000 000 000 = 1000^3

How big is a million?

The following examples may give you some idea of the size of a million.

1 A 1 cm by 1 cm square of 1 mm graph paper contains one hundred small 1 mm × 1 mm squares (Fig. 7.1).

1 cm = 10 mm

1 cm

Fig. 7.1 100 small squares

A 1 m by 1 m square of the same graph paper contains 1 million of these small squares (Fig. 7.2).

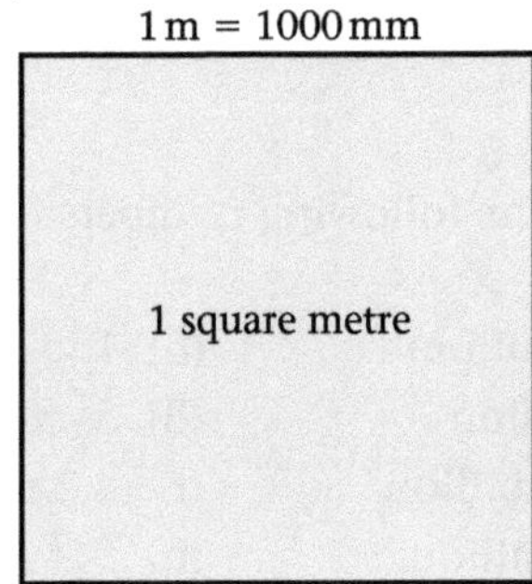

Fig. 7.2. 1000 mm × 1000 mm = 1 000 000 mm² *(not to scale)*

2 A cubic metre measures 100 cm by 100 cm by 100 cm (Fig. 7.3).

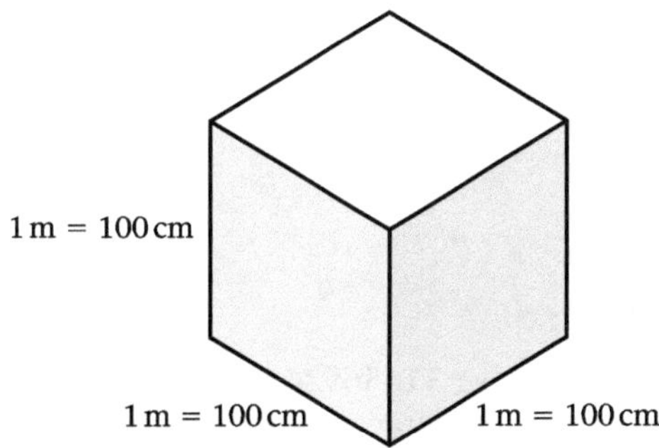

Fig. 7.3 (not to scale)

Volume of cubic metre
$= 100\,\text{cm} \times 100\,\text{cm} \times 100\,\text{cm}$
$= 1\,000\,000\,\text{cm}^3$

Thus 1 million cubic centimetres (cm^3) will exactly fill a 1 cubic metre box.

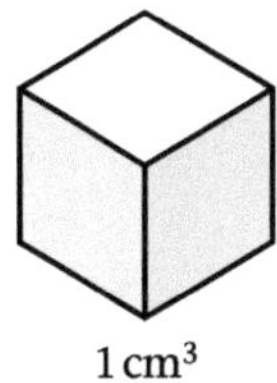

Fig. 7.4 One million of these make 1 cubic metre

Exercise 7a

1. What is the common name for (a) a thousand thousand, (b) a thousand million?

2. A football field measures 80 m by 50 m.
   (a) Change the dimensions to cm and calculate the area of the field in cm^2.
   (b) One face of a cigarette packet measures 8 cm by 5 cm. Calculate how many cigarette packets would be needed to cover the football field.

3. A library has about 4000 books. Each book has about 250 pages. Approximately how many pages are there in the library?

4 How long would it take to count to 1 million if it takes an average of 1 second to say each number? Give your answer to the nearest $\frac{1}{2}$ day.

5 Find out which of the following is nearest to the number of seconds in a year.
(a) 500 000 (b) 1 000 000
(c) 3 000 000 (d) 30 000 000
(e) 2 000 000 000

Writing large numbers

Grouping digits

Read the number 31556926 out aloud. Was it easy to do? It may have been quite difficult. You had to decide, 'Is the number bigger than a million or not?', 'Does it begin 3 million, 31 million or 315 million?'.

It is necessary to write large numbers in a helpful way. It is usual to group the digits of large numbers in threes from the decimal point. A small gap is left between each group.

31556926 should be written 31 556 926. Now it is easy to see that the number begins, 31 million.

Exercise 7b

Write the following numbers, grouping digits in threes from the decimal point.

1 1 million
2 59244
3 721568397
4 2312400
5 8 million
6 3 billion
7 9215
8 14682053
9 108412
10 12345
11 100000000
12 987654

Digits and words

Editors of newspapers know that large numbers sometimes confuse readers. They often use a mixture of digits and words when writing large numbers.

Example 1

What do the numbers in the following headlines stand for?
(a) FOOD IMPORTS RISE TO \$1 BILLION
(b) OIL PRODUCTION NOW 2.3 MILLION BARRELS DAILY
(c) FLOODS IN INDIA: 0.6 MILLION HOMELESS
(d) NEW ROAD TO COST \$$2\frac{1}{4}$MILLION

(a) \$1 billion is short for \$1 000 000 000
(b) 2.3 million $= 2.3 \times 1\,000\,000$
$= 2\,300\,000$
(c) 0.6 million $= 0.6 \times 1\,000\,000$
$= 600\,000$
(d) \$$2\frac{1}{4}$ million = \$2.25 million
= \$2 250 000

Example 2

Express the following in a mixture of digits and words.
(a) \$3 000 000 (b) 6 800 000 000
(c) 240 000 000 (d) \$500 000

(a) \$3 000 000 $= 3 \times 1\,000\,000$
= \$3 million
(b) 6 800 000 000 $= 6.8 \times 1\,000\,000\,000$
= 6.8 billion
(c) 240 000 000 $= 240 \times 1\,000\,000$
= 240 million
or
240 000 000 $= 0.24 \times 1\,000\,000\,000$
= 0.24 billion
(d) \$500 000 = \$0.5 × 1 000 000
= \$ 0.5 million or \$$\frac{1}{2}$ million

Exercise 7c

1 Express the following numbers in digits only.
(a) \$2 million (b) 150 million km
(c) 3 billion (d) $5\frac{1}{2}$ million
(e) \$2.1 billion (f) 4.2 million litres
(g) 0.4 billion (h) \$$1\frac{1}{4}$ million
(i) 0.7 million tonnes (j) \$$\frac{3}{4}$ million
(k) 0.45 million (l) \$0.58 billion

2 Imagine you are a newspaper editor. Write the following numbers using a mixture of digits and words.

(a) 8 000 000 tonnes (b) \$6 000 000
(c) 2 000 000 000 (d) \$3 700 000 000
(e) \$7 400 000 (f) \$1 750 000
(g) 200 000 litres (h) 500 000 000
(i) 300 000 tonnes (j) 250 000
(k) 980 000 barrels (l) 490 000 000

Standard form

Large numbers

Numbers such as 1000, 1 000 000, 1 000 000 000 can be expressed easily as powers of 10. For example,

$$1\,000\,000 = 10 \times 10 \times 10 \times 10 \times 10 \times 10 = 10^6$$

Other numbers can be expressed as a product of two terms, one of which is a power of 10. For example,

$$2\,000\,000 = 2 \times 1\,000\,000 = 2 \times 10^6$$
$$7000 = 7 \times 1000 = 7 \times 10^3$$
$$2\,500\,000 = 2.5 \times 1\,000\,000 = 2.5 \times 10^6$$
$$8\,600\,000\,000 = 8.6 \times 1\,000\,000\,000 = 8.6 \times 10^9$$

The numbers 2×10^6, 7×10^3, 2.5×10^6 and 8.6×10^9 are all in the form $A \times 10^n$ such that A is a number between 1 and 10 and n is a whole number. Such numbers are said to be in **standard form.**

Scientists often use standard form when writing numbers. Standard form makes it easy to tell the size of a number. There is also less chance of making a mistake with zeros when using standard form.

Example 3

Express the following numbers in standard form.

(a) 60 000 (b) 650
(c) 480 000 000 (d) 320 000

(a) $60\,000 = 6 \times 10\,000 = 6 \times 10^4$
(b) $650 = 6.5 \times 100 = 6.5 \times 10^2$
(c) $480\,000\,000 = 4.8 \times 100\,000\,000 = 4.8 \times 10^8$
(d) $320\,000 = 3.2 \times 100\,000 = 3.2 \times 10^5$

Exercise 7d (Oral)

Express the following numbers in standard form.

1 9 000 000
2 4000
3 4 000 000 000
4 600
5 300 000
6 60 000
7 5 000 000 000
8 70
9 20 000 000
10 89 000
11 720 000 000
12 2 300 000
13 55
14 170 000
15 5400
16 25 000 000
17 6300
18 9 400 000 000
19 410 000
20 85 000 000
21 950
22 3600
23 360
24 36

Example 4

Express the following in ordinary form.

(a) 4×10^4 (b) 4.3×10^4 (c) 7.8×10^7

(a) $4 \times 10^4 = 4 \times 10\,000 = 40\,000$
(b) $4.3 \times 10^4 = 4.3 \times 10\,000 = 43\,000$
(c) $7.8 \times 10^7 = 7.8 \times 10\,000\,000 = 78\,000\,000$

Exercise 7e

Change the following from standard form to ordinary form.

1 6×10^4
2 8×10^3
3 9×10^5
4 4×10^2
5 7×10^1
6 3×10^2
7 5×10^7
8 2×10^6
9 6×10^8
10 6.3×10^4
11 8.4×10^3
12 9.8×10^5
13 7.2×10^1
14 3.6×10^2
15 4.4×10^2
16 5.1×10^7
17 2.5×10^6
18 6.7×10^8
19 3.7×10^3
20 5.9×10^7
21 8.5×10^4
22 3.4×10^3
23 3.4×10^2
24 3.4×10^1

Small numbers

Decimal fractions

Decimal fractions also have names.

8 tenths $= 0.8$
8 hundredths $= 0.08$
8 thousandths $= 0.008$
8 ten thousandths $= 0.0008$
8 hundred thousandths $= 0.000\,08$

Notice that digits are grouped in threes from the decimal point as before.

Example 5

Write the following as decimal fractions.

(a) 28 thousandths (b) $\frac{865}{100\,000}$

(c) 350 millionths (d) $\frac{400}{10\,000}$

(a) 28 thousandths $= 1$ thousandth $\times 28$
$= 0.001 \times 28$
$= 0.028$

(b) $\frac{865}{100\,000} = 0.008\,65$
There are 5 zeros in the denominator. The decimal fraction is obtained by moving the digits in the numerator 5 places to the right.

(c) 350 millionths $= 1$ millionth $\times 350$
$= 0.000\,001 \times 350$
$= 0.000\,350 = 0.000\,35$
In a decimal fraction it is not necessary to write any zeros after the last non-zero digit.

(d) $\frac{400}{10\,000} = 0.0400$
$= 0.04$

Exercise 7f

Write the following as decimal fractions.

1. 6 hundredths
2. 4 thousandths
3. 9 tenths
4. 8 millionths
5. 4 ten thousandths
6. 6 hundred thousandths
7. $\frac{3}{1000}$
8. $\frac{9}{100\,000}$
9. $\frac{7}{10\,000}$
10. 16 hundredths
11. 34 thousandths
12. 26 ten thousandths
13. $\frac{28}{100\,000}$
14. $\frac{84}{1000}$
15. $\frac{756}{10\,000}$
16. 27 tenths
17. 65 hundredths
18. 402 thousandths
19. 20 hundredths
20. 240 thousandths
21. 700 thousandths
22. $\frac{620}{100\,000}$
23. $\frac{4020}{100\,000}$
24. 300 ten thousandths
25. $\frac{720}{1000}$
26. $\frac{720}{10\,000}$

Decimal fractions in standard form

Decimal fractions such as 0.001 and 0.000 001 can be expressed as powers of 10. For example,

$$0.000\,001 = \frac{1}{1\,000\,000} = \frac{1}{10^6} = 10^{-6}$$

Any decimal fraction can be expressed in standard form. For example,

$$0.008 = \frac{8}{1000} = \frac{8}{10^3} = 8 \times 10^{-3}$$

$$0.000\,03 = \frac{3}{100\,000} = \frac{3}{10^5} = 3 \times 10^{-5}$$

$$0.000\,25 = \frac{2.5}{10\,000} = \frac{2.5}{10^4} = 2.5 \times 10^{-4}$$

The numbers 8×10^{-3}, 3×10^{-5} and 2.5×10^{-4} are all in standard form $A \times 10^n$, where A is a number between 1 and 10 and n is a whole number. Notice that for decimal fractions, n is a negative whole number.

Example 6

Express the following fractions in standard form.

(a) 0.000 07 (b) 0.075

(c) 0.000 000 022 (d) 0.000 006 3

(a) $0.000\,07 = \frac{7}{100\,000} = \frac{7}{10^5} = 7 \times 10^{-5}$

(b) $0.075 = \frac{7.5}{100} = \frac{7.5}{10^2} = 7.5 \times 10^{-2}$

(c) $0.000\,000\,022 = \frac{2.2}{100\,000\,000} = \frac{2.2}{10^8}$
$= 2.2 \times 10^{-8}$

(d) $0.000\,006\,3 = \frac{6.3}{1\,000\,000} = \frac{6.3}{10^6}$
$= 6.3 \times 10^{-6}$

Exercise 7g (Oral)

Express the following in standard form.

1. 0.005
2. 0.08
3. 0.0006
4. 0.000 004
5. 0.000 02
6. 0.000 000 9
7. 0.3
8. 0.003
9. 0.000 03
10. 0.038
11. 0.0062
12. 0.71
13. 0.000 88
14. 0.000 026
15. 0.000 005 5
16. 0.000 000 91
17. 0.000 000 067
18. 0.000 15
19. 0.0015
20. 0.015

Example 7

Express the following numbers as decimal fractions.
(a) 9×10^{-4} (b) 9.4×10^{-4} (c) 5.3×10^{-7}

(a) $9 \times 10^{-4} = \dfrac{9}{10\,000} = 0.0009$

(b) $9.4 \times 10^{-4} = \dfrac{9.4}{10\,000} = 0.000\,94$

(c) $5.3 \times 10^{-7} = \dfrac{5.3}{10\,000\,000} = 0.000\,000\,53$

Exercise 7h

Express the following as decimal fractions.

1. 2×10^{-4}
2. 8×10^{-6}
3. 5×10^{-3}
4. 4×10^{-2}
5. 7×10^{-1}
6. 3×10^{-5}
7. 6×10^{-3}
8. 9×10^{-5}
9. 2×10^{-7}
10. 2.8×10^{-4}
11. 8.3×10^{-6}
12. 5.1×10^{-3}
13. 4.5×10^{-2}
14. 7.9×10^{-1}
15. 3.3×10^{-5}
16. 6.2×10^{-3}
17. 9.4×10^{-5}
18. 2.6×10^{-4}
19. 1.8×10^{-1}
20. 8.8×10^{-3}

Calculations with numbers in standard form

Adding and subtracting numbers in standard form

Example 8

Find the sum of 6.28×10^3 and 9.5×10^4. Give the sum in standard form.

By factorising:

$$\begin{aligned} 6.28 \times 10^3 + 9.5 \times 10^4 &= 10^3(6.28 + 9.5 \times 10) \\ &= 10^3(6.28 + 95) \\ &= 10^3(101.28) \\ &= 10^3 \times 1.0128 \times 10^2 \\ &= 1.0128 \times 10^5 \end{aligned}$$

or by changing to ordinary form:

$$\begin{aligned} 6.28 \times 10^3 + 9.5 \times 10^4 &= 6\,280 \times 95\,000 \\ &= 101\,280 \\ &= 1.0128 \times 100\,000 \\ &= 1.0128 \times 10^5 \end{aligned}$$

Numbers in standard form can be added or subtracted by taking out the power of 10 which is a common factor. If necessary, the working can be checked by changing the given numbers to ordinary form.

Example 9

Find the value of $2.9 \times 10^6 - 3.8 \times 10^5$. Give the answer in standard form.

By factorising:

$$\begin{aligned} 2.9 \times 10^6 - 3.8 \times 10^5 &= 10^5(2.9 \times 10 - 3.8) \\ &= 10^5(29 - 3.8) \\ &= 10^5 \times 25.2 \\ &= 10^5 \times 2.52 \times 10 \\ &= 2.52 \times 10^6 \end{aligned}$$

Check by changing to ordinary form:

$$\begin{aligned} 2.9 \times 10^6 - 3.8 \times 10^5 &= 2\,900\,000 - 380\,000 \\ &= 2\,520\,000 \\ &= 2.52 \times 10^6 \end{aligned}$$

Special care must be taken when working with small numbers.

Example 10

Express $1.6 \times 10^{-2} - 8.4 \times 10^{-3}$ as a single number in standard form.

By factorising:

$$
\begin{aligned}
1.6 \times 10^{-2} - 8.4 \times 10^{-3} &= 10^{-2}(1.6 - 8.4 \times 10^{-1}) \\
&= 10^{-2}(1.6 - 0.84) \\
&= 10^{-2} \times 0.76 \\
&= 10^{-2} \times 7.6 \times 10^{-1} \\
&= 7.6 \times 10^{-3}
\end{aligned}
$$

Check:

$$
\begin{aligned}
1.6 \times 10^{-2} - 8.4 \times 10^{-3} &= 0.016 - 0.0084 \\
&= 0.0076 \\
&= 7.6 \times 10^{-3}
\end{aligned}
$$

Exercise 7i

Simplify the following. Give all answers in standard form.

1. $3.4 \times 10^3 + 6.2 \times 10^3$
2. $5.7 \times 10^8 + 1.8 \times 10^8$
3. $4.62 \times 10^9 + 3.75 \times 10^9$
4. $8.7 \times 10^4 - 3.5 \times 10^4$
5. $4.3 \times 10^2 - 2.8 \times 10^2$
6. $9.37 \times 10^4 - 6.51 \times 10^4$
7. $9.9 \times 10^5 + 6.8 \times 10^5$
8. $4.1 \times 10^6 + 5.9 \times 10^6$
9. $7.95 \times 10^3 + 3.06 \times 10^3$
10. $5.8 \times 10^4 - 5.2 \times 10^4$
11. $1.75 \times 10^9 - 1.25 \times 10^9$
12. $8.49 \times 10^6 - 8.44 \times 10^6$
13. $3.6 \times 10^{-2} + 4 \times 10^{-2}$
14. $2.9 \times 10^{-4} + 3.5 \times 10^{-4}$
15. $7.8 \times 10^{-3} - 3.4 \times 10^{-3}$
16. $8.65 \times 10^{-5} - 5.76 \times 10^{-5}$
17. $1.7 \times 10^4 + 6.5 \times 10^3$
18. $9.17 \times 10^5 + 7.45 \times 10^6$
19. $6.9 \times 10^{-2} + 5 \times 10^{-3}$
20. $8.31 \times 10^3 - 9.73 \times 10^2$
21. $6.4 \times 10^5 - 1.5 \times 10^4$
22. $5.9 \times 10^{-4} - 4.1 \times 10^{-5}$
23. $3.18 \times 10^{-2} + 9.73 \times 10^{-1}$
24. $1.1 \times 10^{-3} - 8.7 \times 10^{-4}$

Multiplying and dividing numbers in standard form

Use the laws of indices when simplifying powers of 10 that are multiplied or divided:

$10^a \times 10^b = 10^{a+b}$

$10^a \div 10^b = 10^{a-b}$

Example 11

Simplify $(6 \times 10^9) \times (8 \times 10^2)$.

$$
\begin{aligned}
(6 \times 10^9) \times (8 \times 10^2) &= 6 \times 8 \times 10^9 \times 10^2 \\
&= 48 \times 10^{9+2} \\
&= 48 \times 10^{11} \\
&= 4.8 \times 10 \times 10^{11} \\
&= 4.8 \times 10^{12}
\end{aligned}
$$

Example 12

Divide 6×10^3 by 8×10^{-2}.

$$
\begin{aligned}
(6 \times 10^3) \div (8 \times 10^{-2}) &= \frac{6 \times 10^3}{8 \times 10^{-2}} \\
&= \tfrac{6}{8} \times 10^{3-(-2)} \\
&= 0.75 \times 10^5 \\
&= 7.5 \times 10^{-1} \times 10^5 \\
&= 7.5 \times 10^4
\end{aligned}
$$

Example 13

Simplify $(1.4 \times 10^{-5}) \times (2.4 \times 10^6)$.

$$
\begin{aligned}
&(1.4 \times 10^{-5}) \times (2.4 \times 10^6) \\
&= 1.4 \times 2.4 \times 10^{-5} \times 10^6 \\
&= 3.36 \times 10^{-5+6} \\
&= 3.36 \times 10^1 \\
&= 33.6
\end{aligned}
$$

working:

$$
\begin{array}{r}
14 \\
\times\ 24 \\
\hline
56 \\
28 \\
\hline
336
\end{array}
$$

Exercise 7j

Simplify the following. Give all answers in standard form.

1. $(3 \times 10^8) \times (2 \times 10^3)$
2. $(2.8 \times 10^6) \div (1.4 \times 10^2)$
3. $(2 \times 10^{-5}) \times (4 \times 10^{-2})$
4. $(6.3 \times 10^{-2}) \div (2.1 \times 10^4)$
5. $(5 \times 10^2) \times (8 \times 10^5)$
6. $(4.8 \times 10^7) \div (8 \times 10^3)$

7 $(7 \times 10^6) \times (4 \times 10^{-4})$
8 $(3.6 \times 10^2) \div (9 \times 10^{-5})$
9 $(9 \times 10^{-7}) \times (5 \times 10^4)$
10 $(4.2 \times 10^{-9}) \div (7 \times 10^5)$
11 $(6 \times 10^{-3}) \times (6 \times 10^{-3})$
12 $(5.4 \times 10^{-3}) \div (2.7 \times 10^{-7})$
13 $(8.7 \times 10^2) \times (5 \times 10^2)$
14 $(8 \times 10^3) \times (1.5 \times 10^{-3})$
15 $(1.6 \times 10^8) \div (6.4 \times 10^7)$
16 $(1.3 \times 10^{-5}) \times (1.9 \times 10^4)$
17 $(9.1 \times 10^{-2}) \div (1.3 \times 10^{-2})$
18 $(5.5 \times 10^{-6}) \times (4.2 \times 10^{-4})$
19 $(1.92 \times 10^{-6}) \div (1.6 \times 10^{-3})$
20 $(1.05 \times 10^{-7}) \div (1.68 \times 10^{-9})$

Problems involving large and small numbers

In science and astronomy, many measurements are given in very small or very large numbers. For this reason, most scientists prefer to do calculations in standard form. Scientists use standard form so often, it is sometimes called **scientific notation**.

Example 14

A light year is a distance of 9.456×10^{12} km. Express this number to 2 significant figures, then write it out in full.

9.456×10^{12} km $= 9.5 \times 10^{12}$ km to 2 s.f.
9.5×10^{12} km $= 9\,500\,000\,000\,000$ km

Example 15

The density of hydrogen is 8.89×10^{-5} g/cm³.
(a) Find the mass of 1 m³ of hydrogen.
(b) Argon is approximately 20 times as dense as hydrogen. Find the density of argon, giving the answer in standard form correct to 3 s.f. (The density of a substance is mass per unit volume.)

(a) $1\text{ m}^3 = 10^6\text{ cm}^3$
mass of 1 cm³ of hydrogen $= 8.89 \times 10^{-5}$ g
$\therefore$ mass of 1 m³ $= 8.89 \times 10^{-5} \times 10^6$ g
$= 8.89 \times 10^1$ g
$= 88.9$ g

(b) density of argon = mass of 1 cm³
$= 20 \times 8.89 \times 10^{-5}\text{ g/cm}^3$
$= 2 \times 10^1 \times 8.89 \times 10^{-5}\text{ g/cm}^3$
$= 2 \times 8.89 \times 10^{-4}\text{ g/cm}^3$
$= 17.78 \times 10^{-4}\text{ g/cm}^3$
$= 1.778 \times 10^{-3}\text{ g/cm}^3$
$= 1.78 \times 10^{-3}\text{ g/cm}^3$ to 3 s.f.

Example 16

The diameters of the earth and moon are 1.28×10^4 km and 3.5×10^3 km respectively. Find the ratio, diameter of earth : diameter of moon, in the form n : 1 where n is correct to 2 s.f.

$$\frac{\text{diameter of earth}}{\text{diameter of moon}} = \frac{1.28 \times 10^4}{3.5 \times 10^3} = \frac{1.28}{3.5} \times 10 = \frac{128}{35} = \frac{3.65}{1}$$

ratio = 3.7 : 1

working:

```
     3.65
35)128.00
   105
    23 0
    21 0
     2 00
     1 75
       25
```

Example 17

The pages of a book are numbered from 1 to 400. The thickness of the book is 24 mm. Calculate the thickness of 1 leaf (i.e. 2 pages) of the book. Give the answer in metres in standard form.

A book with 400 pages contains 200 leaves of paper.
Thickness of 200 leaves = 24 mm
Thickness of 1 leaf = $\frac{24}{200}$ mm
Thickness of 1 leaf in metres $= \frac{24}{200 \times 1000}$ m
$= \frac{12}{100\,000}$ m
$= \frac{1.2}{10\,000}$ m
$= 1.2 \times 10^{-4}$ m

Exercise 7k

Give all the answers in standard form unless told otherwise.

1 An atom of caesium 133 vibrates 9 192 631 770 times per second. Give this number in standard form correct to 2 s.f.

2. The area of a country is 390 750 km^2. Express this area in standard form correct to 3 s.f.

3. 1 hectare (ha) = 10^4 m^2.
   (a) Find the number of ha in 1 km^2.
   (b) Use the data of question 2 to find the area of the country in hectares in standard form correct to 3 s.f.

4. The distance between two points is 2.54×10^{-2} m. Express this distance in km in standard form.

5. A room measures 4 m by 3 m by $2\frac{1}{2}$ m. Calculate its volume in cm^3 in standard form.

6. The density of air is 1.3×10^{-3} g/cm^3. Calculate the mass of air in the room in question 5. Give your answer in kg in ordinary form.

7. Express 1 hour in seconds in standard form.

8. The velocity of light is approximately 3×10^5 km/s. Use your answer to question 7 to find the distance travelled by light in 1 hour.

9. The height of Blue Mountain Peak in Jamaica is 2.26×10^3 m. The height of Mount Liamuiga in St Kitts is 1.16×10^3 m. Write these heights in ordinary form and find the difference in height between the two mountains.

10. In 1920 the population of the world was 1.81 billion. By 1970 the world's population was 4 billion. Find the increase in world population during those 50 years.

11. The distance of the moon from the earth varies between 3.843×10^5 km and 3.563×10^5 km. Find the difference between these two distances.

12. Mount Everest is the highest point on the earth's surface: 8.848×10^3 m above sea level. The lowest point on the earth's surface is the Marianas Trench: 1.103×10^4 m *below* sea level. Find the vertical distance between the lowest and highest points on the earth's surface.

13. 1 barrel has a volume of 1.65×10^{-1} m^3. Find the total volume, in m^3, of 6.7×10^7 barrels. Give the volume in standard form correct to 2 s.f.

14. A packet of paper contains 500 sheets. The thickness of the packet is 56 mm. Calculate the thickness of 1 sheet of paper. Give your answer in metres in standard form.

15. The pages of a book are numbered 1 to 300.
   (a) How many thicknesses of paper make 300 pages?
   (b) If the thickness of the book is 15 mm, calculate the thickness of one leaf. Give your answer in metres in standard form.

Summary

Very large or very small numbers may be written in **standard form**, that is, in the form $A \times 10^n$,

where A is a number between 1 and 10 and n is a whole number.

For very large numbers, n is positive, e.g. 2.341×10^{12}

For numbers less than 1, n is negative, e.g. 3.215×10^{-9}

When working with numbers written in standard form, the **laws of indices** are applied:

$$b^m \times b^n = b^{m+n}$$
$$b^m \div b^n = b^{m-n}$$
$$(b^m)^n = b^{mn}$$
$$b^{-m} = \frac{1}{b^m}$$

Note: $b^0 = 1$

Practice Exercise P7.1

1. Write each of the following numbers in standard index form.
   (a) 7000 (b) 4 510 000 000
   (c) 30 000 (d) 750 000 000
   (e) 507 000

2. Express each of the following in standard form
   (a) 52.43, 21.045, 2.435, 432.5, 2 543 271
   (b) 0.0314, 0.004 13, 0.134, 17.504, 0.000 005 043
   (c) 36, 0.3, 0.003, 43 505 643, 0.000 63, 0.000 000 105, 7320

3 Write each of the following numbers in ordinary form

(a) 3.7×10^5 (b) 1.3×10^7
(c) 9×10^{12} (d) 2.07×10^5
(e) 9.06×10^8

4 Express each of the following as decimal numbers

(a) 5.24×10^2, 1.415×10^5, 2.43×10^3, 4.325×10^7
(b) 3.14×10^{-2}, 4.13×10^{-5}, 3.124×10^{-3}, 1.054×10^{-6}
(c) 4×10^2, 2.6×10^{-7}, 1.07×10^{-3}, 5.55×10^8

5 The following numbers are in index form. Put them into **standard index form**.

(a) 56×10^8 (b) 846×10^5
(c) 19.7×10^6 (d) 0.74×10^5
(e) 5002×10^8 (f) 0.2×10^{10}

6 Use your calculator to work these out. Give your answers in standard form.

(a) $(3.7 \times 10^4) + (1.5 \times 10^5)$
(b) $(7.85 \times 10^6) - (3.15 \times 10^5)$
(c) $(6.8 \times 10^7) \times (3.4 \times 10^8)$
(d) $(7.2 \times 10^9) \div (2.4 \times 10^7)$

Practice Exercise P7.2

Write your answers to the following questions in **standard index form**. Round decimals to 2 d.p. if necessary.

1 In 1 gram, there are about 9.2×10^{28} electrons.

(a) How many electrons are there in 1 kg?
(b) What is the weight of a cluster of 5×10^{100} electrons?

2 A reservoir, when full, contained 1.9×10^8 litres of water.

(a) If 1.8×10^7 litres were used without any being put back in, how much water would be left?
(b) Find the volume of water in the reservoir when it is half full.

Practice Exercise P7.3

(a) Key into your calculator 10^{10}.
(b) What is the smallest number you can add to this in order for the total not to be still displayed as 10^{10}?
(c) Repeat parts (a) and (b) for 10^{11}; then 10^{12}, and so on up to 10^{16}
(d) What do you notice?

Chapter 8

Sets (2)

Union, intersection, complement

Pre-requisites
- elements of sets, subsets, Venn diagrams

Sets (revision)

Table 8.1 contains some of the symbols and language of sets that appeared earlier in the course, in Chapter 1 of Book 1.

Table 8.1

symbols	meaning
P = {a, b, c, d}	P is the set a, b, c, d
Q = {1, 2, 3, ...}	... means 'and so on'
∅ or { }	the empty set
U	the universal set
∈	is a member of
∉	is not a member of
⊂	is a subset of
⊄	is not a subset of
⊃	includes
⊅	does not include
n(S)	number of elements in S

Exercise 8a (revision)

1. Make each of the following true by writing either ∈ or ∉ in place of the *.
   (a) 9 * {2, 4, 6, 8, 10}
   (b) 15 * {3, 6, 9, ..., 24}
   (c) 44 * {5, 10, 15, 20, ...}
   (d) R * {a, b, c, d, ..., z}
2. Find n(X) when X =
   (a) {h, o, u, s, e} (b) {toes on your feet}
   (c) {0, 1, 2} (d) {months in a year}
   (e) {5, 5, 6, 6} (f) {11, 12, 13, ..., 22}
3. Give three examples of an empty set.
4. Give three examples of an infinite set.
5. Write down all the subsets of the following.
   (a) {3, 4, 5} (b) {x, y}
   (c) {0, 2} (d) {f, o, u, r}
6. Write down the following using symbols.
   (a) 2 and 6 form a subset of the factors of 18
   (b) {trees} is not a subset of {metal objects}
   (c) {vehicles} contains {buses}
   (d) men are members of the human race
7. If U = (months of the year), F = {first eight months of the year}, Y = {months ending in y}, draw a Venn diagram to show the relationship between F, Y and U.

Union and intersection

The **union** of two sets is a third set which includes all the elements of the first two. For example if A = {1, 2, 3} and B = {2, 3, 4, 5, 6}, then the union of A and B is a third set C, where C = {1, 2, 3, 4, 5, 6}.

In symbols, A ∪ B = C means 'A union B equals C'.

Notice that A ∪ B = B ∪ A.

The **intersection** of two sets is a third set which includes only those elements which are in both of the first two sets. For the sets A and B above, the intersection is D, where D = {2, 3}.

In symbols A ∩ B = D means 'A intersects with B to give D'.

Notice that A ∩ B = B ∩ A.

Union and intersection can be represented on Venn diagrams.

In Fig 8.1(a) the shaded region represents the union of A and B. In Fig. 8.1(b) the shaded region represents the intersection of A and B.

(a)
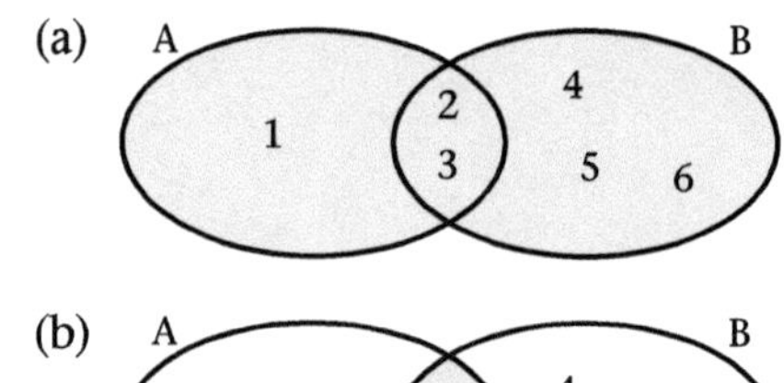

(b)
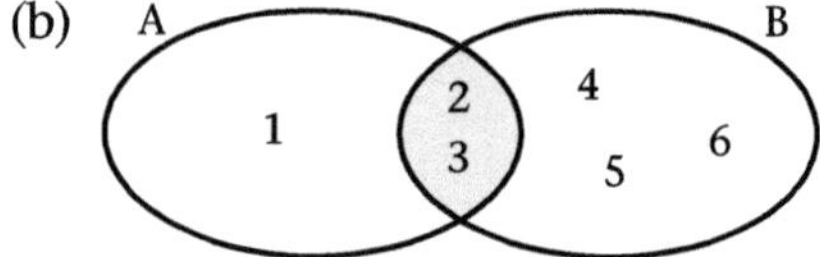

Fig. 8.1 (a) A ∪ B = {1, 2, 3, 4, 5, 6}
(b) A ∩ B = {2, 3}

Example 1

Write down the union and intersection of the following pairs of sets. Represent the results on a Venn diagram.

(a) P = {Friday, Saturday, Sunday},
 Q = {Sunday, Monday}
(b) X = {r, e, l, a}, Y = {t, i, o, n}

(a) $P \cup Q$ = {Friday, Saturday, Sunday, Monday}
 $P \cap Q$ = {Sunday}

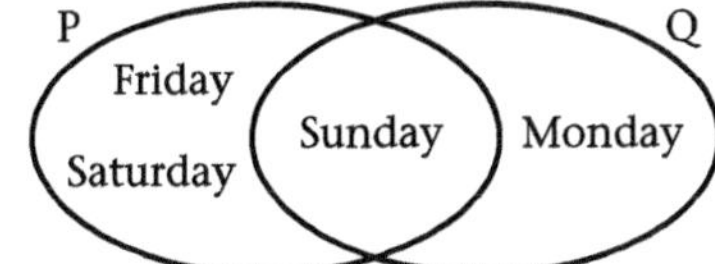

Fig. 8.2

(b) $X \cup Y$ = {r, e, l, a, t, i, o, n}
 $X \cap Y = \varnothing$

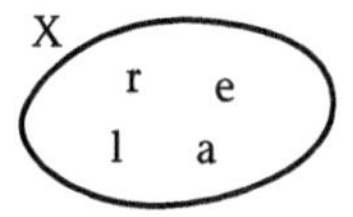

Fig. 8.3

In Example 1 (b), $X \cap Y = \varnothing$. Two sets are said to be **disjoint** if their intersection is the empty set. Hence X and Y are disjoint sets.

Example 2

Given U = {f, o, r, m, u, l, a, e}, A = {f, r, a, m, e} and B = {r, u, l, e},

(a) draw a Venn diagram showing A, B and U;
(b) list the elements of $A \cap B$;
(c) find $n(A \cup B)$.

(a) Fig. 8.4 is the required Venn diagram:

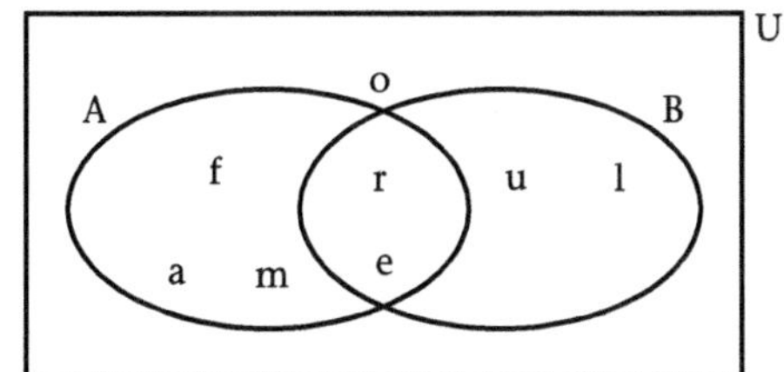

Fig. 8.4

(b) $A \cap B$ = {r, e}
(c) $A \cup B$ = {f, r, a, m, e, u, l}
 $n(A \cup B) = 7$

Exercise 8b

1. If A = {3, 5, 6, 8, 9} and B = {2, 3, 4, 5} write down the sets $A \cup B$ and $A \cap B$. Show A and B on a Venn diagram.

2. If C = {grapefruit, orange, pear} and D = {grapefruit, pear, apple, pawpaw} write down the sets $C \cup D$ and $C \cap D$.

3. What is the union of {January, February, March} and {April, May, June}? What is their intersection? Represent the two sets on a Venn diagram.

4. Name the intersection of the set of all capital cities in Grenada and the set of all towns in Grenada.

5. Fig. 8.5 is a Venn diagram representing a universal set U and the subsets A and B.

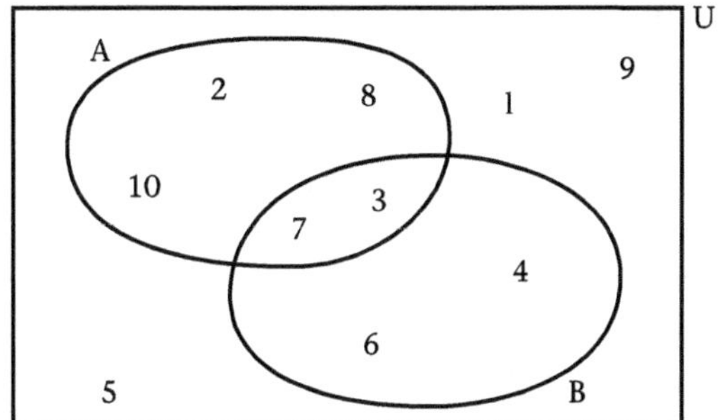

Fig. 8.5

List the elements of the following sets.

(a) $A \cup B$ (b) $A \cap B$
(c) $U \cap B$ (d) $A \cup U$

6. Which of the following pairs of sets are disjoint?
   (a) {even numbers}, {odd numbers}
   (b) {houses in Montego Bay}, {houses in Jamaica}
   (c) {rivers of Africa}, {Thames, Mississippi, Zambesi}
   (d) {letters in pupil}, {letters in teacher}

7. Let U = {students in your class}
   B = {students in the back row}
   S = {students whose names begin with S}
   T = {students taller than the teacher}
   Write down the members of the following.
   (a) $B \cap S$ (b) $B \cup S$ (c) $S \cup T$
   (d) $S \cap T$ (e) $B \cap T$ (f) $U \cap S$

8 Let U = {1, 2, 3, 4, ..., 10}
A = {odd numbers less than 10}
B = {numbers less than 7}
Write out the members of the following.

(a) A ∪ B (b) A ∩ B
(c) A ∪ U (d) A ∩ U

9 The members of a football team form a universal set U, where
U = {Bob, Frank, Kevin, Ned, Andy, Rex, Ron, Sam, Richard, Brian, Tom}
If A = {boys whose names end in *n*},
B = {boys with 3-lettered names},
R = {boys whose names begin with *R*},
list the following sets:

(a) A ∪ B (b) A ∪ R
(c) B ∪ R (d) A ∪ U
(e) B ∪ U (f) R ∪ U
(g) A ∩ B (h) B ∩ R
(i) A ∩ R (j) A ∩ U
(k) B ∩ U (l) R ∩ U

10 Given U = {a, b, c, d, e, f}, X = {a, b, c, d}, Y = {c, d, e} and Z = {b, d, f}, list the following sets. Where brackets are given, do the parts inside the brackets first.

(a) X ∪ Y (b) X ∪ Z
(c) Y ∪ Z (d) X ∩ Y
(e) X ∩ Z (f) Y ∩ Z
(g) X ∪ X (h) Y ∩ Y
(i) Z ∪ Ø (j) X ∪ (Y ∪ Z)
(k) (X ∪ Y) ∪ Z (l) X ∩ (Y ∩ Z)
(m) (X ∩ Y) ∩ Z (n) X ∩ (Y ∪ Z)
(o) (X ∩ Y) ∪ Z (p) X ∪ (Y ∩ Z)
(q) (X ∪ Y) ∩ Z
(r) (X ∩ Y) ∪ (X ∩ Z)
(s) (X ∪ Y) ∩ (X ∪ Z)

Complement of a set

In Fig. 8.6, A is a subset of the universal set U, i.e. A ⊂ U.

The **complement** of A is the set which contains all those elements of U which are *not* members of A. A′ is short for the complement of A. In Fig. 8.6 the shaded region represents A′.

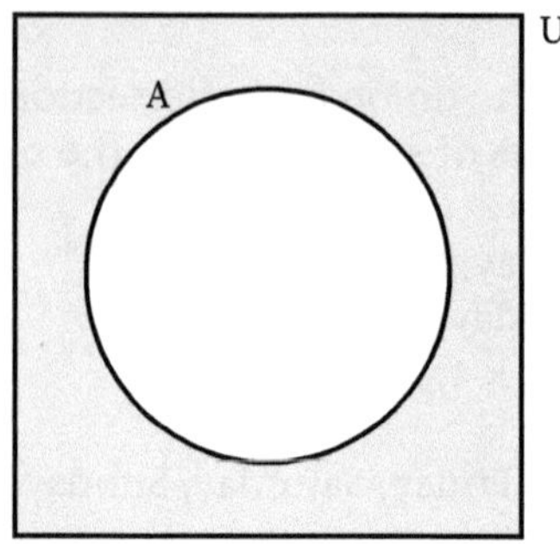

Fig. 8.6

Example 3

If U = {1, 2, 3, 4, 5},
A = {1, 3} and
B = {3, 4}
find (a) A′, (b) B′, (c) (A ∩ B)′, (d) (A ∪ B)′.

(a) A′ = {2, 4, 5}
(b) B′ = {1, 2, 5}
(c) A ∩ B = {3}
(A ∩ B)′ = {1, 2, 4, 5}
(d) A ∪ B = {1, 3, 4}
(A ∪ B)′ = {2, 5}

Example 4

U is the set of all cars, R is the set of all red cars and F is the set of all cars with four doors. Show on a Venn diagram the set of all red cars which do not have four doors.

F = {cars with four doors}
F′ = {cars which do *not* have four doors}.
R ∩ F′ = {red cars which do not have four doors}

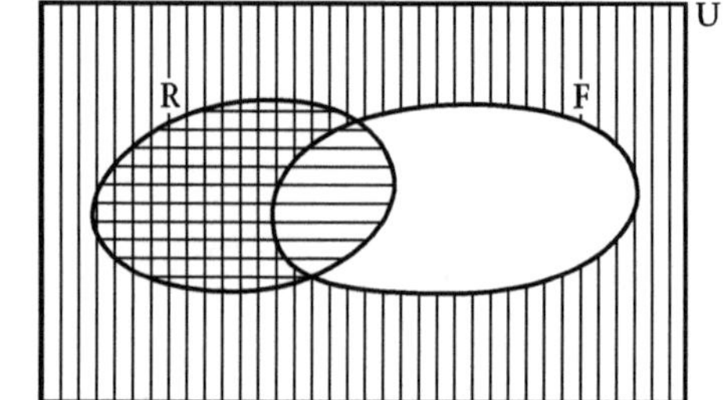

Fig. 8.7

In Fig. 8.7, the region with horizontal shading represents R. The region with vertical shading represents F′. The region which is cross-shaded represents R ∩ F′, the set of all red cars which do not have four doors.

Exercise 8c

1. Given U = {1, 2, 3, 4, 6, 8}, write down the complements of the following sets.
   (a) {1, 2, 3} (b) {4, 6, 8}
   (c) {2, 4} (d) {1, 3, 6, 8}
   (e) {6} (f) {1, 2, 3, 4, 6}
   (g) Ø (h) U
   (i) {1, 2, 3, 6} ∪ {1, 2, 8}
   (j) {1, 2, 3, 6} ∩ {1, 2, 8}

2. If U = {a, b, c, d, e}, A = {a, c, e} and B = {b, e}, list the members of the following sets.
   (a) A′ (b) B′ (c) A′ ∪ B
   (d) A ∪ B′ (e) A′ ∩ B (f) A ∩ B′
   (g) A′ ∪ A (h) B′ ∩ B (i) (A ∪ B)′
   (j) A′∪ B′ (k) (A ∩ B)′ (l) A′ ∩ B′

3. Compare the answers to parts (i) and (l) of question 2. What do you notice?

4. Compare the answers to parts (j) and (k) of question 2. What do you notice?

5. If U = {1, 2, 3, 4, 5, ..., 10}, A = {1, 2, 5, 7}, B = {1, 3, 6, 7}, write down the sets A′, B′, (A ∩ B)′ and (A ∪ B)′.

6. Using the sets of question 5, show that A′ ∪ B′ = (A ∩ B)′ and A′ ∩ B′ = (A ∪ B)′.

7. Use Venn diagrams to illustrate the results of question 6.

8. If U = {days of the week}, S = {words which contain the letter *s*} and N = {words which contain six letters},
   (a) list the members of the sets S, N, S′, N′;
   (b) list the members of the sets (i) (S ∪ N)′, (ii) (S ∩ N)′;
   (c) hence, without further working, list the members of the sets (i) S′ ∩ N′, (ii) S′ ∪ N′.

9. Use a Venn diagram to show that if A ⊂ B then B′ ⊂ A′, and vice versa.

10. If U = {all teachers}, M = {mathematics teachers} and W = {teachers who are women}, show on a Venn diagram the set of all mathematics teachers who are men.

Set-builder notation

A = {integer $x : x < 4$} is short for 'A is the set of integers, x, such that x is less than 4'. We say that A is written in **set-builder notation.** In set-builder notation the colon (:) stands for 'such that' and $x < 4$ describes the condition that the members of the set must satisfy. If set-builder notation is expressed properly it is always possible to list the elements of the given set. For example:

A = {... −2, −1, 0, 1, 2, 3}

Notice that the elements of A are restricted to integral or whole-number values. The letter Z is often reserved for the set of integers:

Z = {... −3, −2, −1, 0, +1, +2, +3, ...}

Hence set A can be rewritten:

A = {x: $x < 4$; $x \in Z$}

i.e. 'A is the set of values of x such that x is less than 4, where x is an integer.'

Example 5

List the members of the set {x: $x \geqslant -2$; $x \in Z$}.

{x: $x \geqslant -2$; $x \in Z$} = {−2, −1, 0, 1, ...}

Remember the meanings of the following symbols:

$>$ is greater than
$\geqslant$ is greater than or equal to
$<$ is less than
$\leqslant$ is less than or equal to

Chapter 16 contains further treatment of inequalities.

Example 6

Given that x is an integer, list the members of {x: $x < 4$} ∩ {x: $x \geqslant -2$}.

If $x \in Z$,
{x: $x < 4$} = {... −3, −2, −1, 0, 1, 2, 3}
{x: $x \geqslant -2$} = {−2, −1, 0, 1, 2, 3, 4, ...}
{x: $x < 4$} ∩ {x: $x \geqslant -2$} = {−2, −1, 0, 1, 2, 3}

Notice, in Example 6, that the expression {x: $x < 4$} ∩ {x: $x \geqslant -2$} can be shortened to {x: $-2 \leqslant x < 4$}, i.e. 'the set of values, x, such that x is greater than or equal to −2 and less than 4'. The expression $-2 \leqslant x < 4$ is a **range of values** of x.

Example 7

List the members of the set $\{y: 1 < y \leqslant 6; y \in Z\}$.

$\{y: 1 < y \leqslant 6; y \in Z\} = \{2, 3, 4, 5, 6\}$

Example 8

If A = {0, 1, 2, 3, 4}, rewrite A using set-builder notation.

either $A = \{x: 0 \leqslant x \leqslant 4; x \in Z\}$
or $A = \{x: 0 \leqslant x < 5; x \in Z\}$
or $A = \{x: -1 < x < 5; x \in Z\}$
or $A = \{x: -1 < x \leqslant 4; x \in Z\}$

Exercise 8d

1. List the elements of the following sets.
   (a) $\{t: t$ is a subject on your timetable$\}$
   (b) $\{f: f$ is a factor of both 24 and 30$\}$
   (c) $\{p: p$ is a prime factor of 42$\}$
   (d) $\{$integer $x: x > 8\}$
   (e) $\{$integer $y: y < -5\}$
   (f) $\{x: 3 < x < 5; x \in Z\}$
   (g) $\{x: 3 < x \leqslant 5; x \in Z\}$
   (h) $\{x: -2 < x < -1; x \in Z\}$

2. If U = {positive integers}, list the members of the following sets. *Note*: 0 (zero) is neither positive nor negative.
   (a) $\{x: x > 7\}$ (b) $\{x: x > -3\}$
   (c) $\{x: x < -1\}$ (d) $\{x: x < 10\}$
   (e) $\{x: x - 3 = 0\}$ (f) $\{x: x + 3 = 9\}$
   (g) $\{x: x + 2 = 0\}$ (h) $\{x: 2x = 7\}$

3. Assuming that x is an integer, state the lowest and highest values of x for each of the following ranges.
   (a) $-2 < x < 5$ (b) $-8 \leqslant x \leqslant 9$
   (c) $-5 < x \leqslant 0$ (d) $2 \leqslant x < 10$
   (e) $4 < x < 6$ (f) $-8 < x < -1$
   (g) $-7 \leqslant x < 1$ (h) $-5\frac{2}{3} < x < 1\frac{3}{4}$

4. Using the inequality symbol <, write the following in set-builder notation.
   (a) {0, 1, 2, 3} (b) {−1, 0, 1}
   (c) {−3, −2, −1, 0} (d) {−9, −8, ..., 3, 4}
   (e) {9, 10} (f) {−8, −7, ..., −1}

5. If $x \in U$ and U = {1, 2, 3, ..., 10}, list the members of the following subsets of U.
   (a) $\{x: x$ is an odd number$\}$
   (b) $\{x: x + 2 \geqslant 7\}$
   (c) $\{x: 3x - 2 = 19\}$
   (d) $\{x: x$ is a factor of 24$\}$

Venn diagrams, problem solving

Venn diagrams can sometimes be used to store numerical information. In such cases it is also possible to use the Venn diagrams to solve problems arising from the data.

Example 9

Thirty students were asked if they liked cricket or football. All the students liked either cricket or football. 20 said they liked cricket and 25 liked football. How many students liked both cricket and football?

Since 20 + 25 = 45 (45 > 30) some students like both cricket and football. Draw a Venn diagram to show the given information (Fig. 8.8).

Let C = {students who like cricket}
F = {students who like football}

You are required to find $n(C \cap F)$.
Let $n(C \cap F) = x$

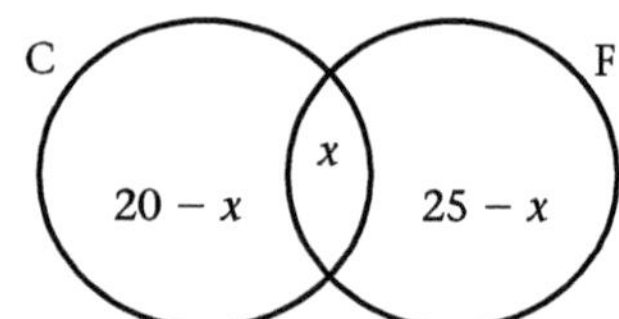

Fig. 8.8

The total of the regions must add up to 30.

$(20 - x) + x + (25 - x) = 30$
$45 - x = 30$
$x = 15$

Check:
No. who liked football only = 25 − 15 = 10
No. who liked cricket only = 20 − 15 = 5
Total = 10 + 5 + 15 = 30

Example 10

50 students were asked what they did last night. 16 said they read a book, 41 said they watched

television. If 7 said they did neither, how many did both?

Since 41 + 16 + 7 = 64 (64 > 50), some students read a book and watched television.

Let U = {all students}, B = {book readers} and T = {television watchers}. You are required to find $n(B \cap T)$. Let $n(B \cap T) = x$.

Fig. 8.9 is a Venn diagram containing the given information. The numbers in the regions of the Venn diagram represent the numbers of elements in the regions.

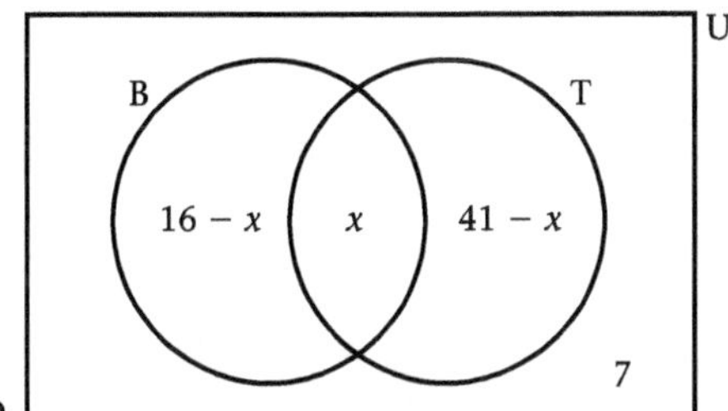

Fig. 8.9

Since $n(T) = 41$, $n(B' \cap T) = 41 - x$
Similarly $n(B \cap T') = 16 - x$
and $n(B \cup T)' = 7$

The totals for the regions must add up to the number of people in the universal set:

$$x + (16 - x) + (41 - x) + 7 = 50$$
$$64 - x = 50$$
$$x = 14$$

14 students read a book *and* watched television.
Note: The 16 who read a book includes the 14 who also watched television.

In general:
1 Identify the sets, including the universal set.
2 Draw a Venn diagram.
3 Enter the data on the Venn diagram, starting with the region where all the sets intersect.
4 Form an equation using the fact that the total number of elements in the regions equals the number of elements in the universal set.

In Exercise 8e the number of subsets will be restricted to two.

Exercise 8e

Draw a suitable Venn diagram in each question.

1 In Fig. 8.10 the numbers of elements in each region of the Venn diagram are as given.

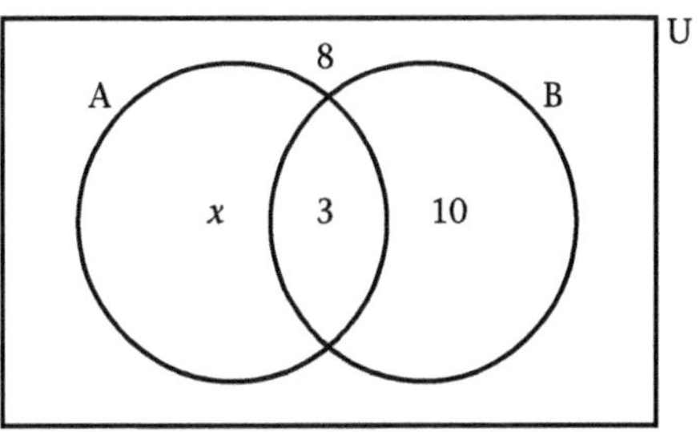

Fig. 8.10

If n(U) = 30 find
(a) $n(A \cap B')$, (b) $n(A \cup B)$.

2 In the Venn diagram in Fig. 8.11 the numbers of elements are as shown.

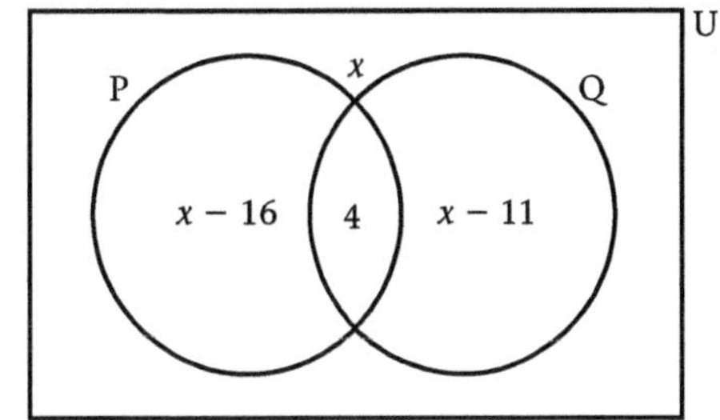

Fig. 8.11

Given that $n(P \cup Q) = n(P \cup Q)'$, find x and hence find n(U).

3 A company employs 100 people, 65 of whom are men. 60 people, including all the women, are paid weekly. How many of the men are paid weekly?

4 In a town all the people speak patois or English or both. If 97% speak patois and 64% speak English, what percentage speak both languages?

3 20 people apply for a job. 12 have school certificates and 10 have diplomas. If 2 have no qualifications, how many have both a school certificate and a diploma?

4 In a class of 35 students, 19 take History and 12 take Economics. If 5 take both subjects, how many take neither?

5 In a school of 750 students, 320 are girls. 559 students do some kind of sport. If 101 girls do no sport, how many boys also do no sport?

8. In a survey of students it is found that 20% of them have visited Grenada and 25% have visited Barbados. If 5% have been to both places, what percentage have visited neither place?

9. Out of 25 teachers, 16 are married and 15 are women. If 6 of the men are married, how many of the women are not married?

10. In a class of 36 students, 29 do Mathematics and 20 do Chemistry. If 5 students do neither, how many students do Chemistry but not Mathematics?

Summary

Given two sets A and B which are subsets of a **universal set** U, then A ∩ B is the **intersection** of sets A and B, i.e. the set whose elements are members of both A and B.

A ∪ B is the **union** of sets A and B, i.e. the set whose elements are members of A or B or both A and B.

If p and q are elements of both A and B, there is no need to write them down twice.

A′ is the **complement** of the set A, i.e. the set whose elements are *not* members of the set A. Similarly, for B′.

The elements of a set may be defined using **set-builder notation**, for example,

$A = \{n: n \text{ is an odd number}, 3 < n < 15\}$

where the colon ':' stands for 'such that'. The above example is read as

'A is the set of n such that n is an odd number greater than 3 and less than 15'

Note that the set A may also be described by listing the elements, that is,

$A = \{5, 7, 9, 11, 13\}$

$3 < n < 15$ is the **range** of values of n.

Venn diagrams may be used to solve problems when numerical information is given for the universal set and for subsets.

Practice Exercise P8.1

1. (a) Using the Venn diagram, list the members of

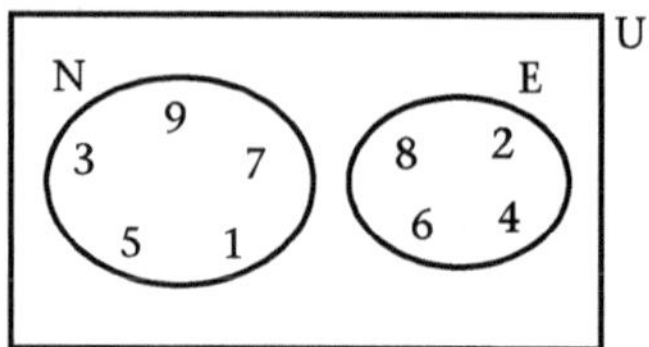

Fig. 8.12

(i) N (ii) N′ (iii) E (iv) E′

(b) Describe with a phrase the members of
(i) N (ii) E

2. Given that U = {real numbers}, list the members of the following sets
(a) $A = \{x : 2x + 10 = 15, x \in Q\}$
(b) B = {factors of 24}
(c) $C = \{x : x \text{ is odd}, x > 20\}$
(d) $D = \{x: x \leqslant 10, x \text{ is a multiple of } 11\}$
(e) $E = \{x : x \text{ is prime}, 7 < x < 20\}$

3. U = {1, 2, 3, 4, 5, ..., 10}
A = {1, 2, 3, 4, 9, 10}
(a) List the members of
(i) A′ (ii) A ∩ A′
(b) What do you notice about the results you got?

4. (a) List three members of each of the subsets of the given universal sets.
(b) List three members of the complement of the subset.
(i) U = {numbers from −5 to +5}
N = {positive numbers}
(ii) U = {8 school subjects}
F = {favourite subjects}
(iii) U = {letters of the alphabet from d to m}
K = {consonants}.

Practice Exercise P8.2

Andy, Bert, David, Jean, Ken, Lisa, Mark, Nina, Roland, Sylvia and Tessa are children in a playground. Andy, Ken, Nina, Roland and Sylvia are cousins.

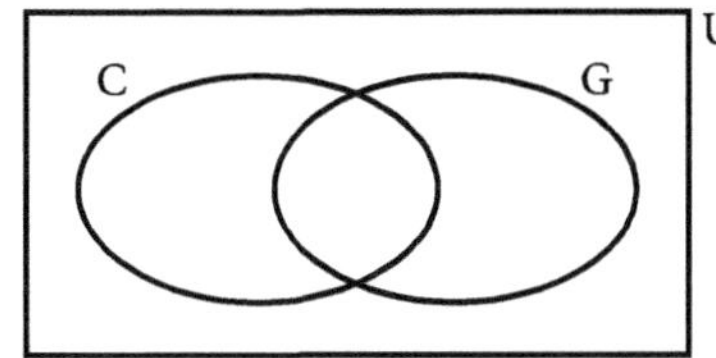

Fig. 8.13

If U = {children in the playground},
C = {cousins}, G = {girls},

(a) using the first letter of the names, complete the Venn diagram for the given information.
(b) list the members of
(i) C ∪ G (ii) C ∩ G

Practice Exercise P8.3

For the following sets,
(i) A = {letters of the word *ladder*}
(ii) B = {letters of the word *dealer*}
(iii) C = {letters of the word *retreat*}
(iv) D = {letters of the word *dress*}
(v) E = {letters of the word *sadder*}

1. List the members of each set
2. State the number of members in each set
3. Describe the following pairs of sets as
   V : equivalent but not equal
   Q : equal
   or N : neither equivalent nor equal
   (a) A and B (b) A and C
   (c) D and E (d) A and D
   (e) C and D (f) B and E

Practice Exercise P8.4

1. Given that
   U = {a, b, c, d, e, f, g, h, m, n, p, r, s, t, w, x, y}
   A = {a, c, f, g, t, w, x, y}
   B = {a, b, c, d, e, f, g}
   C = {a, c, f}
   (a) draw a Venn diagram to show the sets listed above
   (b) list the members of
   (i) A ∩ B (ii) A ∩ C (iii) B ∩ C
   (iv) A ∪ B (v) A ∪ C (vi) B ∪ C

2. For each of the following pairs of sets, find the number of members
   (a) in the intersection of the sets
   (b) in the union of the sets.
   (i) P = {2, 4, 6, 8, 10, 12}
   M = {2, 4, 8, 16, 32}
   (ii) X = {j, k, m, n, p}
   Y = {k, m}
   (iii) T = {3, 6, 9, 12, 15, 18}
   F = {6, 12, 18}
   (iv) S = { 1, 3, 5, 7}
   V = {2, 4, 6}

Practice Exercise P8.5

1. U = {students in Form 2}.
   W = {students who walk to school} and
   P = {students who play games}.
   Some students who walk to school also play games.
   (a) Draw the Venn diagram to show the given sets.
   (b) Show by shading the set of students who both walk to school and play games.
   (c) Insert in the diagram
   (i) *Gabby* who rides to school and plays games
   (ii) ***Rob*** who walks to school and does not play games
   (iii) *Ashley* who rides to school and does not play games.

2. 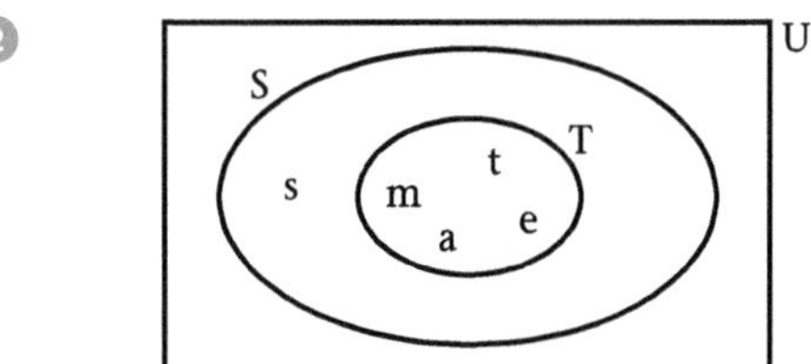

Fig. 8.14

The Venn diagram shows the subsets, S and T, of the universal set,

U = {letters of the word mathematics}

(a) Complete the Venn diagram.
(b) List the members of the following regions:
(i) U (ii) S (iii) T (iv) S ∪ T

Revision exercises and tests

Chapters 1–8

Revision exercise 1 (Chapters 1, 7)

1. Find the next four terms in the following patterns.
   (a) 6, 13, 20, 27, 34, ...
   (b) 5, 6, 8, 11, 15, 20, ...
   (c) 3, 6, 12, 24, ...
   (d) 0, 2, 6, 12, 20, ...
2. Find the smallest number by which 350 must be multiplied to give a perfect square.
3. Express the following numbers in standard form.
   (a) 3 500 000 (b) 5700 (c) 28
   (d) 0.47 (e) 0.085 (f) 0.000 003
4. Simplify the following.
   (a) $10^2 \times 10^5$ (b) $3x^2 \times 5x^3$
   (c) $9a^3 \times 2a^{-2}$ (d) $10^7 \div 10^3$
   (e) $16y^5 \div 8y^3$ (f) $42x^{-3} \div 14x^{-7}$
5. If $\dfrac{0.0001 \times 1.11}{0.1 \times 10^4} = A \times 10^n$ where A is a number between 1 and 10 and n is an integer, find the values of A and n.
6. Simplify the following, giving the answers in standard form.
   (a) $(5.2 \times 10^2) + (6.24 \times 10^3)$
   (b) $(7 \times 10^{-3}) - (8 \times 10^{-4})$
   (c) $(4 \times 10^{-4}) \times (8 \times 10^5)$
   (d) $(5.4 \times 10^2) \div (9 \times 10^{-3})$
7. The populations of two countries are 5.9×10^7 and 8.7×10^6 respectively. Find the total population of the two countries correct to 2 s.f.
8. Using your calculator, find the value of $696 \div 245$. Give your answer correct to 2 decimal places.
9. Complete each of the following as indicated.
   (a) $64 + (59 + 73) = \ldots$ (Associative law)
   (b) $97 + x = \ldots$ (Commutative law)
   (c) $70 \times 86 = \ldots$ (Distributive law)
10. Divide 2.647 by 0.9 and give the answer correct to 2 d.p.

Revision test 1 (Chapters 1, 7)

1. The next term in the sequence 2, 3, 5, 8, 12, 17, ... is
   A 19 B 22 C 23 D 24
2. Which of the following is a member of the sets W, Z and N?
   A 0 B -2 C $2\frac{1}{2}$ D 3
3. $59 \times 23 + 21 \times 23 =$
   A $(59 \times 23) + 21$ B $(59 + 21) \times 23$
   C $59 \times 44 \times 23$ D $59 \times (23 + 44)$
4. Without using tables, calculate $82.5 \div 0.025$, expressing the answer in standard form.
   A 3.3×10^3 B 3.3×10^2
   C 3.3×10^{-2} D 3.3×10^{-3}
5. $25^{-2} =$
   A -23 B $\frac{1}{23}$ C $\frac{2}{25}$ D $\frac{1}{25^2}$
6. Given $1 \times 2 = 2^2 - 2$
   $2 \times 3 = 3^2 - 3$
   $3 \times 4 = 4^2 - 4$
   $4 \times 5 = 5^2 - 5$
   (a) write an expression for $n \times (n + 1)$
   (b) use this to calculate $1001^2 - 1001$.
7. Calculate the value of 108×53, using the distributive law.
8. Simplify the following. Express the answers in standard form.
   (a) $(6.3 \times 10^{-3}) + (3 \times 10^{-4})$
   (b) $(7.42 \times 10^4) - (6.8 \times 10^3)$
   (c) $(4 \times 10^5) \times (3.9 \times 10^{-1})$
   (d) $(9.1 \times 10^{-5}) \div (7 \times 10^{-3})$
9. The following numbers are in standard form. Change them to ordinary form.
   (a) 9×10^2 (b) 3.6×10^5
   (c) 6.1×10^7 (d) 8×10^{-4}
   (e) 6×10^{-1} (f) 3.4×10^{-3}

10 The pages of a dictionary are numbered from 1 to 1322. The dictionary is 7 cm thick (neglecting the covers).
(a) How many thicknesses of paper make 1322 numbered pages?
(b) Find the thickness of 1 sheet of paper. Give your answer in metres in standard form correct to 2 s.f.

Revision exercise 2 (Chapters 2, 6)

1 Use a ruler and set square to construct a pair of parallel lines which are 57 mm apart.

2 Construct the triangle sketched in Fig. R1.

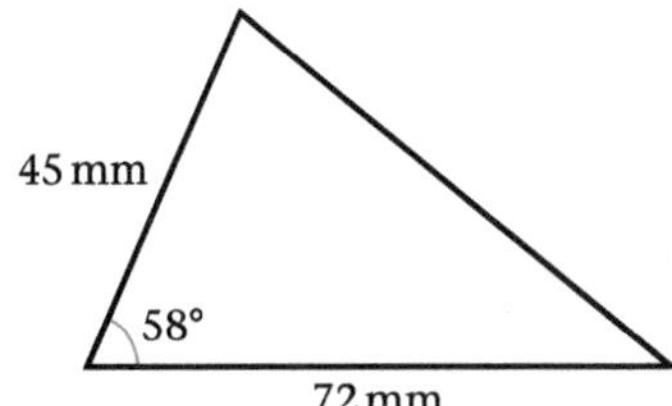

Fig. R1

Measure the third side of the triangle.

3 Make a *freehand* sketch of △ABC in which AB = 6 cm, $B\hat{A}C = 54°$ and $A\hat{C}B = 69°$. Calculate the third angle of the triangle and show this on your sketch.
Make an accurate construction of △ABC.

4 Construct △ABC in which AB = 3.5 cm, BC = 4.5 cm and AC = 5.5 cm. Measure the perpendicular distance of A from BC and hence calculate the area of the triangle.

5 Calculate the volume of a cuboid measuring 12 cm by 10 cm by 6 cm. How many 2 cm by 2 cm by 2 cm cubes would this cuboid contain?

6 How many cm^3 of water does a 5 m by 10 m by 2 m tank hold?

7 The area of the end face of a beam is 24 cm^2. Calculate the volume of a 5 m length of the beam.

8 A triangular prism has a volume of 142 cm^3. If the prism is 8 cm long, calculate the area of one of its triangular faces.

9 Take π to be 3.1 and calculate the volume of a cylinder of height 10 cm and radius 4 cm.

10 Use the value 3 for π to estimate the capacity in litres of a cylindrical drum 49 cm in diameter and 84 cm high.

Revision test 2 (Chapters 2, 6)

1 The following are given as dimensions of triangle ABC.
For which set of dimensions is it impossible to construct the triangle?
A AB = 11 cm, BC = 3 cm, CA = 7 cm
B AB = 8 cm, $A\hat{B}C = 50°$, BC = 3 cm
C AB = 5 cm, $A\hat{B}C = 100°$, AC = 8 cm
D AB = 6 cm, BC = 6.5 cm, CA = 9.5 cm

2 In △XYZ, XY = 5 cm, $X\hat{Y}Z = 40°$ and $X\hat{Z}Y = 60°$. Which one of the sketches in Fig. R2 shows this information correctly?

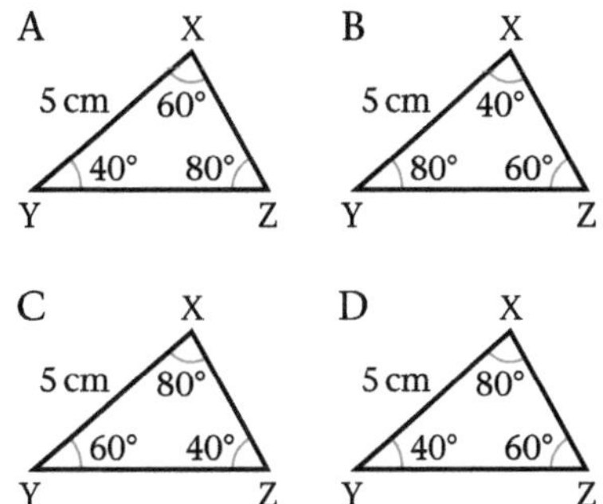

Fig. R2

3 The number of cm^3 in $1 m^3$ is
A 100 B 10 000
C 100 000 D 1 000 000

Use the following information to answer questions 4 and 5.
The capacity of a container is 5 litres. The area of the base of the container is 160 cm^2. 2000 cm^3 of water is poured into the container.

4 5 litres =
A $5 \times 10 cm^3$ B $5 \times 10^2 cm^3$
C $5 \times 10^3 cm^3$ D $5 \times 10^4 cm^3$

5 The height of the water in the container is
A 1.25 cm B 2.5 cm
C 12.5 cm D 25 cm

6 A floor 4 m long by $2\frac{1}{2}$ m wide is concreted to a thickness of 10 cm. Calculate the volume of the concrete.

7 62 litres of water are poured into a cylinder of diameter 40 cm. Use the value 3.1 for π to find how deep the water is in the cylinder.

8 Use a ruler and set square to construct a rectangle 6 cm by 2.5 cm. Measure one of its diagonals.

9 Construct $\triangle$ABC in which BC = 4 cm, $A\widehat{B}C = 50°$ and AB = 6 cm. Measure AC.

10 Construct a parallelogram ABCD in which AB = 4 cm, $A\widehat{B}C = 70°$ and BC = 5 cm. Measure the distance between one pair of parallels and hence calculate the area of ABCD.

Revision exercise 3 (Chapters 3, 5)

1 Simplify the following.
(a) $-4 - 9$ (b) $5 - (-12)$
(c) $-8 - (-3)$ (d) $-6 - (-9)$

2 Simplify the following.
(a) $18 \div (-6)$ (b) $(-8) \div (-4)$
(c) $(-30) \div (+10)$ (d) $\frac{-19}{-1}$
(e) $\frac{-100}{25}$ (f) $\frac{48}{-8}$

3 Place the following in order of size, from lowest to highest.
(a) $(-6) \times (+4)$ (b) $(-8) \div (-2)$
(c) $(+25) \times (-1)$ (d) $(+100) \div (-5)$

4 Simplify the following.
(a) -6×8 (b) $(-2\frac{1}{2}) \times (-2\frac{2}{5})$
(c) $\frac{7}{8}$ of $(-5\frac{1}{3})$ (d) $38 \div (-2)$
(e) $-3.6 \div (-9)$ (f) $-3\frac{2}{3} \div 6\frac{3}{5}$

5 In the year AD 21 a man was 36 years old. In what year was he 12 years old?

6 Remove brackets and simplify the following.
(a) $3x - 2(2x - y)$
(b) $3(3x - 7) + (5x - 9)$
(c) $7(x - 2y) - 5(x - 3y)$
(d) $8(x + y) - 7(2x - y)$

7 Simplify the following.
(a) $\frac{7x}{m} - \frac{3x}{m}$ (b) $\frac{4x}{3} - \frac{2x}{9}$
(c) $\frac{5}{a} + \frac{8}{3a}$ (d) $\frac{1}{p} - \frac{1}{q}$

8 Expand the following.
(a) $(a + b)(x + y)$
(b) $(2p + q)(3r - 5s)$
(c) $(2c - 5)(c - 3)$
(d) $(2a - 9b)(4a + 5b)$
(e) $(b - 5)^2$
(f) $(2x + 1)^2$

9 Factorise the following.
(a) $10 + 15b$ (b) $x^2 - ax$
(c) $4ab - 2a^2$ (d) $27x^2y - 36xy^2$

10 Simplify the following.
(a) $2\frac{2}{3} - 4\frac{1}{2}$
(b) 9.8 °C − 18 °C

Revision test 3 (Chapters 3, 5)

1 $3 - (-8) - 5 =$
A −16 B −10
C +6 D +16

2 The difference between temperatures of 17 °C above zero and 12 °C below zero is
A 5 °C B 17 °C
C 22 °C D 29 °C

3 $(-3) \div (-24) =$
A + 8 B $+\frac{1}{8}$ C $-\frac{1}{8}$ D − 8

4 Evaluate $\frac{x - y}{x + y}$ if $x = -5$ and $y = -15$.
A $-\frac{1}{2}$ B + 1 C $+\frac{1}{2}$ D − 2

5 Simplify $\frac{-5a^2}{15a}$
A $\frac{5a}{3}$ B $-\frac{2}{3}$ C − 2 D $-\frac{a}{3}$

6 Expand the following.
(a) $(v - 4)(v - 9)$
(b) $(b + 4)(3b + 2)$
(c) $(5x + 2)(2x - 3)$
(d) $(4m - n)(3m - 2n)$

7 Factorise the following.
(a) $18 + 9c$ (b) $2\pi rh + \pi r^2$
(c) $28x^2y^2 - 21x^3y$ (d) $2\pi r^3 - \frac{1}{3}\pi r^2h$

8 Simplify the following
(a) $\frac{7}{2a} - \frac{3}{5a}$
(b) $\frac{1}{u} + \frac{1}{v}$
(c) $\frac{9a - 5}{5} - \frac{3a - 2}{2}$
(d) $\frac{3(y - 1)}{4} + \frac{2(5 - 3y)}{7}$

9 A girl is x cm tall. Her father is $1\frac{1}{4}$ times her height. Her sister is $\frac{9}{10}$ of her height.
(a) Express the heights of the father and sister in terms of x.
(b) Write down an expression for the difference in height between the father and sister.
(c) Simplify the expression.

10 Simplify the following.
(a) $(-6) \times (+5)$ (b) $(+7) \times (-0.9)$
(c) $(-10)^2$ (d) $(-5.4) \div (+0.6)$

Revision exercise 4 (Chapters 4, 8)

1 Which of the following pairs of sets are disjoint?
If the sets are not disjoint write down two members of the intersection.
(a) {prime factors of 24}, {prime factors of 55}
(b) {multiples of 5}, {multiples of 7}
(c) {Tobago, Antigua, Australia}, {islands in the Caribbean}
(d) {letters of *bull*}, {letters of *cow*}

2 U = {2, 4, 6, 8, …, 20}
M = {multiples of 3}
L = {numbers less than 14}
(a) Write down the members of the following sets.
(i) $M \cup L$ (ii) $M \cap L$
(iii) $M \cup U$ (iv) $U \cap M$
(v) L' (vi) $(M \cup L)'$
(b) Draw a Venn diagram to represent the data above.
(c) Hence find (i) $n(M \cup L)$, (ii) $n(M \cap L)$, (iii) $n(M' \cap L')$.

3 Make 4 copies of the Venn diagram in Fig. R3.

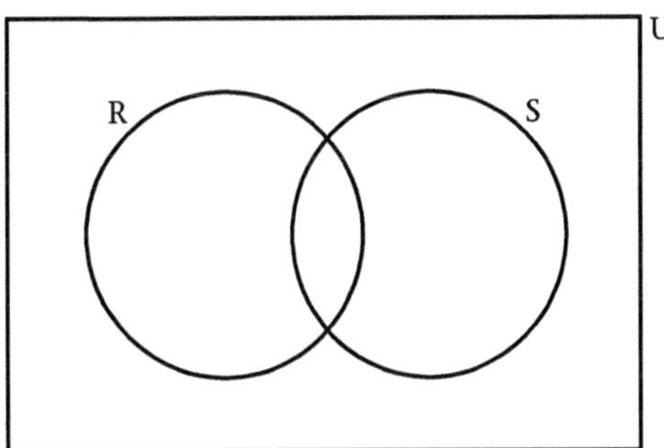

Fig. R3

Using one copy for each of the following, shade the region which represents the set
(a) $(R \cup S)'$ (b) $(R \cap S)'$
(c) $R' \cup S'$ (d) $R' \cap S'$.

4 Consider the set of numbers 1, 2, 3, … 9.
(a) Draw an arrow diagram showing the relation 'is the square root of'. What kind of relation is this?
(b) Write the information as a set of ordered pairs

5 Draw an arrow diagram to show a mapping that is not a one-to-one mapping

6 List the members of the following sets.
(a) {integer x: $-5 < x < 0$}
(b) {x: $2x = 7$}
(c) {x: $3x = -5$, $x \in Z$}
(d) {y: $5y > -9$, $y \in Z$}

7 Out of 83 cattle 39 have been de-horned and 55 have been vaccinated. If 28 of the cattle with horns have been vaccinated, find the number of cattle which have neither been de-horned nor vaccinated.

8 Which of the following mappings are one-to-one for the set of students in your class?

(a) Student → student's chair
(b) Student → student's desk
(c) Student → bag for books
(d) Student → subject liked most

9 John collects stamps at the rate of 7 stamps per week.

(a) Make a table of the mapping: number of weeks → number of stamps collected. Complete the table for 12 weeks
(b) Write the mapping as a set of ordered pairs.

10 Choose 8 of your classmates. Record how many questions in this exercise each of them answered correctly. Draw a mapping of the information. State the rule of the mapping.

Revision test 4 (Chapters 4, 8)

1 Which of the following diagrams does *not* show a one-to-one mapping:

A

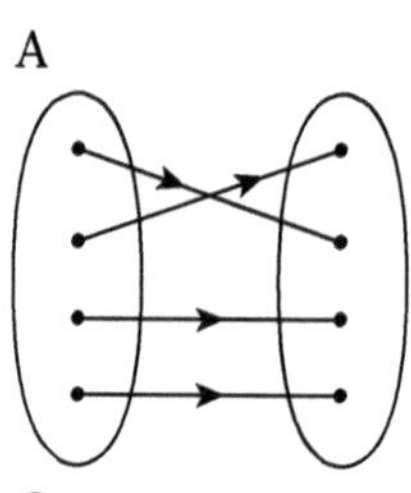

B

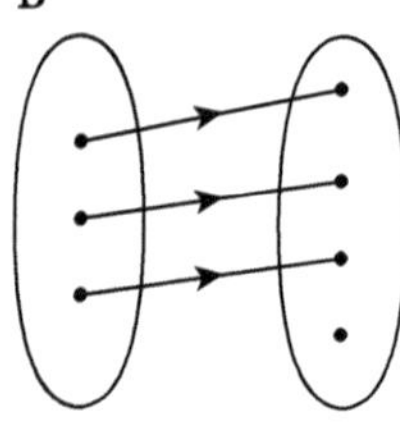

C

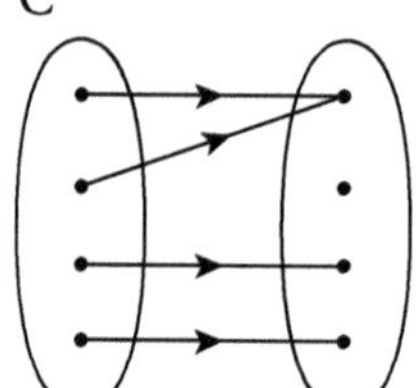

D

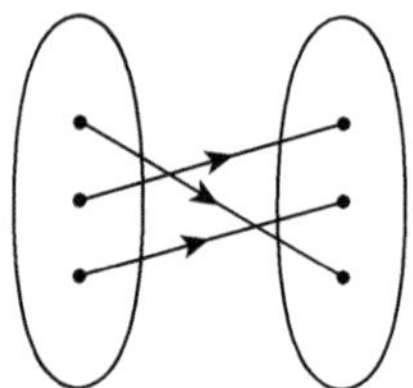

Fig. R4

2 Which set of ordered pairs represents a one-to-one mapping?

A {(0, 1), (1, 3), (2, 3), (3, 4)}
B {(2, 3), (3, 4), (3, 5), (4, 6)}
C {(2, 3), (3, 4), (4, 5), (5, 6)}
D {(0, 2), (2, 4), (3, 5), (3, 6)}

3 Fig. R5 is a Venn diagram showing a universal set, U, with subsets X and Y.

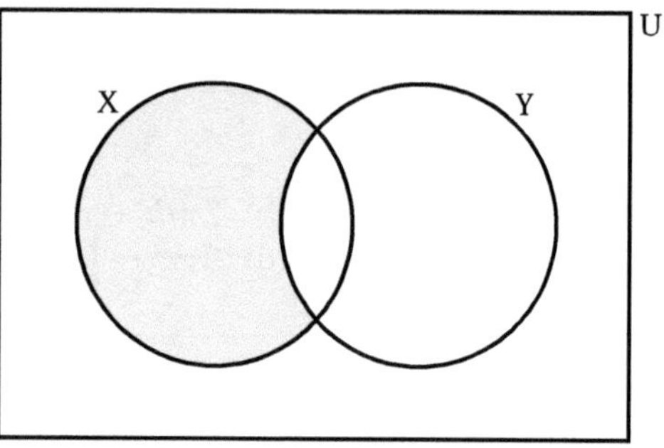

Fig. R5

Which one of the following represents the shaded region of Fig. R5?

A $X' \cup Y'$ B $X' \cap Y$
C $X' \cap Y'$ D $X \cap Y'$

4 U = {3, 6, 9, 12, ..., 30} and $x \in$ U. Which of these lists the members of the subset {x: $x < 19$} ∩ {x: x is a factor of 30}

A {1, 2, 3, 5, 6, 10, 15, 30}
B {3, 6, 9, 12, 15, 18}
C {3, 6, 15, 30}
D {3, 6, 15}

5 Which of the following statements is true for a one-to-one mapping for two sets?

A More than one arrow goes from a member of the first set to a member of the second set.
B One arrow goes from each member of the first set to a member of the second set.
C The number of members in the second set is more than the number of members in the first set.
D The number of members in the first set is more than the number of members in the second set.

6 The scale of a map is 1 cm to 3 km. Make a table of the mapping: distance in centimetres on the map → distance in kilometres on the ground. Complete the table up to 10 cm.

7 There are approximately 250 words on each page of a book. Make a table of the mapping: number of pages → number of words. Complete the table for 9 pages.

8 Which of the following are one-to-one mappings for the set of students in your class?
(a) student → marks in last mathematics test
(b) student → home address
(c) student → size of shoes
(d) student → height

9 A worker is paid \$25 per hour for overtime work. Make a table of the mapping: number of hours → amount earned in overtime pay. Complete this table for a week in which the workers clock up to 15 hours overtime.

10 A motorist is charged a fine of \$50 for every km he exceeds the speed limit through a town. Make a table of the mapping: number of km in excess of speed limit → fine charged. Complete the table as far as 12 km in excess of the speed limit.

General revision test A (Chapters 1–8)

1 What is the next term in the sequence 1, 3, 7, 15, 31, ...?
A 47 B 51 C 59 D 63

2 Find the least number by which 112 must be multiplied to give a perfect square.
A 2 B 5 C 7 D 11

3 Simplify $(a - (-a)) + b$.
A $2a - b$ B $2a + b$
C $2a$ D b

4 If $r = 3$, $s = -2$, $t = -1$, then $r^2 + s^3 - t^2 =$
A 4 B 1 C 0 D -1

5 Express 0.000 263 in standard form.
A 2.63×10^{-5} B 2.63×10^{-4}
C 2.63×10^{4} D 2.63×10^{5}

6 Which of the following sets of ordered pairs show a one-to-one mapping?
A (1,3), (2,3), (3,3) B (1,3), (1,4), (1,5)
C (1,3), (2,6), (3,9) D (1,3),(2,5),(3,5)

7 If U = {a, b, c, d, e} and X = {a, b, e}, then X′ =
A {c, d} B {a, b, c,} C {a} D ∅

8 The LCM of $5ab^2$ and $3a^2b$ is
A ab B $\frac{5b}{3a}$
C $15a^2b^2$ D $15a^3b^3$

9 Fig. R6 is a Venn diagram showing the elements of the sets U, X, Y.

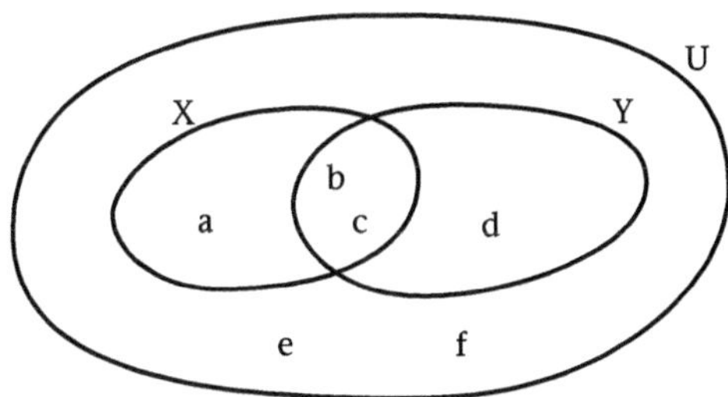

Fig. R6

X ∪ Y =
A {a, b, c, d, e, f} B {a, b, c, d}
C {e, f} D {b, c}

10 Simplify $(y + z)^2 - 2yz$.
A $y^2 + z^2 + 4yz$ B $y^2 - z^2$
C $y^2 + z^2$ D $y^2 + z^2 - 2yz$

11 Given

$$\begin{aligned} 1 &= 1^2 \\ 1 + 3 &= 2^2 \\ 1 + 3 + 5 &= 3^2 \\ 1 + 3 + 5 + 7 &= 4^2 \end{aligned}$$

(a) what will be the value of x if $1 + 3 + 5 + \ldots + x = 23^2$?
(b) what will be the value of y if $1 + 3 + 5 + \ldots + 99 = y^2$?

12 Construct △ABC in which BC = 6 cm, $A\hat{B}C = 30°$ and $B\hat{A}C = 100°$. Measure the perpendicular distance of A from BC and hence calculate the area of the triangle which you have drawn.

13 A cylindrical rain barrel has a base area of 0.38 m². It contains rainwater to a depth of 91 cm. How many times can a bucket holding 14×10^3 cm³ be filled from the barrel?

14 (a) Expand the following.
   (i) $(x - 5)(x - 6)$
   (ii) $(2p - 3q)(2p - 5q)$
   (iii) $(m + 4)(m - 4)$
   (iv) $(t - 8)^2$
(b) Factorize the following.
   (i) $8xy + 2x^2$
   (ii) $mn^2 + m^2n$

15 Simplify the following.
(a) $\frac{3a + 1}{2} + \frac{a + 2}{3}$
(b) $\frac{2b - 3}{3} - \frac{2b - 5}{4}$
(c) $\frac{5a + 2}{2} - \frac{2(3a + 1)}{3}$

16 A salesman keeps a record of the distance, in km, travelled after each hour:
(1, 48), (2, 96), (3, 144), (4, 192), (5, 240).
(a) Represent the information in a mapping table.
(b) How long did the trip last?
(c) What was the distance travelled after 3 hours?
(d) What was the total distance travelled?

17 A household uses water from a tank at the rate of 250 litres per day.
(a) Make a table of the mapping: number of days → number of litres of water used.
(b) How much water is left in the tank after 7 days if the tank held 1805 litres at the beginning?

18 Make a table of the mapping: radius of circle → circumference of circle. Complete the table for the following values of radii: $3\frac{1}{2}$ cm, 7 cm, $10\frac{1}{2}$ cm, 14 cm, $17\frac{1}{2}$ cm, 21 cm. (Use the value $\frac{22}{7}$ for π.)

19 Simplify the following, giving the answers in standard form.
(a) $(5.8 \times 10^4) + (7.7 \times 10^4)$
(b) $(8.2 \times 10^{-2}) - (4 \times 10^{-2})$
(c) $(5.1 \times 10^{-3}) \times (3 \times 10^{-2})$
(d) $(5 \times 10^5) \div (8 \times 10^{-2})$

20 In a class of 32 students, 15 are boys. 20 of the students walk to school each day. 8 of the girls do not walk to school. How many boys walk to school each day?

Chapter 9
The Cartesian plane

Pre-requisites
- directed numbers; ordered pairs; mapping

Points on a line

The number line is a **graph**, or picture, of all the positive and negative numbers (Fig. 9.1).

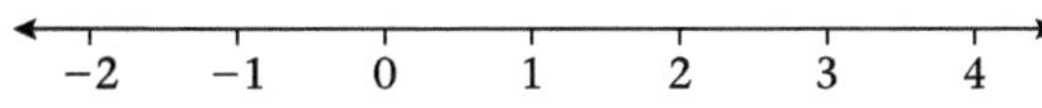

Fig. 9.1

If we draw points on the number line, we can say exactly where they are on the line.

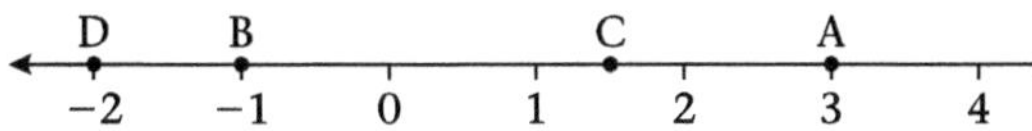

Fig. 9.2

In Fig. 9.2, A is 3 units to the right of zero and B is 1 unit to the left of zero. We can shorten this to A(3) and B(−1). In the same way, C is the point C($1\frac{1}{2}$) and D is the point D(−2).

A(3) and B(−1) give the positions of A and B. Notice that we are using brackets in a different way from the way we use them in algebra and arithmetic.

Exercise 9a

1. In Fig. 9.3, P(2) gives the position of P and Q(− 3) gives the position of Q. Give the positions of R, S, T, U and V in the same way.

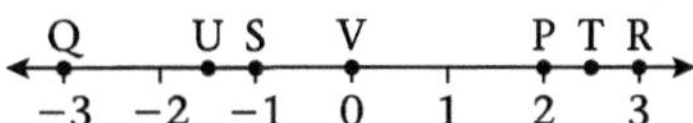

Fig. 9.3

2. In Fig. 9.4, A(0.7) describes the position of A. Describe the positions of B, C, D, E, F and G in the same way.

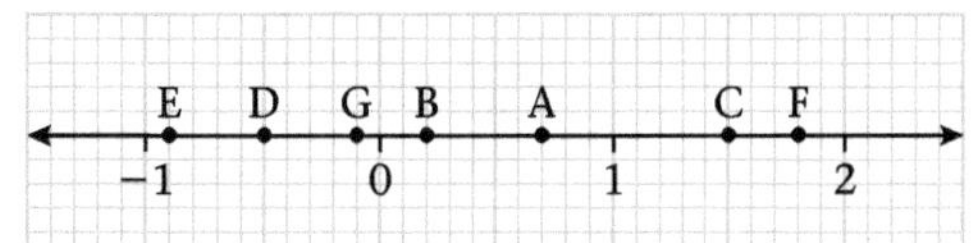

Fig. 9.4

3. Draw a number line from −10 to 10. On the line, mark the points A(6), B(3), C(−4), D(−8), E(9), F(−9), G(0), H($7\frac{1}{2}$) and I($-6\frac{1}{2}$).
4. Use graph paper to draw a number line like that of Fig. 9.4. On the line, mark the points P(0.8), Q(1.3), R(0.4), S(−0.4), T(−0.7), U(1.9) and V(1.0).

Points on a plane

Exercise 9b (Oral)

1. Try to describe the positions of points P, Q and R in Fig. 9.5.

   Hint: one way is to measure the distances of P, Q and R from the edges of the page.

×
Q

R×

Fig. 9.5

2 Fig. 9.6 shows the same points on a cm square grid. Starting at the cross at O, describe how to get to P, Q and R. Does this make it easier to describe the positions of the points?

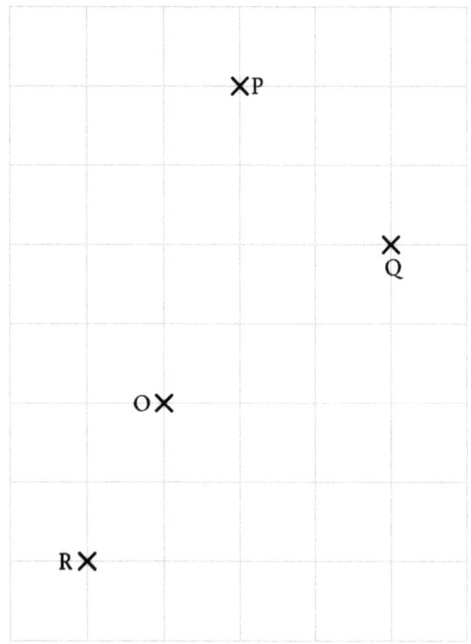

Fig. 9.6

Cartesian plane

The positions of points on a line are found by using a number line. The positions of points on a plane surface are found by using *two* number lines, usually at right angles. See Fig. 9.7.

In Fig. 9.7, starting from the zero point, P is in position 1 unit to the *right* and 4 units *up*;
Q is in position 3 units to the *right* and 2 units *up*;
R is in position 1 unit to the *right* and 2 units *down*;
S is in position 1 unit to the *left* and 1 unit *down*;
T is in position 2 units to the *left* and 3 units *down*.

We can shorten this to P(1, 4), Q(3, 2), R(1, −2), S(−1, −1) and T(−2, −3).

The position of each point is represented by a pair of numbers.

Fig. 9.7 is a graph, or picture, of the five points P, Q, R, S, T. In a graph like this, the number lines are called **axes**. They cross at the zero-point of each axis. This point is called the **origin**, O. The axis going across from left to

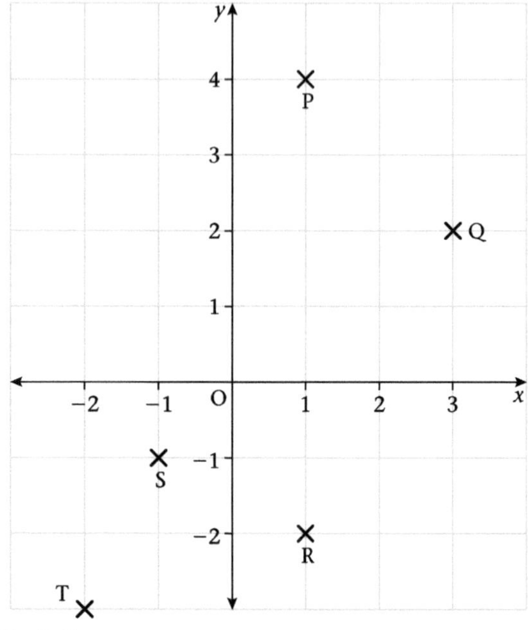

Fig. 9.7

right is called the **x-axis**. It has a positive scale to the right of O and a negative scale to the left of O. The axis going up the page is called the ***y*-axis**. It has a positive scale upwards from O and a negative scale downwards from O.

A plane surface with axes drawn on it, such as Fig. 9.7 and Fig. 9.8, is called a **Cartesian plane**. It is named after the French philosopher and mathematician, Descartes. His work made it possible to represent geometry in a numerical way.

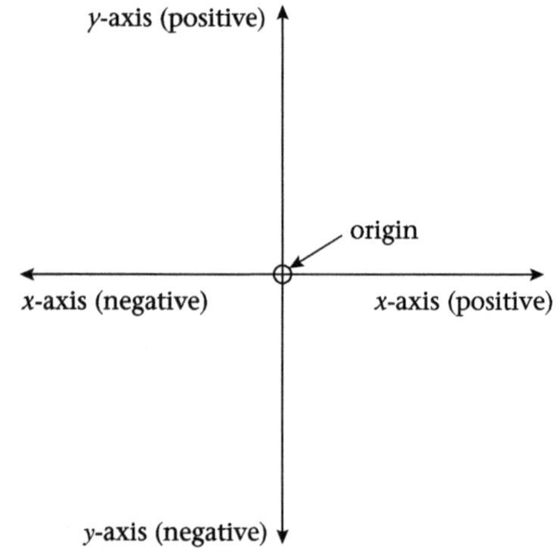

Fig. 9.8

Coordinates

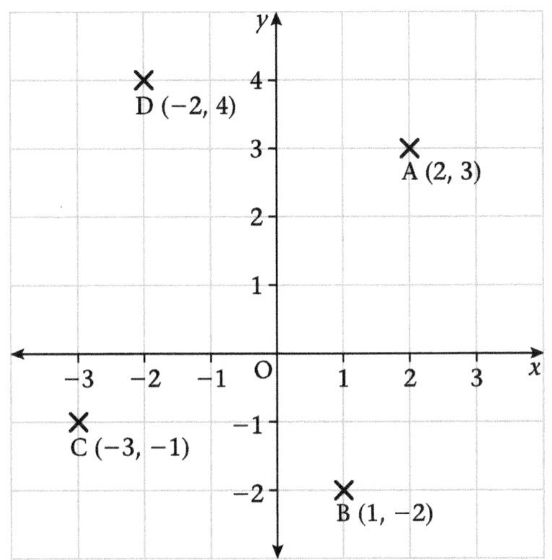

Fig 9.9

Fig. 9.9 shows a Cartesian plane with points A, B, C and D drawn on it.

The position of A is found by moving 2 units to the *right* of the origin and then 3 units *up* the page parallel to the y-axis. We can shorten this to A(2, 3). C is found by moving 3 units to the *left* of the origin and then 1 unit *down* the page. Its position is C(−3, −1). In the same way, B and D are the points B(1, −2) and D(−2, 4).

The position of each point is given by an **ordered pair of numbers**. These are called the **coordinates** of the point. The first number is called the **x-coordinate**. The x-coordinate gives the distance of the point in a direction parallel to the x-axis. The second number is called the **y-coordinate**. The y-coordinate gives the distance of the point in a direction parallel to the y-axis. The coordinates are separated by a comma and enclosed in brackets (Fig. 9.10).

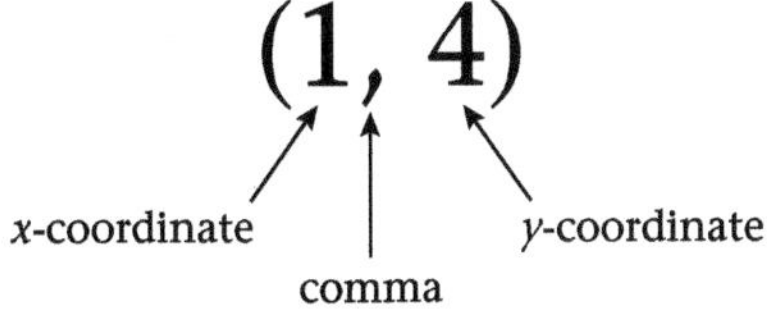

Fig. 9.10

The *order* of the pair of numbers is very important. For example, the point (1, 4) is not the same as the point (4, 1). This is shown in Fig. 9.11.

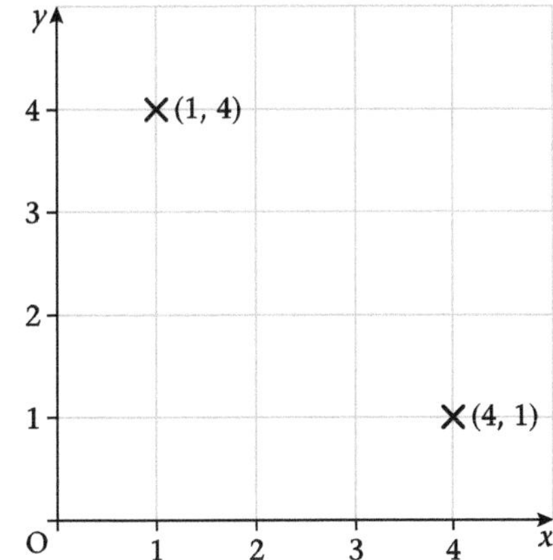

Fig. 9.11

Example 1

Write down the coordinates of the vertices of triangle ABC and parallelogram PQRS in Fig. 9.12.

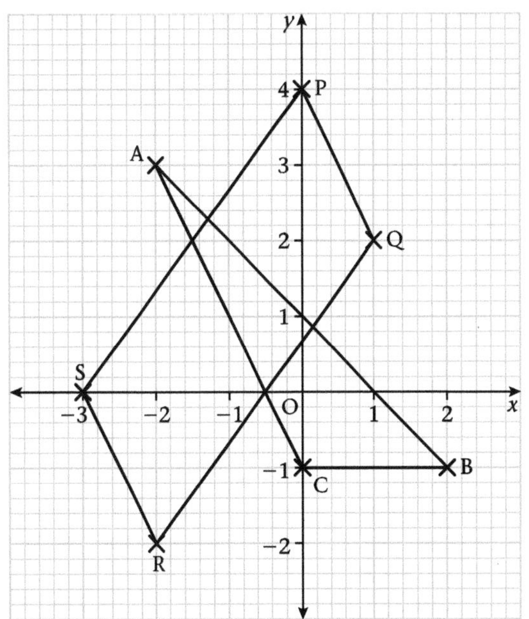

Fig. 9.12

The vertices of triangle ABC are A(−2, 3), B(2, −1) and C(0, −1). The vertices of parallelogram PQRS are P(0, 4), Q(1, 2), R(−2, −2) and S(−3, 0).

Notice that C and P are on the y-axis. Their x-coordinate is 0 (zero). S is on the x-axis. Its y-coordinate is 0.

Exercise 9c

1. What are the coordinates of the points A, B, C, D, E, F, G, H, I and J in Fig. 9.13?

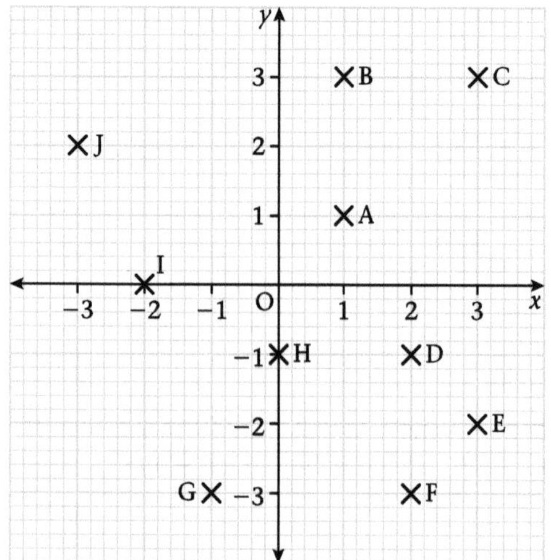

Fig. 9.13

2. What are the coordinates of the vertices of the 'elephant' in Fig. 9.14? Start where shown and work clockwise round the figure.

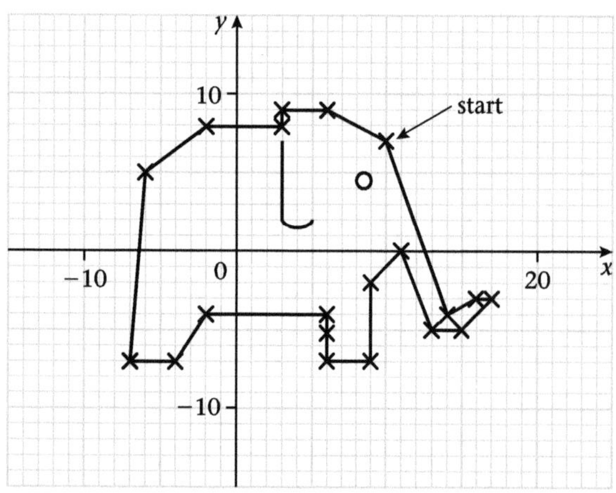

Fig. 9.14

3. In Fig. 9.15 name the points which have the following coordinates.

(a) (9, 5) (b) (5, −8)
(c) (−15, −10) (d) (−5, 8)
(e) (12, 0) (f) (0, 12)
(g) (0, −7) (h) (−7, 0)
(i) (14, −11) (j) (−13, 15)
(k) (−4, −12) (l) (14, 14)

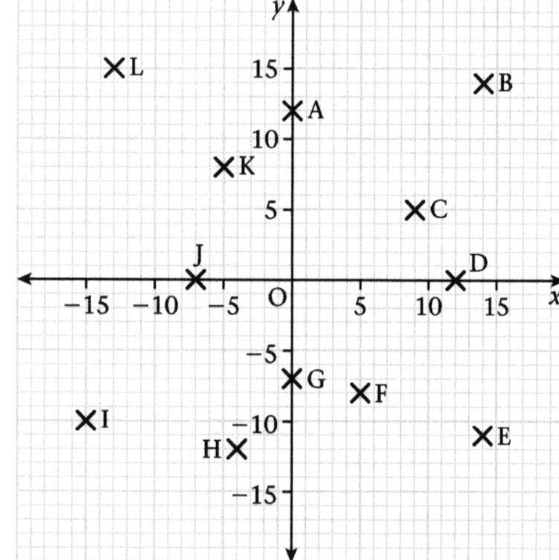

Fig. 9.15

4. What are the coordinates of the vertices T, U, V, W, X, Y and Z of the shape in Fig. 9.16?

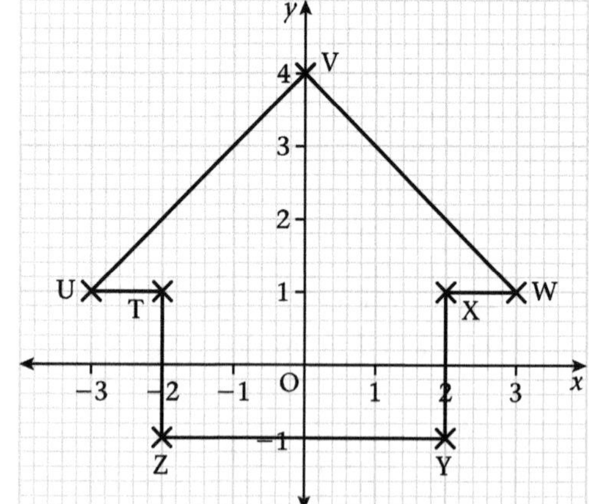

Fig. 9.16

5 Fig. 9.17 shows part of a map drawn on a Cartesian plane.

Find the coordinates of

(a) the big tree
(b) the garage
(c) the farm
(d) the manhole
(e) the hospital
(f) the top of the hill
(g) the point where the railway line crosses the road
(h) the point where the railway line crosses the river
(i) the point where the road crosses the river
(j) the point where the road branches to the right

Find the coordinates of any 4 points on the railway line. What do you notice?

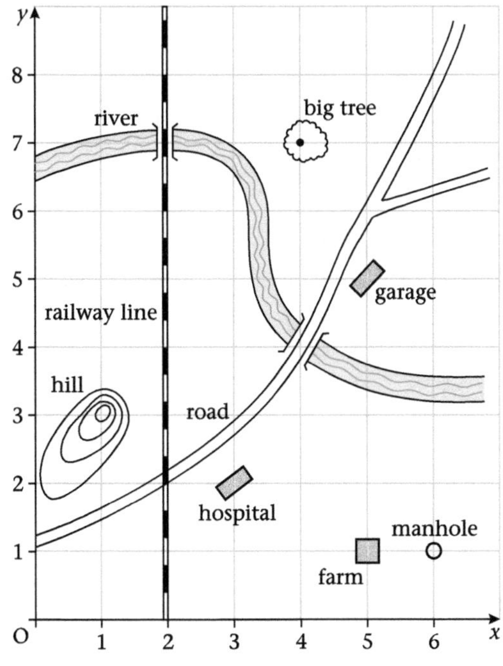

Fig. 9.17

6 Fig. 9.18 is the graph of lines l and m.

(a) Write down the coordinates of the points marked ✕ on line l. What do you notice?

(b) Write down the coordinates of the points marked ✕ on line m. What do you notice?

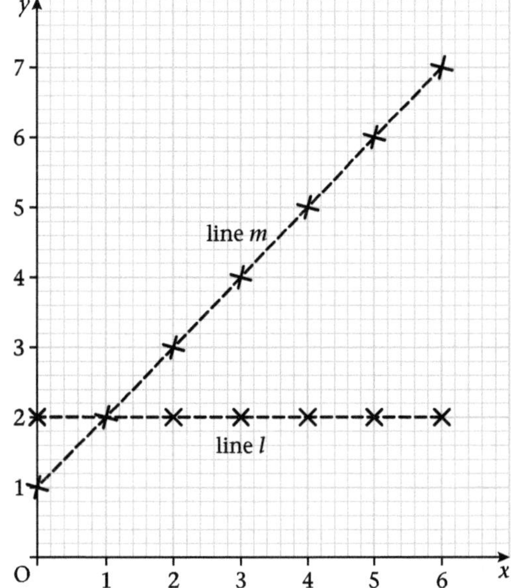

Fig. 9.18

Plotting points

To **plot** a point means to draw its position on a Cartesian plane.

The easiest way to plot a point is as follows.

1 Start at the origin.

2 Move along the x-axis by an amount and in a direction given by the x-coordinate of the point.

3 Move up or down parallel to the y-axis by an amount and in a direction given by the y-coordinate.

Example 2

Plot the points (−1, 2) and (2.6, −1.8) on a Cartesian plane.

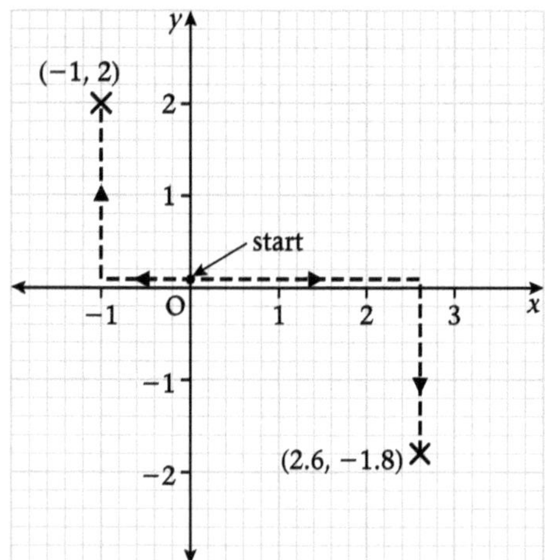

Fig. 9.19

The dotted arrows in Fig. 9.19 show the method of plotting.

For (−1, 2):

Start at the origin. The x-coordinate is −1. Move 1 unit to the left along the x-axis. The y-coordinate is 2. Move 2 units up parallel to the y-axis. Plot the point.

For (2.6, −1.8):

Start at the origin. The x-coordinate is 2.6. Move 2.6 units to the right on the x-axis. The y-coordinate is −1.8. Move 1.8 units down parallel to the y-axis. Plot the point.

Notes:

1 The dotted arrows in Fig. 5.19 are not normally put on the graph. They are given here to show the method only.

2 Use a small cross (×) to plot points.

Example 3

The vertices of quadrilateral PQRS have coordinates P(−3, 18), Q(15, 14), R(11, −4) and S(−7, 0). A and B are the points A(−3, −7) and B(3, 0).

(a) Using a scale of 2 cm to represent 10 units on both axes, plot points P, Q, R, S, A and B.

(b) Join the vertices of quadrilateral PQRS. What kind of quadrilateral is it?

(c) Find the coordinates of the point where the diagonals of PQRS cross.

(d) What do you notice about the points A, B and Q?

(a) The scale is given. The highest x-coordinate is 15 and the lowest is −7. The x-axis must include these numbers. A scale from −10 to 20 on the x-axis will be suitable. The highest y-coordinate is 18 and the lowest is −7. A scale from −10 to 20 on the y-axis will be suitable. The points are plotted in Fig. 9.20.

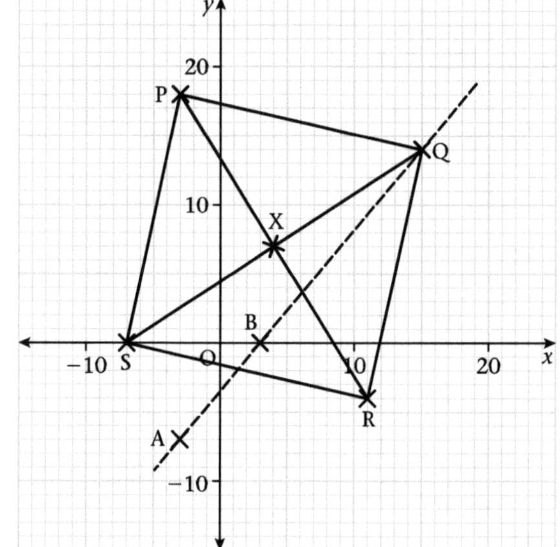

Fig. 9.20

(b) PQRS is a square.

(c) The diagonals of PQRS cross at X(4, 7).

(d) B, A and Q lie on a straight line (broken in Fig. 9.20).

When drawing Cartesian graphs, *always*:

1 draw the axes;

2 label the origin O;

3 label the axes, x and y;

4 write the scales along each axis.

Exercise 9d

Work on graph paper in this exercise.

1. Draw the origin, O, near the middle of a clean sheet of graph paper. Use a scale of 1 cm to represent 1 unit on both axes. Plot the following points:
A(8, 10), B(−8, −10), C(3, −5), D(−6, 9), E(−4, −7), F(1, 8), G(2, 0), H(0, −6), I(−2.4, 5.2), J(−4, 3.8), K(0, 6.6), L(0.8, −7.8).

2. Draw the origin, O, near the middle of a sheet of graph paper. Use a scale of 2 cm to represent 1 unit on both axes. Plot the following points then join each point to the next in alphabetical order.
A(0, 1), B(1, 2), C(1, 1), D(2, 1), E(1, 0), F(2, −1), G(1, −1), H(1, −2), I(0, −1), J(−1, −2), K(−1, −1), L(−2, −1), M(−1, 0), N(−2, 1), P(−1, 1), Q(−1, 2).
Finally, join Q to A.

3. Draw the origin, O, near the middle of a sheet of graph paper. Use a scale of 2 cm to represent 5 units on both axes. Plot the following points. Join each point to the next in the order they are given.
START (−10, −5), (−5, 10), (0, 15), (5, 17), (3, 14), (3, 12), (15, 6), (14, 3), (11, 3), (13, 2), (5, 3), (6, −6) FINISH
What does your graph show a picture of?

4. Take O near the middle of your graph paper and let 2 cm represent 1 unit on both axes.
   (a) Plot the points P(4, 3), Q(4, 1), R(−1, 1), S(−1, 3), W(1, 2), X(1, −1), Y(−2, −1) and Z(−2, 2).
   (b) Find the areas, in unit^2, of rectangle PQRS, square WXYZ, triangle SXY, triangle PYZ.

5. As in question 4, but plot the points A(0, 4), B(−3, −1), C(−2, −4), D(1, 1), E(3, 2), F(4, 0) and G(2, −1).
   (a) Draw quadrilateral ABCD. What kind of quadrilateral is it? Let its diagonals cross at X. Find the coordinates of X.
   (b) What do you notice about points B, X, D and E?
   (c) Draw quadrilateral DEFG. What kind of quadrilateral is it? Let its diagonals cross at Y. Find the coordinates of Y.

Graphs of mappings

In Chapter 4 we saw that a mapping may be shown as

(i) an arrow diagram
(ii) a table
(iii) a set of ordered pairs.

We now notice that points on a Cartesian plane are identified by ordered pairs of numbers. A graph of any mapping may be drawn.

Example 4

Using the set of numbers {0, 1, 2, 3, 4},

(a) make a table of the mapping:
number → twice its value
(b) write the mapping as a set of ordered pairs
(c) using a scale of 1 cm to 1 unit on both axes of your graph papers plot each pair of numbers in (b) as a point.
Join the points with a straight edge.

(a)

0	1	2	3	4
↓	↓	↓	↓	↓
0	2	4	6	8

(b) {(0, 0), (1, 2), (2, 4), (3, 6), (4, 8)}

(c)

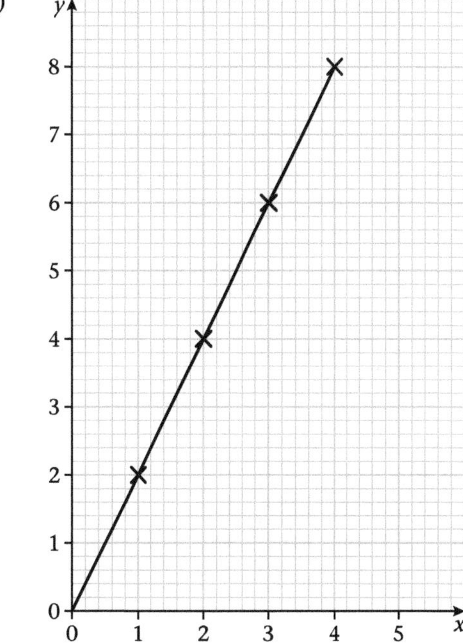

Fig. 9.21

Example 5

Fig. 9.22 is a graph of a mapping. Make a table of the mapping. State the rule of the mapping.

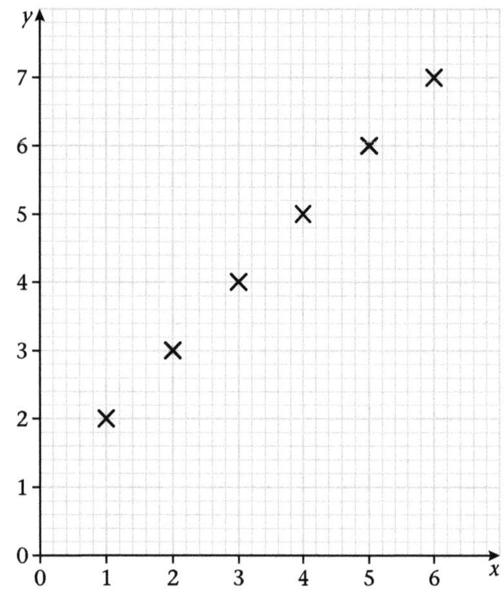

Fig. 9.22

0	1	2	3	4	5	6
1	2	3	4	5	6	7

rule of mapping:

number → one more than

Note that

1 The first number of the ordered pair is shown on the horizontal axis.

2 The second number of the ordered pair is shown on the vertical axis.

The rule of the mapping above could be written as: $x \rightarrow x + 1$

Example 6

Make a table of the mapping: $x \rightarrow 2x + 1$. Draw a graph of the mapping. Use a scale of 2 cm to 5 units on the vertical axis.

x	0	1	2	3	4	5
$2x + 1$	1	3	5	7	9	11

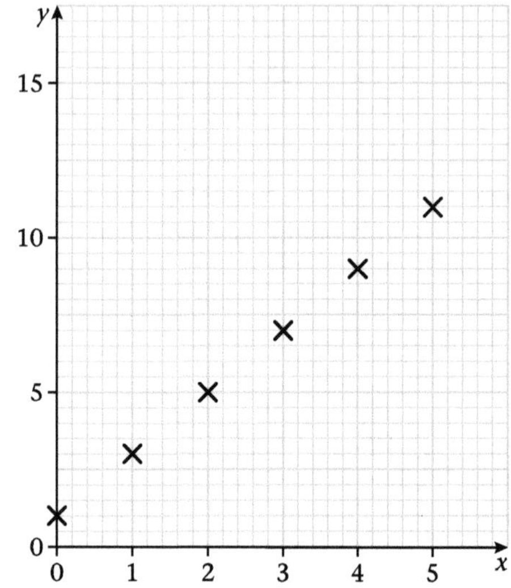

Fig. 9.23

Note that to make the table, $x \rightarrow (2 \times x) + 1$

Exercise 9e

1 Figs. 9.24, 9.25 and 9.26 are graphs of some mappings. Make a table of the mapping shown by each graph. State the rule of the mapping in the form: $x \rightarrow$

(a)

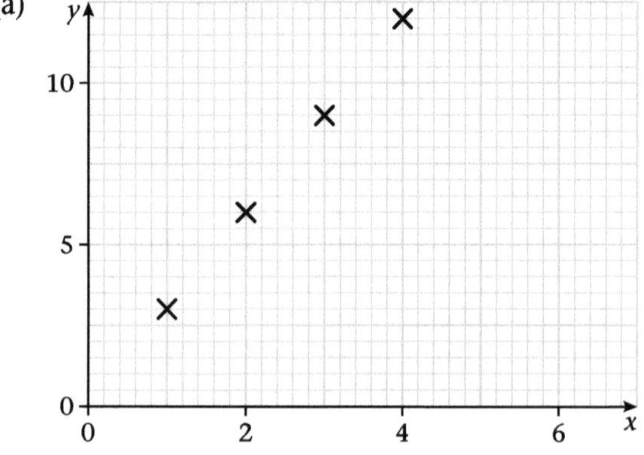

Fig. 9.24

(b)

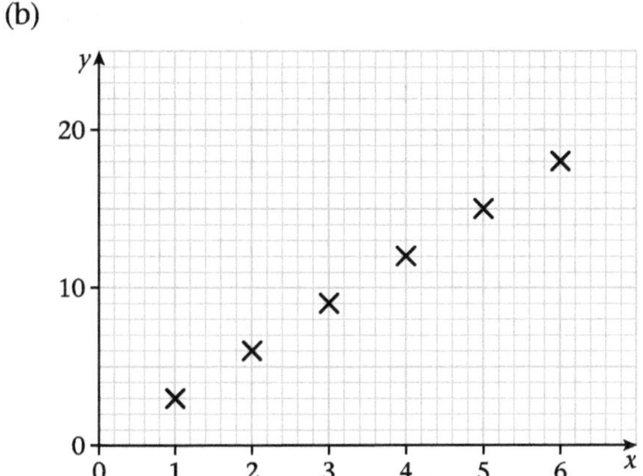

Fig. 9.25

(c)

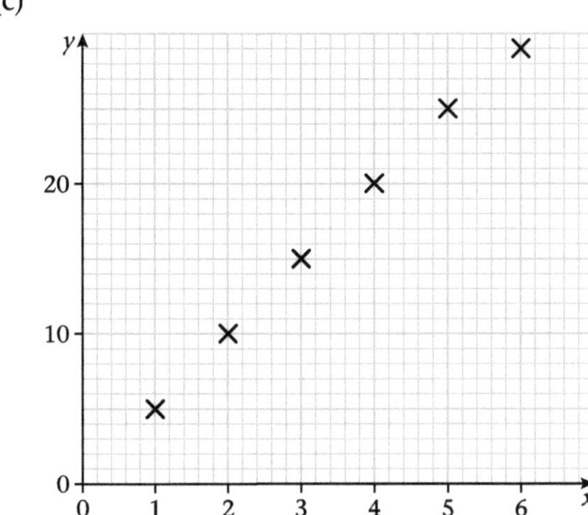

Fig. 9.26

2. An hourly paid worker is paid at the rate of \$50 per hour.
   - (a) Make a table of the mapping: number of hours → wages. Complete the table for 8 hours.
   - (b) Draw a graph of the mapping. Use 1 cm to represent 50 units on the vertical axis.

   Join the points with a straight edge.

3. Fig. 9.27 shows the diagonals of a quadrilateral and a pentagon.

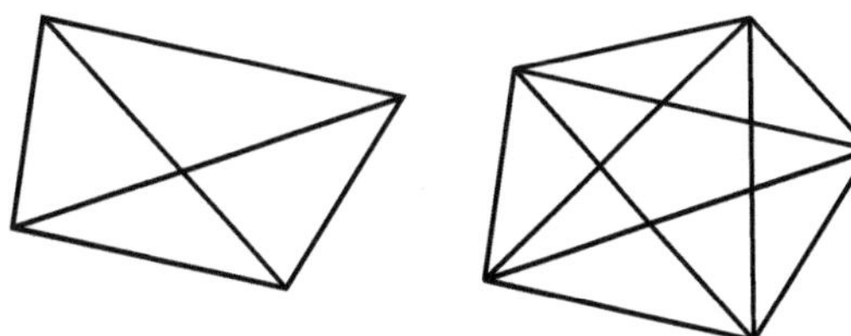

Fig. 9.27

   Make a table of the mapping: number of sides → number of diagonals. Complete the table for up to 10 sides. Draw a graph of the mapping.

4. Fig. 9.28 shows the number of lines that can be drawn through two points drawn as shown.

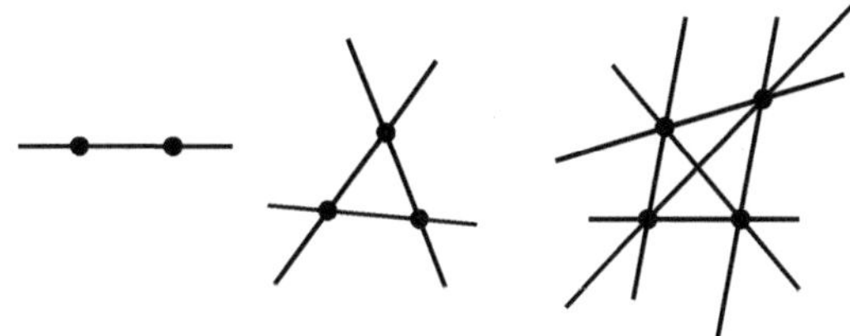

Fig. 9.28

   Make a table of the mapping: number of points → number of lines joining them. Complete the table for up to 8 points. Draw a graph of the mapping.

5. Make a table of the mapping: $x \to 3x - 1$ Complete the table for $x = (1, 2, \ldots 6)$ Draw a graph of the mapping.

6. (a) Complete the ordered pairs in the following pattern: (0, 0), (1, 1), (2, 4), (3, 9), (4, 16), (5, 25), (6,), (7,), (8,), (9,), (10,).
   - (b) Draw the origin, O, at the bottom left-hand corner of a sheet of graph paper. Draw an x-axis with a scale of 1 cm to 1 unit. Draw a y-axis with a scale of 1 cm to 5 units.
   - (c) Plot the points in part (a).
   - (d) Join the points you have plotted by drawing as smooth a curve as you can.
   - (e) Use your graph to find $(8.4)^2$, $(6.5)^2$, $\sqrt{20}$, $\sqrt{90}$.

Summary

The number line is a **graph** or picture of all the positive and negative numbers.

Two number lines, usually at right angles and intersecting at the zero-point on each line, may be used to fix the positions of points on a plane surface.

The point of intersection is called the **origin** O; the number lines are called **axes**. When the plane surface is a **Cartesian plane**, the x-axis goes from left to right (the horizontal axis) and the y-axis upwards (the vertical axis) through O.

The position of a point in the Cartesian plane is defined by its **coordinates**, that is, its distance from O measured in given units in the x-direction (the **x-coordinate**) and in the y-direction (the **y-coordinate**). Positive x-units lie to the right and negative x-units to the left of O; positive y-units lie upwards and negative y-units downwards from O. Coordinates are written as an ordered pair of numbers in brackets, the x-coordinate being stated first, that is (x, y).

A mapping can also be represented as points on a Cartesian plane (graph). When drawing a graph you should choose the scale carefully.

Practice Exercise P9.1

1. For each of the following

A	$x = -3$	B	$x = 4.5$
C	$y = -3$	D	$y = 2.5$
E	$y = x + 3$	F	$y = -x$
G	$y = 3x$	H	$y = x - 3$

(a) Complete the following table

Table 9.1

x	−2	0	1	3
A				
B				
C				
D				
E				
F				
G				
H				

(b) Write the mapping as a set of ordered pairs

(c) On the same axes, draw the graph of the mappings, using values on the x-axis from −4 to +5, and on the y-axis from −10 to +10

2. (a) Complete Table 9.2 for the function $y = x + 1$.

Table 9.2

x	−2	0	1	2
1	1	1		
$x + 1$	−1			

(b) Draw a set of coordinate axes, labelling the x-axis from −3 to 3, and the y-axis from −8 to 8 and use the table to draw the graph of $y = x + 1$

(c) Complete Table 9.3 and use it to draw the graph of $y = 2x + 3$ on the same axes as (b).

Table 9.3

x	−2	0	1	2
$2x$				
3				
$2x + 3$				

(d) Use your graph to find the coordinates of the point that lies on both lines: $y = x + 1$ and $y = 2x + 3$.

3. (a) Draw the next two dot patterns.

Pattern #	1	2	3	4	5	6
	·	· ·	· · ·	· · · ·		
	·	· ·	· · ·	· · · ·		
	·	· ·	· · ·	· · · ·		
No. of dots	3					

Fig. 9.29

(b) Write the number of dots underneath each pattern.

(c) List the set of ordered pairs.

(d) State the rule of the mapping.

(e) Draw a graph to show this relation, using a scale of 1 cm to 1 unit on the horizontal axis and 1 cm to 1 unit on the vertical axis.

(f) Use your graph to find the number of dots in the 10th pattern.

Chapter 10

Algebraic equations (2)

Word problems

Pre-requisites

- linear equations; lowest common multiple

We can use algebraic equations to solve many word problems.

Sometimes we may use a diagram or a picture to help us to change the given information into an algebraic equation.

Exercise 10a (Oral)

1. Use the information in Fig. 10.1 to find the value of d.

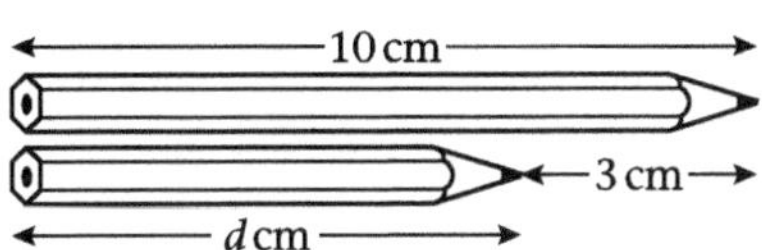

Fig. 10.1

2. When new, the box of matches in Fig. 10.2 contained m matches. Find m if 4 are used and 42 remain.

Fig. 10.2

3. The three cows in Fig. 10.3 are worth x dollars. Find x if each cow is worth \$340.

Fig. 10.3

4. The height of the blocks in Fig. 10.4 is h cm. If each block is 20 cm thick, find the number that h stands for.

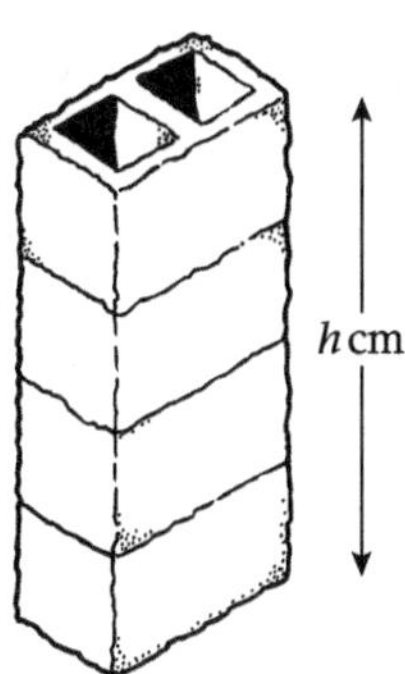

Fig. 10.3

If we have not been given a letter to represent the unknown number, we first have to identify the unknown and the units to measure its value. We then write the equation and solve the equation by the balance method (See Examples 1 and 2). We should always check that our solution, that is, the value we have found for the unknown, makes the statement true. Finally, we should give our answer in written form as it was asked in the problem. Remember that the letter replaces only the number and not the units. The units are *not* written in the equation.

Example 1

I think of a number. I multiply it by 5. I add 15. The result is 100. What is the number I thought of?

Let the number be n.
I multiply n by 5: $5n$.
I add 15: $5n + 15$
The result is 100: $5n + 15 = 100$
Subtract 15 from both sides.

$$5n + 15 - 15 = 100 - 15$$
$$5n = 85$$

Divide both sides by 5.

$$\frac{5n}{5} = \frac{85}{5}$$
$$n = 17$$

The number is 17.

Check: $17 \times 5 = 85$; $85 + 15 = 100$

Remember that we may use a flow chart to represent the steps needed in writing an equation and in solving the equation. Using Example 1 on page 81:

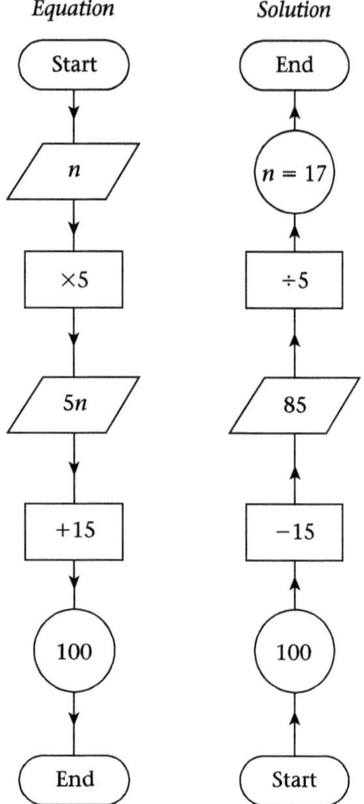

Fig. 10.5

Example 2

A rectangle is 8 m long and its perimeter is 30 m. Find the breadth of the rectangle.

Let the breadth of the rectangle be b metres.

Perimeter $= 8 + b + 8 + b$ metres
$= 16 + 2b$ metres

Thus $16 + 2b = 30$

Subtract 16 from both sides.

$$2b = 30 - 16 = 14$$

Divide both sides by 2.

$$b = \frac{14}{2} = 7$$

The breadth of the rectangle is 7 metres.
Check: 8 m + 7 m + 8 m + 7 m = 30 m

Using a flow chart, we have

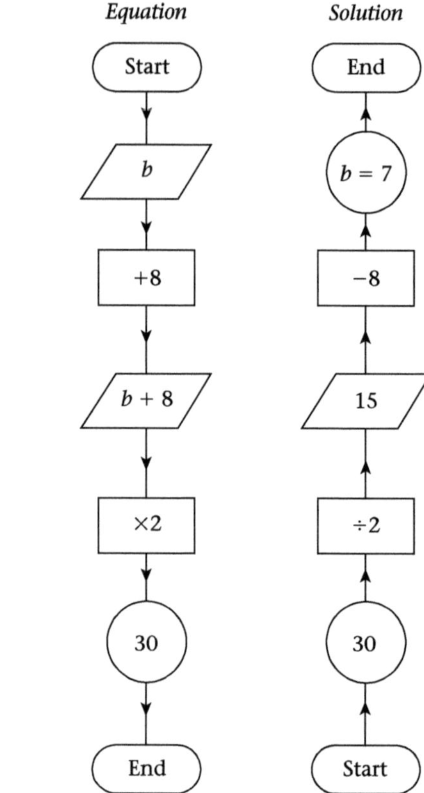

Fig. 10.6

Notice the method in Examples 1 and 2.

1 Choose a letter for the unknown.
2 Write down the information of the question in algebraic form.
3 Make an equation.
4 Solve the equation.
5 Give the answer in written form.
6 Check the result against the information of the question.

Exercise 10b will give you practice in changing written information into algebraic form.

Exercise 10b

1. Write the following information in algebraic form and simplify your answer, if possible:
   (a) a number x is doubled
   (b) a number n is multiplied by 6
   (c) a number m is multiplied by 6 and then 4 is added
   (d) a number y is doubled and then 5 is taken away
   (e) a number is 3 less than a
   (f) a number d is added to another number 4 times as big
   (g) Rudy has h cents and Peter has twice as much
   (h) a number t is trebled and 7 is subtracted
   (i) one girl has k cents and another girl has 9 cents less
   (j) team X scores g goals and team Y scores 23 goals more than X

2. What is the perimeter of
   (a) a square of side x cm?
   (b) a triangle with sides $2a$ metres, a metres and 4 metres?
   (c) a regular hexagon of side c cm?
   (d) a rectangle of breadth b metres and length 3 times as long?
   (e) a rectangle of length 10 metres and breadth h metres?
   (f) an isosceles triangle with two sides of length $2t$ cm and one side of length t cm?

Exercise 10c

The questions in this exercise correspond in order to the questions of Exercise 10b.

1. John thinks of a number. He doubles it. His result is 58. What number did John think of?

2. 6 boys each have the same number of sweets. The total number of sweets is 78. How many sweets did each boy have?

3. A number is multiplied by 6 and then 4 is added. The result is 34. Find the first number.

4. A man has two boxes of matches. He uses 5 matches and has 75 matches left. How many matches were in each box?

5. I am thinking of a number. I take away 3. The result is 14. What number did I think of?

6. When a number is added to another number 4 times as big, the result is 30. Find the first number.

7. Rudy and Peter share 21 cents so that Peter gets twice as much as Rudy. How much does Rudy get?

8. Find the number such that when it is trebled and 7 is subtracted, the result is 8.

9. One girl has 9 cents less than another girl. They have 29 cents between them. How much does each girl have?

10. During a football season, one team scored 23 goals more than another. Between them they scored 135 goals. How many goals did each team score?

11. A square has a perimeter of 32 m. Find the length of one side of the square.

12. A triangle is such that the first side is twice the length of the second side. The third side is 4 m long. If the perimeter of the triangle is 13 m, find the lengths of the first and second sides.

13. A regular hexagon has a perimeter of 90 cm. Find the length of one side of the hexagon.

14. A rectangle is three times as long as it is broad. If the perimeter of the rectangle is 40 m, find its length and breadth.

15. A rectangle is 10 m long and its perimeter is 26 m. Find the breadth of the rectangle.

16. An isosceles triangle has 2 long sides and 1 short side. The short side is half the length of a long side. If the perimeter of the triangle is 15 cm, find the length of the short side.

It is possible to use operations with directed numbers when solving equations.

Example 3

Solve $25 - 9x = 2$.

$25 - 9x = 2$

Subtract 25 from both sides.

$25 - 25 - 9x = 2 - 25$

$-9x = -23$

Divide both sides by -9.

$\frac{-9x}{-9} = \frac{-23}{-9}$

$x = \frac{23}{9} = 2\frac{5}{9}$

Check: When $x = \frac{23}{9}$,

LHS $= 25 - 9 \times \frac{23}{9} = 25 - 23 = 2 =$ RHS

If an equation has unknown terms on both sides of the equals sign, collect the unknown terms on one side and the number terms on the other side.

Example 4

Solve $5x - 4 = 2x + 11$.

$5x - 4 = 2x + 11$

Subtract $2x$ from both sides.

$5x - 2x - 4 = 2x - 2x + 11$

$3x - 4 = 11$

Add 4 to both sides.

$3x - 4 + 4 = 11 + 4$

$3x = 15$

Divide both sides by 3.

$x = 5$

Check: When $x = 5$,

LHS $= 5 \times 5 - 4 = 25 - 4 = 21$

RHS $= 2 \times 5 + 11 = 10 + 11 = 21 =$ LHS

Exercise 10d

Solve the following equations and check the solutions.

1. $13 - 6a = 1$
2. $12 + 5a = -3$
3. $4b + 24 = 0$
4. $0 = 25 - 15x$
5. $12 = 9 - 3a$
6. $9 - 8y = 3$
7. $5 - 4n = 8$
8. $7 = 9 - 3m$
9. $7a = 3a + 20$
10. $20 - 2t = 3t$
11. $5n = 12 - n$
12. $7c - 6 = c$
13. $10q = 3q - 7$
14. $3x = 18 - 3x$
15. $3m + 8 = m$
16. $9x + 1 = 7x$
17. $4h - 2 = h + 7$
18. $5a + 6 = 2a + 20$
19. $18 - 5f = 2f + 4$
20. $11 - 3e = 2e - 19$
21. $6x + 1 = 26 - 2x$
22. $4x + 7 = 5x + 6$
23. $x + 7 = 19 + 2x$
24. $11 + 9n = 6n + 13$

Equations with brackets

Always remove brackets before collecting terms.

Example 5

Solve $3(3x - 1) = 4(x + 3)$.

$3(3x - 1) = 4(x + 3)$

Remove brackets.

$9x - 3 = 4x + 12$

Subtract $4x$ from and add 3 to both sides.

$9x - 4x - 3 + 3 = 4x - 4x + 12 + 3$

$5x = 15$

Divide both sides by 5.

$x = 3$

Check: When $x = 3$,

LHS $= 3(3 \times 3 - 1) = 3(9 - 1)$

$= 3 \times 8 = 24$

RHS $= 4(3 + 3) = 4 \times 6 = 24 =$ LHS

Example 6

Solve $5(x + 11) + 2(2x - 5) = 0$.

$5(x + 11) + 2(2x - 5) = 0$

Remove brackets.

$5x + 55 + 4x - 10 = 0$

Collect like terms.

$9x + 45 = 0$

Subtract 45 from both sides.

$9x = -45$

Divide both sides by 9.

$x = -5$

Check: When $x = -5$,

LHS $= 5(-5 + 11) + 2(2 \times (-5) - 5)$

$= 5 \times 6 + 2(-10 - 5)$

$= 30 + 2 \times (-15) = 30 - 30 = 0$

$=$ RHS

Exercise 10e

Solve the following equations and check the solutions.

1. $2(x + 5) = 18$
2. $15 = 3(x - 3)$
3. $55 = 5(2a - 1)$
4. $2(3y + 1) = 14$
5. $4(x + 7) + 12 = 0$
6. $0 = 7(x - 3)$
7. $6(2s - 7) = 5s$
8. $4b = 3(3b + 15)$
9. $3(f + 2) = 2 - f$
10. $3x + 1 = 2(3x + 5)$
11. $5(a + 2) = 4(a - 1)$
12. $5(b + 1) = 3(b + 3)$
13. $7(2e + 3) = 3(4e + 9)$
14. $8(2d - 3) = 3(4d - 7)$
15. $5(x + 1) = 7(2 - x)$
16. $2(4 - x) = 3(2 - x)$
17. $2(y - 2) + 3(y - 7) = 0$
18. $5(y + 8) + 2(y + 1) = 0$
19. $5(x - 4) - 4(x + 1) = 0$
20. $3(2x + 3) - 7(x + 2) = 0$
21. $5(5z - 2) - 9(3z - 2) = 2$
22. $3(6 + 7y) + 2(1 - 5y) = 42$
23. $5(v + 2) + 3(v + 5) = 1$
24. $4(3 - 5n) - 7(5 - 4n) + 3 = 0$

Word problems involving brackets

Example 7

I subtract 3 from a certain number, multiply the result by 5 and then add 9. If the final result is 54, find the original number.

Let the original number be x.
I subtract 3: this gives $x - 3$
I multiply by 5: this gives $5(x - 3)$
I add 9: this gives $5(x - 3) + 9$
The result is 54.
Thus, $5(x - 3) + 9 = 54$

Clear brackets.

$$5x - 15 + 9 = 54$$

Collect terms.

$$5x = 54 + 15 - 9 = 60$$
$$x = 60 \div 5 = 12.$$

The original number is 12.

Example 8

Jimmy and Ivy sell ballpoint pens at the same price. One day Jimmy increases his price by 12 cents and Ivy reduces her price by 14 cents. Jimmy sells 6 pens and Ivy sells 9 pens. If they both take in the same amount of money, what was the original price of a pen?

Let the original price of a pen be x cents.

$$\begin{aligned} \text{Jimmy's new price} &= (x + 12) \text{ cents} \\ \text{Ivy's new price} &= (x - 14) \text{ cents} \\ \text{Jimmy's income} &= 6(x + 12) \text{ cents} \\ \text{Ivy's income} &= 9(x - 14) \text{ cents} \end{aligned}$$

They both take in the same money, thus,

$$6(x + 12) = 9(x - 14)$$

Clear brackets.

$$6x + 72 = 9x - 126$$

Collect terms.

$$\begin{aligned} 72 + 126 &= 9x - 6x \\ 198 &= 3x \\ 66 &= x \end{aligned}$$

The original price of a pen was 66 cents.

Exercise 10f

1. I add 12 to a certain number and then double the result. The answer is 42. Find the original number.
2. I subtract 8 from a certain number. I then multiply the result by 3. The final answer is 21. Find the original number.
3. I think of a number. I multiply it by 5. I then subtract 19. Finally, I double the result. The final number is 22. What number did I think of?
4. Find two consecutive whole numbers such that 5 times the smaller number added to 3 times the greater number makes 59. (*Hint*: let the numbers be x and $(x + 1)$.)
5. Find two consecutive odd numbers such that 6 times the smaller added to 4 times the greater comes to 138. (*Hint*: let the numbers be x and $(x + 2)$.)

6 A rectangular room is 2 m longer than it is wide. Its perimeter is 70 m. If the width of the room is w m, express the length of the room in terms of w. Hence find the width of the room.

7 A man has a body mass of m kg. He is 30 kg heavier than each of his twin children. Express the body mass of each child in terms of m. If the mass of the father and 2 children comes to 156 kg, find the mass of the father.

8 Black pencils cost 45c each and coloured pencils cost 61c each. If 24 mixed pencils cost \$13.36, how many of them were black? (*Hint*: let there be x black pencils. Thus there are $(24 - x)$ coloured pencils. Work in cents.)

9 A worker gets \$6 an hour for ordinary time and \$9 an hour for overtime. If she gets \$324 for a 50-hour week, how many hours were overtime?

10 The cost of petrol rises by 12c a litre. Last week a man bought 20 litres at the old price. This week he bought 10 litres at the new price. Altogether, the petrol cost \$55.20. What was the old price for 1 litre?

11 A trader bought some oranges at 75c each. She finds that 6 of them are rotten. She sells the rest at \$1 each and makes a profit of \$31.50. How many oranges did she buy? (*Hint*: let the number of oranges be x. Work in cents.)

12 In 1996 an egg cost 7c less than in 1997. In 1998 an egg cost 8c more than in 1997. The cost of 11 eggs in 1996 was the same as the cost of 8 eggs in 1998. Find the cost of an egg in 1997.
(*Hint*: let the 1997 cost of an egg be n cents. Express the 1996 and 1998 costs in terms of n.)

Equations with fractions

Always clear fractions before collecting terms. To clear fractions, multiply both sides of the equation by the LCM of the denominators of the fractions.

Example 9

Solve the equation $\frac{4m}{5} - \frac{2m}{3} = 4$.

$$\frac{4m}{5} - \frac{2m}{3} = 4$$

The LCM of 5 and 3 is 15.
Multiply both sides of the equation by 15, i.e. multiply *every* term by 15.

$$15 \times \left(\frac{4m}{5}\right) - 15 \times \left(\frac{2m}{3}\right) = 15 \times 4$$
$$3 \times (4m) - 5 \times (2m) = 15 \times 4$$
$$12m - 10m = 60$$
$$2m = 60$$

Divide both sides by 2.

$$m = 30$$

Check: When $m = 30$

$$\text{LHS} = \frac{4 \times 30}{5} - \frac{2 \times 30}{3}$$
$$= 4 \times 6 - 2 \times 10$$
$$= 24 - 20 = 4 = \text{RHS}$$

Example 10

Solve the equation $\frac{3x - 2}{6} - \frac{2x + 7}{9} = 0$.

The LCM of 6 and 9 is 18.
Multiply both sides of the equation by 18.

$$\frac{18(3x - 2)}{6} - \frac{18(2x + 7)}{9} = 18 \times 0$$
$$3(3x - 2) - 2(2x + 7) = 0$$

Clear brackets.

$$9x - 6 - 4x - 14 = 0$$

Collect terms.

$$5x - 20 = 0$$

Add 20 to both sides.

$$5x = 20$$

Divide both sides by 5.

$$x = 4$$

Check: When $x = 4$,

$$\text{LHS} = \frac{3 \times 4 - 2}{6} - \frac{2 \times 4 + 7}{9}$$
$$= \frac{12 - 2}{6} - \frac{8 + 7}{9}$$
$$= \frac{10}{6} - \frac{15}{9} = \frac{5}{3} - \frac{5}{3} = 0 = \text{RHS}$$

Exercise 10g

Solve the following equations.

1. $\frac{x}{3} = 5$
2. $\frac{x}{5} = \frac{1}{2}$
3. $4 = \frac{a}{9}$
4. $\frac{7a}{2} - 21 = 0$
5. $\frac{4}{3} = \frac{2z}{15}$
6. $1\frac{1}{2} - \frac{3x}{4} = 0$
7. $\frac{x-2}{3} = 4$
8. $\frac{5+a}{4} = 6$
9. $\frac{2-a}{5} = 1$
10. $5 = \frac{2y-3}{7}$
11. $\frac{3n+1}{8} = 2$
12. $4 = \frac{9+5a}{6}$
13. $\frac{x+18}{2} = 5x$
14. $x = \frac{x-24}{9}$
15. $\frac{5x-8}{2} = 2x$
16. $\frac{22-3x}{4} = 2x$
17. $\frac{4-z}{7} = z$
18. $\frac{2(8x+7)}{3} = 5x$
19. $\frac{x}{2} - \frac{x}{3} = 2$
20. $\frac{x}{2} + \frac{3x}{4} = 5$
21. $\frac{3m}{5} - \frac{m}{3} = \frac{8}{5}$
22. $\frac{3x}{7} = \frac{2x}{3} - \frac{1}{3}$
23. $\frac{x-5}{2} = \frac{x-4}{3}$
24. $\frac{4t+3}{5} = \frac{t+3}{2}$
25. $\frac{5e-1}{4} - \frac{7e+4}{8} = 0$
26. $\frac{2d+7}{6} + \frac{d-5}{3} = 0$

Fractions with unknowns in the denominator

Example 11

Solve $2\frac{3}{4} + \frac{33}{2x} = 0$.

$2\frac{3}{4} + \frac{33}{2x} = 0$

Express $2\frac{3}{4}$ as an improper fraction.

$\frac{11}{4} + \frac{33}{2x} = 0$

The denominators are 4 and $2x$. Their LCM is $4x$. Multiply each term in the equation by $4x$.

$$4x\left(\frac{11}{4}\right) + 4x\left(\frac{33}{2x}\right) = 4x \times 0$$
$$11x + 66 = 0$$
$$11x = -66$$
$$x = -6$$

Check: When $x = -6$,

LHS $= 2\frac{3}{4} + \frac{33}{-12} = \frac{11}{4} - \frac{11}{4} = 0 =$ RHS

Example 12

Solve $\frac{1}{3a} + \frac{1}{2} = \frac{1}{2a}$

The denominators are $3a$, 2 and $2a$. Their LCM is $6a$. Multiply each term in the equation by $6a$.

$$6a \times \left(\frac{1}{3a}\right) + 6a \times \frac{1}{2} = 6a \times \left(\frac{1}{2a}\right)$$
$$2 + 3a = 3$$
$$3a = 1$$
$$a = \frac{1}{3}$$

Check: When $a = \frac{1}{3}$,

LHS $= \frac{1}{3 \times \frac{1}{3}} + \frac{1}{2} = 1 + \frac{1}{2} = 1\frac{1}{2}$

RHS $= \frac{1}{2 \times \frac{1}{3}} = \frac{3}{2} = 1\frac{1}{2} =$ LHS

Examples 11 and 12 show that when unknowns, such as x or a, appear in the denominator, they are treated in the same way as known numbers. Clear fractions by multiplying each term of the equation by the LCM of the denominators of the fractions. The equation can then be solved in the usual way.

Exercise 10h

Solve the following equations.

1. $\frac{1}{x} = \frac{1}{5}$
2. $\frac{1}{9} = \frac{1}{r}$
3. $\frac{1}{m} - \frac{1}{4} = 0$
4. $\frac{1}{y} = \frac{2}{7}$
5. $2\frac{1}{2} = \frac{1}{s}$
6. $2\frac{3}{4} + \frac{1}{n} = 0$
7. $\frac{2}{t} = \frac{6}{11}$
8. $\frac{9}{10} = \frac{3}{z}$
9. $\frac{4}{9} - \frac{2}{p} = 0$
10. $\frac{10}{3} = \frac{5}{a}$
11. $\frac{13}{x} = 5\frac{1}{5}$
12. $\frac{8}{q} + 3\frac{3}{7} = 0$
13. $\frac{1}{5b} = \frac{1}{30}$
14. $\frac{1}{40} = \frac{1}{8y}$
15. $\frac{1}{3r} - \frac{1}{24} = 0$
16. $\frac{7}{3c} = \frac{21}{2}$

17 $\dfrac{3}{10} = \dfrac{9}{2d}$

18 $\dfrac{16}{9} + \dfrac{4}{3s} = 0$

19 $3\frac{3}{5} = \dfrac{12}{25z}$

20 $\dfrac{33}{2r} = 3\frac{1}{7}$

21 $3\frac{3}{4} - \dfrac{5}{2t} = 0$

22 $\dfrac{1}{x} = \dfrac{1}{4} + \dfrac{1}{12}$

23 $\dfrac{1}{y} + \dfrac{1}{5} = \dfrac{1}{3}$

24 $\dfrac{1}{9} = \dfrac{1}{d} - \dfrac{1}{18}$

25 $\dfrac{1}{f} + \dfrac{1}{2} = \dfrac{5}{6}$

22 $2 = \dfrac{7}{2x} - \dfrac{1}{3}$

Word problems involving fractions

Example 13

I add 55 to a certain number and then divide the sum by 3. The result is 4 times the first number. Find the number.

Let the number be n.

I add 55 to n: this gives $n + 55$

I divide the sum by 3: this gives $\dfrac{n + 55}{3}$

The result is $4n$.

Thus $\dfrac{n + 55}{3} = 4n$

Multiply both sides by 3.

$$\frac{3(n + 55)}{3} = 3 \times 4n$$
$$n + 55 = 12n$$

Collect terms.

$$55 = 12n - n$$
$$55 = 11n$$
$$n = 5$$

The number is 5.

Example 14

The mass of a package A is x kg. The masses of two other packages B and C are $\frac{5}{6}$ and $\frac{4}{5}$ that of A. (a) Express the masses of B and C in terms of x. (b) If the difference between the masses of B and C is 2.3 kg, find the mass of A.

(a) Mass of B is $\frac{5}{8}$ of x kg $= \dfrac{5x}{6}$ kg

Mass of C is $\frac{4}{5}$ of x kg $= \dfrac{4x}{5}$ kg

(b) $\dfrac{5x}{6} - \dfrac{4x}{5} = 2.3$

The LCM of 5 and 6 is 30.
Multiply both sides by 30.

$$30 \times \frac{5x}{6} - 30 \times \frac{4x}{5} = 30 \times 2.3$$
$$5 \times 5x - 6 \times 4x = 69$$
$$25x - 24x = 69$$
$$x = 69$$

The mass of A is 69 kg.

Example 15

The students in a class have a total mass of 1717 kg. If the average mass per student is $50\frac{1}{2}$ kg, find the number of students in the class.

The number of students is the unknown. Let there be n students in the class. Then,

average mass per student $= \dfrac{1717}{n}$ kg
(from first sentence in question).

Thus, $50\frac{1}{2} = \dfrac{1717}{n}$
(from second sentence in question).

$$\frac{101}{2} = \frac{1717}{n}$$

Multiply throughout by $2n$

$$101n = 2 \times 1717$$
$$n = \frac{2 \times 1717}{101} = 2 \times 17$$
$$n = 34$$

There are 34 students in the class.

The problem in Example 15 could have been solved by simple arithmetic. The algebraic method was not really necessary. However, Example 16 shows a problem which is best solved by using algebra.

Example 16

A cow costs 7 times as much as a goat. For $5040 I can buy 18 more goats than cows. How much does a goat cost?

The cost of a goat is the unknown. Let a goat cost $\$h$. Thus a cow costs $\$7h$ (from first sentence in question). For $5040 I can buy

$\dfrac{5040}{h}$ goats or $\dfrac{5040}{7h}$ cows.

Thus, $\frac{5040}{h} - \frac{5040}{7h} = 18$
(from second sentence).

Multiply throughout by $7h$.

$$7h \times \frac{5040}{h} - 7h \times \frac{5040}{7h} = 7h \times 18$$
$$7 \times 5040 - 1 \times 5040 = 7 \times 18 \times h$$
$$6 \times 5040 = 7 \times 18 \times h$$
$$\frac{6 \times 5040}{7 \times 18} = h$$
$$h = \frac{720}{3}$$
$$= 240$$

Thus a goat costs \$240 (and a cow costs \$1680).

When solving problems of this kind:

1 find out what the unknown is;
2 choose a letter to stand for the unknown quantity;
3 change the statements in the question into algebraic expressions;
4 make an equation;
5 solve the equation, leaving any numerical simplifying until the last step of the working.

In Exercise 10i below, some questions give a letter for the unknown; in other questions you must choose a letter for yourself.

Exercise 10i

1. I think of a number. I double it. I divide the result by 5. My answer is 6. What number did I think of?

2. I subtract 17 from a certain number and then divide the result by 5. My final answer is 3. What was the original number?

3. I add 9 to a certain number and then divide the sum by 16. Find the number if my final answer is 1.

4. A fisherman catches n fish. Their total mass is 15 kg.
   (a) Write down the average mass of a fish in terms of n.
   (b) If the average mass of a fish was $\frac{3}{4}$ kg, find the number of fish caught.

5. A car travels 120 km at a certain average speed. If the journey takes $2\frac{1}{2}$ h, find the average speed.

6. A pencil costs x cents and a notebook costs $4x$ cents. I spend \$10.80 on pencils and \$10.80 on notebooks.
   (a) Write down the number of pencils I get in terms of x.
   (b) Write down the number of notebooks I get in terms of x.
   (c) If I get 15 more pencils than notebooks, how much does a pencil cost?

7. A table costs 5 times as much as a chair. For \$6000 a trader can buy 20 more chairs than tables. Find the cost of a chair.

8. The price of a packet of salt goes up by 9 cents. The old price is $\frac{4}{5}$ of the new price. Find the old and new prices.
   (*Hint*: let the old price be n cents. Thus the new price is $n + 9$ cents.)

9. A man's weekly pay is \$$x$. He spends $\frac{1}{2}$ of his pay on food and $\frac{1}{3}$ on rent.
   (a) Express the amount he spends on food in terms of x.
   (b) Express the amount he spends on rent in terms of x.
   (c) Find his weekly pay if he spends a total of \$400 on food and rent.

10. The distance between two villages is d km.
    (a) Express $\frac{4}{5}$ of that distance in terms of d.
    (b) Express $\frac{3}{4}$ of the distance in terms of d.
    (c) If the difference between these distances is 1.5 km, find the value of d.

11. A girl walks 8 km at v km/h. She then cycles 15 km at $2v$ km/h. In terms of v, write down the time taken in hours (a) when walking, (b) when cycling. (c) If the total time for the journey is 2 h 35 min, find the girl's walking speed.

12. Kathy is y years old.
    (a) How old was she 3 years ago?
    (b) How old will she be in 4 years time?
    (c) Find her age, if $\frac{1}{2}$ of what she was 3 years ago is equal to $\frac{1}{3}$ of what she will be in 4 years time.

13. A car travels for 15 km in a city at a certain speed. Outside the city it travels 72 km at twice its former speed. If the total travelling time is 1 h 8 min, find the average speed in the city.

14. Mary has n oranges of total mass 14.5 kg. Ann has $2n$ oranges of total mass 21 kg.
    (a) What is the average mass of one of Mary's oranges in terms of n?
    (b) What is the average mass of one of Ann's oranges in terms of n?
    (c) If the average mass of Ann's oranges is 0.1 kg less than the average mass of Mary's oranges, find the number of oranges that Mary has.

15. A man caught 15 kg of fish on Monday and 23 kg on Tuesday. On Tuesday there were twice as many fish as on Monday, but their average mass was $\frac{1}{8}$ kg less. How many fish did the man catch on Monday?

Summary

To **solve an equation** is to find the value of the **unknown** which makes the equation true. In a word problem, the unknown must be identified and defined by a letter; consideration is given to any required units.

The information given in the problem is then written as an algebraic equation. The rules which apply, for example, in removing brackets, in clearing fractions, in expanding expressions, are used to solve the equation. The solution of the problem is given as a statement. The value(s) of the solution of the equation must be checked against the information given in the word problem.

Practice Exercise P10.1

Solve the following equations. Write down all the steps and check your answers.

1. $5x + 6 = 2x$
2. $2a + 3 = 5a$
3. $4c = c - 15$
4. $\frac{5u}{8} - 3 = \frac{u}{4}$
5. $5x + 4 = 2x + 10$
6. $b - 4 = 3b - 2$
7. $3(4m - 1) - (m - 3) = 0$
8. $2(n - 3) = 5(n + 3)$
9. $3a + 2(2a + 11) = 1$
10. $3(2y - 2) = 4(y + 7)$
11. $\frac{m}{5} - 2 = \frac{m}{2} + 10$
12. $\frac{1}{2}f - \frac{1}{4}f = 2$
13. $\frac{y}{2} - \frac{y}{3} = \frac{2}{3}$
14. $\frac{3d}{2} - \frac{1}{2} = 3 - \frac{d}{4}$
15. $\frac{2b}{3} = \frac{b - 1}{2}$
16. $\frac{x + 1}{4} = \frac{2x - 3}{3}$
17. $\frac{h + 1}{h} = \frac{h - 1}{2h}$
18. $\frac{1 - n}{4 + n} = \frac{n + 2}{3 - n}$

Practice Exercise P10.2

Remove the brackets and solve the following equations.

1. $-\frac{1}{3}(3k) = 5k - 9$
2. $4(m - 2) - 2(2m - 3) = m$
3. $4(u - 2) = 3(2u - 3)$
4. $6y - 3(y - 2) = 0$
5. $4 - 2(1 + 2y) - 3(2 - 3y) = 11$
6. $4(-2d + 1) + 3(d - 2) = -7$

Practice Exercise P10.3

Write the following as equations using n to represent the number,

1. I think of a number, add 4 and the answer is 12.
2. I think of a number, subtract 5 and the answer is 13.
3. I think of a number and multiply it by 3. The answer is 15.
4. I think of a number, divide by 7 and the answer is 3.
5. When 2 is added to a number the answer is 13.
6. If 9 is added to a number the answer is 35.
7. If 15 is subtracted from a number, the answer is 20.
8. When 14 is taken away from a number the answer is 23.
9. 5 times a number gives 45.
10. When 10 is multiplied by a number, you get 120.
11. When a number is divided by 4, you get 12.
12. When a number is doubled, then 3 is added to the result, the answer is 9.

Practice Exercise P 10.4

Solve the following problems. Write down all the steps and check your answers.

1. Balloons cost b cents each. Whistles cost 72 cents each.
   (a) Write an expression for the total cost of 4 balloons and 1 whistle.
   (b) The total cost is $2.92. Write an equation to show this information.
   (c) Solve the equation.
   (d) How much does 1 balloon cost?
2. The capacity of a cup is C ml. A mug holds twice as much as a cup, and a jug holds 150 ml more than four cups.
   (a) Write expressions in terms of C for the capacity of
      (i) the mug
      (ii) the jug.
   (b) If the jug fills three cups and a mug, write an equation to show this information.
   (c) Solve the equation for C.
   (d) State the capacity of
      (i) the cup
      (ii) the mug
      (iii) the jug.

Chapter 11

Scale drawing (1)

Scale, use of scales in drawing

Pre-requisites

- ratio; basic units of measurement; geometrical constructions

Scale

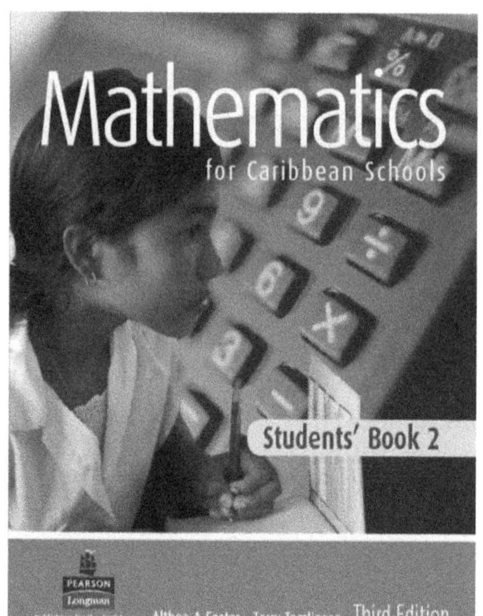

Fig. 11.1

Fig. 11.1 is a **scale drawing** of the front cover of this book. The only difference between Fig. 11.1 and the front cover is size. The scale drawing is smaller.

Check the following measurements:
width of Fig. 11.1 ≃ 6.2 cm
width of front cover ≃ 18.6 cm
where ≃ means 'is approximately equal to'.

Also
length of Fig. 11.1 ≃ 8.0 cm
length of front cover ≃ 24.0 cm

We can write the ratio of the measurements as follows:

ratio of two widths is
6.2 cm : 18.6 cm = 1 : 3

ratio of two lengths is
8 cm : 24 cm = 1 : 3

We say that the **scale** of Fig. 11.1 is **1 to 3** or **1 cm to 3 cm** or **1 cm represents 3 cm**.

The scale of a drawing is found by comparing a length on the drawing with the corresponding length on the object which has been drawn.

$$\text{Scale} = \frac{\text{any length on the scale drawing}}{\text{corresponding length on the actual object}}$$

All corresponding measurements when compared are in the same ratio. So that in Fig. 11.1
ratio of the two widths is 1 : 3
ratio of the two lengths is 1 : 3

Example 1

Fig. 11.2(a) is a scale drawing of Fig. 11.2(b). Use measurement to find the scale of the drawing.

Width of Fig. 11.2(a) = 20 mm
Width of Fig. 11.2(b) = 40 mm

$$\text{Scale} = \frac{20\text{ mm}}{40\text{ mm}} = \frac{1}{2}$$

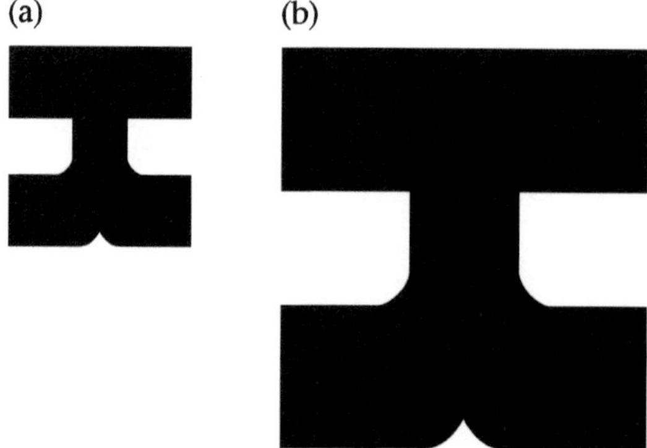

Fig. 11.2

The scale is 1 to 2 or 1 : 2,
1 cm to 2 cm or 1 cm : 2 cm

The result in Example 1 can be checked by comparing a different pair of corresponding lengths, e.g. the heights of the figures. Notice that the measurements must be in the same units.

Example 2

A plan of a school is drawn to a scale of 1 cm represents 5 m. (a) If the football field is 80 m by 53m, find its length and breadth on the drawing. (b) If the scale drawing of the hall is a 7 cm by 3.2 cm rectangle, find its actual length and breath.

In scale drawings, a **plan** is a drawing of the view from above the object.

(a) 5 m is represented by 1 cm
1 m is represented by $\frac{1}{5}$ cm
80 m is represented by $80 \times \frac{1}{5}$ cm $= 16$ cm
53 m is represented by $53 \times \frac{1}{5}$ cm $= 10.6$ cm
On the drawing, the football field will be 16 cm long and 10.6 cm wide.

(b) 1 cm represents 5 m
7 cm represents 7×5 m $= 35$ m
3.2 cm represents 3.2×5 m $= 16$ m
The hall is 35 m long and 16 m wide.

Notice, in Example 2, that the scale is given in mixed units. 1 cm represents 5 m is the same as 1 cm represents 500 cm, 1 to 500 or 1 : 500.

Exercise 11a

1 In each part of Fig. 11.3, the smaller diagram is a scale drawing of the larger diagram. Measure two corresponding lengths and give the scales of the drawings.

(a)

(b)

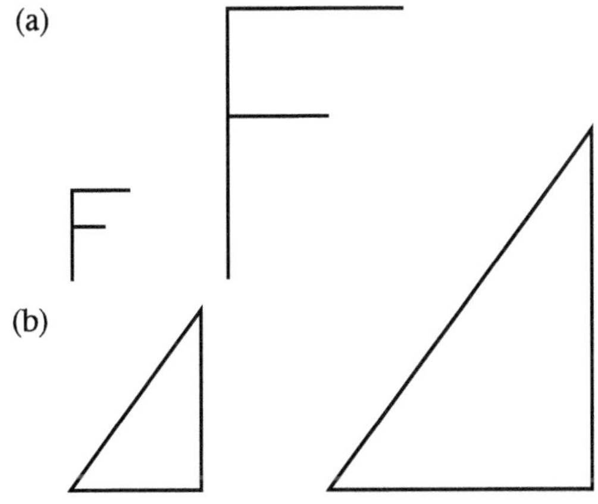

(c)

2 Copy and complete Table 11.1. The first part has been done.

Table 11.1

	True length	Scale	Length on drawing
(a)	90 m	1 cm to 10 m	9 cm
(b)	20 m	1 cm to 5 m	
(c)	8 m	1 cm to 2 m	
(d)	73 m	1 cm to 10 m	
(e)	65 m	1 cm to 5 m	
(f)	3 km	1 cm to 200 m	
(g)	450 m	1 cm to 100 m	
(h)	375 km	1 cm to 50 km	
(i)	1.53 km	10 cm to 1 km	
(j)	2.86 km	5 cm to 1 m	

3 Copy and complete Table 11.2. The first part has been done.

Table 11.2

	Length on drawing	Scale	True length
(a)	6 cm	1 cm to 10 m	60 m
(b)	11 cm	1 cm to 5 m	
(c)	5 cm	1 cm to 2 m	
(d)	7.5 cm	1 cm to 10 m	
(e)	8.2 cm	1 cm to 100 m	
(f)	9.3 cm	1 cm to 2 m	
(g)	8.6 cm	1 cm to 50 km	
(h)	14.8 cm	2 cm to 1 km	
(i)	11.3 cm	5 cm to 1 m	

4 Each part of Fig. 11.4, on page 94, is a scale drawing.

(a) For each part, use the given dimensions and a ruler to complete the statement, 'Scale: 1 cm represents _____'.

(b) Hence find the dimensions w, h, d, t.

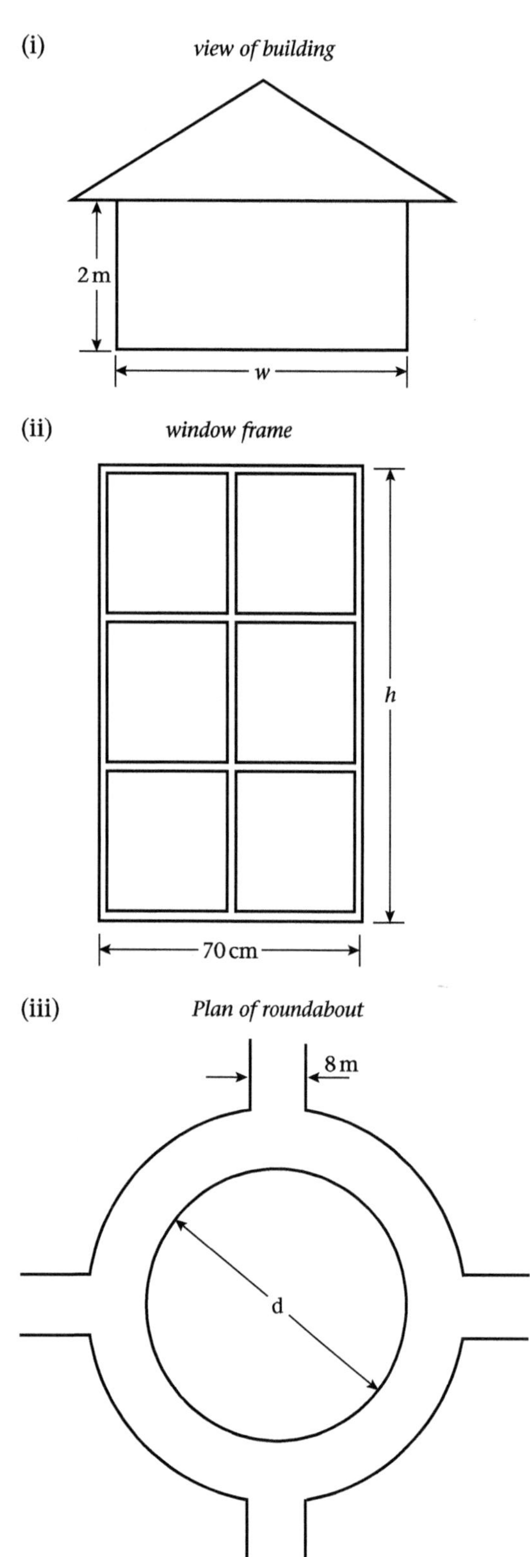

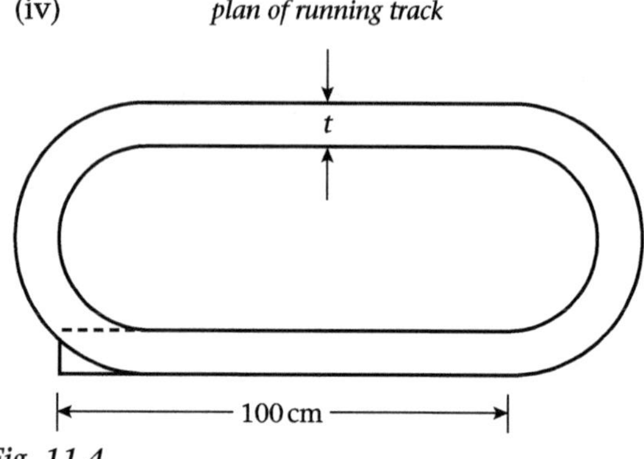

Fig. 11.4

Scale drawing

Example 3

A rectangular field measures 45 m by 30 m. Draw a plan of the field. Use measurement to find the distance between opposite corners of the field.

First, make a rough sketch of the plan. Enter the details on the rough sketch as in Fig. 11.5.

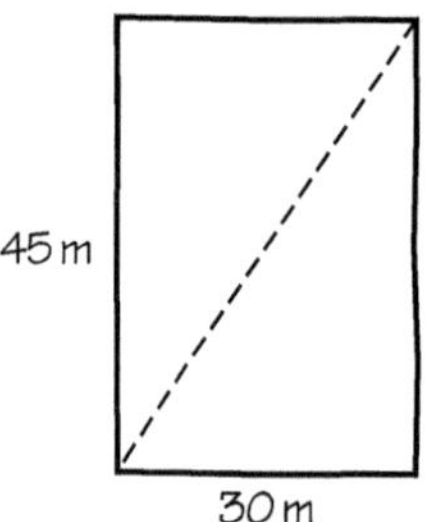

Fig. 11.5

Second, choose a suitable scale. As with graphs, the scale must suit the size of the page.

A scale of 1 cm to 1m will give a 45 cm × 30 cm rectangle. This will be too big for the page.

A scale of 1 cm to 5m will give a 9 cm by 6 cm rectangle. This will be suitable.

Third, make an accurate drawing of the plan. This is shown in Fig. 11.6

The distance between opposite corners of the field is represented by the broken line.

Length of broken line ≃ 10.8 cm

Actual distance $\simeq 10.8 \times 5$ m
$= 54$ m (to nearest metre)

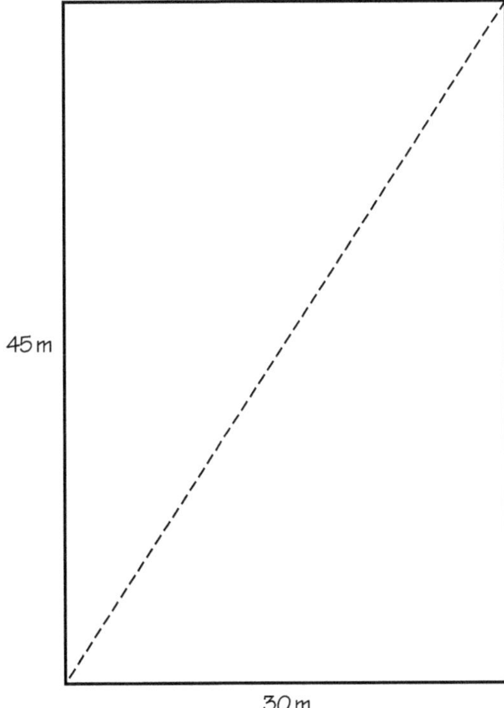

Fig. 11.6 Plan of field Scale: 1 cm to 5 m

Notice the following points.

1. Scale drawings should be made on plain paper.
2. Mathematical instruments are needed. For example, a pencil, a ruler and a set-square were used to draw Fig. 11.6.
3. The drawing has a title and the scale is given.
4. The dimensions of the actual object are written on the drawing.

Group work

Students are advised to have available a tape measure/rule, mathematical instruments, a well-sharpened pencil, and so on.

The class divides into groups of four students. The students in each group with the teacher's guidance choose either in the classroom or in the school compound a large object having a simple shape (preferably rectangular, square or circular), for example, the door, window, wall, desk-top, netball/football court, waste basket, text-book, and so on. The four students in each group work in two pairs. Each pair makes a scale drawing of a surface on the object, each pair deciding independently on the scale to be used. When both pairs in a group have finished the scale drawing, they discuss together the completed drawing, paying close attention to the scale used, the title of the drawing, the accuracy and the general appearance of the drawing. Questions to be asked may include the following:

Was a sketch done before the accurate drawing? Are all the construction lines visible? Are they 'sharp' and not 'fuzzy and thick'?

Example 4

Fig. 11.7 shows a sketch of two paths AX and BX. Points A and B are 178 m and 124 m from X respectively. The distance between A and B is 108 m. Make a scale drawing of the paths and find the angle between the paths at X.

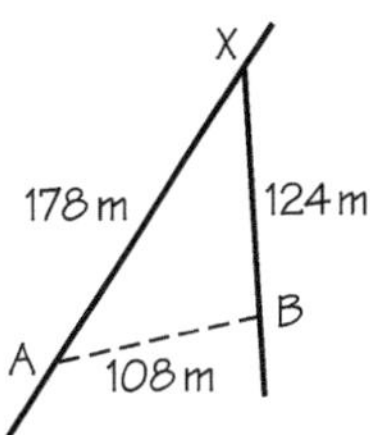

Fig. 11.7

It is necessary to construct a scale drawing of triangle AXB. Using a scale of 1 cm to 20 m, the sides of the triangle in the scale drawing will be as follows.

$$AX = \frac{178}{20}\text{ cm} = 8.9\text{ cm}$$

$$BX = \frac{124}{20}\text{ cm} = 6.2\text{ cm}$$

$$AB = \frac{108}{20}\text{ cm} = 5.4\text{ cm}$$

Using the method of constructing a triangle given its three sides, Fig. 11.8 (on page 96) is the required scale drawing.

Using a protractor, $A\hat{X}B = 37°$ (to the nearest degree). The angle between the paths is 37°.

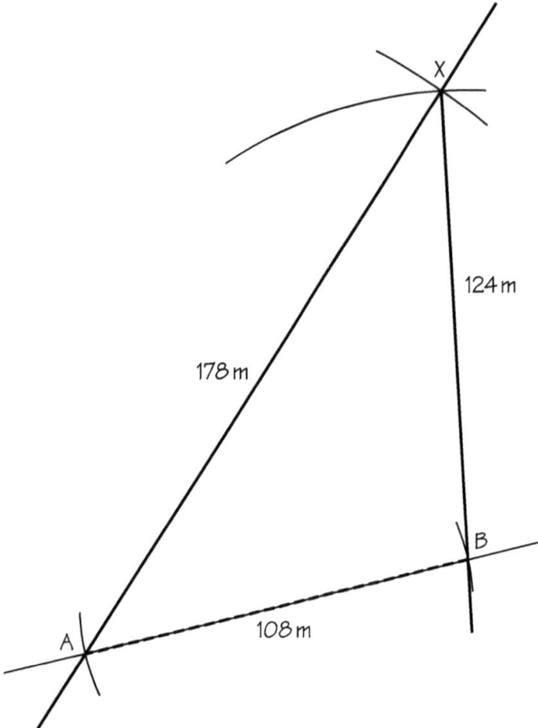

Fig. 11.8 *Plan of paths AX and BX*
Scale: 1 cm to 20 m

Exercise 11b

Make sketches where none is given. Choose suitable scales where none is given. All questions should be answered by taking measurements from an accurate scale drawing.

1. Find the distance between the opposite corners of a rectangular room which is 12 m by 9 m. Use a scale of 1 cm to 1 m.

2. A rectangular field measures 55 m by 40 m. Draw a plan of the field. Use measurement to find the distance between opposite corners of the field.

3. Measure the length and breadth of the top of your desk. Draw a plan of the top of your desk. Find the length of a diagonal from your drawing. Check your work by measuring the actual diagonal on your desk.

4. Measure the length and breadth of your classroom. Draw a plan of your classroom. Find a way of showing that your drawing is accurate.

5. A square field is 300 m × 300 m. Draw a plan of the field. Find the distance of the centre of the field from one of its corners.

6. Fig. 11.9 shows the plan of a room ABCD. PQ and XY are windows. HK is a door.

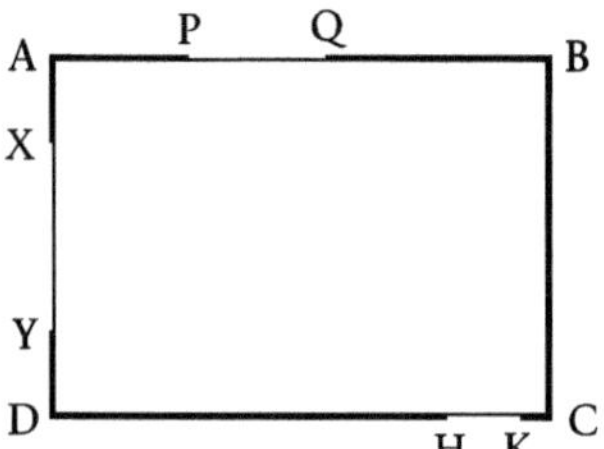

Fig. 11.9

AB = 10 m, BC = 7 m, AX = 1.5 m,
XY = 4 m, AP = 3 m, PQ = 2.5 m,
KC = 0.75 m, HK = 1.5 m.

Draw a plan on a scale of 1 cm to 1 m. Find the distances AC, XK, PH and QY.

7. A football field measures 104 m by 76 m. Use a scale of 1 cm to 10 m to draw a plan of the field. Find the distance from the centre spot to a corner flag.

8. Fig. 11.10 is a sketch of a cross-section of a hut. Use the dimensions on the figure to make an accurate scale drawing.

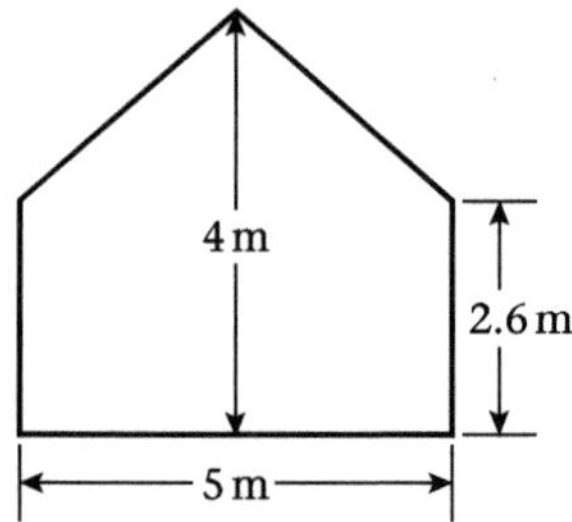

Fig. 11.10

Find the angle at the vertex of the roof.

9 Fig. 11.11 is a sketch of the end view of a house.

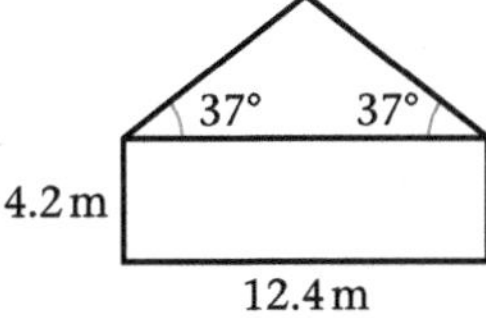

Fig. 11.11

Make a scale drawing and find the greatest height of the house.

10 A triangular plot, ABC, is such that AB = 120 m, BC = 80 m and CA = 60 m. P is the middle point of AB. Find the length of PC. Use a scale of 1 cm to 10 m.

Reading scale drawings

Many professions use scale drawings. The most common scale drawings are **maps** and **technical drawings**. Surveyors and cartographers make maps. Maps are used by geographers, navigators, planners, the police and soldiers. Engineers and draughtsmen make technical drawings. Technical drawings are used by builders, mechanics, electricians and skilled tradesmen. A scale drawing is an accurate way of storing and giving information. It is important to be able to read scale drawings.

Technical drawings

Fig. 11.12 is the plan of the wiring for an electric plug. It is drawn full size.

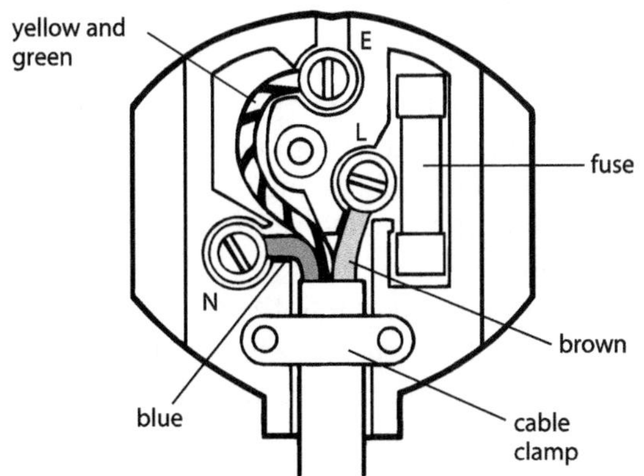

Fig. 11.12 Electric plug: wiring diagram (full size)

Fig. 11.13 is the ground plan of a house. Some of the important dimensions are given on the drawing. All such dimensions are in mm.

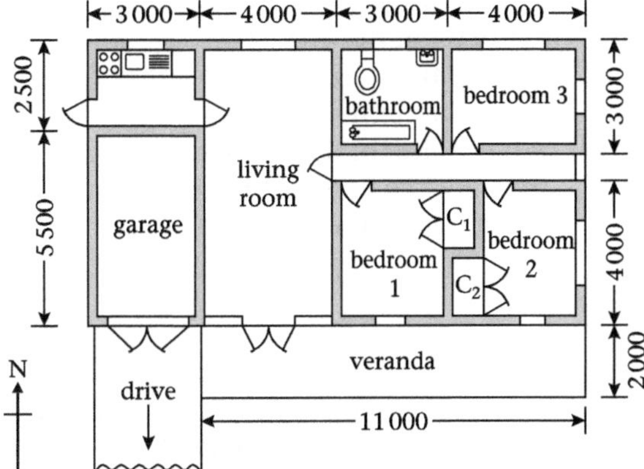

Fig. 11.13 House: ground plan

Exercise 11c

1 Use the wiring diagram in Fig. 11.12 to answer the following.
  (a) Each wire is connected to a terminal. How many terminals are there?
  (b) What is the colour of the wire which is connected to the terminal marked L?
  (c) What is the colour of the wire which is connected to the terminal marked E?
  (d) What is the colour of the wire which is connected to the terminal marked N?
  (e) Which terminal is next to the fuse?
  (f) How many screws are on the cable clamp?
  (g) Find the greatest width of the plug.

2 Use the ground plan in Fig. 11.13 to answer the following.
  (a) How many rooms has the house? (Do not count the garage as a room.)
  (b) Which is the biggest room in the house?
  (c) Which is the smallest room in the house?
  (d) If a person walked from the kitchen to the bathroom, how many doors would he pass through?
  (e) Which room has most windows?
  (f) How many windows does the house have altogether?
  (g) Which room(s) is (are) north of bedroom 2?

(h) Which room(s) is (are) west of the bathroom?
(i) What do you think C_1 and C_2 stand for?
(j) What is the length and breadth of the living room in metres?
(k) What is the length and breadth of the garage in metres?
(l) What is the total area that the house covers? Give your answer in m^2 and include the garage and veranda.

Maps

A map is a scale drawing of a piece of land. Distances on the map represent the horizontal distances between points on the land. Fig. 11.14 is a small-scale map of Barbados.

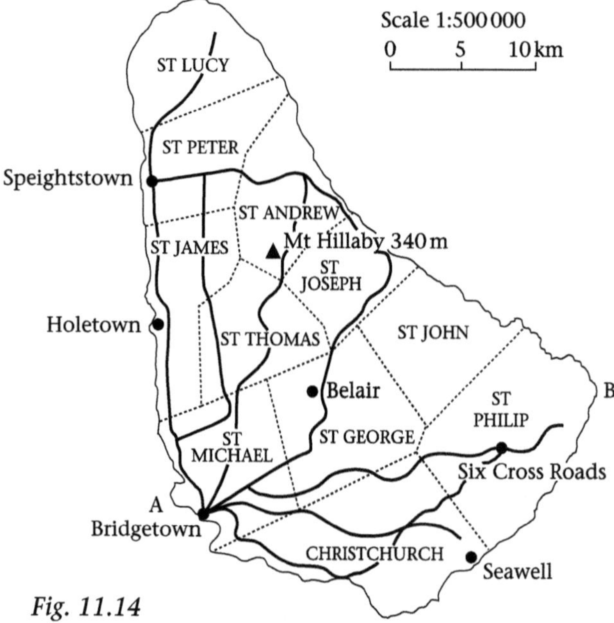

Fig. 11.14

Fig. 11.15 is a large-scale map of an imaginary town, Zama. Look at the way the scale is given in Fig. 11.15. Measure the scale. You will find that 1 cm represents 500 m.

Exercise 11d

1. Use a ruler and the map in Fig. 11.14. Find the following distances to the nearest km.
   (a) Speightstown to Holetown
   (b) Bridgetown to Holetown
   (c) Bridgetown to Belair
   (d) Mt. Hillaby to Holetown
   (e) Mt. Hillaby to Bridgetown
   (f) Mt. Hillaby to Six Cross Roads

2. Use the map in Fig. 11.15. Which roads would you travel on if you took the best route between the following? (Main roads only.)
   (a) The Mayor's house and the hospital
   (b) The station and the airport
   (c) The Shell garage and the BP garage
   (d) The bank and the Post Office
   (e) The two primary schools
   (f) Government College and the dispensary
   (g) The church and the airport
   (h) The Police Station and the airport
   (i) The hospital and the airport
   (j) Government Buildings and the Post Office

3. Use the map in Fig. 11.15. Find the following distances as accurately as possible.
   (a) From the Mayor's house to the church
   (b) From the prison to the bank
   (c) From Government Buildings to Eastside Primary School
   (d) From the hospital to the dispensary
   (e) From the Shell garage to the market
   (f) The length of Moortown Road
   (g) The length of Palm Avenue
   (h) The width of the river
   (i) The width of the airport's runway
   (j) The length of Church Road

4. Use the map in Fig. 11.15. Find the following distances, (i) in a straight line, (ii) by going on the main roads, taking the best route.
   (a) From the Post Office to the Mayor's house
   (b) From the station to the airport
   (c) From the dispensary to the Post Office
   (d) From Southside Primary School to Government Buildings
   (e) From Government Buildings to the station

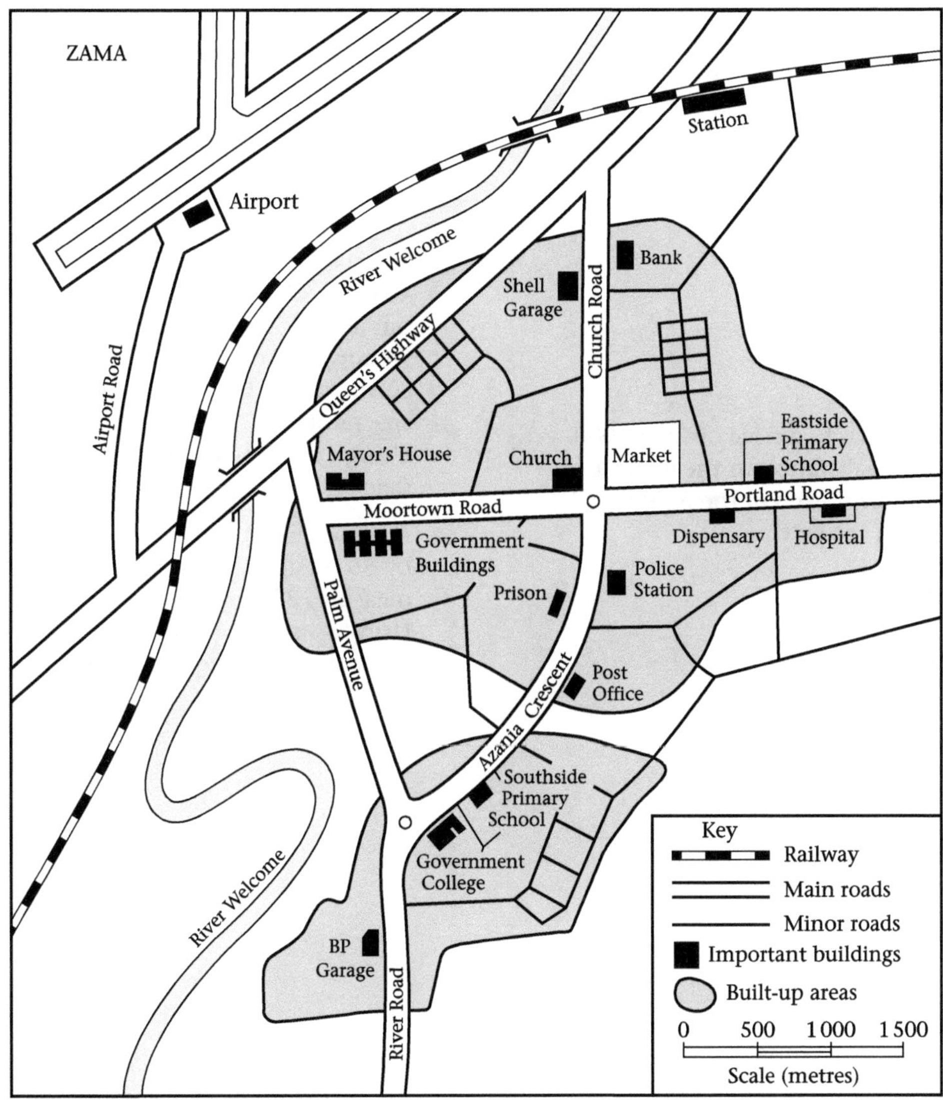

Fig. 11.15 Map of Zama

Summary

A **scale drawing** of a plane figure is such that while the shape remains the same, the size is different.

The **scale** gives the rate or ratio by which the dimension of the original is reduced. Note that a scale is also a ratio. A scale may be written as

1 to 3 or 1 : 3
1 cm to 5 m or 1 : 500

Note that when the scale is written as a ratio, both measurements must be stated in the same units.

Maps usually are drawn to small scales, that is, when the second number is large, the scale is small. e.g. 1 : 250 000 meaning 1 cm to 2.5 km. The actual dimensions of the original and the scale used must be given on the scale drawing.

Practice Exercise P11.1

1. Each of the following distances represent the distance between two places on a map which is drawn to a scale of 1: 200 000. Calculate the actual distance (km) between the two places:
   (a) 3.4 cm
   (b) 10.5 cm
   (c) 12.4 cm
   (d) 12.9 cm
   (e) 15.8 cm
   (f) 25.3 cm

2. A map is drawn to a scale of 1 : 250 000. Calculate the distance on the map that would represent each of the following distances on the earth:
   (a) 24 km
   (b) 36.5 km
   (c) 45.6 km
   (d) 59.4 km
   (e) 112.6 km
   (e) 150 km

3. A rectangle measures 8 m long and 5.4 m wide. Using a scale 1 : 100
   (a) draw a plan of the rectangle
   (b) measure the length of the diagonal of the rectangle
   (c) calculate the actual length of the diagonal of the rectangle

4. A rectangular playing field measures 40 m by 32 m. A circle radius 10 m is marked out at the centre of the field and a smaller rectangle 10 m by 6 m is marked at one corner of the field for a stand for seating a spectators.
   Using a scale of 1 : 500 draw a plan of the field showing the circle and the seating area.

5. A rectangular picture frame measures 35 cm by 56 cm and the space for the picture measures 28 cm by 49 cm.
   Using a scale of 1 cm to 7 cm, draw a plan of the frame showing the space for the picture.

Chapter 12

Angles (2)

Other polygons

Pre-requisites

- properties of plane shapes; measurement of angles

Polygons

A **polygon** is any plane figure with straight sides. Thus a triangle is a 3-sided polygon and a quadrilateral is a 4-sided polygon. Polygons are named after the number of sides they have. Table 12.1 gives the names of the first 8 polygons.

Table 12.1

triangle	3 sides
quadrilateral	4 sides
pentagon	5 sides
hexagon	6 sides
heptagon	7 sides
octagon	8 sides
nonagon	9 sides
decagon	10 sides

Angles in a quadrilateral

Exercise 12a

1. Draw two large quadrilaterals.
   (a) Measure the angles in each quadrilateral.
   (b) Find the sum of the angles in each quadrilateral.
   (c) What do you notice about your results?

2. Draw any large quadrilateral on a sheet of scrap paper. Cut it out carefully along its sides.
   (a) Tear off the four angles of the quadrilateral as in Fig. 12.1.

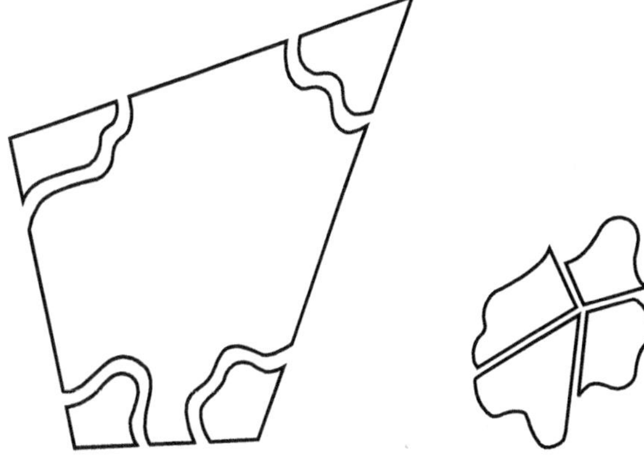

Fig. 12.1 *Fig. 12.2*

   (b) Take the four angles and arrange them so that they are adjacent to each other as in Fig. 12.2.
   (c) What do you notice? What is the sum of the angles at a point?

When working through Exercise 12a you may have noticed that the **sum of the angles of a quadrilateral is 360°**. We can show that this is true as follows.

Any quadrilateral can be divided into two triangles by drawing its diagonals (Fig. 12.3).

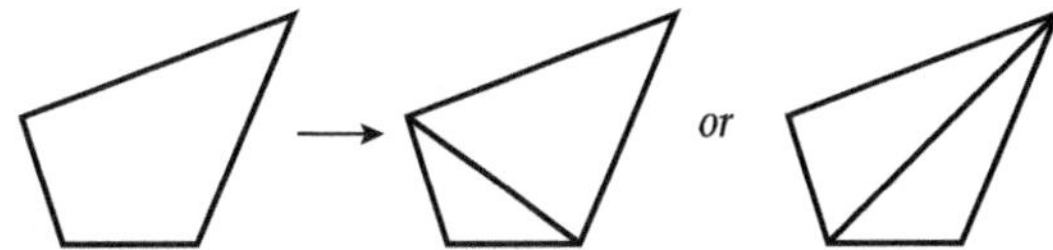

Fig. 12.3

The sum of the angles in each triangle is 180°. Thus the sum of the angles in the quadrilateral is 360°.

Exercise 12b

1. Find the sizes of all the angles of the quadrilaterals in Fig. 12.4.

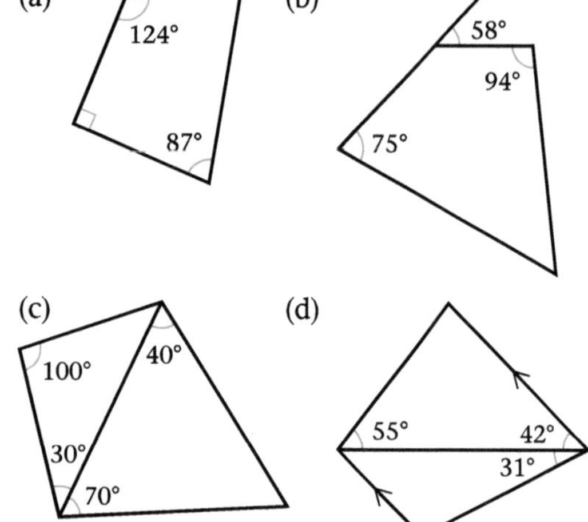

Fig. 12.4

2. Calculate the 4th angle of the quadrilaterals whose other three angles, in order, are:
   (a) 100°, 60°, 80°
   (b) 58°, 117°, 122°
   (c) 95°, 85°, 90°
   (d) 109°, 71°, 109°
   (e) 114°, 95°, 114°

3. Make rough sketches of quadrilaterals (c) and (d) of question 2. What types of quadrilaterals are they?

4. The angles of a quadrilateral are x, $2x$, $3x$ and $4x$ in that order.
   (a) Form an equation in x.
   (b) Find x.
   (c) Find the angles of the quadrilateral.
   (d) Make a rough drawing of the quadrilateral.
   (e) What kind of quadrilateral is it?

5. In Fig. 12.5, first find the value of x, then find the unknown angles of the quadrilaterals.

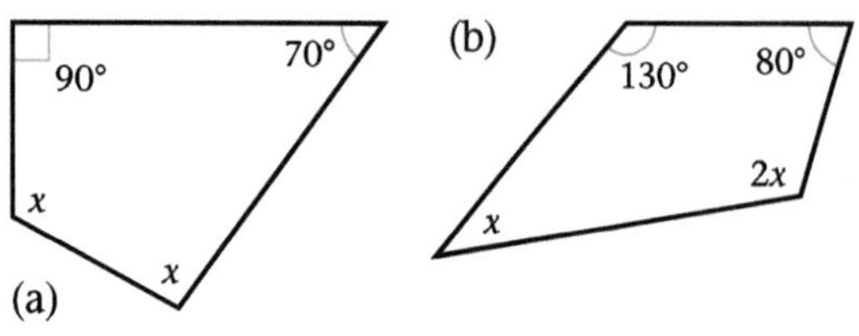

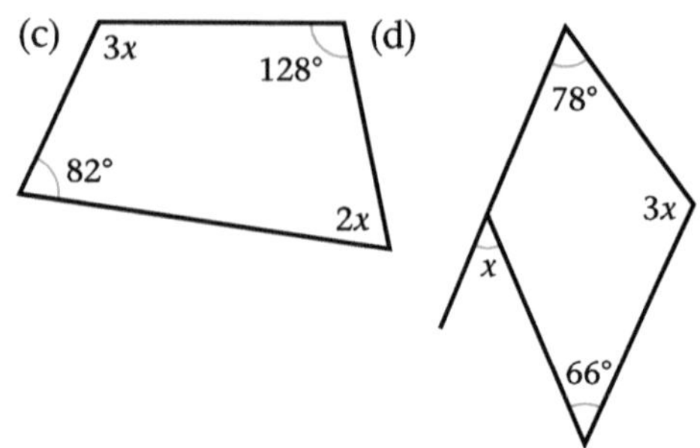

Fig. 12.5

Regular polygons

A regular polygon has all sides and all angles equal. Fig. 12.6 shows the first 6 regular polygons.

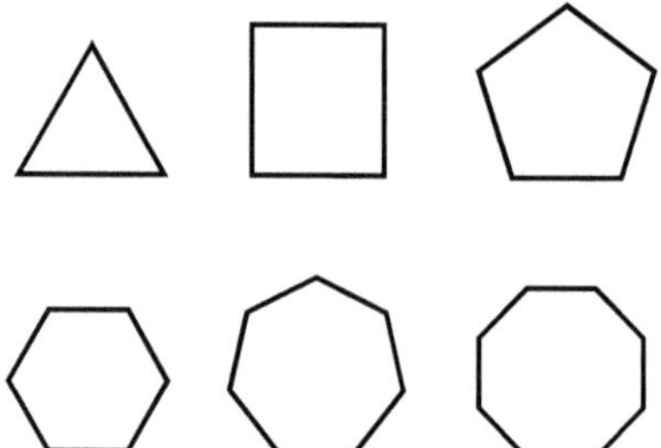

Fig. 12.6

Exercise 12c

1. How many lines of symmetry has each polygon in Fig. 12.6?

2. What is the order of rotational symmetry of each polygon in Fig. 12.6?

3. Fig. 12.7 is a regular pentagon ABCDE. O is the centre of point symmetry.

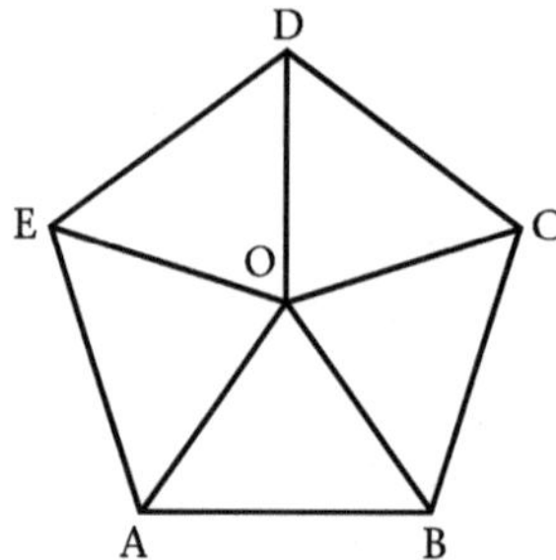

Fig. 12.7

(a) What is the sum of the angles at O?
(b) How many angles are at O?
(c) Calculate the size of each angle at O.
(d) What kind of △ is △AOB?
(e) Calculate $O\widehat{A}B$ and $O\widehat{B}A$.
(f) What is the size of $A\widehat{B}C$?
(g) Calculate the sum of the angles of pentagon ABCDE.

4 Fig. 12.8 shows a pattern made from regular hexagons.

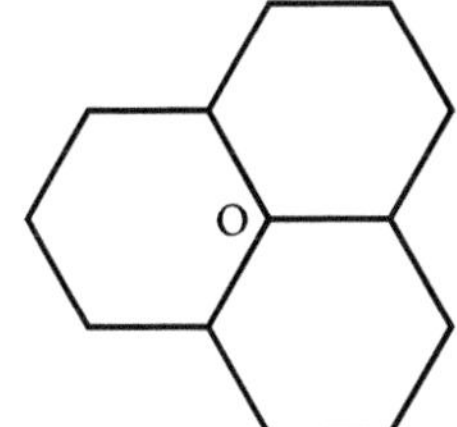

Fig. 12.8

(a) What is the order of symmetry of Fig. 12.8 about O?
(b) What is the size of each angle at O?

5 Calculate the sum of the angles of a regular hexagon.

6 In Fig. 12.9, ABCDEFGH is a regular octagon.
(a) What is the order of symmetry of Fig. 12.9 about O?
(b) What is the size of each angle at O?
(c) Show that $O\widehat{A}B + A\widehat{B}C + O\widehat{C}B = 270°$.
(d) Hence find the sum of the angles of ABCDEFGH.
(e) Hence find the size of each angle of the regular octagon.

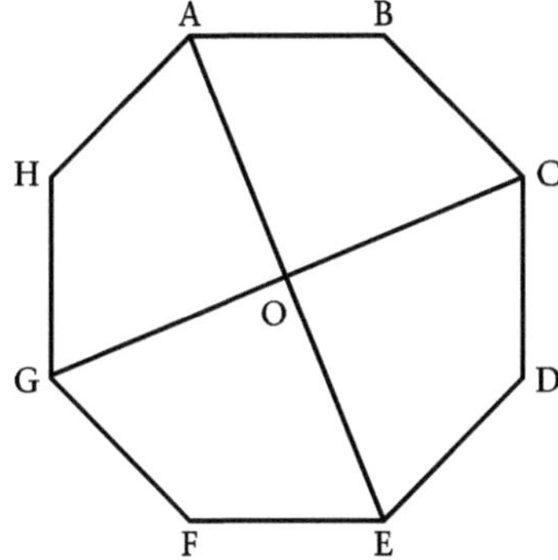

Fig. 12.9

7 Fig. 12.10 shows polygons labelled (a), (b), (c), (d), (e).

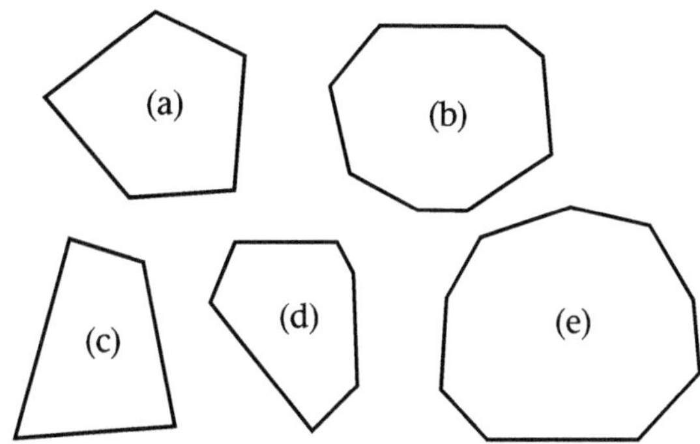

Fig. 12.10

Name each polygon correctly.

8 (a) Draw any large non-regular pentagon. Each side should be more than 5 cm long.
(b) Use a protractor to measure the angles of your pentagon.
(c) Find the sum of the five angles.
(d) Compare your result in (c) with your classmates'. What do you notice?

9 Repeat question 8 with any large non-regular hexagon.

10 Using your results in questions 8 and 9, copy and complete Table 12.2.

Table 12.2

Polygon	Sum of angles
triangle	
quadrilateral	
pentagon	
hexagon	

11 (a) Write your results for the sum of angles in question 10 as a ratio $a : b : c : d$.
(b) Simplify the ratio as far as possible.
(c) Hence guess the sum of the angles of any heptagon.

Interior angles of a polygon

Look at the quadrilateral (a), the pentagon (b) and the hexagon (c) in Fig. 12.11.

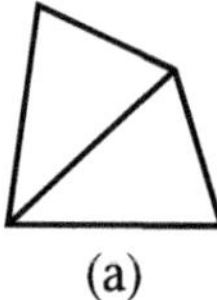
(a)

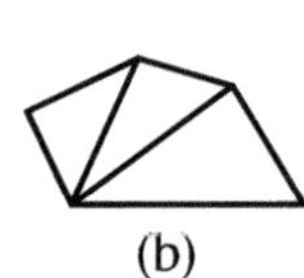
(b)

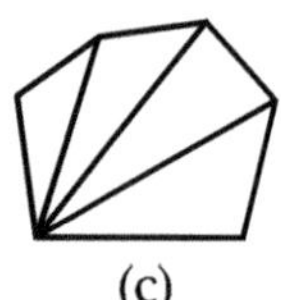
(c)

Fig. 12.11

In each polygon, one vertex is joined to all the other vertices. This divides the polygons into triangles. The number of triangles depends on the number of sides of the polygon. Table 12.3 shows the number of triangles for the polygons in Fig. 12.11. What do you notice?

Table 12.3

Polygon	Number of sides	Number of triangles
quadrilateral	4	2
pentagon	5	3
hexagon	6	4

In each case, the number of triangles is 2 less than the number of sides. For a polygon with n sides there will be $n - 2$ triangles. The sum of the angles of a triangle is 180°. Thus,

the sum of the angles of an n-sided polygon = $(n - 2) \times 180°$

Notice that this formula is true for the polygons we know. In a triangle, $n = 3$.

$$\begin{aligned}\text{Sum of angles} &= (3 - 2) \times 180° \\ &= 1 \times 180° = 180°\end{aligned}$$

In a quadrilateral, $n = 4$.

$$\begin{aligned}\text{Sum of angles} &= (4 - 2) \times 180° \\ &= 2 \times 180° = 360°\end{aligned}$$

Example 1

Calculate the size of each angle of a regular octagon.

A regular octagon has 8 equal sides and 8 equal angles.

Use the formula to find the sum of the angles, that is, sum of angles of polygon = $(n - 2) \times 180°$.

In an octagon, $n = 8$.

$$\begin{aligned}\text{Sum of angles} &= (8 - 2) \times 180° \\ &= 6 \times 180° = 1080°\end{aligned}$$

Since there are 8 equal angles

$$\text{Each angle} = \frac{1080°}{8} = 135°$$

Example 2

The sum of 7 of the angles of a nonagon is 1000°. The other two angles are equal to each other. Calculate the sizes of the other two angles.

A nonagon has 9 sides and 9 angles.
Use the formula:
sum of angles of polygon = $(n - 2) \times 180°$.
In a nonagon, $n = 9$.

$$\begin{aligned}\text{Sum of angles} &= (9 - 2) \times 180° \\ &= 7 \times 180° = 1260° \\ \text{Sum of 7 of the angles} &= 1000° \\ \text{Sum of 2 other angles} &= 1260° - 1000° \\ &= 260° \\ \text{Size of each angle} &= \frac{260°}{2} = 130°\end{aligned}$$

$$\begin{aligned}\text{Notice that } (n - 2) \times 180° &= (n - 2) \times 2 \times 90° \\ &= (2n - 4) \times 90°\end{aligned}$$

Hence **the sum of the angles of a polygon with n sides is $(2n - 4)$ right angles.**

Example 3

The sum of the angles of a polygon is 1980°. How many sides has the polygon?

Let the polygon have n sides.
Use the formula:
sum of angles of polygon

$$\begin{aligned}&= (2n - 4) \text{ right angles} \\ \text{then } (2n - 4) \times 90° &= 1980° \\ 2n - 4 &= 22 \\ 2n &= 26 \\ n &= 13\end{aligned}$$

The polygon has 13 sides.

Exercise 12d

1 In Fig. 12.12, O is any point inside each polygon. O is joined to the vertices of each polygon by straight lines. This divides the polygons into triangles.

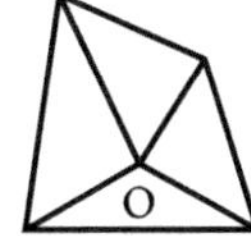

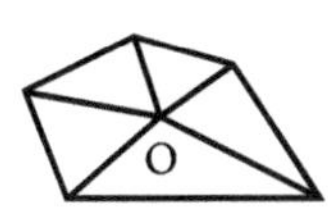

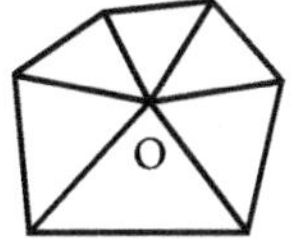

Fig. 12.12

(a) Draw a heptagon (7 sides) and an octagon (8 sides). Divide these into triangles in the same way as shown in Fig. 12.11 and 12.12.

(b) Copy and complete Table 12.4.

Table 12.4

Polygon	Number of sides	Number of triangles	Sum of angles at O	Sum of angles of polygon
quadrilateral	4	4	360°	$4 \times 180° - 360°$
pentagon	5	5	360°	$5 \times 180° - 360°$
hexagon				
heptagon				
octagon				
n-gon	n			

(c) Hence find a formula for the sum of the angles of an n-sided polygon.

2 Calculate the size of each angle of (a) a regular hexagon, (b) a regular decagon.

3 Fig. 12.13 is a regular *pentagram*. A regular pentagram has a regular pentagon at its centre. Calculate the size of each angle in the pentagram.

Fig. 12.13

4 Fig. 12.14 shows a pattern made with squares and regular octagons.

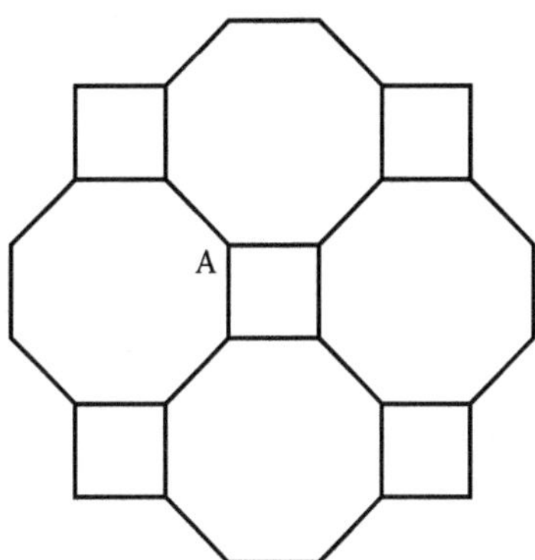

Fig. 12.14

Find the sizes of the three angles at point A, (a) from knowledge of the sum of the angles at a point, (b) by calculating the sizes of the angles of a regular octagon using the formula: $(n - 2) \times 180°$.

5 The pattern in Fig. 12.15 is made from four regular pentagons.

(a) What kind of quadrilateral is shown shaded?

(b) Calculate the sizes of the four angles of the shaded quadrilateral.

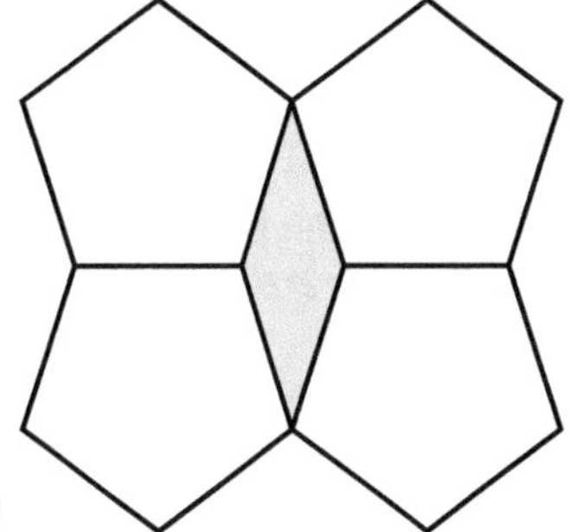

Fig. 12.15

6 In Fig. 12.16, first find the value of x, then find the unknown angles in each polygon.

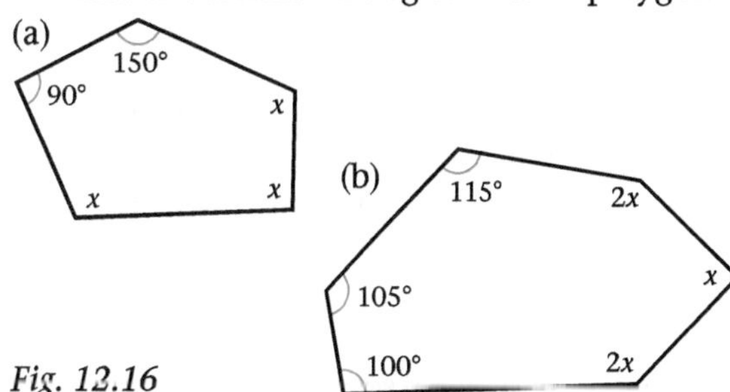

Fig. 12.16

7. The sum of seven of the angles of a decagon is 1170°. The other three angles are all equal to each other. Calculate the sizes of the other three angles.

8. If the angles of a pentagon are $2x$, $3x$, $4x$, $4x$ and $5x$, what is the value of x? Calculate the size of the largest angle.

9. How many sides has a polygon if the sum of its angles is (a) 3240°, (b) 2340°?

10. (a) How many sides has a polygon if the sum of its angles is 3960°
    (b) If the polygon is regular, what is the size of each angle?

Exterior angles of a polygon

Fig. 12.17 represents an n-sided polygon with each side produced.

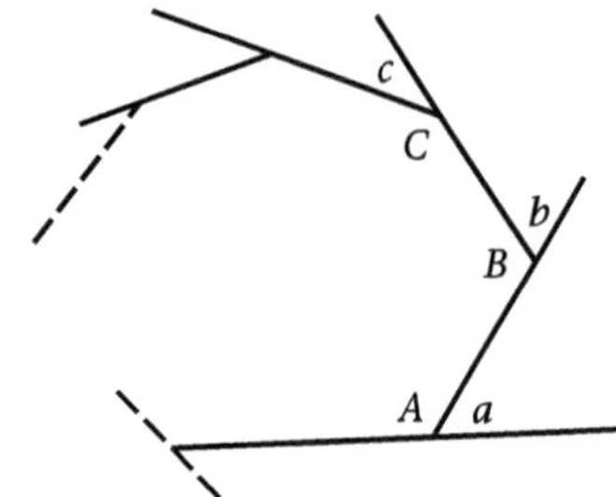

Fig. 12.17

The interior angles of the polygon are A, B, C, ... and the exterior angles are a, b, c, ... It follows that

$(A + a) = 180°$ (*straight angle*)
and $(B + b) = 180°$
etc.

Since there are n sides and n vertices

$$(A + a) + (B + b) + \ldots = n \times 180°$$

that is, $(A + B + \ldots) + (a + b + \ldots) = n \times 180°$
But $(A + B + \ldots) = (n - 2) \times 180°$ (*sum of interior angles*)

Hence, by subtraction,

$$\begin{aligned}(a + b + c + \ldots) &= n \times 180° - (n - 2) \times 180° \\ &= 180n° - 180n° + 360° \\ &= 360°\end{aligned}$$

The sum of the exterior angles of any polygon is 360°.

Example 4

Calculate the size of each interior angle of a regular decagon (10 sides).

A decagon has 10 sides, 10 interior angles and 10 exterior angles.
The 10 exterior angles add up to 360°.
Since the polygon is regular the exterior angles are equal.

Each exterior angle $= \dfrac{360°}{10} = 36°$

Each interior angle $= 180° - 36° = 144°$

Compare the working of Example 4 with your calculation for question 2(b) in Exercise 12d. Both methods are correct.

Example 5

How many sides has a regular polygon if each interior angle is 135°?

Each interior angle $= 135°$
Each exterior angle $= 180° - 135° = 45°$
The sum of the exterior angles $= 360°$
Hence the number of exterior angles $= \dfrac{360°}{45} = 8$

The polygon has 8 sides (since it has 8 exterior angles).

Exercise 12e

1. Use the method of Example 4 to calculate the interior angles of regular polygons with (a) 9, (b) 12, (c) 20 sides.

2. Use the method of Example 5 to find the number of sides that a regular polygon has if its interior angles are (a) 144°, (b) 168°, (c) 156°.

3. The angles of a pentagon are $2x°$, $3x°$, $4x°$, $5x°$, $6x°$.
   (a) Calculate x.
   (b) Hence calculate the angles of the pentagon.

4. In Fig. 12.18, AB, BC, CD are three sides of a regular pentagon ABCDE. Calculate $B\widehat{X}C$.

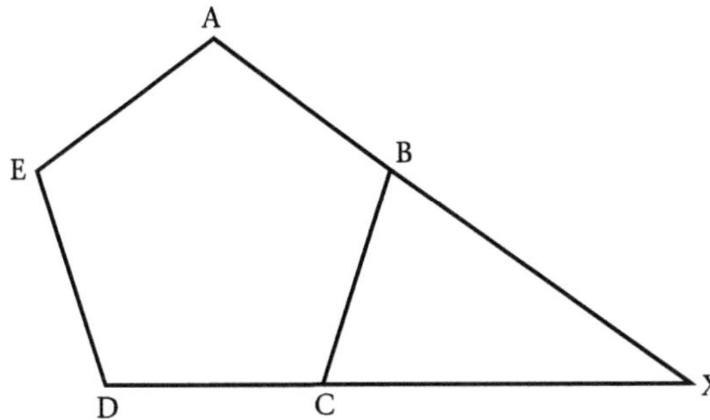

Fig. 12.18

5. Calculate x in Figs. 12.19 (*a*) and (*b*).

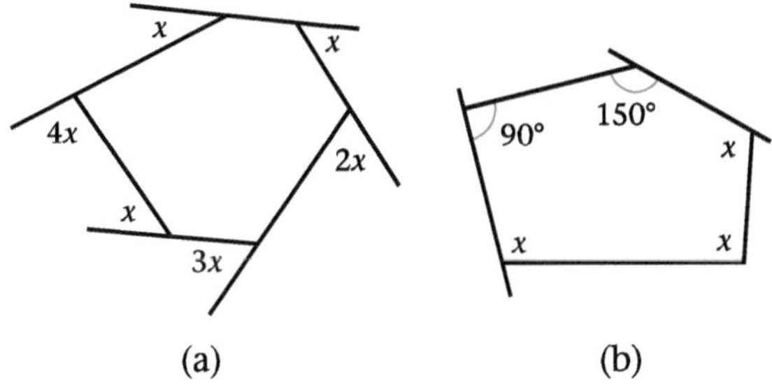

Fig. 12.19

Summary

A **polygon** is any plane figure bound by straight sides. Polygons are named according to the number of sides.

The **sum of the interior angles** of a polygon is *either* $(n - 2) \times 180°$ *or* $(2n - 4) \times 90°$ where n is the number of sides.

The **sum of the exterior angles** of a polygon is 360°.

A **regular polygon** has equal sides and equal angles.

Practice Exercise P12.1

1. The angles of a hexagon are $(x + 30)°$, $(x + 10)°$, x, $2x°$, $(2x + 30)°$, $(2x + 20)°$.
   (a) Write an equation in x for the sum of the angles of a hexagon.
   (b) Solve for x
   (c) Find the size of the angles of the hexagon.

2. Calculate the size of the interior angle of a regular decagon (10 sides)

3. Each interior of a regular polygon is 150° .
   (a) What is the size of each exterior angle?
   (b) Calculate the number of sides of the regular polygon.

4. The size of an exterior angle of a regular polygon is one-half the size of the exterior angle of a square. Calculate the number of sides of this polygon.

5. (a) State the formula for the sum of angles of a polygon.
   If the sum of the angles of a polygon is 1800°
   (b) Write an equation for the sum of the angles of the polygon;
   (c) Find the number of sides in the polygon.

6. Calculate the size of an interior angle of a polygon of 24 sides.

7. The sum of five angles of a decagon is 1105°. What is the size the other angles of the decagon if they are all equal?

8. Three interior angles of an octagon are 120°.
   (a) What is the sum of the angles of an octagon?
   (b) If the other angles are all equal, calculate the size of each remaining angle.

9. The size of an exterior angle of a regular polygon is one-third the size of an interior angle.
   (a) What is the size of the exterior angle?
   (b) Calculate the number of sides of the polygon.

Chapter 13

Straight-line graphs

Pre-requisites

- the Cartesian plane; ratio and rate; basic units of measurement

Continuous graphs

Consider the following example.
If 1m of ribbon costs \$5, then 2 m costs \$10, 3 m costs \$15, and so on. We can show lengths and costs in a **table of values** (Table 13.1). The values in the table form a set of ordered pairs: (1, 5), (2, 10), (3, 15), (4, 20), (5, 25).

Table 13.1

Length (m)	1	2	3	4	5
Cost (\$)	5	10	15	20	25

We can plot the ordered pairs on a Cartesian plane. Fig. 13.1 shows the graph of the 5 points. Length (m) is on the horizontal axis and cost (\$) is on the vertical axis.

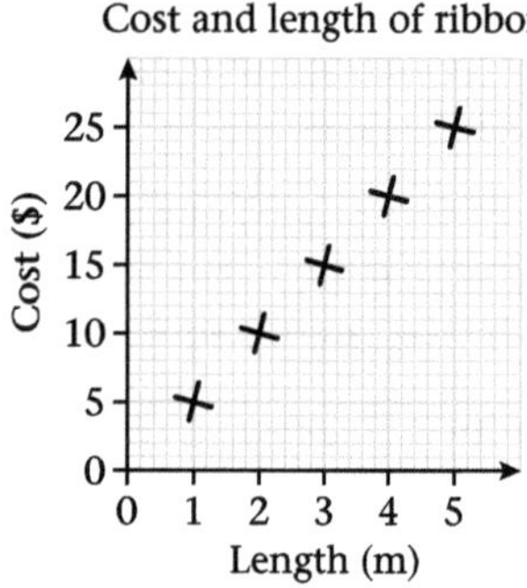

Fig. 13.1

It can be seen that the points in the graph lie in a straight line.

Other lengths of ribbon, such as 10 m, 1.4 m, 3.75 m would have corresponding costs. Thus it is possible to plot more points. Instead of this, we can draw a **continuous line** through the points which have been plotted. Starting at (0, 0) (no ribbon costs nothing!) we can continue the line as far as we want.

Fig. 13.2 shows the graph extended as far as the cost of 10 m of ribbon.

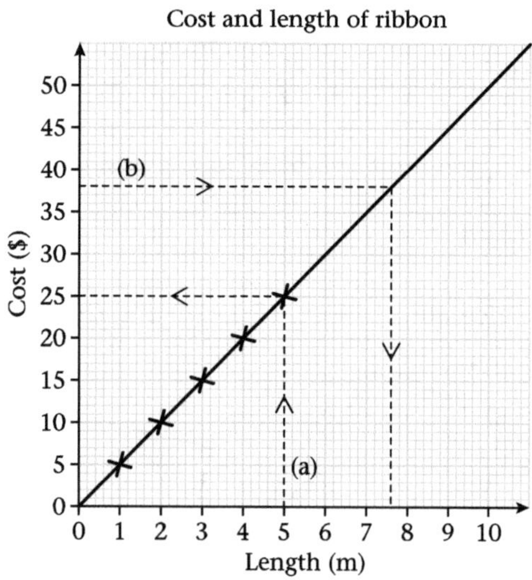

Fig. 13.2

The graph can be used to answer questions like these.

1 How much would 5 m of ribbon cost? Follow the broken line (a) in Fig. 13.2. Start at 5 m on the length axis. Read the cost which corresponds to 5 m: \$25.

2 How much ribbon can be bought for \$38? Follow the broken line (b) in Fig. 13.2. Start at \$38 on the cost axis. Read the length which corresponds to \$38: 7.6 m.

Example 1

A boy walks along a road at a speed of 120 m per minute.

(a) Make a table of values showing how far the boy has walked after 0, 1, 2, 3, 4, 5, minutes.

(b) Using a scale of 1 cm to 1 min on the horizontal axis and 1 cm to 100 m on the vertical axis, draw a graph of this information.

(c) Use the graph to find (i) how far the boy has walked after 2.6 min, (ii) how long it takes him to walk 500 m.

(a) Table 13.2, is the table of values.

Table 13.2

Time (min)	0	1	2	3	4	5
Distance (m)	0	120	240	360	480	600

(b) See Fig. 13.3.

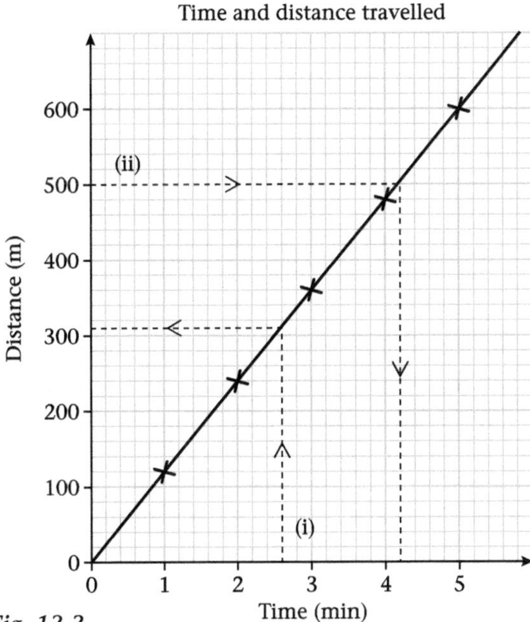

Fig. 13.3

(c) (i) See broken line (i) on Fig. 13.3; 2.6 min corresponds to 310 m (approximately). He has walked about 310 m after 2.6 min.

(ii) See broken line (ii) on Fig. 13.3; 500 m corresponds to 4.2 min (approximately). He takes about 4.2 min to walk 500 m.

Notice that graphs usually only give approximate results.

When drawing graphs, *always:*

(a) draw and name the two axes;

(b) give a title to the graph.

Exercise 13a

1. Use Fig. 13.2 to find the following.
   (a) The cost of 6 m, 3.5 m, 9 m, 8.2 m, 2.6 m, 7.1 m of ribbon.
   (b) How much ribbon can be bought for $35, $40, $21, $9, $12.50, $16.50.

2. A car increases its speed steadily over 6 seconds as shown in Table 13.3.

Table 13.3

Time (s)	0	1	2	3	4	5	6
Speed (km/h)	0	15	30	45	60	75	90

(a) Use a scale of 2 cm to represent 1 second on the horizontal axis and 2 cm to represent 10 km/h on the vertical axis. Draw a graph of the information in Table 13.3.

(b) Use your graph to find, (i) the speed of the car after 2.5 s, (ii) the time taken to reach a speed of 80 km/h.

3. A girl walks along a road at a speed of 100 m per minute.
   (a) Copy and complete Table 13.4.

Table 13.4

Time (min)	0	1	2	3	4	5	6
Distance (m)	0	100	200				

(b) Using a scale of 2 cm to 1 min on the horizontal axis and 2 cm to 100 m on the vertical axis, draw a graph of the information.

(c) Use your graph to find, (i) how far the girl has walked after 5.7 min, (ii) how long it takes her to walk 335 m.

4. Ribbon costs $6 for 1 metre.
   (a) Copy and complete Table 13.5.

Table 13.5

Length (m)	1	2	3	4	5	6
Cost ($)	6	12	18			

(b) Using a scale of 2 cm to 1 metre on the horizontal axis and 2 cm to $5 on the vertical axis, draw a graph of the information.

(c) Use your graph to find, (i) the cost of 3.8 m of ribbon, (ii) how much ribbon can be bought for $14.

5 A car travels 7 km on 1 litre of petrol.

(a) Copy and complete Table 13.6.

Table 13.6

Petrol (litres)	0	10	20	30	40	50
Distance (km)	0	70	140	210		

(b) Using a scale of 2 cm to 10 litres on the horizontal axis and 2 cm to 100 km on the vertical axis, draw a graph of the information.

(c) Use your graph to find, (i) the distance that the car will travel on 22 litres, (ii) how much petrol the car uses in travelling 230 km.

6 1 litre of petrol costs $1.60.

(a) Copy and complete Table 13.7.

Table 13.7

Petrol (litres)	0	10	20	30	40	50	60
Cost ($)	0	16	32	48			

(b) Using a scale of 2 cm to 10 litres on the horizontal axis and 2 cm to $20 on the vertical axis, draw a graph of this information.

(c) Use the graph to find, (i) the cost of 22 litres of petrol, (ii) how much petrol can be bought for $60.

7 Sugar costs $3.20 per kg.

(a) Copy and complete Table 13.8.

Table 13.8

Sugar (kg)	1	2	3	4	5	6
Cost ($)	3.20	6.40	9.60			

(b) Using a scale of 2 cm to 1 kg on the horizontal and 2 cm to $5 on the vertical axis, draw a graph of this information.

(c) Use your graph to find, (i) the cost of $2\frac{1}{2}$ kg of sugar, (ii) how much sugar can be bought for $12.

8 The drill of an oil well drills downwards at a rate of 7.5 m/h.

(a) Copy and complete Table 13.9.

Table 13.9

Time (h)	0	1	2	3	4	5
Cost ($)	0	−7.5	−15	−22.5		

(b) Draw the origin near the top left corner of your graph paper. Using a scale of 2 cm to represent 1 hour on the horizontal axis and 1 cm to represent 5 m on the vertical axis, draw a graph of the information.

(c) Use the graph to find, (i) how long it takes the drill to drill down through 25 m, (ii) the distance of the drill from ground level after 90 min.

9 A car travelling at 90 km/h covers 3 km in 2 min, 6 km in 4 min, 9 km in 6 min, and so on.

(a) Make a table of values showing how far the car travels in 2 min, 4 min, 6 min, 8 min, 10 min.

(b) Using a scale of 1 cm to 1 min on the horizontal axis and 1 cm to 1 km on the vertical axis, draw a graph of the information in your table.

(c) Use the graph to find, (i) how far the car travels in $3\frac{1}{2}$ min, (ii) how long it takes the car to travel 10 km.

10 A man cycles at a speed of 18 km/h.

(a) Make a table of values showing how far the man travels in $\frac{1}{2}$, 1, $1\frac{1}{2}$, 2, $2\frac{1}{2}$, 3 hours.

(b) Using a scale of 2 cm to represent 1 hour on the horizontal axis and 2 cm to represent 10 km on the vertical axis, draw a graph of the information.

(c) Use the graph to find, (i) how far the man cycles in 1.6 hours, (ii) how long it takes him to cycle 40 km.

Discontinuous graphs

Example 2

Glasses cost 90 cents each. (a) Make a table of values showing the cost of 1, 2, 3, 4, 5 glasses. (b) Draw a graph to show this information.

(a) See Table 13.10.

Table 13.10

Number of glasses	1	2	3	4	5
Cost ($)	0.90	1.80	2.70	3.60	4.50

(b) See Fig. 13.4.

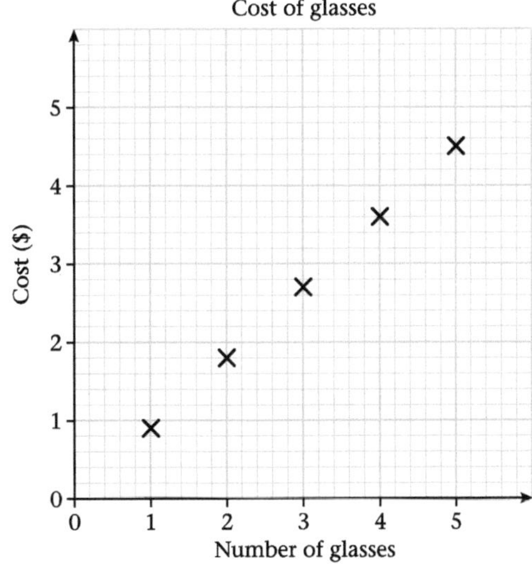

Fig. 13.4

The points in the graph in Fig. 13.4 lie in a straight line. However, we do *not* connect them. This is because it is impossible to buy a fraction of a glass such as $1\frac{1}{2}$ or 3.2. There is no part of the graph which corresponds to fractions. A graph like this is called a **discontinuous graph**.

Exercise 13b

The graphs in this exercise are discontinuous. Plot the points only.

1. Cinema tickets cost $8 each.
   (a) Copy and complete Table 13.11.

Table 13.11

No. of tickets	0	1	2	3	4	5
Cost ($)	0	8	16			

   (b) Using a scale of 2 cm to 1 ticket on the horizontal axis and 1 cm to $5 on the vertical axis, draw a graph of this information.

2. A bottle of 60 pills costs $4.80.
   (a) Copy and complete Table 13.12.

Table 13.12

Number of pills	10	20	30	40	50	60
Cost (cents)	80					480

   (b) Using a scale of 2 cm to represent 10 pills on the horizontal axis and 2 cm to represent 100 cents on the vertical axis, draw a graph to show the information.
   (c) Without doing any more calculation, plot the points representing the cost of a bottle containing 5, 15, 25, 35, 45, 55 pills.
   (d) Hence *estimate* the cost of 17 pills.

3. In Chapter 12 you learned that the sum of the angles of an n-sided polygon is $(n - 2) \times 180°$.
   (a) Use this formula to complete Table 13.13.

Table 13.13

Number of sides of polygon	3	4	5	6	7
Sum of interior angles (degrees)	180	360	540		

   (b) Using a scale of 2 cm to 1 side on the horizontal and 2 cm to 100° on the vertical axis, draw a graph to show the information in the table.

4. Car tyres cost $240 each.
   (a) Make a table of values showing the cost of 1, 2, 3, 4, 5 tyres.
   (b) Using a scale of 2 cm to represent 1 tyre on the horizontal axis and 2 cm to represent $250 on the vertical axis, draw a graph to show this information.

Choosing scales

Most of the graphs shown in this chapter are drawn to a small scale. This is to fit the sizes of the columns in the book. However, it is better to choose a big scale when drawing graphs. The scales given in Exercises 13a and 13b are all of a suitable size.

When choosing scales, first look at your table of values. For example, look at Table 13.14.

Table 13.14

Time (min)	0	10	20	30	40	50
Temperature (°C)	−8	−1	6	13	20	27

This shows that the time scale on the horizontal axis must go from 0 min to 50 min, as shown in Fig. 13.5.

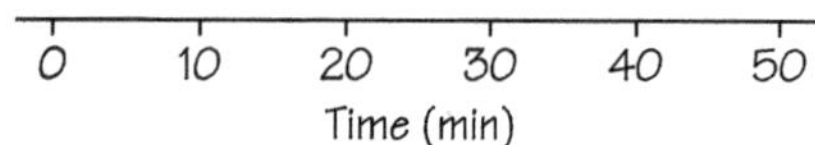

Fig. 13.5

The temperature scale on the vertical axis must go from −8 °C to 27 °C. It is better to round these values to give a range from −10 °C to 30 °C, as shown in Fig. 13.6.

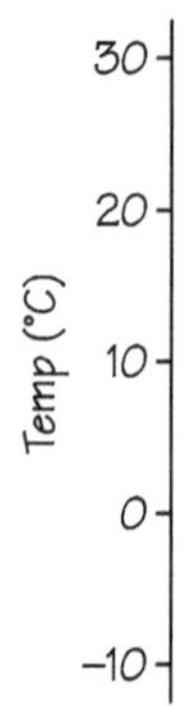

Fig. 13.6

Both axes meet at the origin. A rough sketch of the axes, such as in Fig. 13.7, can be made. Most graph paper is about 24 cm long by 18 cm wide. In this case, a scale of 2 cm to 10 units on both axes will be suitable.

Always look at the data and make a rough sketch as in Fig. 13.7. This will help you to place your graph on your graph paper.

On 2 mm graph paper, it is usual to let 2 cm represent 1, 2, 5, 10, 20, 50, 100, ... units (Fig. 13.8).

The scale that you use will depend on the data. Do not use scales in multiples of 3 or 4.

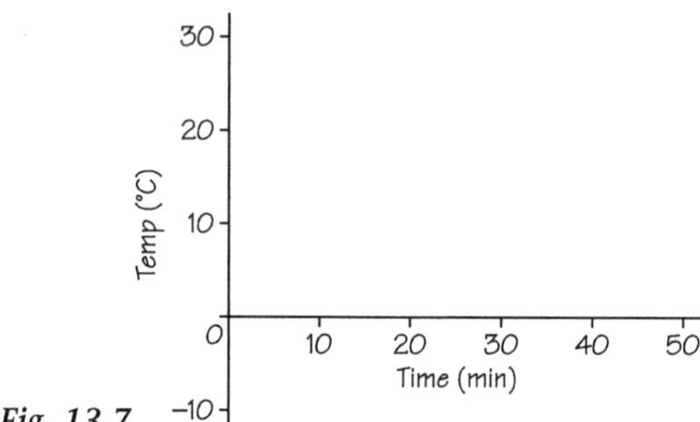

Fig. 13.7

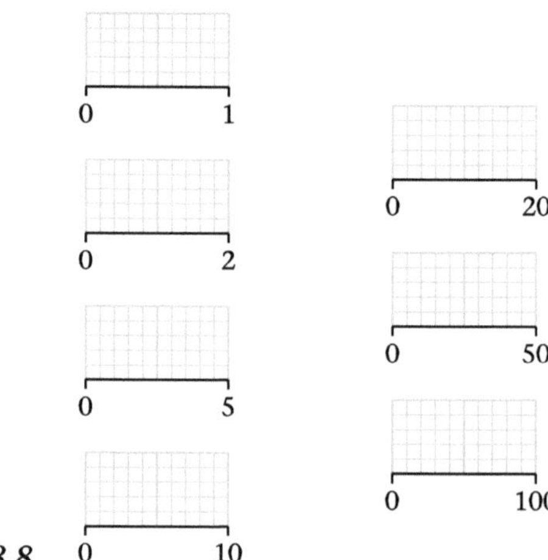

Fig. 13.8

Example 3

The labour charges for repairing a television set consist of: a standing charge of \$50 on all bills and an hourly rate of \$20 per hour.

(a) Make a table showing the total labour charges for jobs which take $\frac{1}{2}$ h, 1 h, 2 h, 3 h, 4 h.

(b) Choose a suitable scale and draw a graph of the information.

(c) Find the total labour charges for a job which takes
(i) $2\frac{1}{2}$ hours, (ii) 24 min.

(a) The total labour charges are shown in Table 13.15.

Table 13.15

Time (hours)	$\frac{1}{2}$	1	2	3	4
Standing charge (\$) Hourly rate (\$)	50 10	50 20	50 40	50 60	50 80
Labour charges (\$)	60	70	90	110	130

(b) Choosing scales:
Time goes from $\frac{1}{2}$ h to 4 h. Use a range 0 to 4 h (Fig. 13.9).

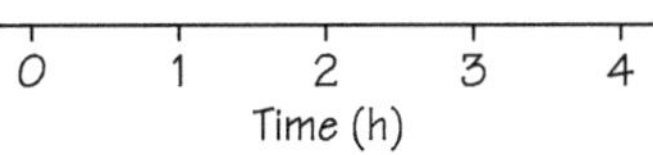

Fig. 13.9

Labour charges go from \$60 to \$130. Use a range 0 to \$150 (Fig. 13.10). Always include the origin if possible.

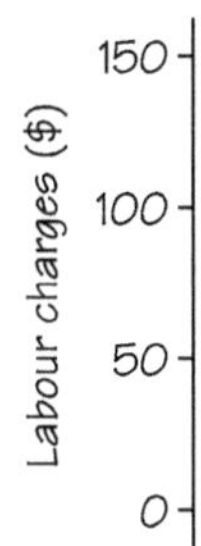

Fig. 13.10

Fig. 13.11 is a sketch of the axes.

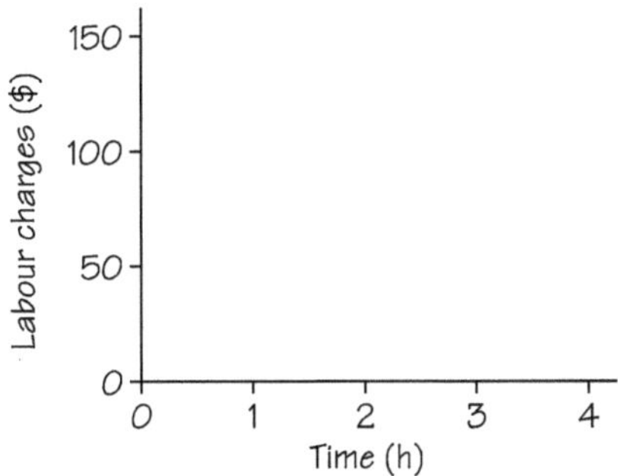

Fig. 13.11

Scales of 2 cm to 1 h on the horizontal axis and 2 cm to \$50 on the vertical axis will be suitable. The graph is given in Fig. 13.12.

(c) (i) Reading up from $2\frac{1}{2}$ h on the time axis corresponds to a charge of \$100 on the labour charges axis.

(ii) 24 min $= \frac{24}{60}$ hour $= 0.4$ h

0.4 h corresponds approximately to \$57.50. The charge for a 24-minute job is about \$57.50.

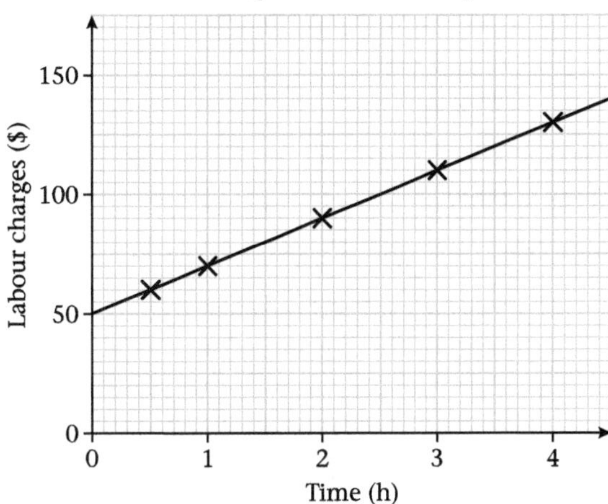

Fig. 13.12

Exercise 13c

1 A piece of meat is taken out of a freezer. Its temperature rises steadily as shown in Table 13.16.

Table 13.16

Time (min)	0	10	20	30	40	50
Temperature (°C)	−6	0	6	12	18	24

(a) Choose a suitable scale and draw a graph of this information.
(b) Use the graph to estimate, (i) the temperature of the meat after 13 min, (ii) the time taken for the meat to reach a temperature of 20 °C.

2 The unstretched length of a rubber band is 120 mm. When masses were hung on the rubber band, its total length changed as given in Table 13.17.

Table 13.17

Mass (g)	0	200	400	600	800	1000
Length (mm)	120	170	220	270	320	370

(a) Choose a suitable scale and draw a graph of the data.
(b) Use your graph to find, (i) the length of the rubber band when a mass of 250 g is hung on it, (ii) the mass which stretches the rubber band to a length of 300 mm.

3 A baby was 3.4 kg when he was born. For his first 6 weeks, his mass increased by about 0.3 kg per week.

(a) Copy and complete Table 13.18.

Table 13.18

Week number	0	1	2	3	4	5	6
Mass (kg)	3.4	3.7	4.0	4.3			

(b) Choose a suitable scale and draw a graph of this information.

(c) Approximately how many days old was the baby when his mass was 5 kg?

4 An oil company sells petrol to garages at the rate of \$840 per kilolitre. There is also a delivery charge of \$180 on all orders.

(a) Copy and complete Table 13.19.

Table 13.19

Petrol (kℓ)	1	2	3	4	5
Del. charge (\$)	180	180	180	180	180
Unit charge (\$)	840	1680	2520		
Total cost (\$)	1020	1860	2700		

(b) Choose a suitable scale and draw a graph of this information.

(c) Use the graph to find, (i) the cost of 4500 litres of petrol, (ii) how much petrol is delivered for \$3000.

5 The unit cost of window glass is \$16 per m^2. There is also a handling and cutting charge of \$4 on all orders.

(a) Copy and complete Table 13.20.

Table 13.20

Area of glass (m^2)	$\frac{1}{2}$	1	$1\frac{1}{2}$	2	$2\frac{1}{2}$
Handling/cutting (\$)	4	4	4	4	4
Unit charge (\$)	8	16	24		
Total cost (\$)	12	20	28		

(b) Choose a suitable scale and draw a graph of this information.

(c) A man buys 9 panes of glass. Each pane is 25 cm by 50 cm. (i) Calculate the total area, in m^2, of the glass that he buys. (ii) Use the graph to find out how much the glass costs.

Information from graphs

Conversion graphs

A **conversion graph** changes one set of units into another. Fig. 13.13 is a conversion graph for changing Dollars into Pounds Sterling and Pounds Sterling into Dollars.

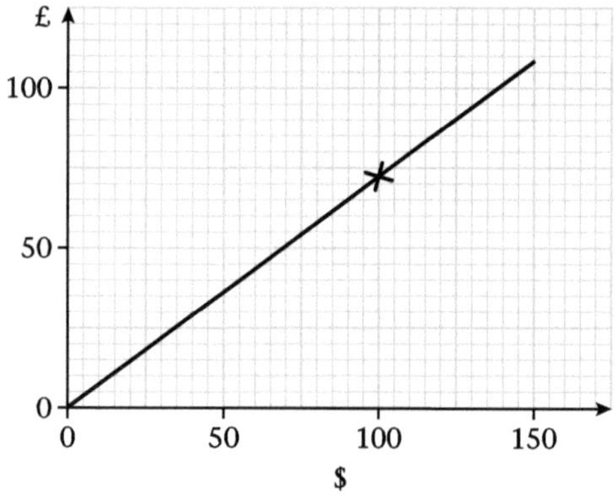

Fig. 13.13

The conversion graph was drawn using the exchange rate. The exchange rate for Fig. 13.13 is \$1 = £0.72. this is equivalent to \$100 = £72. The graph can be extended as far as we like. Fig. 13.13 can be used for values as high as \$150 and £108.

Example 4

Use the conversion graph in Fig. 13.13 to find the following:

(a) The Sterling equivalent of (i) \$50, (ii) \$140.

(b) The Dollar equivalent of (i) £25, (ii) £90.

From the graph

(a) (i) \$50 is approximately equivalent to £36.
(ii) \$140 is approximately equivalent to £100.

(b) (i) \$25 is approximately equivalent to \$35.
(ii) \$90 is approximately equivalent to \$125.

A bigger scale would give a graph which could be read more accurately.

Distance-time graphs

Example 5

Sam and Ned leave home at the same time to walk to school 3 km away. Their journeys are shown in Fig. 13.14.

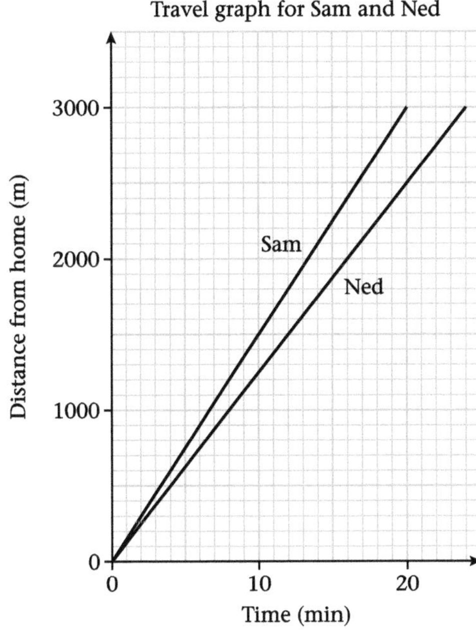

Fig. 13.14

(a) How long did Sam take to walk to school?
(b) How long did Ned take to walk to school?
(c) Find Sam's and Ned's speeds in m/min.

From the graph:
(a) Sam took 20 min.
(b) Ned took 24 min.
(c) In each case the journey was 3000 m.
Sam took 20 min to walk 3000 m.
In 1 min Sam walked $\frac{3000}{20}$ m = 150 m.
Sam's speed = 150 m/min.
Ned took 24 min to walk 3000 m.
In 1 min Ned walked $\frac{3000}{24}$ m = 125 m.
Ned's speed = 125 m/min.

A speed is a rate of change. It is the rate of change of distance with time. There is another way to find the speeds in Example 5. Fig. 13.15 is the graph of Ned's journey.

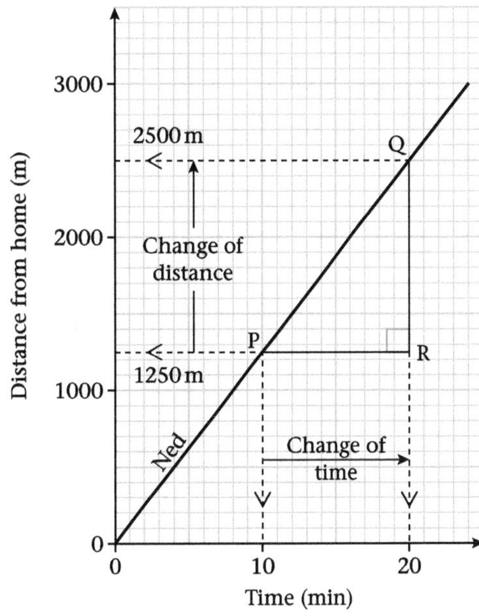

Fig. 13.15

A right-angled triangle PQR has been drawn on part of Ned's line. See Fig. 13.15. In going from P to Q, the time changes from P to R and the distance changes from R to Q.

$$\text{Ned's speed} = \frac{\text{distance travelled}}{\text{time taken}}$$
$$= \frac{\text{change of distance RQ}}{\text{change of time PR}}$$
$$= \frac{(2500 - 1250)\,\text{m}}{(20 - 10)}\text{min}$$
$$= \frac{1250\,\text{m}}{10\,\text{min}}$$
$$= 125\ \text{m/min}$$

Any right-angled triangle can be drawn. The bigger the triangle, the more accurate the result.

Exercise 13d

1. Use Fig. 13.13 to find the Sterling equivalent of the following.
(a) \$70 (b) \$105 (c) \$132.50 (d) \$17.50

2. Use Fig. 13.13 to find the Dollar equivalent of the following.
(a) £40 (b) £65 (c) £82.50 (d) £37.50

3. Refer to Fig. 13.14 to answer the following.
   (a) If Sam arrived at school just when the first lesson started, how many minutes late was Ned?
   (b) At the moment Sam arrived at school, how much further did Ned have to walk?
   (c) When Sam was halfway to school how far was Ned (i) from home, (ii) from school?
   (d) How far had Sam walked after 14 min?
   (e) How long did it take Ned to walk 1.8 km?

4. Use Fig. 13.16 to change the following marks out of 30 to percentages. Give your answers to the nearest whole per cent.

(a) 15	(b) 12	(c) 27	(d) 18
(e) 10	(f) 20	(g) 5	(h) 25
(i) 4	(j) 13	(k) 16	(l) 29
(m) $7\frac{1}{2}$	(n) $22\frac{1}{2}$	(o) $11\frac{1}{2}$	(p) $24\frac{1}{2}$

5. Use Fig. 13.16 to change the following percentages to marks out of 30. Give your answers to the nearest whole mark.

(a) 80%	(b) 20%	(c) 30%	(d) 70%
(e) 53%	(f) 47%	(g) 63%	(h) 13%

A test is marked out of 30. Fig. 13.16 is a conversion graph for changing the marks into percentages.

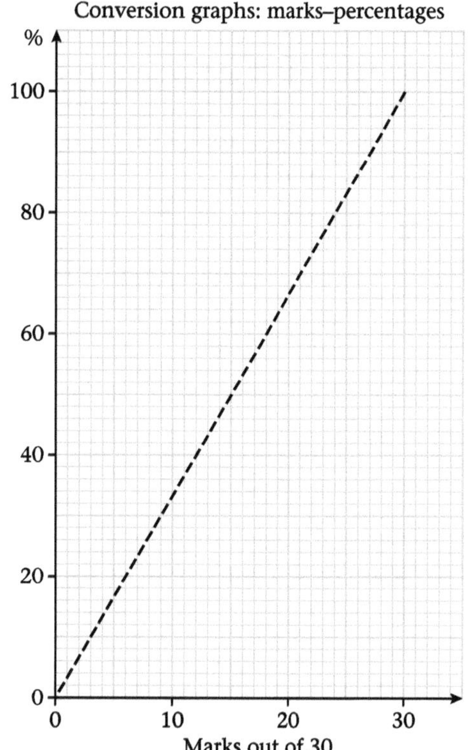

Fig. 13.16

Car A and car B leave Robley at the same time. They travel 200 km to Manga. Their journeys are shown in Fig. 13.17.

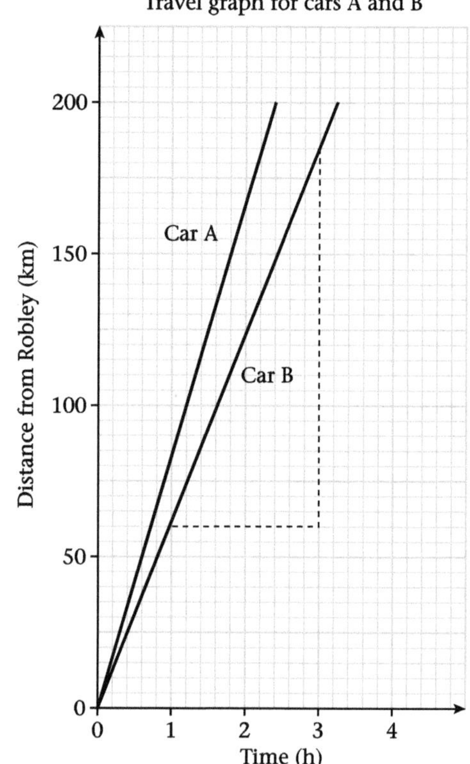

Fig. 13.17

6 Use Fig. 13.17 to answer the following.
(a) How long did car A take to get to Manga?
(b) How long did car B take to get to Manga?
(c) Find the speeds of car A and car B.
(d) Use the triangle (dotted) to check your result for car B.

7 Use Fig. 13.17 to answer the following.
(a) When car A reached Manga how far behind was car B?
(b) After 1 hour, how far was each car from Manga?
(c) After 2 hours, how far were the cars apart?
(d) How long did it take each car to travel the first 50 km?

Fig. 13.18 is a graph of the journeys of two students, Mary and Guy. They leave their university at different times and travel 6 km to hospital. Mary walks. The line ABCD shows her journey. Guy cycles. The line FD shows his journey.

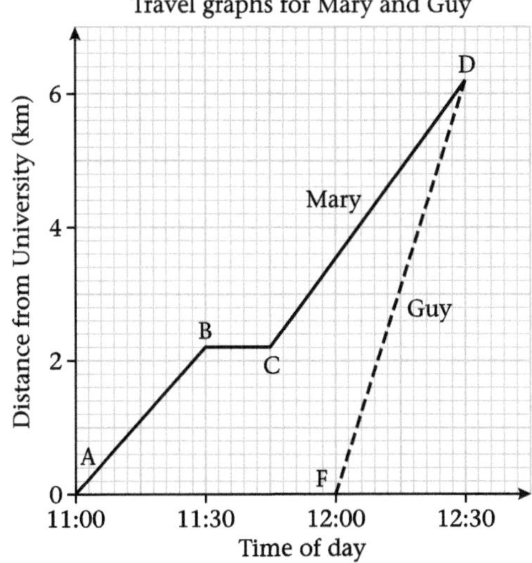

Fig. 13.18

8 Use Fig. 13.18 to answer the following.
(a) What time did Mary leave the university?
(b) What time did Mary arrive at the hospital?
(c) Mary stopped for a rest during her journey. How long did she stop for?
(d) During part AB of Mary's journey, how far did she walk?
(e) How long did part AB take?
(f) During part CD of Mary's journey, how far did she walk?
(g) How long did part CD take?
(h) How far was Mary from the hospital at 11:15?

9 Use Fig. 13.18 to answer the following.
(a) What time did Guy leave the university?
(b) What time did Guy arrive at the hospital?
(c) When Guy left the university, how far was Mary from the hospital?
(d) How far apart were the students at 12:15?

10 Use your answers to questions 8 and 9 to find the following speeds.
(a) Mary's speed between A and B.
(b) Mary's speed between C and D.
(c) Mary's average speed for the whole journey.
(d) Guy's average speed for the whole journey.

Summary

Continuous straight-line graphs may be used to represent the relation between a set of ordered pairs when the variables can take any of the values between the maximum and minimum values in the range.

Discontinuous graphs indicate that only certain values are possible in the given range.

It is important to choose a suitable scale when drawing a graph.

Practice Exercise P13.1

For each of the following functions
(a) complete the table of values
(b) sketch the graph of the function.

1 *Table 13.21*

x	2	0	−4
5			
$x+5$			

$y = x + 5$

2 *Table 13.22*

x	1	0	−2
$3x$			
−1			
$3x-1$			

$y = 3x - 1$

3 *Table 13.23*

x	−5	4	8
−4			
$x-4$			

$y = x - 4$

4 *Table 13.24*

x	−2	−1	2
1			
$-4x$			
$1-4x$			

$y = 1 - 4x$

5 *Table 13.25*

x	−7	0	5
4			
$x+4$			

$y = x + 4$

6 *Table 13.26*

x	−10	0	6
$-x$			
−4			
$-x-4$			

$y = -4 - x$

7 *Table 13.27*

x	−1	6	8
7			
$-2x$			
$7-2x$			

$y = 7 - 2x$

8 *Table 13.28*

x	0	2	4
$4x$			
−8			
$4x-8$			

$y = 4x - 8$

Practice Exercise P13.2

For each of the following functions
(a) draw up a table of values
(b) use suitable scales to represent 1 unit on each axis
(c) plot the points on graph paper
(d) draw the graph of the function.

1 $y = 3x - 2, -4 < x < 5$

2 $y = -x - 2, -5 \leqslant x \leqslant 4$

3 $f : x \rightarrow 2x + 5, -5 \leqslant x \leqslant 3$

4 $G : x \rightarrow -3x - 1, -4 \leqslant x \leqslant 3$

Practice Exercise P13.3

1 A cyclist starts at 09:00 h from O to travel a distance of 39 km. He travels at a constant speed of 12 km/h for $1\frac{1}{4}$ h, then rests for $\frac{1}{2}$ h and continues at a speed of 18 km/h.

A second cyclist leaves from a place P that is at a distance 30 km from O at 10:09 h and travels towards O at a constant speed of 14 km/h.

(a) Using a scale of 1 cm: 15 min on the *time-axis* and 1 cm : 2 km on the *distance-axis*, show these journeys on a distance–time graph.
(b) Find the distance from O when they pass each other and the time when they do.
(c) Find the time when each cyclist reaches his destination.

2 Fig. 13.19 shows the relationship between the speed and the time for a journey.

(a) Describe the information given by each straight line (OA, AB, ...) of the graph. *[Hint: State the values of any acceleration or deceleration, any constant speed, the maximum speed reached, and the time periods for each motion.]*
(b) Calculate
  (i) the total distance travelled
  (ii) the average speed for the whole journey

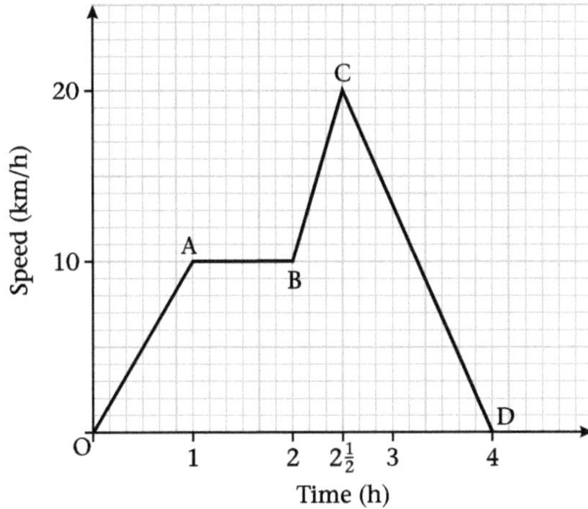

Fig. 13.19

3 *Table 13.29*

	Bds$	EC$	Ja$	TT$	Pesos	US$	UK£
Bds$ 1.00	1	1.35	30.87	3.10	5.72	0.50	0.28
EC$ 1.00	0.74	1	22.85	2.30	4.23	0.37	0.20
Ja$ 1.00	0.03	0.04	1	0.10	0.19	0.02	0.01
TT$ 1.00	0.32	0.44	9.96	1	1.85	0.16	0.09
Pesos 1.00	5.72	0.24	5.40	0.54	1	0.09	0.05
US$ 1.00	2.00	2.70	61.59	6.19	11.41	1	0.55
UK£ 1.00	3.63	4.91	112.20	11.27	20.79	1.82	1

Using the exchange rate table (Table 13.29),
(a) state suitable scales
(b) draw the conversion graphs
to calculate the amount received for the following transactions
(i) TT$120 to Ja$
(ii) EC$630 to US$.

Chapter 14

Solving triangles (1)

The right-angled triangle – sides

Pre-requisites

- properties of right-angled triangles; units of measurement; squares and square roots

To **solve** a triangle means to find the sizes of its sides and angles by calculation.

In any triangle, the sum of the angles is 180°. Thus if two angles are known, it is easy to calculate the third angle.

In any right-angled triangle, if two of the sides are known, it is possible to calculate the length of the third side. Work carefully through Exercise 14a.

Exercise 14a

1. The longest side of a right-angled triangle is called the **hypotenuse**. The hypotenuse is opposite the right angle.

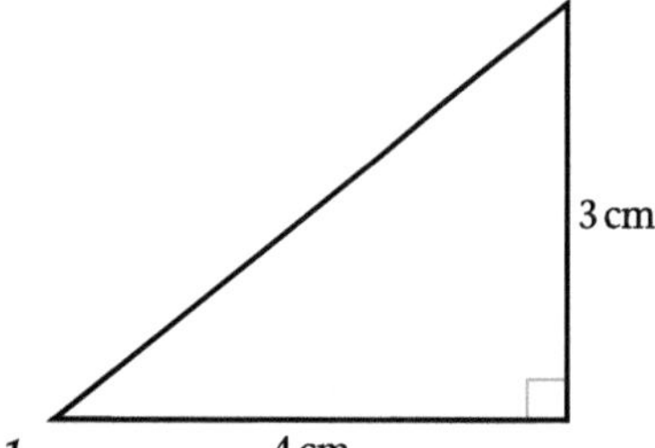

Fig. 14.1

Measure the length of the hypotenuse of the triangle in Fig. 14.1

2.

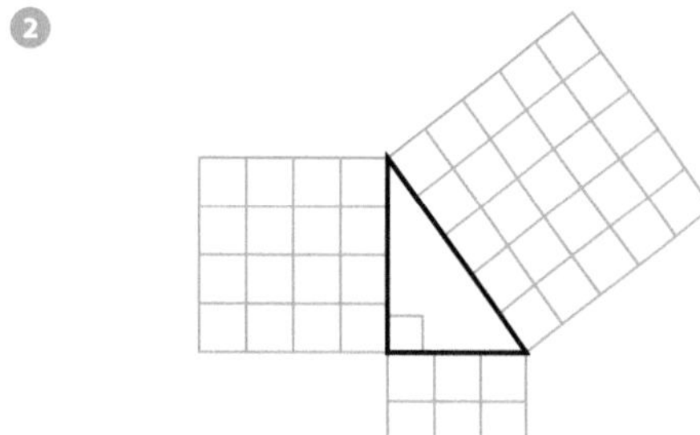
Fig. 14.2

Fig. 14.2 shows a right-angled triangle with sides of 3, 4 and 5 units. A square has been drawn on each side of the triangle. Each square is divided into small squares of area 1 unit^2.

(a) Count the number of unit^2 in the square on the hypotenuse.
(b) Count the number of unit^2 in the squares on the other two sides. Add these together.
(c) What do you notice?

3. Repeat question 2 with the triangle in Fig. 14.3.

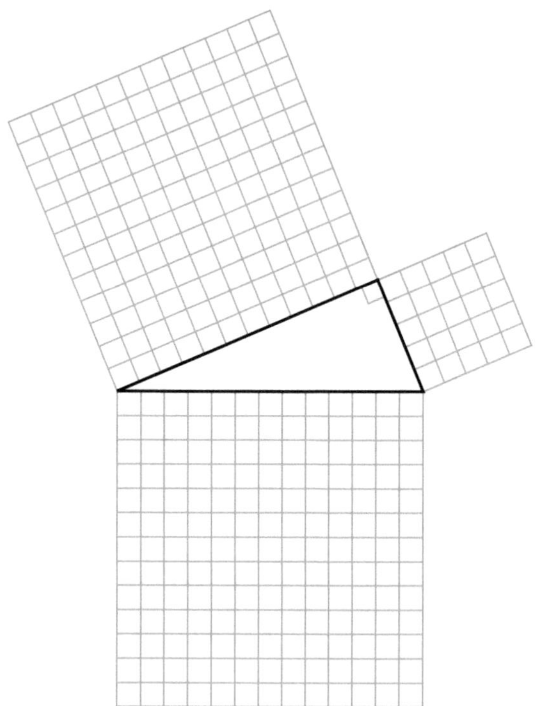
Fig. 14.3

4. (a) On a large sheet of paper, draw a right-angled triangle such that the sides containing the right angle are 8 cm and 6 cm.

(b) Measure the hypotenuse.
(c) Draw squares on the three sides of the triangle as in Figs. 14.2 and 14.3.
(d) Divide the squares into 1 cm^2 small squares. Count the small squares as in question 2.

5 Repeat question 4 for a right-angled triangle such that the sides containing the right angle are 8 cm and 15 cm.

Pythagoras' theorem

From the work of Exercise 14a, it can be seen that the square on the hypotenuse is equal in area to the sum of the squares on the other two sides.

This rule is very famous. It is called **Pythagoras' theorem**. Pythagoras was a Greek philosopher who lived about 2500 years ago. The theorem was proved by Pythagoras and his friends at that time. However, the rule was used long before then and can be written in different ways. For example, in Fig. 14.4, $AB^2 = BC^2 + AC^2$.

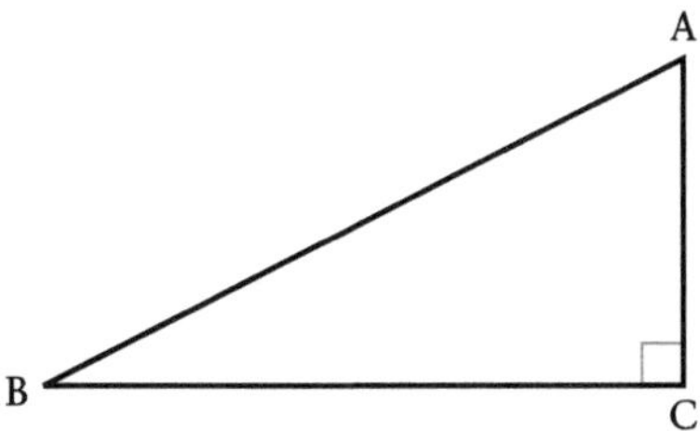

Fig. 14.4

AB^2 means the square on the line segment AB, and the measurement of the area of the square is the distance AB squared.

In Fig. 14.5, X, Y, and Z are squares drawn on the sides AB, BC, CA of the right-angled triangle ABC. Pythagoras' rule can be written as

Area of X = Area of Y + Area of Z

Look at triangle ABC in Fig. 14.5. c is the hypotenuse of $\triangle ABC$ and a and b are its other two sides. See Fig. 14.6. Note that the sides a, b, c are opposite the vertices A, B, C in that order.

For any right-angled triangle with hypotenuse c and other sides a and b, $c^2 = a^2 + b^2$.

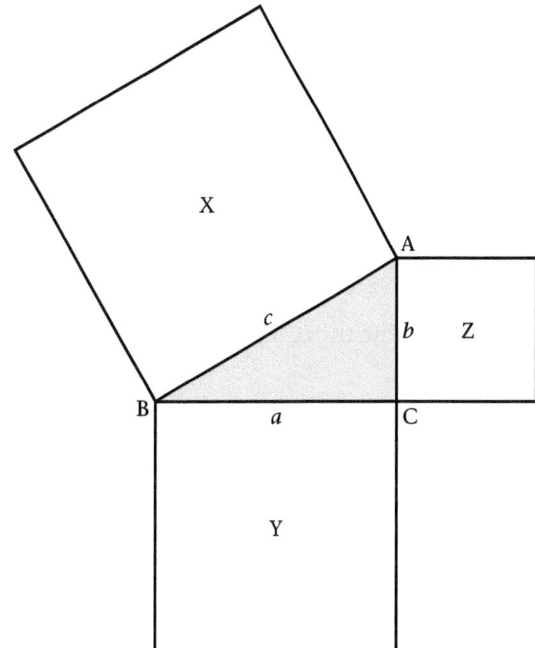

Fig. 14.5

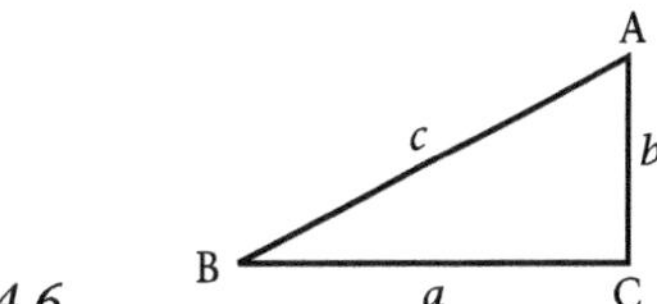

Fig. 14.6

Example 1

Given the data of Fig. 14.7, calculate the value of c.

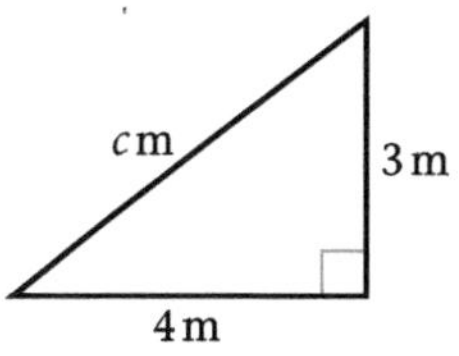

c m
3 m
4 m

Fig. 14.7

Using Pythagoras' theorem.

$c^2 = 3^2 + 4^2$
$= 9 + 16$
$c^2 = 25$
$c = \sqrt{25} = 5$

The hypotenuse is 5 m long.

Example 2

Calculate the length of the third side of the triangle in Fig. 14.8, on page 122.

Using Pythagoras' theorem,

$13^2 = a^2 + 5^2$
$169 = a^2 + 25$

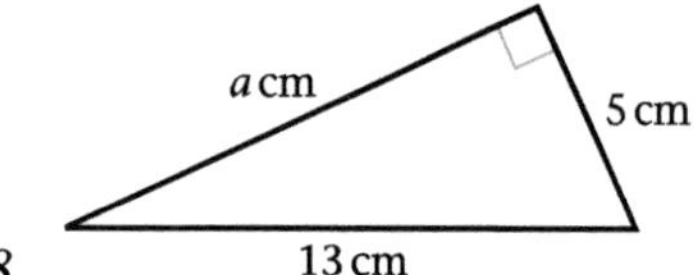

Fig. 14.8

Subtract 25 from both sides.

$$169 - 25 = a^2$$
$$a^2 = 144$$
$$a = \sqrt{144} = 12$$

The length of the third side of the triangle is 12 cm.

Example 3

In Fig. 14.9, calculate the length of PS.

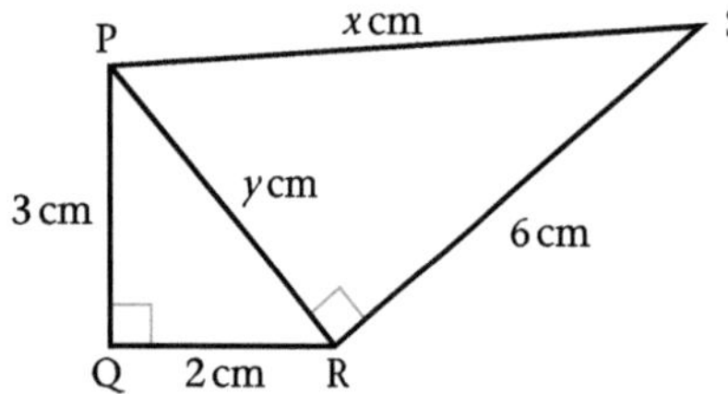

Fig. 14.9

PS is in right-angled triangle PRS.
Let PR be y cm.
In $\triangle$PQR, $y^2 = 3^2 + 2^2$
$= 9 + 4$
$= 13$

Let PS be x cm.
In $\triangle$PRS, $x^2 = y^2 + 6^2$
$= 13 + 36$
$= 49$
$x = \sqrt{49} = 7$
PS = 7 cm

Example 4

In Fig. 14.10, calculate the length of AD.

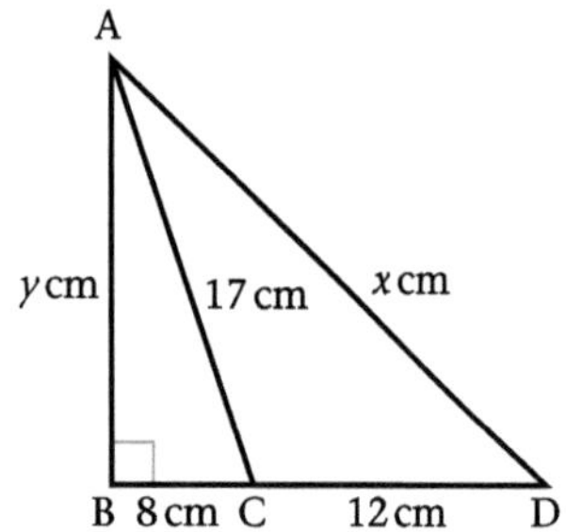

Fig. 14.10

AD is in right-angled triangle ABD.
Let AB be y cm.
In $\triangle$ABC, $y^2 = 17^2 - 8^2 = 289 - 64 = 225$
Let AD be x cm.
In $\triangle$ABD, $x^2 = y^2 + (8 + 12)^2$
$= 225 + 20^2 = 225 + 400$
$= 625$
$x = \sqrt{625} = 25$
AD = 25 cm

In Examples 3 and 4, notice that the sides labelled y are *intermediate* sides. We do not have to find their values. When y^2 has been found, there is no need to find y, since y^2 is used in the second part of the working.

Exercise 14b

1. ABC is a triangle in which $\widehat{B} = 90°$. In each of the following, draw and label a sketch then calculate the length of the third side of the triangle.
   (a) AB = 6 m, BC = 8 m
   (b) AB = 9 cm, BC = 12 cm
   (c) AB = 5 m, BC = 12 m
   (d) AB = 15 cm, BC = 8 cm
   (e) AC = 25 m, BC = 24 m
   (f) AC = 25 cm, BC = 20 cm
   (g) AC = 100 m, AB = 80 m
2. In triangle PQR, angle PQR is a right angle. If PQ = 20 cm and QR =21 cm, what is the area of the square drawn on side PR?
3. Find the value of x in each of the figures in Fig. 14.11. It will be necessary to find a value for y^2 before finding x.

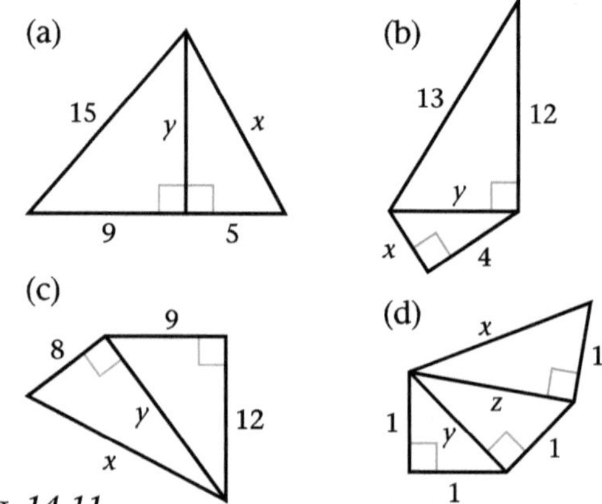

Fig. 14.11

Pythagorean triples

The sides of the triangle in Example 1 are 3 m, 4 m and 5 m. We call this a 3, 4, 5 triangle. The numbers (3, 4, 5) are called a **Pythagorean triple.** A Pythagorean triple is a set of 3 whole numbers which give the lengths of the sides of right-angled triangles. (5, 12, 13), (7, 24, 25), (8, 15, 17) are some other common Pythagorean triples. You discovered these and others in Exercise 14b.

(6, 8, 10) and (30, 40, 50) are multiples of (3, 4, 5). They are also Pythagorean triples.

Example 5

Which of the following is a Pythagorean triple?

(a) (33, 56, 65) (b) (15, 30, 35)

(a)
$$33^2 + 56^2 = 1089 + 3\,136$$
$$= 4225$$
$$65^2 = 4225$$

Hence, $33^2 + 56^2 = 65^2$

(33, 56, 65) is a Pythagorean triple.

(b)
$$15^2 + 30^2 = 225 + 900$$
$$= 1125$$
$$35^2 = 1225$$

Hence $15^2 + 30^2 \neq 35^2$

(15, 30, 35) is *not* a Pythagorean triple.

The results of Example 5 show that a triangle with sides of length 33, 56 and 65 units will be right-angled, whereas one with sides of length 15, 30 and 35 will *not* be right-angled. This method, therefore, can be used as a test for right-angled triangles.

Exercise 14c

1. Write down four multiples of each of the following Pythagorean triples.
   (a) (3, 4, 5) (b) (5, 12, 13)
   (c) (7, 24, 25) (d) (8, 15, 17)

2. Find out which of the following are Pythagorean triples.
   (a) (20, 21, 29) (b) (15, 22, 27)
   (c) (28, 45, 53) (d) (11, 60, 61)

3. Try to complete the following pattern of Pythagorean triples.
   (3, 4, 5) → $3^2 = 4 + 5$
   (5, 12, 13) → $5^2 = 12 + 13$
   (7, 24, 25) → $7^2 = 24 + 25$
   (9, …, …) → $9^2 =$
   (11, …, …) → $11^2 =$
   Hint: notice that the difference between the last two numbers of each triple is 1.

4. Try to complete the following pattern of Pythagorean triples.
   (6, 8, 10) → $\frac{1}{2}$ of $6^2 = 8 + 10$
   (8, 15, 17) → $\frac{1}{2}$ of $8^2 = 15 + 17$
   (10, 24, 26) → $\frac{1}{2}$ of $10^2 = 24 + 26$
   (12, …, …) → $\frac{1}{2}$ of $12^2 =$
   (14, …, …) → $\frac{1}{2}$ of $14^2 =$
   Hint: notice that the difference between the last two terms of each triple is 2.

5. Try to extend the patterns of questions 3 and 4 for five more terms.

Everyday use of Pythagoras' theorem

So far the exercises in this chapter have been arranged so that when a square root of a number was needed, it could be found exactly. However, this does not often happen with numbers in everyday situations. More often we have to find squares and square roots from tables.

Using tables of squares

The table of squares on page 224 can be used to find the squares of 3-digit numbers.

Example 6

Use the table of squares to find 4.16^2.

The digits 4.1 appear in the left-hand column of the table of squares. 6 is the third digit. Look for the column headed 6. Find the number which is across from 4.1 and under 6. See Fig. 14.12 on page 124.

The number is 17.31.
Hence $4.16^2 = 17.31$.
This result is correct to 4 s.f.

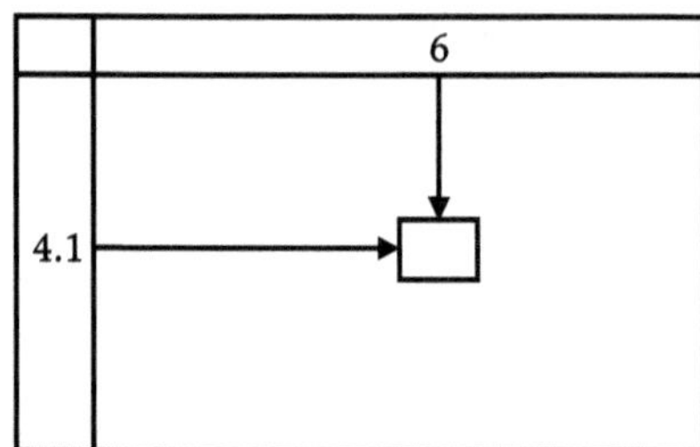

Fig. 14.12

The table of squares on page 224 shows the squares of numbers from 1 to 10. Sometimes you may need to find the square of a number greater than 10. Read Example 7 carefully.

Example 7

Use the table of squares to find (a) 19^2, (b) 190^2.

(a) $19 = 1.9 \times 10$
$19^2 = (1.9 \times 10)^2$
$= 1.9^2 \times 10^2$

From the table of squares,
$1.9^2 = 3.61$
Hence $19^2 = 3.61 \times 100$
$= 361$

(b) $190 = 1.9 \times 100$
$190^2 = (1.9 \times 100)^2$
$= 1.9^2 \times 100^2$
$= 3.61 \times 10\,000$
$= 36\,100$

In Example 7, notice that
$1.9^2 = 3.61$
$19.0^2 = 361$
$190.0^2 = 36\,100$

When a number is multiplied by increasing powers of 10, its square is multiplied by increasing powers of 100.

Exercise 14d

Use the table of squares on page 224 in this exercise.

1. Find the values of the following.
   (a) 1.4^2 (b) 2.3^2 (c) 6.8^2
   (d) 7.2^2 (e) 4.9^2 (f) 8.6^2
   (g) 5.63^2 (h) 9.08^2 (i) 3.15^2
   (j) 1.88^2 (k) 5.71^2 (l) 4.54^2

2. Find the values of the following.
   (a) 18^2 (b) 31^2
   (c) 32^2 (d) 15^2
   (e) 29^2 (f) 44^2
   (g) 70.5^2 (h) 20.6^2
   (i) 62.7^2 (j) 59.8^2
   (k) 81.3^2 (l) 90.9^2

3. Round off the following to 3 s.f. and then find the approximate square of each number.
   (a) 1.733 (b) 2.808 (c) 78.65
   (d) 52.14 (e) 96.47 (f) 49.57
   (g) 632.6 (h) 805.3 (i) 303.6

4. Find the values of the following.
   (a) 130^2 (b) 410^2
   (c) 870^2 (d) 504^2
   (e) 2700^2 (f) 8350^2

5. Look at the following pattern:
   $1.5^2 = 2.25 = 1 \times 2 + 0.25$
   $2.5^2 = 6.25 = 2 \times 3 + 0.25$
   $3.5^2 = 12.25 = 3 \times 4 + 0.25$
   Find out if the pattern continues in the same way.

Square root tables

Tables of square roots are given on pages 225 and 226. Notice that there are two tables.

Example 8

Use square root tables to find (a) $\sqrt{5.7}$ (b) $\sqrt{57}$.

(a) 5.7 lies between 1 and 9.99. Use the first table (page 225).
$\sqrt{5.7} = 2.39$

(b) 57 lies between 10 and 9.99. Use the second table (page 226).
$\sqrt{57} = 7.55$

The square root tables gives results rounded to 3 s.f. For example $\sqrt{5.7} = 2.39$ to 3 s.f. However, if 2.39^2 is worked out, the result is 5.712 1, *not* 5.7. Nevertheless, 3 significant figures are accurate enough for most purposes.

Example 9

Use square root tables to find (a) $\sqrt{875}$ (b) $\sqrt{3827}$.

(a) $875 = 8.75 \times 100$

$\sqrt{875} = \sqrt{8.75 \times 100}$

$= \sqrt{8.75} \times \sqrt{100}$

$= \sqrt{8.75} \times 10$

From the first table $\sqrt{8.75} = 2.96$

Hence $\sqrt{875} = 2.96 \times 10 = 29.6$ to 3 s.f.

(b) $3827 = 3830$ to 3 s.f.

$= 38.3 \times 100$

$\sqrt{3827} = \sqrt{38.3 \times 100}$

$= \sqrt{38.3} \times \sqrt{100}$

$= \sqrt{38.3} \times 10$

From the second table $\sqrt{38.3} = 6.19$

Hence $\sqrt{3827} = 6.19 \times 10 = 61.9$ to 3 s.f.

Notice again that the final results are not exact. For example, $61.9^2 = 3831.61$, *not* 3827.

Exercise 14e

Use the square root tables on pages 225 and 226 in this exercise.

1. Find the square roots of the following.
   (a) 9 (b) 90 (c) 2.8 (d) 28
   (e) 4.7 (f) 47 (g) 5.04 (h) 50.4
   (i) 36.2 (j) 3.62 (k) 25.7 (l) 2.57

2. Find the square root of the following.
   (a) 7 (b) 70 (c) 700 (d) 7000
   (e) 2.9 (f) 29 (g) 290 (h) 2900
   (i) 38.2 (j) 382 (k) 3820 (l) 38 200
   (m) 10 (n) 100 (o) 1000 (p) 10 000

3. Round off the following to 3 s.f. Then find their approximate square roots.
   (a) 9.286 (b) 78.23 (c) 463.8
   (d) 59.03 (e) 5.806 (f) 5003
   (g) 500.3 (h) 63 945 (i) 1982

4. Find out if $\sqrt{10}$ is a good approximation for π.

5. (a) Use square root tables to find m if $m = \sqrt{40}$.
   (b) Using the value of m found in part (a), find the value of m^2 from the table of squares.
   (c) What do you notice? Explain.

Use of the calculator

One advantage of the calculator is the speed with which accurate calculations are carried out.

Exercise 14f

Using the calculator check the accuracy of the results for Exercise 14d, question 1.

Approximation using the calculator

In question 3, Exercise 14d you were required to round off to 3 s.f. and then find the approximate square of each number. Rounding off would not be necessary when using the calculator.

Example 10

Find the values of the following giving your answer to 3 s.f.

(a) 1.733^2 (b) 2.08^2 (c) 3.5^2

(a) 3.00 (b) 4.33 (c) 12.3

The calculator shows more digits and you must follow the rules for approximation.

Example 11

Use the calculator to find the square root of

(a) 5.7 (b) 875 (c) 39.02

(a) $\sqrt{5.7} = 2.39$ to 3 s.f.

(b) $\sqrt{875} = 29.6$ to 3 s.f.

(c) $\sqrt{39.02} = 6.25$ to 3 s.f.

Exercise 14g

Use the calculator in this exercise. Give your answers to 3 s.f.

1. Find the square roots of the the following.
   (a) 5 (b) 24 (c) 3.9
   (d) 26.1 (e) 4.8 (f) 5.04
   (g) 22.9 (h) 17.2 (i) 502
   (j) 0.258 (k) 9.1 (l) 3.94

2. (a) Find the square root of 3.9 to 3 s.f.
   (b) Multiply this value by 10.
   (c) Find the square of the result in (b).
   (d) Write this value as the square of a number, n.
   (e) Investigate the range of the values of n.

Group work

Find the square root of the numbers in Exercise 14e, question 3. Compare your answers with the answers in Exercise 14e, question 3.

Investigate the range of values of m for which $\sqrt{m} = 1.84$

Example 12

Fig. 14.13 shows a ladder leaning against a wall. The ladder is 7.3 m long and the foot of the ladder is 1.8 m from the wall. Find how far up the wall the ladder reaches.

Draw a sketch of the right-angled triangle which contains the ladder (Fig. 14.14).

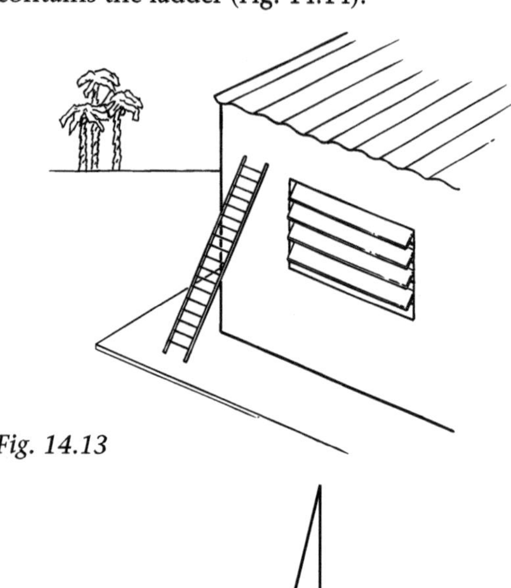

Fig. 14.13

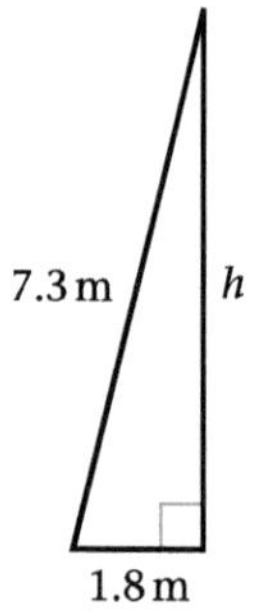

Fig. 14.14

Either

(a) By calculation, using Pythagoras' theorem

$$h^2 = 7.3^2 - 1.8^2$$
$$= 53.3 - 3.24 \text{ (from squares table)}$$
$$= 50.06$$
$$h = \sqrt{50.06} \simeq \sqrt{50.1}$$
$$= 7.08$$

The ladder reaches about 7.1 m up the wall.

or

(b) By scale drawing

Fig. 14.15, is a scale drawing of the data on a scale 1 cm to 1 m.

Measure the distance h on Fig. 14.15.

From the drawing, $h = 7.1$ cm

Height of ladder up the wall $\simeq 7.1 \times 1$ m
$\simeq 7.1$ m

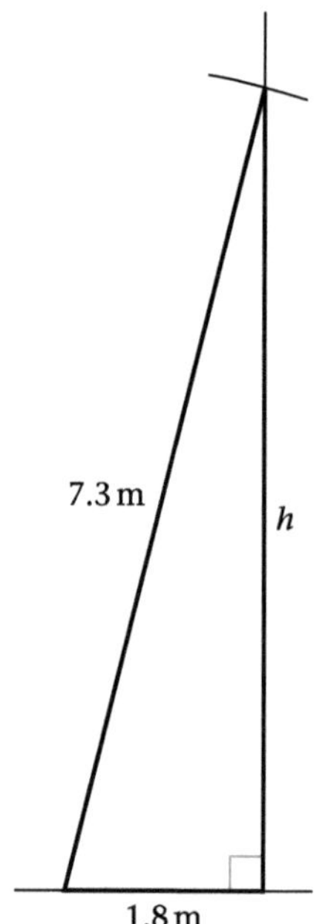

Fig. 14.15
Ladder and wall,
Scale: 1 cm to 1 m

Exercise 14h

In each question, sketch the right-angled triangle which contains the unknown. Either use Pythagoras' theorem or make a scale drawing to solve the triangle.

1. A pencil which has been sharpened at each end just fits along the diagonal of the base of a box. See Fig. 14.16.

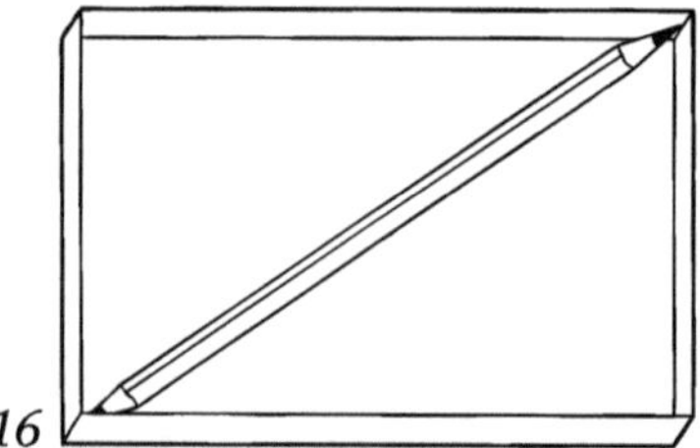

Fig. 14.16

If the box measures 14 cm by 8 cm, find the length of the pencil.

2 A telegraph pole is supported by a wire as shown in Fig. 14.17.

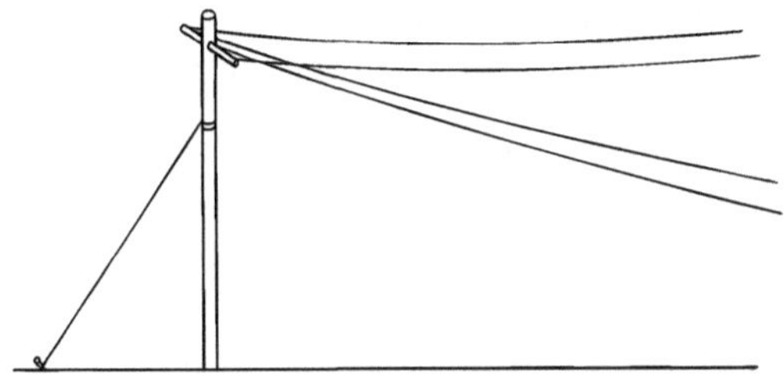

Fig. 14.17

The wire is attached to the pole 6 m above the ground and to a point on the ground 2.5 m from the foot of the pole. Calculate the total length of wire needed if an extra 0.8 m of wire is needed for the attachments.

3 Fig. 14.18 shows a straight pipe which carries water from a reservoir at R to a tap at T.

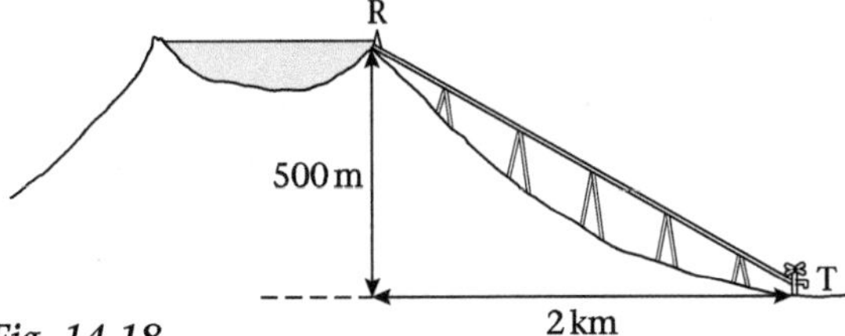

Fig. 14.18

R and T are 2 km apart horizontally and R is 500 m above the level of T. Find the length of the pipe.

4 Find the length of the longest straight line which can be drawn on a rectangular blackboard which measures 2.2 m by 1.2 m.

5 A plane flies northwards for 430 km. It then flies eastwards for 380 km. How far is it from its starting point? (Neglect its height above the ground.)

6 Fig. 14.19 shows a simple bridge over a ditch.

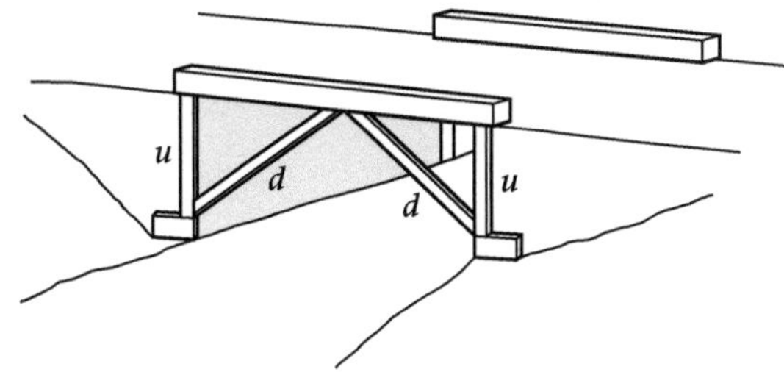

Fig. 14.19

The bridge is supported by uprights, u, and diagonals, d. Find the length of a diagonal support, if the upright supports are 4 m long and the bridge is 20 m long.

7 A ladder 7m long leans against a wall as in Fig. 14.20

Fig. 14.20

Its foot is 2 m from the wall. Calculate how far up the wall the ladder reaches.

8 A rectangular window frame measures 72 cm by 1.32 m. Calculate the length of a wooden strip that is tacked along the diagonal of the frame. Give your answer correct to 3 significant figures.

9 A graph page measures 24 cm by 15 cm. Calculate the length of the longest straight line that can be drawn on the graph page.

10 The pages of a book are each of length 16.5 cm and width 9.5 cm. Calculate the length of the longest straight line that can be drawn across two consecutive pages seen together. Give your answer correct to 3 significant figures.

Summary

To **solve** a triangle is to find the size of the angles and the length of the sides of the triangle.

Pythagoras' theorem can be used to find the length of the sides of a right-angled triangle. Note that Pythagoras' theorem only solves for the *sides* of the triangle.

Pythagoras' theorem states that in any right-angled triangle the square on the hypotenuse (the longest side) equals the sum of the squares on the other two sides.

A **Pythagorean triple** (a, b, c) is a set of numbers which are equivalent to the lengths of the sides of a right-angled triangle ABC.

Using square and square root tables and the calculator for numbers greater than 1 simplifies the calculations. The calculator has the advantage of speed for calculations.

Practice Exercise P14.1

1. For each set of numbers, find if they could be the lengths in cm, of sides of a right-angled triangle:

   (a) 5, 12, 13 (b) 6, 8, 10
   (c) 8, 10, 14 (d) 7, 24, 25
   (e) 8, 17, 25 (f) 9, 12, 15

2. Three sides of a triangle are of lengths 20 cm, 21 cm and 29 cm. Show that the triangle is right-angled.

3. Triangle ABC is an isosceles triangle If one of the equal angles measures 45°, how else can triangle ABC be described. Give a reason.

4. In the right-angled triangle PQR, the length of the hypotenuse PQ, is 24 cm and QR is 16 cm. Calculate the length of PR giving your answer correct to 1 decimal place.

5. A pencil case is 19 cm long and 7 cm wide. Calculate the length of the longest pencil that will fit flat, in the case. Give your answer correct to 3 significant figures.

6. A pane of glass measures 64 cm by 45 cm. Two strips of sticking tape are pasted along the diagonals of the pane of glass for protection during a hurricane. Calculate to the nearest centimetre, the length of one strip of tape.

7. A rectangular metal gate measuring 1.5 m by 1.2 m is strengthened by a bar of metal nailed across a diagonal of the gate. Calculate the length of this piece of metal.

8. Calculate the length of the line of symmetry along the diagonal of a square of side 15 cm. Give your answer correct to one decimal place.

9. The area beneath a set of stairs is a right-angled triangle of height 4 m and base 5.4 m. Fig.14.21.

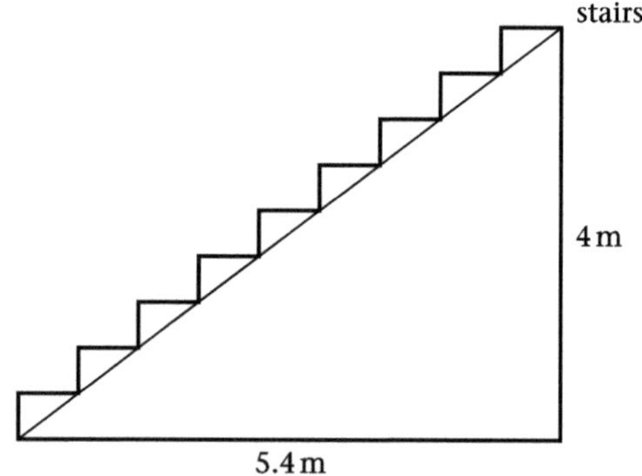

Fig.14.21

Calculate the length of the staircase. Give your answer correct to 2 significant figures.

Chapter 15
Geometrical transformations
Congruencies

Pre-requisites
- geometrical constructions; the Cartesian plane

Congruency

Look at the patterns in Fig. 15.1.

(a)

(b)

(c)

(d)

Fig. 15.1 (a) brick wall (b) wall pattern (c) cloth pattern (d) print

The patterns in Fig. 15.1 all have something in common. They are made by taking a **basic shape** and repeating it to build up the pattern.

Look at the patterns in Fig. 15.2.

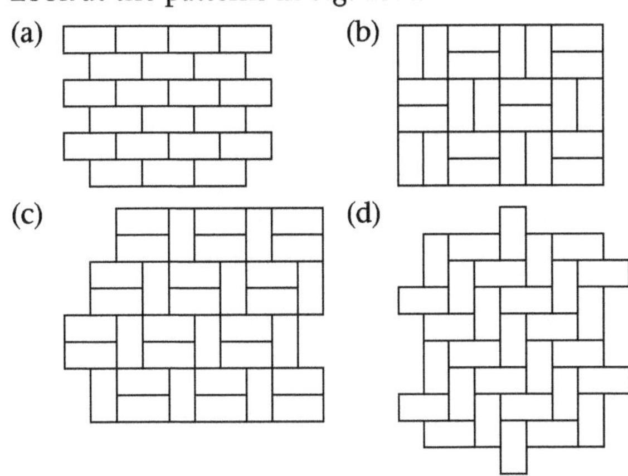

Fig. 15.2

The basic shape which makes each pattern is a 2×1 rectangle. The patterns appear different because the rectangles have been arranged in different ways.

Exercise 15a

You will need graph paper and a ruler for this exercise.

1. Using a rectangle 2 cm by 1 cm as your basic shape, make repeated patterns of your own design on your graph paper (Fig. 15.2.).

2 (a) Name the basic shapes which make the patterns in Fig. 15.3.
  (b) Copy each pattern on to graph paper. Draw more shapes until each pattern is about 8 cm wide and 6 cm long.

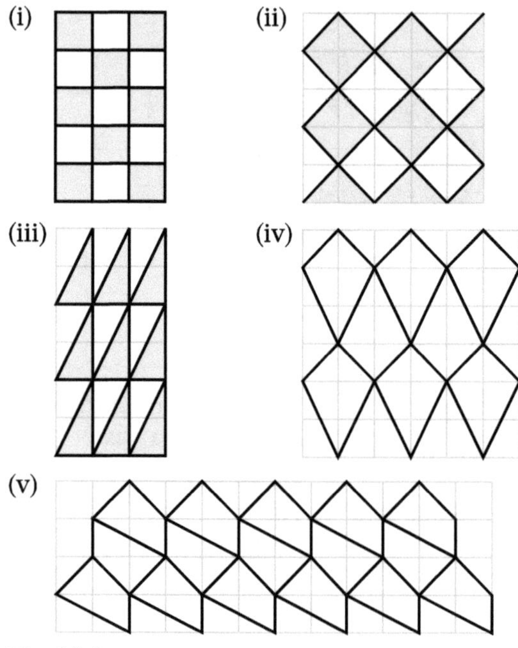

Fig. 15.3

When the position or dimensions (or both) of a shape changes, we say that it is **transformed**. The **image** of a shape is the figure which results after transformation (Fig. 15.4).

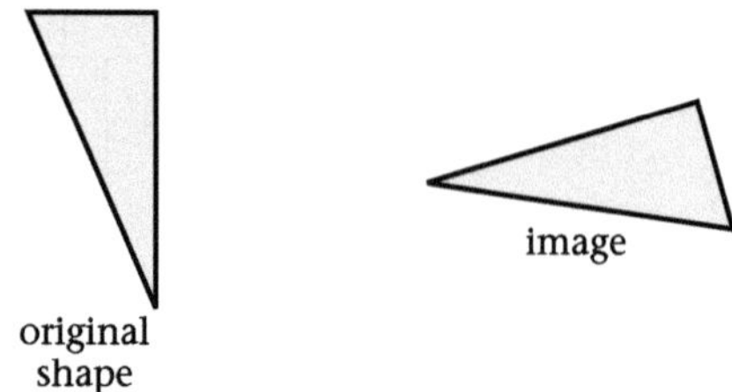

Fig. 15.4

The patterns in Exercise 15a were made by building up images of basic shapes on a plane surface. In every case, each image has the same dimensions as the original given shape. Transformations of this kind are called **congruencies**. Two shapes are congruent if their corresponding dimensions are identical. There are three basic congruencies: translations, reflections and rotations. The size of the image remains the same when the shape is transformed through translation, reflection or rotation.

Translation

A **translation** is a movement in a straight line. Fig. 15.5 shows the letter p being given translations of 1 cm steps across the page.

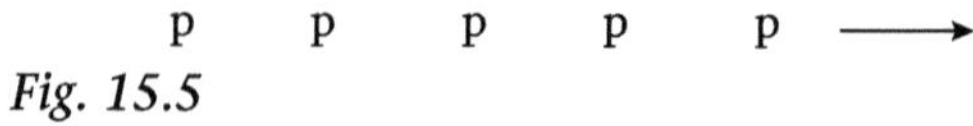

Fig. 15.5

Translations may be in any direction (Fig. 15.6).

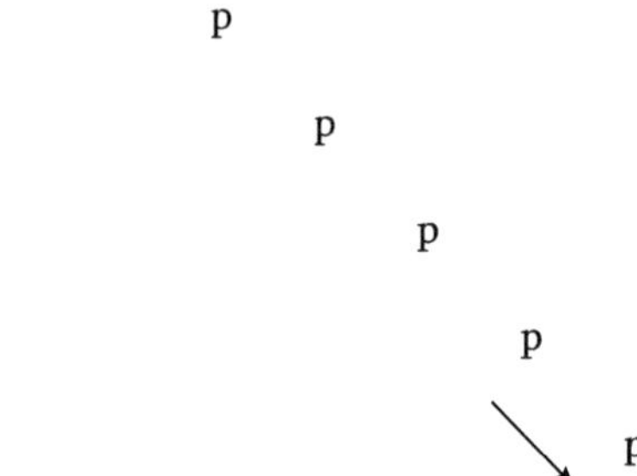

Fig. 15.6

Two or more translations of a basic shape may give a pattern which fills the plane (Fig. 15.7).

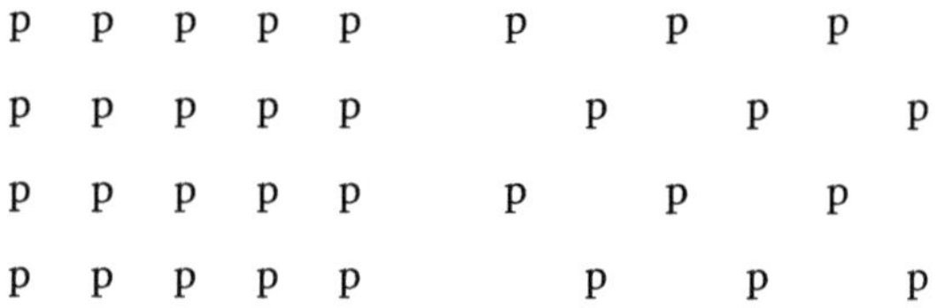

p p p p p   p   p   p
p p p p p   p   p   p
p p p p p   p   p   p
p p p p p   p   p   p

(a)   (b)

Fig. 15.7

Exercise 15b

1 Which of the patterns in Figs. 15.1, 15.2, 15.3 are translation patterns?

2 Copy each pattern in Fig. 15.8. Use translation to draw at least 6 more basic shapes on each pattern. (The direction of translation is shown by the arrows.)

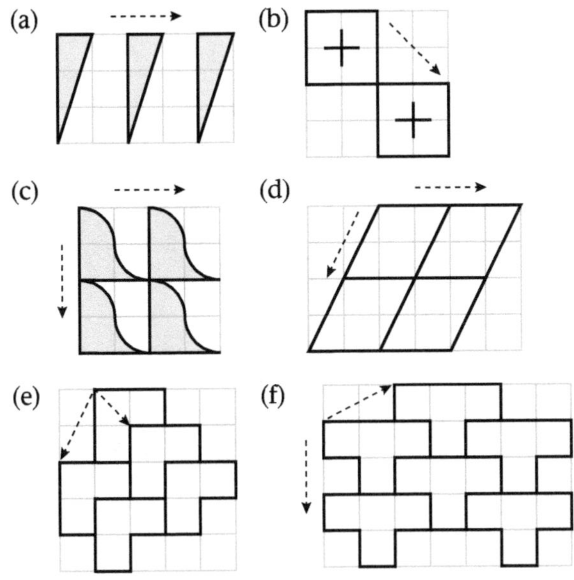

Fig. 15.8

3 Fig. 15.9, shows △ABC drawn on the Cartesian plane.

(a) State the coordinates of A, B and C.

(b) Draw the new position of △ABC when the triangle is translated 4 units to the left. Find the coordinates of the images of A, B and C. What do you notice?

(c) △ABC is translated so that the image of C is the point (−3, −2). Find the images of A and B.

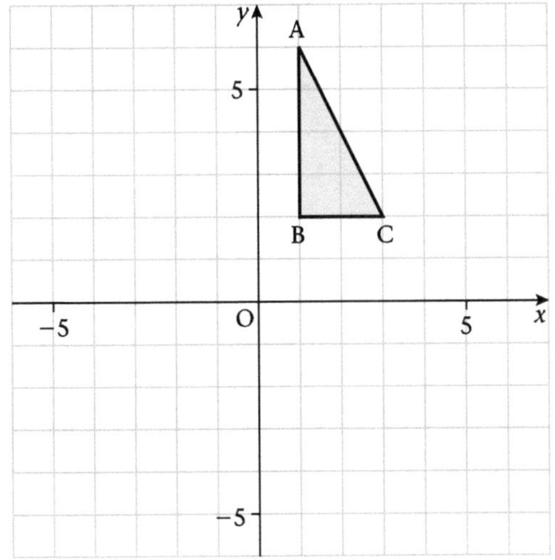

Fig. 15.9

Note that in a translation every point in a shape moves through the same distance in the same direction.

Translation vectors

The movement of each shape in Fig. 15.8 (b)−(e) was in two directions across the page and down the page (to the right and down).

In Fig. 15.10 △ABC is translated to △PQR.

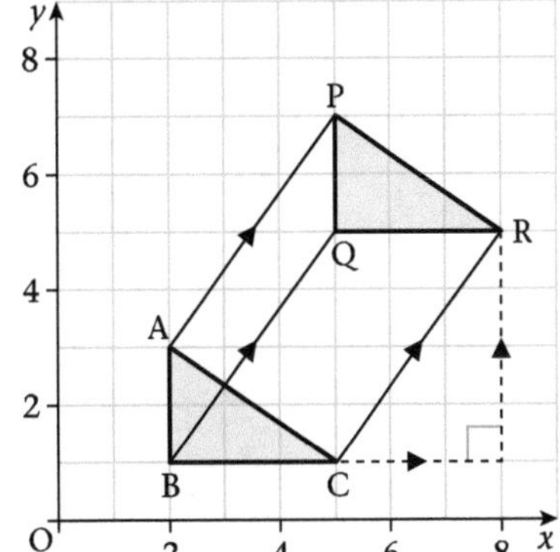

Fig. 15.10

Point A moves to point P. This movement can be written as $\overrightarrow{AP}$. $\overrightarrow{AP}$ is a **translation vector**. A **vector** is any quantity which has direction as well as size. The other translation vectors in Fig. 15.10 are $\overrightarrow{BQ}$ and $\overrightarrow{CR}$.

$\overrightarrow{BQ}$ and $\overrightarrow{CR}$ have the same size and direction as $\overrightarrow{AP}$.

Hence $\overrightarrow{AP} = \overrightarrow{BQ} = \overrightarrow{CR}$

Any one of these vectors describes the translation that takes △ABC to the position shown by △PQR.

In Fig. 15.10 the broken lines show that the vector $\overrightarrow{CR}$ is equivalent to a movement of 3 units to the right followed by a movement of 4 units upwards. These movements can be combined as a single column matrix or **column vector**:

$$\text{CR} = \begin{pmatrix} 3 \\ 4 \end{pmatrix}$$

In general any translation of the Cartesian plane can be written as a column vector $\begin{pmatrix} x \\ y \end{pmatrix}$ where x represents a movement parallel to the x-axis and

y represents a movement parallel to the y-axis. Movements to the right and movements upwards are positive. Movements to the left and movements downwards are negative.

Example 1

In Fig. 15.11 the line segments represent vectors $\overrightarrow{AB}$, $\overrightarrow{CD}$, ..., $\overrightarrow{IJ}$

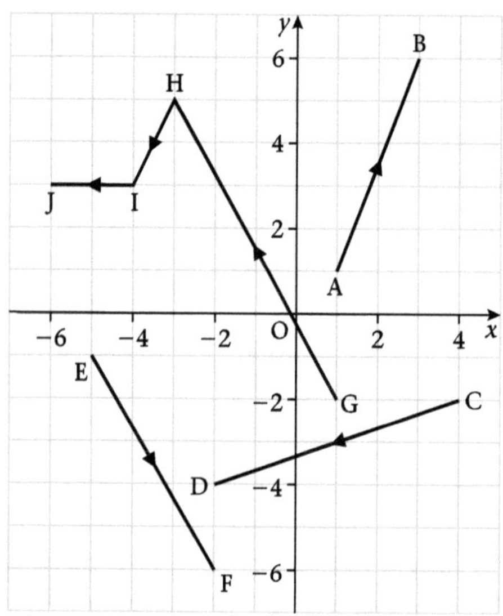

Fig. 15.11

Write these vectors in the form $\begin{pmatrix} x \\ y \end{pmatrix}$.

$\overrightarrow{AB} = \begin{pmatrix} 2 \\ 5 \end{pmatrix}$ $\overrightarrow{CD} = \begin{pmatrix} -6 \\ -2 \end{pmatrix}$

$\overrightarrow{EF} = \begin{pmatrix} 3 \\ -5 \end{pmatrix}$ $\overrightarrow{GH} = \begin{pmatrix} -4 \\ 7 \end{pmatrix}$

$\overrightarrow{HI} = \begin{pmatrix} -1 \\ -2 \end{pmatrix}$ $\overrightarrow{IJ} = \begin{pmatrix} -2 \\ 0 \end{pmatrix}$

Example 2

A triangle ABC has coordinates A(2, 1), B(5, 1), C(3, 6). The triangle is translated by a vector $\begin{pmatrix} 2 \\ 4 \end{pmatrix}$ to triangle DEF. Find the coordinates of D, E and F.

Fig. 15.12 shows the translation from $\triangle$ABC to $\triangle$DEF. The coordinates of D, E, F are D(4, 5), E(7, 5), F (5, 10).

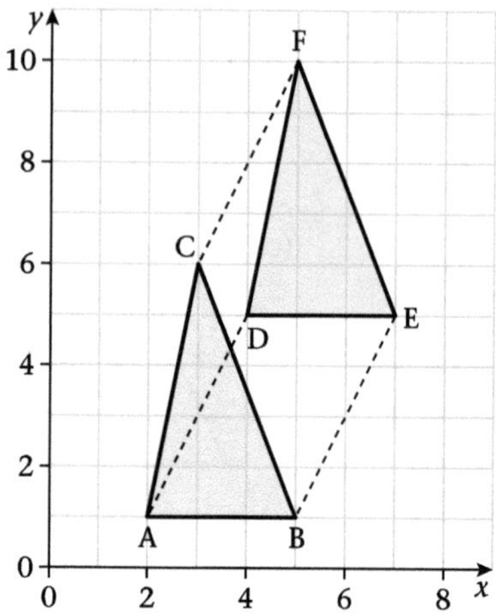

Fig. 15.12

Example 3

A square OABC has coordinates O(0, 0), A(3, 0), B(3, 3), C(0, 3). It is translated by vector $\begin{pmatrix} 2 \\ -4 \end{pmatrix}$ to square PQRS. Find the coordinates of P, Q, R and S.

Fig. 15.13 shows the translation from square OABC to square PQRS.

Notice that each point on the original square moves through the same vector.

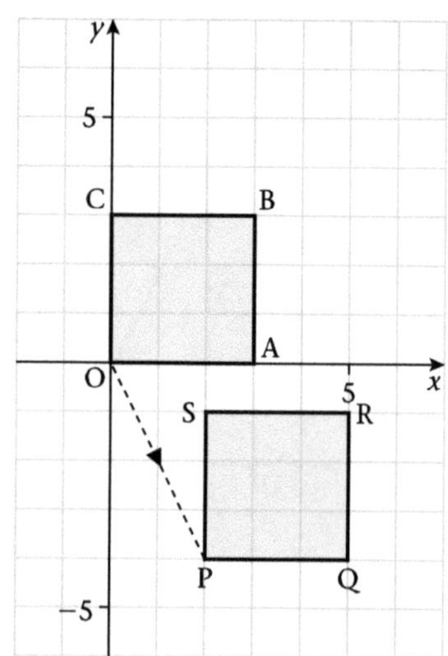

Fig. 15.13

The coordinates of PQRS are P(2, −4), Q(5, −4), R(5, −1) and S(2, −1).

Exercise 15c

Use graph paper when answering questions 3–10. Use a scale of 1 cm to 1 unit throughout.

1. In Fig. 15.14 the line segments represent vectors $\overrightarrow{AB}$, $\overrightarrow{CD}$, ..., $\overrightarrow{KL}$. Write these vectors in the form $\begin{pmatrix} x \\ y \end{pmatrix}$.

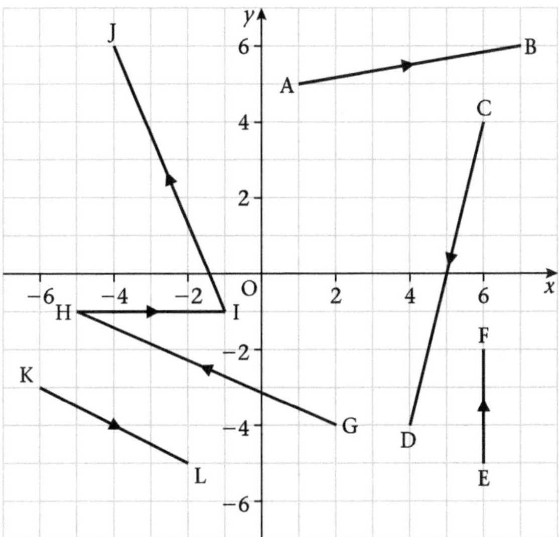

Fig. 15.14

2. In Fig. 15.15, $\triangle$ABC is translated to $\triangle$PQR.
   (a) Write $\overrightarrow{AP}$ in the form $\begin{pmatrix} x \\ y \end{pmatrix}$.
   (b) What is the vector of translation?

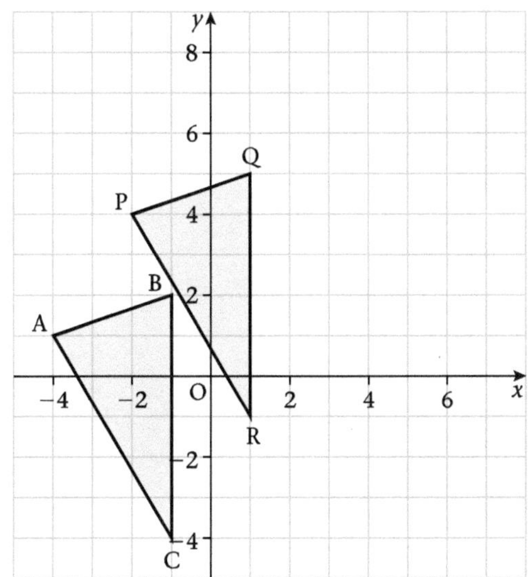

Fig. 15.15

3. $\triangle$OAB has coordinates O(0, 0), A(4, 2), B(2, 5). Find the coordinates of the images of O, A and B when the triangle is translated by the vector
   (a) $\begin{pmatrix} 3 \\ 6 \end{pmatrix}$, (b) equal to $\overrightarrow{OA}$, (c) $\begin{pmatrix} -1 \\ 2 \end{pmatrix}$,
   (d) equal to $\overrightarrow{AB}$,
   (e) equal to $\overrightarrow{BA}$.

4. The coordinates of the vertices of rectangle ABCD are A(1, 2), B(1, 5), C(2, 5) and D(2, 2). Find the coordinates of the images of A, B, C, D when the rectangle is translated by the vector
   (a) $\begin{pmatrix} -1 \\ 3 \end{pmatrix}$, (b) $\begin{pmatrix} 3 \\ 1 \end{pmatrix}$, (c) equal to $\overrightarrow{AC}$.

5. A Cartesian plane is translated so that (4, 5) is the image of the point (3, 1).
   (a) What is the vector of translation?
   (b) What are the coordinates of the image of the origin?
   (c) What are the coordinates of the point whose image is (−1, 2)?

6. A quadrilateral has vertices P(1, 4), Q(5, 7), R(9, 4), S(5, 1).
   (a) What kind of quadrilateral is PQRS?
   (b) Find the coordinates of the images of P, Q, R, S after translation of PQRS through
   (i) $\begin{pmatrix} 3 \\ -1 \end{pmatrix}$, (ii) $\begin{pmatrix} -5 \\ -4 \end{pmatrix}$, (iii) $\overrightarrow{PR}$.

Reflection

A reflection is the image you see when you look in a mirror. We have already seen in Book 1 that a line of symmetry acts like a mirror line.
Fig. 15.16 shows the letter p and its reflection in a line.

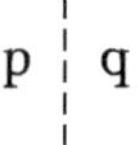

Fig. 15.16

Two or more mirror lines will reflect a shape many times to give a pattern which may fill the plane (Fig. 15.17).

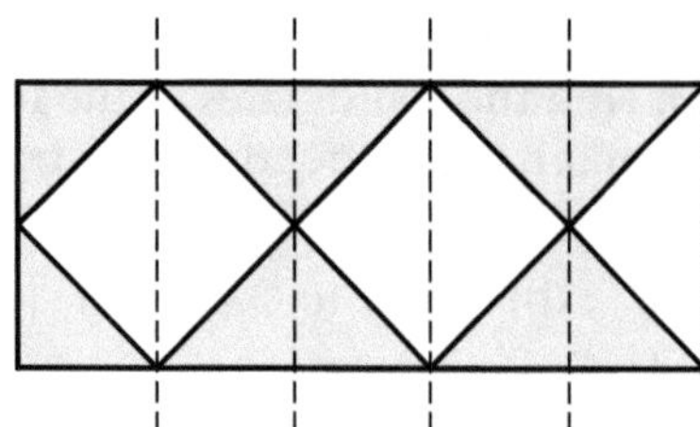

Fig. 15.17

Every line of symmetry in Fig. 15.17 is a mirror line for the whole pattern. Place a mirror along any of the broken lines and look into it. You will see that it continues the pattern.

Exercise 15d

1. Which of the patterns in Figs. 15.1, 15.2, 15.3 are reflection patterns? If possible, use a mirror to help you to decide.

2. In each of the following, copy the given figure on to graph paper. Take broken lines to be mirror lines.

   (a) Reflect the word HAND (i) across the page, (ii) down the page (Fig. 15.18).

(i)

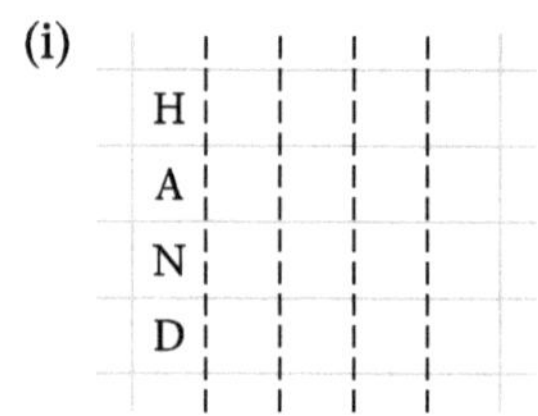

(ii)

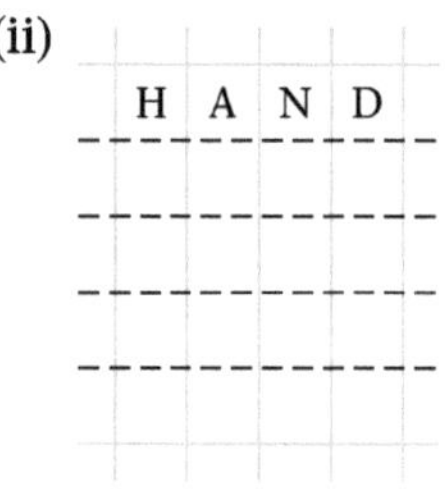

Fig. 15.18

   (b) In Fig. 15.19, use the given shape and mirror lines to make a reflection pattern.

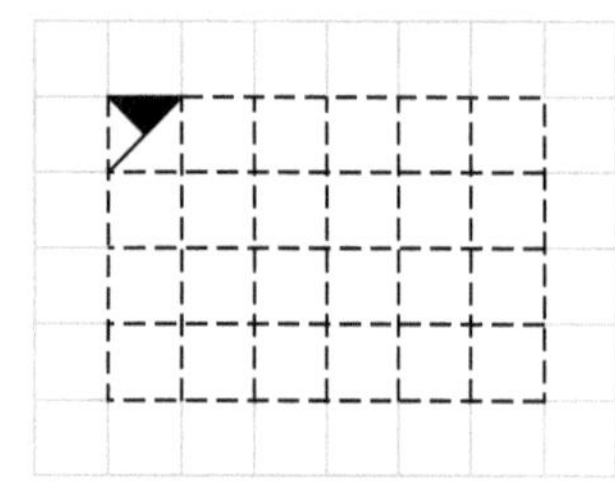

Fig. 15.19

3. Copy Fig. 15.20 on to graph paper. By folding the paper about the x-axis and the y-axis draw the image of $\triangle PQR$ after reflection in (a) the x-axis (b) the y-axis.

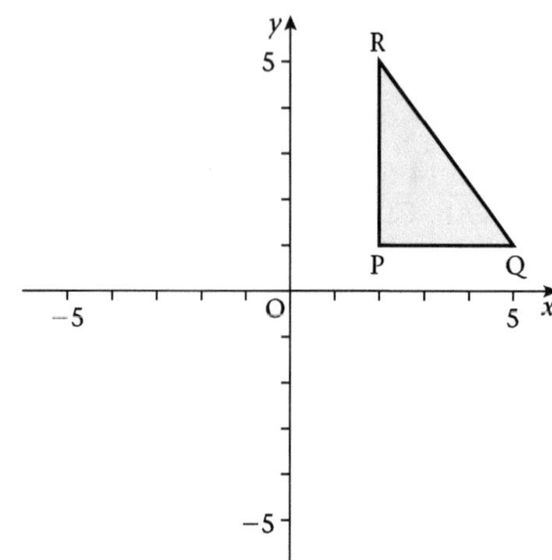

Fig. 15.20

   (i) What is the distance of the points P, Q, R from the x-axis?
   (ii) What is the distance of the images of the points, P, Q, R from the x-axis?
   (iii) What do you notice?
   (iv) What is the distance of the points P, Q, R from the y-axis?
   (v) What is the distance of the image of the points P, Q, R from the y-axis?
   (vi) What do you notice?

4. In Fig. 15.9 on page 131 state the coordinates of the vertices A, B and C after reflection in (a) the x-axis, (b) the y-axis.

Rotation

When a shape turns about a point, we say it **rotates**.

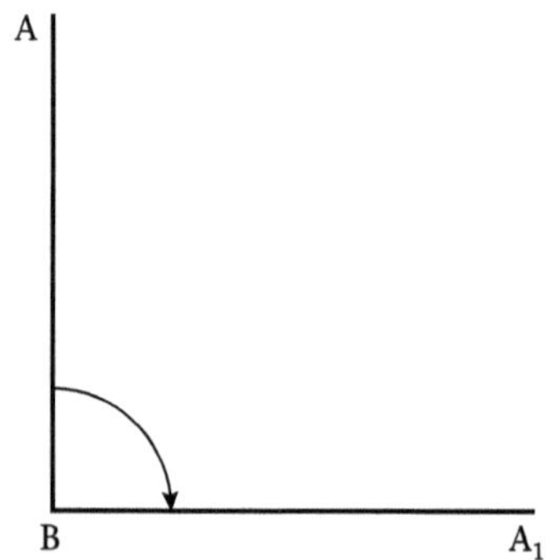

Fig. 15.21

In Fig. 15.21 the line AB is rotated through 90° **clockwise** about the point B. Note that clockwise is the direction in which the hands of the clock move.

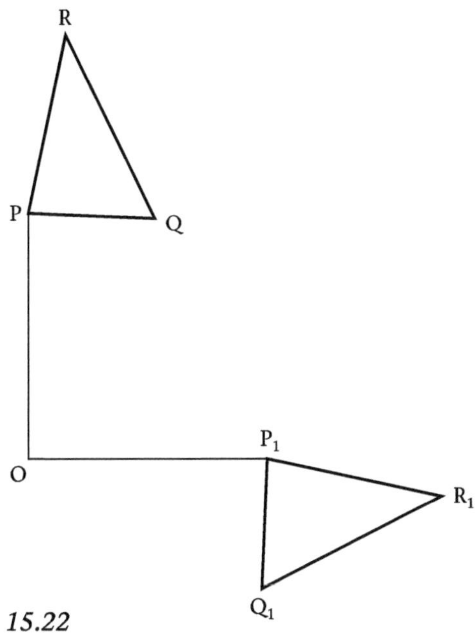

Fig. 15.22

In Fig. 15.22 the △PQR has been rotated through 90° clockwise about the point O. This point O is called the **centre of rotation**.

Exercise 15e

1. Copy Fig. 15.22 on to a sheet of plain paper. Measure the angle POP_1.
   (a) Join OQ, OQ_1 and measure angle QOQ_1. What do you notice?
   (b) Join OR, OR_1 and measure angle ROR_1. What do you notice?
   (c) What can you say about the angle through which the points P, Q, R are rotated?
   (d) What can be said about the lengths of the lines PR and P_1R_1?
   (e) Is the same true of QR and Q_1R_1? PQ and P_1Q_1?
   (f) What can be said of the triangles PQR and $P_1Q_1R_1$?
2. Draw a triangle ABC with AB = 3 cm, BC = 4 cm and AC = 6 cm.
   (a) Mark any point O outside the triangle. Join OA and rotate OA through 50° anticlockwise. Mark A_1 on this line such that $OA_1 = OA$.
   (b) Repeat for OB and OC to obtain B_1 and C_1.
   (c) What is the size of the angle BOB_1? Why?
   (d) What can you say about the size of angle COC_1? Why?
   (e) What can be said of triangles ABC and $A_1B_1C_1$?
3. In Fig. 15.9 on page 131 state the coordinates of the vertices A, B and C after a clockwise rotation about the origin of
   (a) 90°,
   (b) 180°,
   (c) 270°.
4. As question 3, with the point (−2, 1) as the centre of rotation

Summary

Transformation means change. Under a **geometrical transformation** the position and/or dimensions of a shape are changed.

The **image** of a shape is the figure that results after a transformation.

A **translation** is a movement of a shape along a straight line. Every point in the shape moves through the same distance. A translation can be described in terms of a vector of translation.

A **reflection** is a mirror image of the shape. A point and its image are at equal distances from the mirror line. The shape is laterally reversed or the sense of the shape is changed.

A **rotation** is a movement about a fixed point through an angle in a clockwise or an anti-clockwise direction. The fixed point is called the **centre of rotation**. Every line in the shape is turned through the same angle. The sense of the shape is not changed.

Under translation, reflection and rotation the dimensions of the shapes remain unchanged. The shapes and their images are **congruent**.

Practice Exercise P15.1

1. Using a scale of 1cm to 1 unit, draw on graph paper, the triangle ABC with A(2, 4), B(5, 6), C(6, 1). Find the coordinates of the images of A,B and C when the triangle is translated by the vector
   (a) $\begin{pmatrix} 2 \\ 3 \end{pmatrix}$ (b) $\begin{pmatrix} -4 \\ -3 \end{pmatrix}$
   Draw the images of triangle ABC.

2. Using a scale of 1 cm to 1 unit, draw on graph paper, triangle PQR with P(−2, −5), Q(−6, 0) and R(−2, 3). Translate the triangle PQR by the vector $\begin{pmatrix} 3 \\ 4 \end{pmatrix}$ to triangle WXY. Draw triangle WXY.

3. Using a scale of 2 cm to 1 unit
   (a) draw rectangle PQRS with P(2, 2), Q(4, 2), R(4, 4) and S(2, 4)
   (b) translate each point of the rectangle by the vector $\begin{pmatrix} 1 \\ 1 \end{pmatrix}$
   (c) draw the image of rectangle PQRS and name it WXYZ.

4. Draw a triangle PQR with P(2, 1), Q(6, 1) and R(5, 6). Draw the reflection of the triangle PQR in the line $x = 7$.

5. Using a scale of 1 cm to 1 unit, draw triangle PQR with P(2, 1), Q(5, 2) and R(2, 6). State the coordinates of the points P, Q and R after a clockwise rotation of (a) 90° (b) 180° about the origin.

6. Using a scale of 1 cm to 1 unit, draw on the same axes, triangle ABC with A(−2, 1), B(0, 6) and C(3, 0) and triangle DEF with D(2, 4), E(4, 9) and F(7, 3).
   Is triangle DEF a translation of triangle ABC? If so, what is the translation vector?

Chapter 16

Inequalities (1)

Linear inequalities in one variable

Pre-requisites

- number line; algebraic expressions

Greater than, less than

In mathematics, we probably use the equals sign, =, more than any other. For example, $5 + 3 = 8$.

However, many quantities are *not* equal:

$5 + 5 \neq 8$

where $\neq$ means *is not equal to*. We can also write:

$5 + 5 > 8$

where $>$ means *is greater than*. Similarly we can write the following:

$3 + 3 \neq 8$

$3 + 3 < 8$

where $<$ means *is less than*.

$\neq$, $>$ and $<$ are **inequality symbols**. They tell us that quantities are not equal. The $>$ and $<$ symbols are more helpful than $\neq$. They tell us more. For example, $x \neq 0$ tells us that x does not have the value 0; x can be any positive or negative number. However, $x < 0$ tells us that x is less than 0; x must be a negative number.

Exercise 16a

1. Rewrite each of the following, using either $>$ or $<$ instead of the words.
   (a) 6 is less than 11
   (b) -1 is greater than -5
   (c) 0 is greater than -2.4
   (d) -3 is less than $+3$
   (e) x is greater than 12
   (f) y is less than -2
   (g) 4 is greater than a
   (h) a is less than 4
   (i) 15 is less than b
   (j) b is greater than 15

2. State whether each of the following is true or false.
   (a) $13 > 5$ (b) $19 < 21$
   (c) $-2 < -4$ (d) $-15 > 7$
   (e) $3 + 9 < 10$ (f) $0 > -4 - 3$
   (g) $14 - 6 > 8$ (h) $30 > -50 + 20$

3. Find which symbol, $>$ or $<$, goes in the box to make each statement true.
   (a) $9 + 8 \square 10$ (b) $7 - 2 \square 7$
   (c) $6 \square 12 - 5$ (d) $0 \square 3 - 6$
   (e) $16 \square 2 \times 10$ (f) $29 \div 7 \square 3.6$
   (g) $13 \square 3 \times 4.9$ (h) $(-5)^2 \square (2)^2$

The symbols $>$ and $<$ can be used to change English statements into algebraic statements.

Example 1

The distance between two villages is over 18 km. Write this as an algebraic statement.

If the distance between the villages is d km, then, $d > 18$.

A statement like $d > 18$ is called an **inequality**.

Example 2

I have x cents. I spend 20 cents. The amount I have left is less than 5 cents. Write an inequality in x.

I spend 20c out of x cents.
Thus I have $x - 20$ cents left.
Thus $x - 20 < 5$.

Example 3

The area of a square is less than 25 cm^2. What can be said about (a) the length of one of its sides, (b) its perimeter?

(a) Let the length of a side of the square be a cm.
Then, $a^2 < 25$
$\sqrt{a^2} < \sqrt{25}$
Since $a > 0$,
$a < 5$

(b) Perimeter $= 4a$

$$a < 5$$
$$4a < 4 \times 5$$
$$4a < 20$$

The length of a side of the square is less than 5 cm. Its perimeter is less than 20 cm.

Exercise 16b

1. For each of the following, write an inequality in terms of the given unknown.
   (a) The height of the building, h m, is less than 5 m.
   (b) The mass of the boy, m kg, is less than 50 kg.
   (c) The cost of the meal, $\$x$, was over \$5.
   (d) The time taken, t min, was under 5 minutes.
   (e) The number of pages, n, was less than 24.
   (f) The mass of the letter, m g, was under 20 g.
   (g) The cost of the stamp, s cents, was less than \$1.
   (h) The time for the journey, t min, was over 2 hours.

2. For each of the following, choose a letter for the unknown and write an inequality.
   (a) The boy is less than 1.5 m tall.
   (b) The volume of the space is less than 800 m^3.
   (c) The book cost more than \$12.
   (d) The girl got over 60% in the exam.
   (e) The car used more than 28 litres of petrol.
   (f) Her mass was less than 55 kg.
   (g) The light was on for over 6 hours.

3. 7 lorries each carry a load of over 4 tonnes. The total mass carried by the lorries is m tonnes. Write an inequality in m.

4. A boy saved over \$5. His father gave him \$2. The boy now had $\$y$ altogether. Write an inequality in y.

5. In 3 years time a girl will be over 18 years of age. If her age now is x years, write an inequality in x.

6. There are x goats on a field of area 3 ha. There are less than 20 goats on each hectare. Write an inequality in x.

7. A square has an area of more than 36 cm^2. What can be said about (a) the length of one of its sides, (b) its perimeter?

8. The perimeter of a square is less than 28 cm. What can be said about (a) the length of one of its sides, (b) its area?

Not greater than, not less than

In most towns there is a speed limit of 50 km/h.

If a car, travelling at s km/h, is within the limit, then s is not greater than 50. If $s < 50$ or if $s = 50$, the speed limit will not be broken. This can be written as one inequality:

$$s \leqslant 50$$

where $\leqslant$ means *is less than or equal to*. Thus, not greater than means the same as less than or equal to.

In most countries, voters in elections must not be less than 18 years of age. If a person of age a years is able to vote, then a is not less than 18. The person can vote if $a > 18$ or if $a = 18$. This can be written as one inequality:

$$a \geqslant 18$$

where $\geqslant$ means *is greater than or equal to*. Thus, not less than means the same as greater than or equal to.

Example 4

Notebooks cost 60 cents each. David has d cents. It is not enough to buy a notebook. Tracy has t cents. She is able to buy a notebook. What can be said about the values of d and t?

The question tells us that: d is less than 60.
$d < 60$
If Tracy gets no change, then $t = 60$
If Tracy gets some change, then $t > 60$
It is not known if Tracy gets change or not, so, $t \geqslant 60$.
Finally, Tracy clearly has more money than David.
Thus, $t > d$

Similarly, $d < t$
In conclusion, the following can be said about d and t:

$d < 60$ $\quad t \geqslant 60$
$t > d$ $\quad d < t$

Notice that $t > d$ and $d < t$ are two ways of saying the same thing. It is like saying $9 > 4$ and $4 < 9$.

Exercise 16c

1. For each of the following, write an inequality in terms of the given unknown.
   (a) The age of the girl, a years, was 12 years or less.
   (b) The number of goals, n, was 5 or more.
   (c) The temperature, t °C, was not greater than 38 °C.
   (d) The selling price, $\$s$, was not less than \$24.
   (e) The number of students, n, was less than 36.
   (f) The speed of the car, v km/h, was never more than 120 km/h.

2. For each of the following, choose a letter for the unknown and then write an inequality.
   (a) The lorry can carry a load of not more than 7 tonnes.
   (b) My car cannot go faster than 140 km/h.
   (c) To join the police force, you must be not less than 160 cm tall.
   (d) The taxi cannot carry more than 5 passengers.
   (e) The hole must be not less than 6 m deep.
   (f) Each cow needs at least 100 m² of grazing land.

3. The pass mark in a test was 27. One person got x marks and failed. Another got y marks and passed. What can be said about x and y?

4. Pencils cost 42 cents each. a cents is not enough to buy a pencil. A person with b cents is able to buy a pencil. Write down 3 different inequalities in terms of:
   (i) a (ii) b (iii) a and b.

5. The radius of a circle is not greater than 3 m. What can be said about (a) its circumference, (b) its area? Give any inequalities in terms of π.

Graphs of inequalities

Line graphs

The inequality $x < 2$ means that x can have any value less than 2. We can show these values on the number line Fig. 16.1.

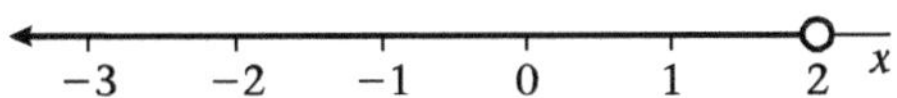

Fig. 16.1

The heavy arrowed line in Fig. 16.1 shows the range of values that x can have. The empty circle at 2 shows that the value 2 is not included. x can have any value to the left of 2. The inequality $x \geqslant -1$ means that x can have the value -1 or any value greater than -1. Its graph is given in Fig. 16.2.

Fig. 16.2

The shaded circle in Fig. 16.2 shows that the value -1 is included. x can have the value -1 and any value to the right of -1.

Example 5

Fig. 16.3 shows the graph of an inequality. What is the inequality?

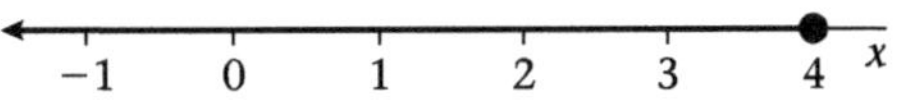

Fig. 16.3

The shaded circle above 4 shows that the value $x = 4$ is included. The heavy line to the left of 4 shows that x can have values in the range $x < 4$. Thus Fig. 16.3 is the graph of $x \leqslant 4$.

Exercise 16d

1 Write down the inequalities in the following graphs.

(a) (b) (c)

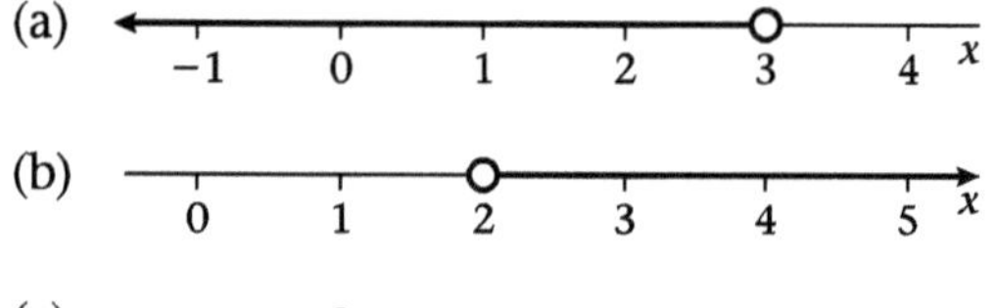

(d)

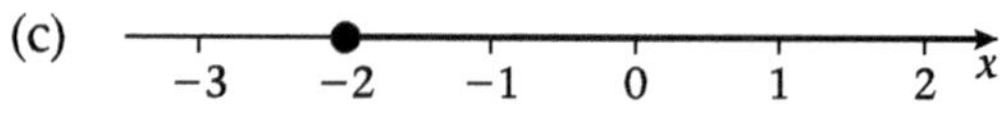

(e)

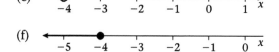

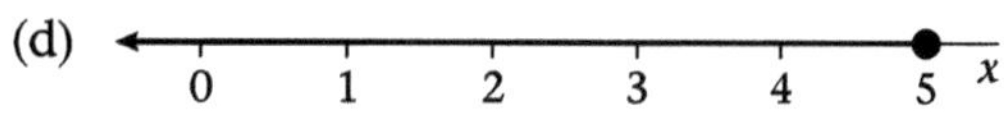

(f) −5 −4 −3 −2 −1 0 x

Fig. 16.4

2 Sketch graphs of the following inequalities.

(a) $x > 1$ (b) $x < -2$

(c) $x \geqslant -3$ (d) $x \leqslant 0$

(e) $x < 3$ (f) $x \geqslant -2$

(g) $x \geqslant 4$ (h) $x < -1$

Cartesian graphs

(x, y) represents any point on the Cartesian plane which has coordinates x and y. If (x, y) is such that $x \geqslant 2$ then (x, y) may lie anywhere in the unshaded region in Fig. 16.5.

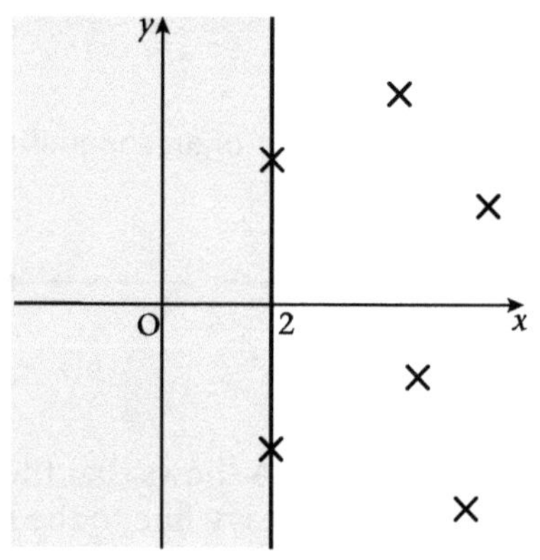

Fig. 16.5

Small crosses show some possible positions of (x, y). The region on the left is shaded to show that it is *not* wanted.

In Fig. 16.5 the boundary of the shaded region is a line through the x-axis where $x = 2$. $x = 2$ at every point on this line. We say that the **equation of the line** is $x = 2$.

The points shown by crosses are members of a **set of points**, P, where $P = \{(x, y): x \geqslant 2\}$, i.e. P is the set of points (x, y) such that $x \geqslant 2$. similarly the line $x = 2$ is composed of the set of points, L, where $L = \{(x, y): x = 2\}$.

Example 6

On a Cartesian plane, sketch the region which contains the set of points Q where $Q = \{(x, y): y > -3\}$.

The unshaded part of Fig. 16.6 is the required region.

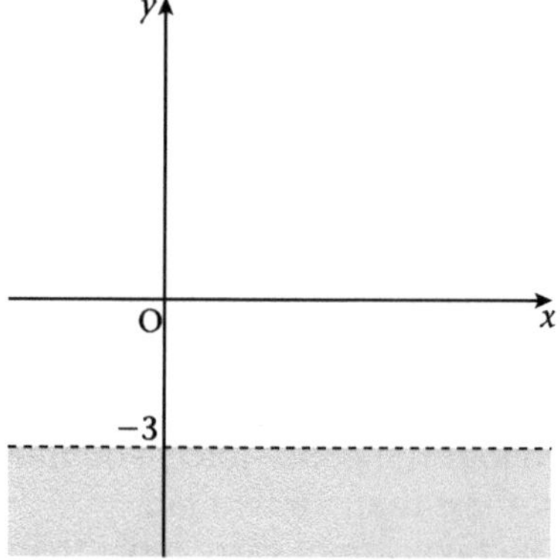

Fig. 16.6

Note: The boundary line has the equation $y = -3$. In Fig. 16.6 this line is broken to show that the points on the line are *not* included in the required region.

Example 7

Combine Fig. 16.5 with Fig. 16.6 to show the region which represents the set

$$\{(x, y): x \geqslant 2\} \cap \{(x, y): y > -3\}.$$

In Fig. 16.7 the unshaded part is the required region. The other parts are shaded to show that they are not wanted.

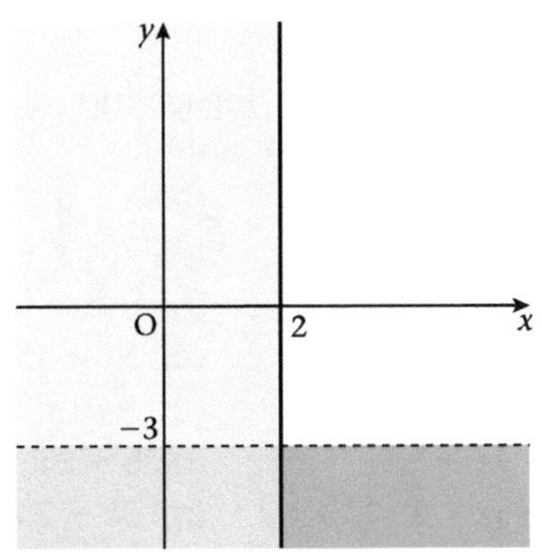

Fig. 16.7

Note: Fig. 16.7 shows the intersection of sets P and Q. The unshaded region contains those points which belong to both P and Q.

Exercise 16e

1. On a Cartesian plane, sketch and label the lines represented by the following equations and sets of points.

   (a) $x = 3$ (b) $\{(x, y): y = 5\}$
   (c) $y = -4$ (d) $\{(x, y): x = -2\}$
   (e) $x = 0$ (f) $\{(x, y): y = 0\}$
   (g) $y = -1$

2. In Fig. 16.8 the points belonging to the shaded region are not wanted. State the set of points represented by the unshaded region.

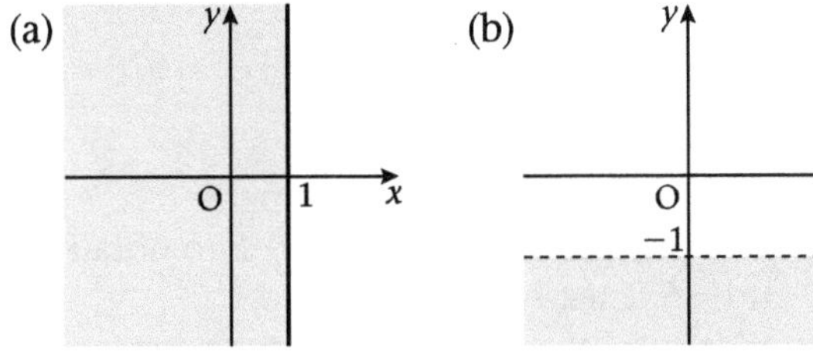

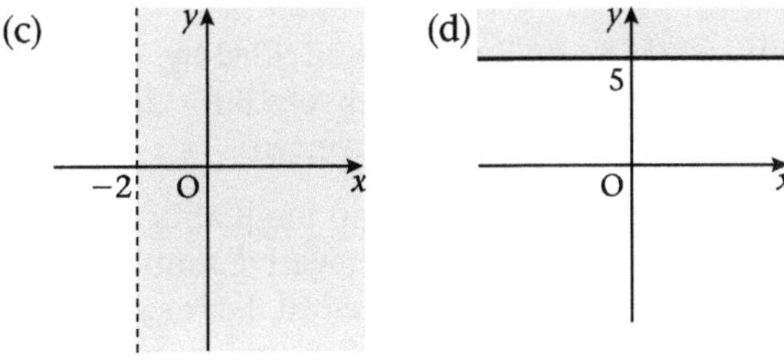

Fig. 16.8

3. On a Cartesian plane sketch the region which represents the following sets of points (a)–(g). *Use the rule that shaded regions contain unwanted points and that continuous boundary lines mean that the equality is included.*

   (a) $x \geqslant -1$ (b) $x < 2$
   (c) $y > -2$ (d) $y \leqslant 3$
   (e) $x \leqslant 0$ (f) $y \geqslant 0$
   (g) $y < 6$

4. In each part of Fig. 16.9 the unshaded region represents the intersection of two sets of points. Use set notation to write out the sets.

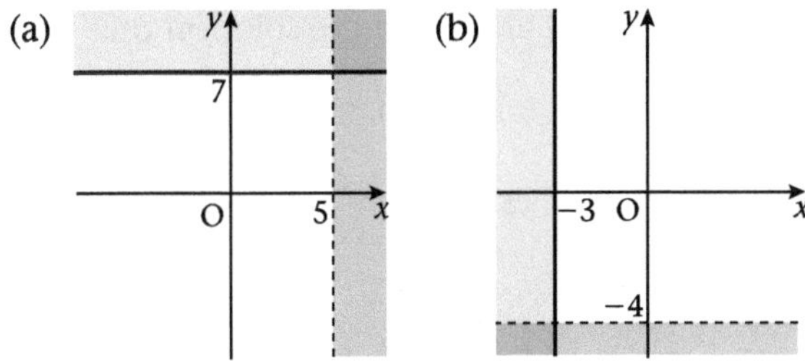

Fig. 16.9

5. On a Cartesian plane sketch the region which represents the region containing the sets of points in each of the following intersections.

   (a) $\{(x, y): x < 2\} \cap \{(x, y): y \geqslant 5\}$
   (b) $\{(x, y): x > 1\} \cap \{(x, y): y < -1\}$

Solution of inequalities

Solution sets

Consider a bus which can hold 46 people. At any one time there may be x people in the bus. If the bus is full, then $x = 46$. This is an equation.

If the bus is not full, then $x < 46$. This is an inequality.

The equation has only one solution: $x = 46$.

The inequality has many solutions: if $x < 46$, then x can have the values 0, 1, 2, 3, ..., 44, 45. Notice that negative and fractional values of x are impossible in this example. The set of values of x which make $x < 46$ true is called the **solution set** of the inequality.

If $x < 46$, then $x \in \{0, 1, 2, 3, ..., 45\}$. The solution set is $\{0, 1, 2, 3, ..., 45\}$.

Equations may also have solution sets. The solution set of $x = 46$ is $\{46\}$.

Inequalities are solved in much the same way as equations. We can use the balance method. However, there is one important exception; this will be shown in Examples 10 and 11. Meanwhile, read the following examples carefully.

Example 8

Solve $6 \leqslant 2x - 1$ and show the solution on a line graph.

$6 \leqslant 2x - 1$

Add 1 to both sides.

$7 \leqslant 2x$

Divide both sides by 2.

$\frac{7}{2} \leqslant \frac{2x}{2}$

$3\frac{1}{2} \leqslant x$

If $3\frac{1}{2} \leqslant x$, then $x \geqslant 3\frac{1}{2}$.

$x \geqslant 3\frac{1}{2}$ is the solution of $6 \leqslant 2x - 1$. Fig. 16.10 is the graph of the solution set.

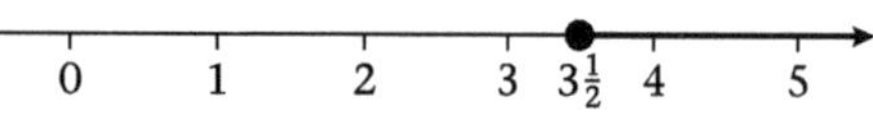

Fig. 16.10

Note: Normally we arrange for the unknown to be on the left-hand side of the inequality. This makes it easier to sketch the graph.

Example 9

Given that x is an integer, find the solution set of $3x - 3 > 7$.

$3x - 3 > 7$

Add 3 to both sides.

$3x > 10$

Divide both sides by 3.

$x > 3\frac{1}{3}$

Since x is an integer,

$x \in \{4, 5, 6 ...\}$

$\{4, 5, 6, ...\}$ is the solution set of $3x - 3 > 7$.

Exercise 16f

1 Solve the following inequalities. Sketch a line graph of each solution.

(a) $x - 2 < 3$ (b) $x + 3 \geqslant 6$
(c) $2 > x - 4$ (d) $7 < x + 2$
(e) $x + 9 \leqslant 3$ (f) $0 > x + 5$
(g) $2x < 6$ (h) $5x \geqslant 45$
(i) $12 \geqslant 3x$ (j) $-10 < 5x$
(k) $4x \geqslant -9$ (l) $8 \leqslant 3x$
(m) $3x + 1 < 13$ (n) $5x - 2 \geqslant 8$
(o) $-5 > 4x + 15$ (p) $3 \leqslant 17 + 2x$
(q) $4x - 2 > 19$ (r) $3 \leqslant 3x + 5$

2 Find the solution sets of the following, given that x is an integer in each case.

(a) $2x > 9$ (b) $3x < 7$
(c) $4x < -11$ (d) $4x > -14$
(e) $3 > 5x$ (f) $-8 < 3x$
(g) $2x + 1 < 12$ (h) $5x - 7 > 9$
(i) $7x > 5x - 9$ (j) $8x < 5x - 10$
(k) $6 > 4x + 1$ (l) $3x + 20 > 4$
(m) $2x + 4 \leqslant 2$ (n) $5x - 8 \geqslant 12$
(o) $1 \geqslant 6x - 11$ (p) $3x - 8 \leqslant 5x$
(q) $8x + 16 \leqslant 0$ (r) $x + 4 \geqslant 10x - 23$

Multiplication and division by negative numbers

Consider the following true statement: $5 > 3$. Multiply both the 5 and the 3 by -2 to obtain -10 and -6. Clearly $-10 < -6$, so multiplication by a negative number reverses the inequality.

Similarly if 5 and 3 are both divided by -2, they become $-2\frac{1}{2}$ and $-1\frac{1}{2}$ respectively. But $-2\frac{1}{2} < -1\frac{1}{2}$, so again the inequality is reversed.

In general, if both sides of an inequality are multiplied or divided by a negative number, the inequality sign must be reversed. For example, if $-2x > 10$ is true, then, on division throughout by -2, $x < -5$ will be true.

Example 10

Solve $5 - x > 3$.

Either:

$$5 - x > 3$$

Subtract 5 from both sides

$$-x > -2$$

Multiply both sides by -1 and reverse the inequality.

$$(-1) \times (-x) < (-1) \times (-2)$$

$$x < 2$$

or:

$$5 - x > 3$$

Add x to both sides.

$$5 > 3 + x$$

Subtract 3 from both sides

$$2 > x$$

$$x < 2$$

The second method in Example 10 shows that the same result is obtained as that obtained by using the rule of reversing the inequality sign when multiplying by a negative number.

Example 11

Solve $19 \geqslant 4 - 5x$.

$$19 \geqslant 4 - 5x$$

Subtract 4 from both sides.

$$15 \geqslant -5x$$

Divide both sides by -5 and reverse the inequality sign.

$$\frac{15}{-5} \leqslant \frac{-5x}{-5}$$

$$-3 \leqslant x$$

$$x \geqslant -3$$

Exercise 16g

Solve the following inequalities.

1. $-2x < 8$
2. $-3a < -6$
3. $12 \geqslant -4m$
4. $40 \leqslant -5d$
5. $3 - y \leqslant 7$
6. $5 - z \geqslant 1$
7. $5 - 2a > 1$
8. $2 - n \leqslant 3$
9. $2r \geqslant 5r + 6$
10. $9 \geqslant 3 - 4t$

Summary

An algebraic statement in which quantities are **not equal** is called an **inequality**. The symbols which connect the left-hand side and the right-hand side of an inequality are $\leqslant, <, \geqslant, >, \neq$.

Linear inequalities in one variable may be represented graphically on a number line or on a Cartesian plane. An inequality separates the Cartesian plane into a region which contains the set of points which satisfy the inequality, and a region which does not. The boundary between these regions is the line whose equation is found by replacing the inequality symbol (one of $\leqslant, <, \geqslant, >$) by the equal sign $(=)$.

When an inequality is multiplied by a **negative number**, the inequality sign is **reversed**.

Practice Exercise P16.1

In each of the following, replace $*$ with the symbol for 'greater than' or 'less than' to make each statement true:

1. $8 * 3$
2. $-8 * -3$
3. $5 \times 3 * 4 + 10$
4. $3 \times 5 + 7 * 3 \times (5 + 7)$
5. $18 \div 3 * 12 \div 4$
6. $\frac{1}{2} + \frac{3}{4} * \frac{1}{2} \times \frac{3}{4}$
7. $6 \div \frac{2}{3} * 6 \times \frac{2}{3}$
8. $(-3)^2 * (2)^3$
9. $(5)^2 * \frac{1}{(5)^2}$
10. $\left(\frac{2}{3}\right)^2 * \frac{1}{\left(\frac{2}{3}\right)^2}$

Practice Exercise P16.2

The diagrams below are the graphs of linear inequalities in x. For each inequality, write the inequality using the symbols: $<$, $>$, $\leqslant$, or $\geqslant$

1. 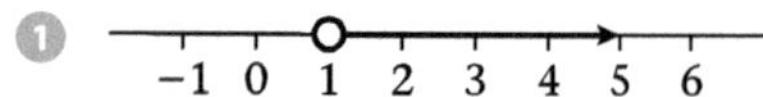

2. −3 −2 −1 0 1 2 3 4 5

3. −2 −1 0 1 2 3 4 5 6

4. −4 −3 −2 −1 0 1 2 3 4 5

5. −1 0 1 2 3 4 5 6 7 8

6. −2 −1 0 1 2 3 4 5 6 7

Fig.16.11

Practice Exercise P16.3

For each of the following write the inequality using the symbols : $<$, $>$, $\leqslant$, or $\geqslant$ and n to represent the number

1. A number which is less than 2.
2. A number which is not greater than 2.
3. A number which is not less than 2.
4. I think of a number, subtract 5 and the answer is greater than 1.
5. I think of a number and multiply it by 5. The answer is not less than 15.
6. I think of a number, divide by 7 and the answer is less than 3.
7. When 2 is added to the number, the answer is not more than 8
8. When the number is divided by 4, the result is more than 1.
9. If 5 is subtracted from the number, the answer is less than 0.
10. When 4 is taken away from the number, the answer is greater than 3.
11. 5 times the number gives an answer which is not greater than 15.
12. When 10 is multiplied by the number, the result is less than 60.
13. When the number is doubled, and 3 is added to the result, the answer is not less than 5.
14. If 9 is added to the number, the answer is not more than 16 but more than 12.
15. When the number is multiplied by itself, the result is greater than 4.

Practice Exercise P16.4

For each problem under P16.3,

(a) solve the inequality to find the value(s) of n

(b) draw the graph to illustrate the solution.

Practice Exercise P16.5

Show on Cartesian axes the region which represents the set of points:

1. $\{(x, y) : x \geqslant -1\}$
2. $\{(x, y) : y < 1\}$
3. $\{(x, y) : y \geqslant -1\}$.
4. $\{(x, y) : x \leqslant -3\}$

[*Shade the region where the points do* **not** *satisfy the inequality.*]

Practice Exercise P16.6

In each of the following graphs, state the set of points represented by the unshaded area.

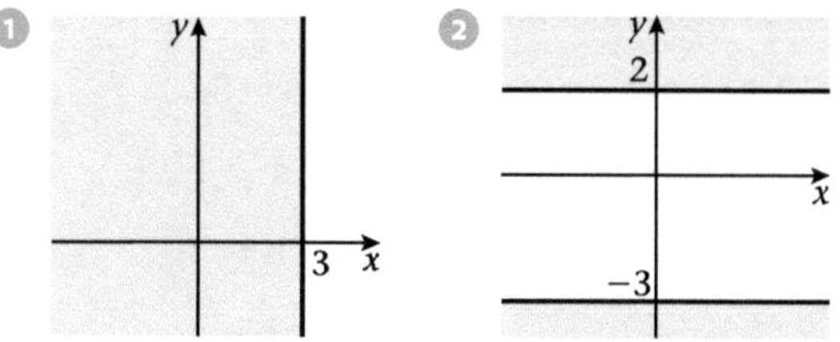

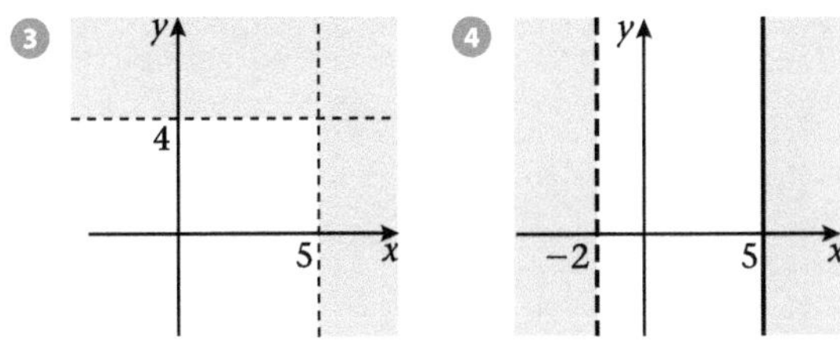

Fig. 16.12

Practice Exercise P16.7

On a Cartesian plane draw a graph showing the region that contains the set of points in each of the following intersections:

1. $\{(x, y) : x \geqslant -3\} \cap \{(x, y) : x < 2\}$
2. $\{(x, y) : y \geqslant -2\} \cap \{(x, y) : y < 2\}$
3. $\{(x, y) : x < -2\} \cap \{(x, y) : y \geqslant -1\}$

Practice Exercise P16.8

Solve the following inequalities:

1. $-5x < 10$
2. $6v + 3 \geqslant 8v$
3. $-3r \leqslant -21$
4. $6 > 3d$
5. $4m - 1 < 5m + 3$
6. $3 + 2(n - 3) > 5(n - 3)$
7. $3a - 2(2a + 11) \geqslant 1$
8. $3(y - 2) < 6(y + 1)$
9. $4(m + 2) \leqslant 5(m - 2) + 2m$
10. $(a + 2) \leqslant 3(a + 2)$

Revision exercise and tests

Chapters 9–16

Revision exercise 5 (Chapters 9, 13)

1. Draw a number line from −10 to 10. On the line mark the points A(8), B(2), C(−5), D(−2), E(0), F($3\frac{1}{2}$), G($-7\frac{1}{2}$), H($5\frac{1}{2}$).

2. Write down the coordinates of the points P, Q, R, S, T, U, V, W, X, Y, Z in Fig. R7.

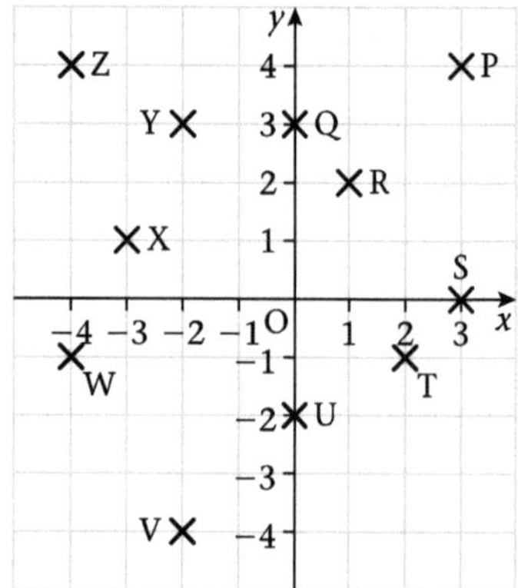

Fig. R7

3. In Fig. R8 name the points which have coordinates (3, −4), (−7, −3), (15, 6), (3, 4), (−4, 3), (6, 9), (−9, 8), (17, −2).

4. (a) Choose a suitable scale and plot the points P(−2, 1), Q(0, 2), R(2, 0) and S(0, −1).
   (b) What kind of quadrilateral is PQRS?
   (c) Find the coordinates of the point where the diagonals of PQRS cross each other.

5. Make a mapping table for the relation $n \rightarrow 75n$ for the values of n 1, 2, 3, 4, 5, 6 and 7.

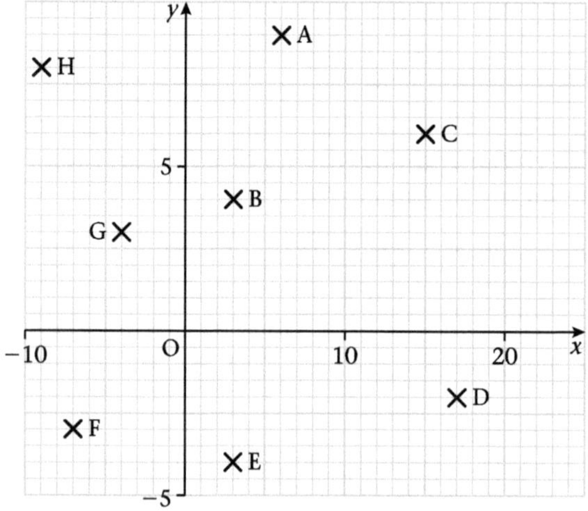

Fig. R8

6. A new candle is 15 cm long. When lit it burns steadily as shown in Table R1.

Table R1

Burning time (h)	0	1	2	3	4	5
Length of candle remaining (cm)	15	13.8	12.6	11.4	10.2	9

Choose a suitable scale and draw a graph of the information in Table R1. Use your graph to answer the following.
   (a) Find the length of the candle remaining after it has burned for 2.6 h.
   (b) How long does it take to burn 5 cm of candle?
   (c) Hence estimate how long the candle will last altogether.

7 Fig. R9 is a sketch of a graph which shows the distances from A and the times for a car going from A to B and back again. Find.
(a) the distance from A to B,
(b) how long the car stops at B,
(c) the average speed in km/h from (i) A to B, and (ii) B to A.

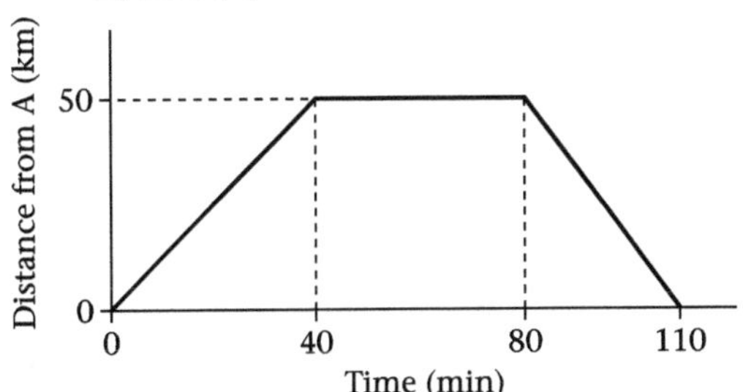

Fig. R9

8 The sum of the angles on a straight line is 180°. In Fig. R10, $x + y = 180$.

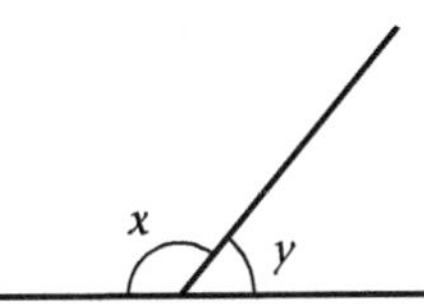

Fig. R10

(a) Copy and complete Table R2.

Table R2

x	0	45	90	135	180
y	180	135			

(b) Choose a suitable scale and draw a graph of the information in your table.
(c) Use your graph to find (i) y when $x = 40$, (ii) x when $y = 128$.

9 The cost of car insurance is \$60 per \$1000 worth of insurance. To this is added a standing charge of \$50.
(a) Copy and complete Table R3.

Table R3

Value insured (× \$1000)	1	2	3	4	5
Standing charge (\$) Basic rate (\$)	50 60	50 120	50 180	50	50
Total cost (\$)	110	170	230		

(b) Choose a suitable scale and draw a graph of this information.

10 Use the graph you drew in question 9 to answer the following.
(a) What is the cost of insuring a car worth \$3800?
(b) A sports car costs \$260 to insure. How much does it cost to insure a car half the value of the sports car?

Revision test 5 (Chapters 9, 13)

1 The position of point X is given by X(−0.9). Which of the points A, B, C, D, in Fig. R11 is in the same position as X?

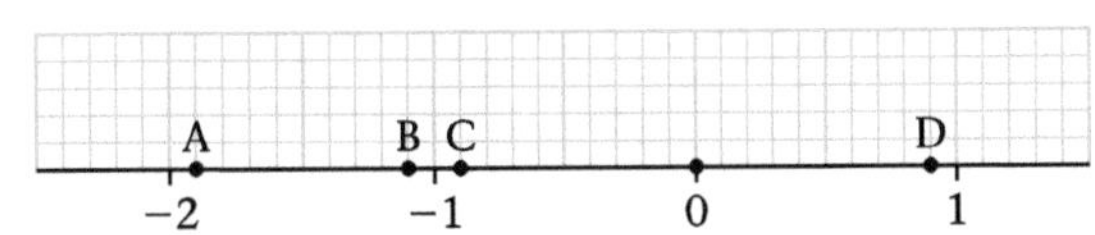

Fig. R11

2 Which of the points A, B, C, D in Fig. R12 has coordinates (−2, 3)?

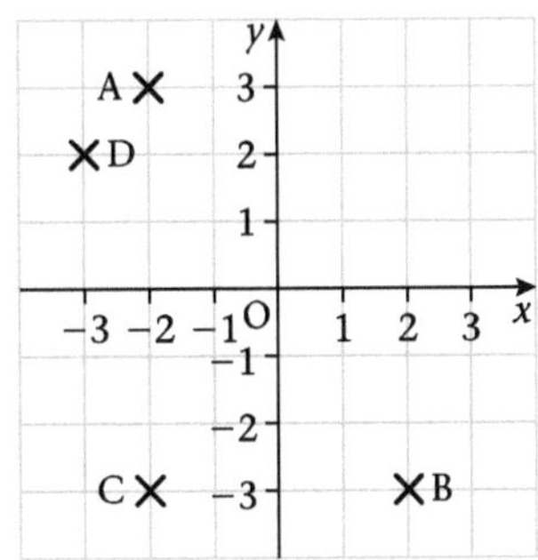

Fig. R12

3 A straight line PO joins the point P(5, −3) to the origin, O. PO is extended to Q so that PO = OQ. What are the coordinates of Q?
A (−5, −3) B (−5, 3)
C (−3, 5) D (3, −5)

4 The point (−2, 7) is reflected in the y-axis. What are the coordinates of its image?
A (2, 7) B (7, −2)
C (2, −7) D (−2, −7)

Fig. R13 is a discontinuous graph which gives the postal rates in cents (c) for letters up to 500 g in mass.

Each step in the graph is in the form of a line with a cross-mark at the right-hand end. Take this to mean that the value at the right-hand end is included but that the value at the left-hand end is *not* included.

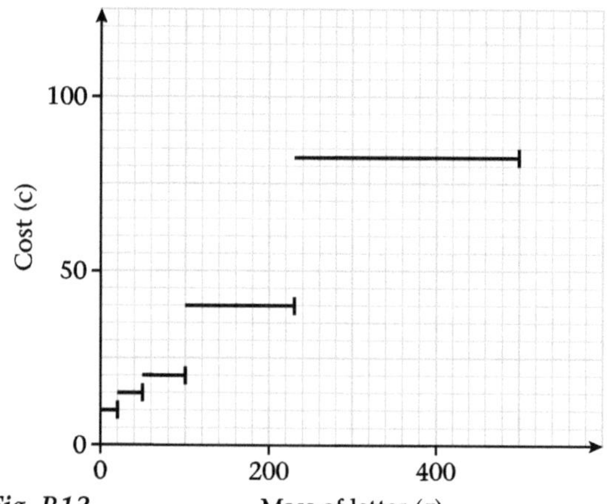

Fig. R13

Use Fig. R13 to answer questions 5 and 6.

5 How much does it cost to send a letter of mass 80 g?
A 10c B 15c C 20c D 40c

6 A letter of mass x g can be sent for 40c. Express the range of values of x in the form $a < x \leqslant b$.

7 Choose a suitable scale and plot the points A(1,3), B(6, 3), C(3, −1), D(−2, −1).
(a) AC and BD cross at P. Find the coordinates of P.
(b) What is the size of $\widehat{APD}$?
(c) What kind of quadrilateral is ABCD?

8 A man walks with a speed of 6 km/h.
(a) Make a table of values to show how far the man walks in 0.25, 0.5, 0.75, 1, 1.25, 1.5, 1.75, 2 hours.
(b) Choose a suitable scale and draw a graph of the information in the table.

9 Use the graph drawn in question 8 to find:
(a) how far the man walks in 69 min,
(b) how long it takes the man to walk 10 km.

10 The exchange rate between Eastern Caribbean currency and Barbados currency is EC\$1.00 = Bds \$0.74.
(a) Find how many Bds\$ are equivalent to (i) EC\$10, (ii) EC\$100.
(b) Hence draw a conversion graph for changing up to EC\$100 to Bds\$.
(c) Read off the equivalent amounts in the other currency of (i) EC\$80, (ii) Bds\$44.

Revision exercise 6 (Chapters 10, 16)

1 A car costs \$$c$ when new. It was sold for four-fifths of its cost price. If \$14 000 was lost on the sale of the car, calculate the cost price of the car when new.

2 (a) Solve the following.
(i) $-3x = 12$ (ii) $8n = -32$
(iii) $-5d = -35$ (iv) $-2\frac{1}{3}x = -14$
(b) Solve the following.
(i) $12 - 5a = 2$
(ii) $11 - 5x = 4x - 16$
(iii) $2(3x + 1) = 4(x + 5)$
(iv) $5(2a - 3) - 3(2a + 1) = 0$

3 One man earns \$17 more than another man. Between them they earn a total of \$105. How much does each man earn?

4 Solve the following.
(a) $\frac{1}{6}m = \frac{5}{36}$ (b) $\frac{3a}{2} = \frac{9}{14}$
(c) $\frac{3r}{5} - 15 = 0$ (d) $\frac{3x}{2} - 2 = \frac{2x}{3}$
(e) $\frac{m-3}{2} + \frac{m-8}{3} = 0$
(f) $\frac{3(2x-1)}{4} = \frac{4(x+2)}{3} - 3$
(g) $\frac{18}{2x-1} = 3$ (h) $\frac{7}{a-4} = \frac{5}{a-2}$

5 The length of a rectangular sheet of paper is twice its width.
(a) If the width is d cm, write an expression in d for the total length of the edges of the sheet of paper.
(b) Given that the total length is 84 cm, find
(i) the value of d
(ii) the area of the sheet of paper.

6. Theresa has \$50. She buys x blouses and y skirts. A blouse costs \$9 and a skirt costs \$13. If she gets change, write down an inequality in x and y.

7. On a Cartesian plane sketch the region which represents the following set of points.
   (a) $\{(x, y): x < 7\}$ (b) $\{(x, y): y \geqslant -1\}$
   (c) $\{(x, y): x > -2\} \cap \{(x, y): y \leqslant 0\}$

8. Find the range of values of x for which
   (a) $x - 3 < 2$
   (b) $2 - x > 5$
   (c) $2x - 2 < \frac{x + 2}{2}$
   (d) $\frac{x + 2}{5} \geqslant \frac{x - 3}{3} + 1$

9. If $x \in$ {integers}, find the solution sets of the following.
   (a) $x - 7 < -4$ (b) $3 - 2x \leqslant 15$
   (c) $\frac{4 - 7x}{8} \geqslant -3$ (d) $13 - 10x > 1 - 4x$

10. n is an integer. 3 times n is subtracted from 38. The result is less than 20.
    (a) Make an inequality in n.
    (b) Find the four lowest values of n.

Revision test 6 (Chapters 10, 16)

1. Given that $\frac{x + 2}{3} + 2x = 10$, $x =$
   A $9\frac{1}{3}$ B $4\frac{2}{3}$ C 4 D $1\frac{1}{7}$

2. If $\frac{6}{5} = \frac{3}{d}$ then $d =$
   A $\frac{5}{18}$ B $\frac{2}{5}$ C $2\frac{1}{2}$ D $3\frac{3}{5}$

3. If x is an integer, the highest value of x in the range $-6\frac{3}{4} < x < 2\frac{2}{3}$ is
   A 3 B 2
   C −6 D −7

4. A boy has more than \$9. He spends \$3 and has \$$m$ left. Which one of the following is the correct inequality in m?
   A $m > 6$ B $m < 6$
   C $m < 12$ D $m > 12$

5. Which of the lines in Fig. R14 has the equation $y = -2$?

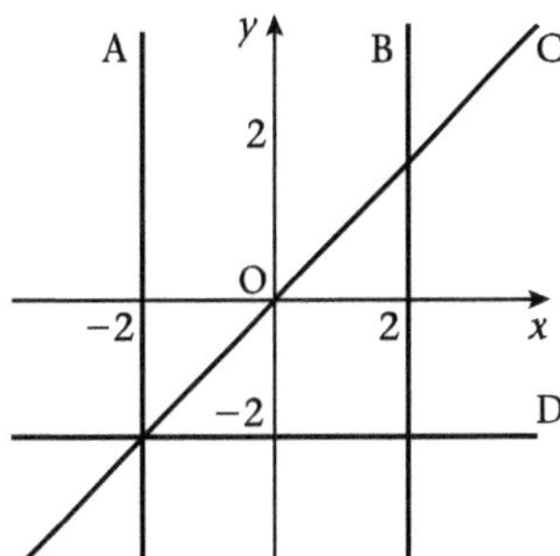

Fig. R14

6. Solve the following.
   (a) $-3x = 18$ (b) $\frac{1}{5}a = \frac{7}{15}$
   (c) $-\frac{7m}{10} = -2\frac{4}{5}$ (d) $\frac{5}{8}y = -1\frac{1}{4}$
   (e) $9 - 4x = 11 - 7x$
   (f) $6a - 3 = 25 + 5a$
   (g) $17y - 2(6y + 1) = 8$
   (h) $\frac{7z}{2} - 18 = \frac{z}{2}$ (i) $\frac{2}{x - 10} + \frac{1}{3} = 0$
   (j) $\frac{23 - 3x}{x + 1} = \frac{4}{3}$

7. A sum of \$30 is shared equally between x girls. One of the girls spends \$2 and has \$1.75 remaining. Find the value of x.

8. Use number lines to draw graphs of the solutions of the following.
   (a) $x - 2 \leqslant 0$ (b) $2x + 5 \leqslant 3$

9. Find the solution sets of the following, given that x is an integer.
   (a) $5 + x \geqslant 7$ (b) $3 - 7x \leqslant 59$
   (c) $\frac{2x + 3}{3} > 5$ (d) $\frac{3x + 2}{5} < \frac{x - 8}{3}$

10. A rectangle is of length x cm and breadth 5 cm. Its perimeter is p cm where $14 \leqslant p \leqslant 32$. Find the corresponding range of values of x.

Revision exercise 7 (Chapters 11, 12)

1. A map is drawn to a scale of 1 cm to 20 km. On the map a river is 6.3 cm long. What is the true length of the river?

2. A map is drawn to a scale of 5 cm to 1 km. A straight road is 1.58 km long. How long will the road appear on the map?

3. Use measurement to find the scale of △ (a) to △ (b) in Fig. R15.

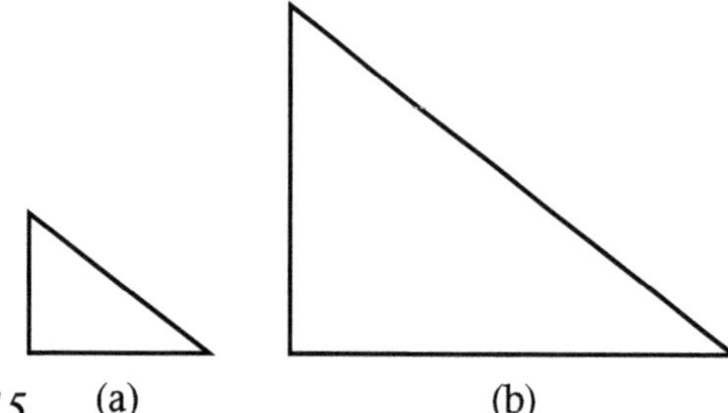

Fig. R15 (a) (b)

4. Use Fig. 11.14 on page 98 to estimate the distance of point A near Bridgetown to point B on the opposite side of the island.

5. A triangular plot PQR is such that PQ = 100 m, QR = 70 m and RP = 50 m. X is the mid-point of side PQ.
Use a scale of 1 cm to 10 m to make a scale drawing of the plot. Hence find the true length of XR.

6. Sketch a hexagon which has rotational symmetry of order 2 and (a) 2, (b) 0 lines of symmetry.

7. How many sides has a regular polygon if each interior angle is 171°?

8. Calculate the size of each angle of a regular 15-sided polygon.

9. (a) Find the sum of the angles of a hexagon.
(b) Hence find the value of x in Fig. R16.

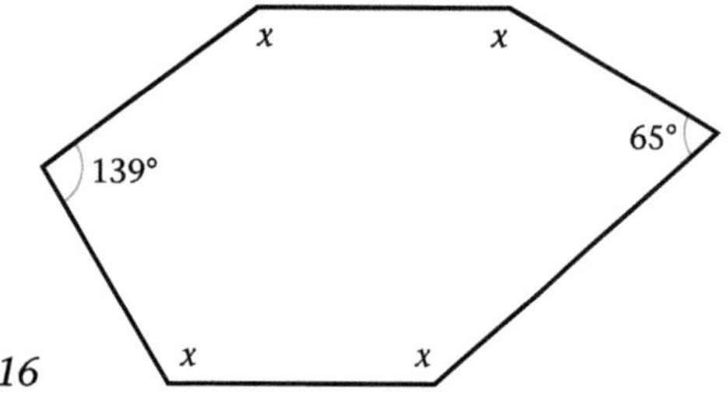

Fig. R16

10. The sum of 3 of the angles of a nonagon (9 sides) is 462°. The other 6 angles are all equal to each other. Calculate the size of each of the other angles.

Revision test 7 (Chapters 11, 12)

1. A road is 90 km long. On a map it appears as a line 6 cm long. The scale of the map is 1 to
A 15 B 1500 C 15 000 D 1 500 000

2. A map is drawn to a scale of 2 cm to 100 km. If the distance between two towns is 374 km, what is this distance as measured on the map?
A 1.87 cm B 3.74 cm
C 7.48 cm D 18.7 cm

3. Refer to the map in Fig.11.15 on page 99. Which one of the following is on Azania Crescent?
A Bank B Prison
C Dispensary D Church

4. How many sides has a polygon if the sum of its angles is 1620°?
A 7 B 9 C 11 D 16

5. The exterior angle of a regular decagon (ten sides) is
A 36° B 30° C 20° D 18°

6. A photograph measures 8 cm by 10 cm. It is enlarged so that the shorter side becomes 12 cm. What is the length of the longer side?

7. A plan is drawn on a scale of 1 cm to 5 m.
(a) If a wall is 24 m long, find its length on the plan.
(b) If the scale drawing of the floor of a tower is a circle of diameter 0.7 cm, find the actual diameter of the tower.

8. A rectangular sheet of paper measures 60 cm by 40 cm. Use a scale of 1 cm to 10 cm and make a scale drawing of the sheet of paper. Hence find the length of a diagonal of the sheet of paper.

9. (a) Find the sum of the angles of a pentagon.

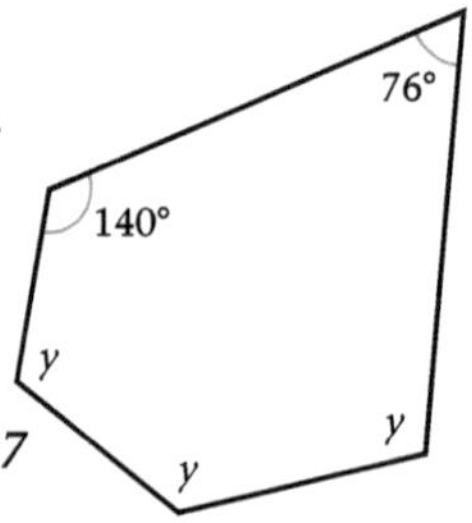

76°
140°
y
y
y

Fig. R17

(b) Hence calculate y in Fig. R17.

10 The sum of four of the angles of a heptagon (7 sides) is 600°. The other three angles are of sizes $4x°$, $5x°$ and $6x°$. Find the value of x and hence find the sizes of those three angles.

Revision exercise 8 (Chapters 14, 15)

1 A rectangle measures 8 cm by 15 cm. Make a sketch and use Pythagoras' theorem to calculate the length of one of its diagonals.

2 In Fig. R18 calculate XY.

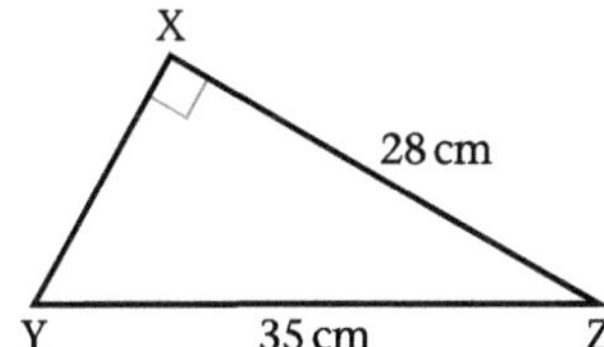

Fig. R18

3 Which of the following are Pythagorean triples?

(a) (10, 24, 26) (b) (12, 29, 31)
(c) (14, 49, 50) (d) (16, 30, 34)

4 A ladder 9.6 m long leans against a wall. It touches the wall at a point 9 m above the ground. Find the distance of the foot of the ladder from the wall.

Fig. R19 shows △PQR drawn on the Cartesian plane. Use it to answer questions 5, 6, 7 and 8.

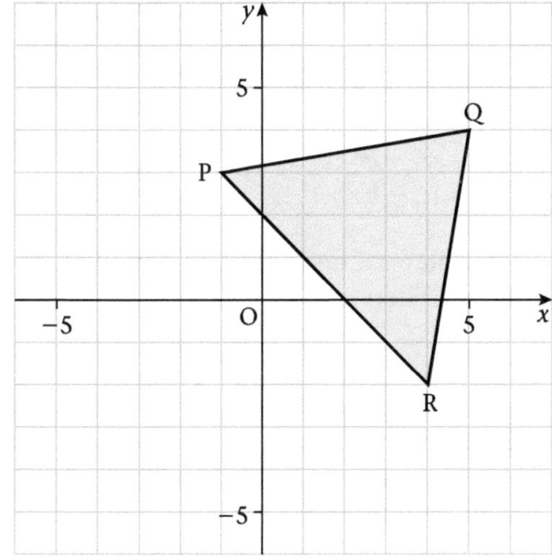

Fig. R19

5 (a) State the coordinates of P, Q and R.
(b) Find the coordinates of the images of P, Q and R if the triangle is translated 5 units downwards.
(c) △PQR is translated so that the image of R is at (1, −3). Find the images of P and Q.

6 State the coordinates of the images of the vertices of △PQR after reflection in (a) the x-axis, (b) the y-axis.

7 Find the coordinates of the images of P, Q and R after a clockwise rotation of (a) 180°, (b) 270° about the origin.

8 As question 7, with the point (0, 3) as the centre of rotation.

9 A triangle PQR has coordinates P(0, −2), Q(5, 3), R(−3, 4). The triangle is translated by $\begin{pmatrix} -2 \\ 3 \end{pmatrix}$ to form triangle TUV. State the coordinates of the vertices of the triangle TUV.

10 On graph paper:
(a) draw triangle PQR as in question 9.
(b) State the coordinates of the vertices P, Q and R after a clockwise rotation of 90° about the origin.

Revision test 8 (Chapters 14, 15)

1 Which of the following are Pythagorean triples?
I (3, 4, 5), II (5, 12, 13), III (8, 13, 17)

A I only B I and II only
C II only D II and III only

2 PQRS is a rectangle with sides 3 cm and 4 cm. If its diagonals cross at O, calculate the length of PO.

A 2.5 cm B 3.5 cm C 5.0 cm D 6.0 cm

3 The diagonals of a rhombus measure 8 cm by 6 cm. What is the length of a side of the rhombus?

A 5 cm B 6 cm C 7 cm D 8 cm

4 The coordinates of the point (2, 3) after reflection in the line $x = 0$ are

A (−2, −3) B (2, −3) C (−2, 3) D (3, 2)

5 The coordinates of the point $(-3, 4)$ after translation by a vector $\begin{pmatrix} 1 \\ -3 \end{pmatrix}$ are

A $(-2, 7)$ B $(2, 7)$ C $(4, -7)$ D $(-2, 1)$

6 Use tables to find the value of the following.

(a) 4.8^2 (b) 48^2 (c) 480^2

(d) $\sqrt{8}$ (e) $\sqrt{80}$ (f) $\sqrt{8520}$

Fig. R20 shows trapezium PQRS drawn in Cartesian axes. Use it to answer questions 7, 8 and 9.

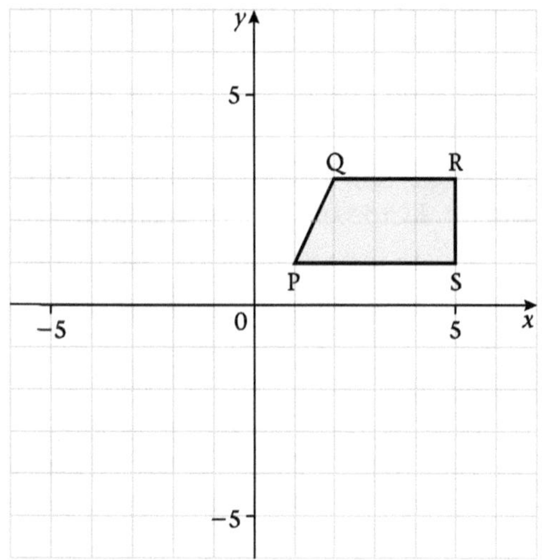

Fig. R20

7 In Fig. R20 PQRS is reflected first in the x-axis. Its image is then reflected in the y-axis. What are the coordinates of the final image of Q?

8 PQRS is translated so that the image of S is the point $(-3, -1)$. Find the images of P, Q and R.

9 Find the coordinates of the image of point R if PQRS is given an anticlockwise rotation of 90° about (a) the origin, (b) point P, (c) point $(4, -5)$.

10 A ladder leans against a wall. The ladder reaches 5 m up the wall and its foot is 2 m from the wall. If the foot of the ladder is placed 1 m further from the wall, calculate how far up the wall the ladder then reaches. Give your answer to 3 significant figures.

General revision test B (Chapters 9–16)

1 Find the value of $\sqrt{94}$ to 2 s.f.

A 3.1 B 9.7 C 31 D 47

2 In Fig. R21, what are the coordinates of the point where the diagonals of kite PQRS cross?

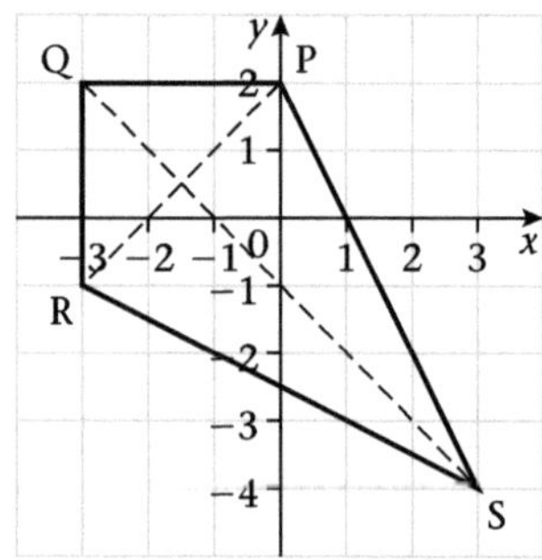

Fig. R21

A $(-1\frac{1}{2}, -\frac{1}{2})$ B $(-1\frac{1}{2}, \frac{1}{2})$

C $(\frac{1}{2}, -1\frac{1}{2})$ D $(1\frac{1}{2}, -\frac{1}{2})$

3 In Fig. R22, which one of the following equations gives the value of a^2?

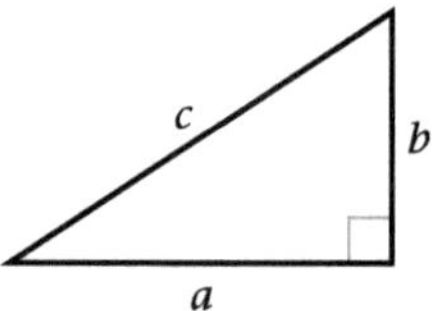

Fig. R22

A $a^2 = b^2 - c^2$

B $a^2 = b^2 + c^2$

C $a^2 = c^2 - b^2$

D $a^2 = (c - b)^2$

4 A quadrilateral has angles of 128°, 91°, n° and $2n$°. Then $n =$

A 47 B 73

C 89 D 141

5 A map is drawn to a scale of 1:40 000. What distance, in km, will a line on the map 2.8 cm long represent?

A 11.2 km B 7 km

C 1.12 km D 0.7 km

Fig. R23 is a graph giving the cost in $ of ribbon in metres. Use it to answer questions 6 and 7.

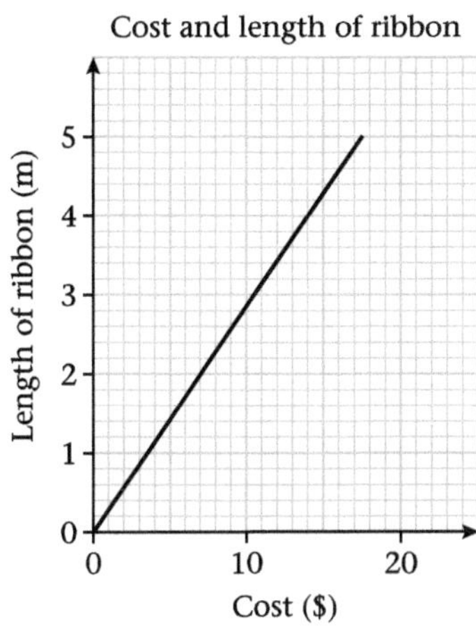

Fig. R23

6 What is the cost of 3 m of ribbon?

A $0.90 B $1.80
C $10.50 D $15

7 Approximately what length of ribbon can be bought for $4.50?

A 1.3 m B 2.5 m
C 15.75 m D 17.5 m

8 The range of values of a for which $11 - 2a \geqslant 1$ is

A $a \geqslant 5$ B $a \geqslant -5$
C $a \leqslant 5$ D $a \leqslant -5$

9 What is the length of the longest rod that can lie flat in a rectangular box which is 2 m long and $1\frac{1}{2}$ m wide?

A $5\frac{1}{2}$ m B 5 m C $3\frac{1}{2}$ m D $2\frac{1}{2}$ m

10

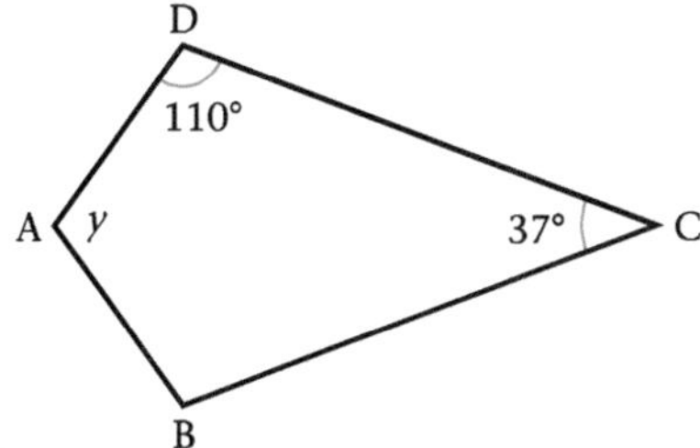

Fig. R24

In Fig. R24 ABCD is a kite with angles as shown. y =

A 37° B 53° C 103° D 143°

11 Solve the following.

(a) $3(x - 6) = 4(1 - 2x)$

(b) $\frac{c - 5}{4} - \frac{6 - c}{8} = 1$

(c) $\frac{15}{k} = \frac{3}{k} + 2$ (d) $\frac{15}{k} = \frac{3}{k + 2}$

(e) $\frac{1}{x + 5} - \frac{1}{4x - 7} = 0$

12 Choose a suitable scale and plot the following points:

A (11, −15), B (−4, 0),
C (−7, −3), D (−10, 0),
E (−4, 6), F (5, 6),
G (−2, 2), H (13, −13).

Join the points in alphabetical order and join H to A. What have you drawn a picture of?

13 A girl walks 6 km at a speed of v km/h.

(a) Write the time taken in hours in terms of v.

(b) Find v if her journey takes 1 h 20 min.

14 Find the value of k in Fig. R25.

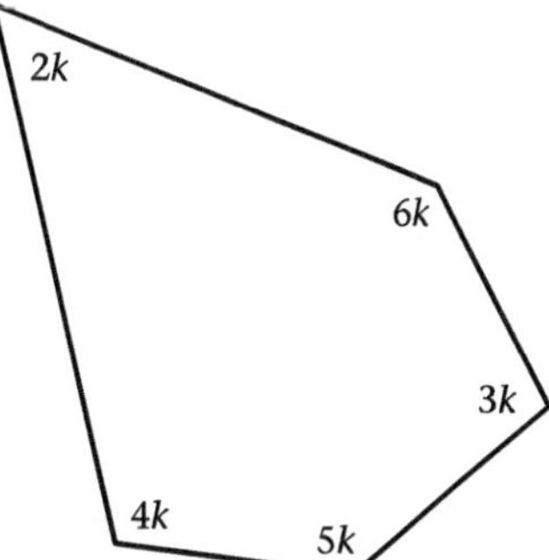

Fig. R25

15 Use number lines to draw graphs of the solutions of the following.

(a) $x + 2\frac{1}{2} > 0$ (b) $9 \geqslant 1 - 2x$

16 Fig. R26 is a sketch of a simple bridge. M is the mid-point of the bridge.

(a) Use a scale of 1 cm to 1m to draw an accurate scale drawing of the bridge.

(b) Find the length of the support AM.

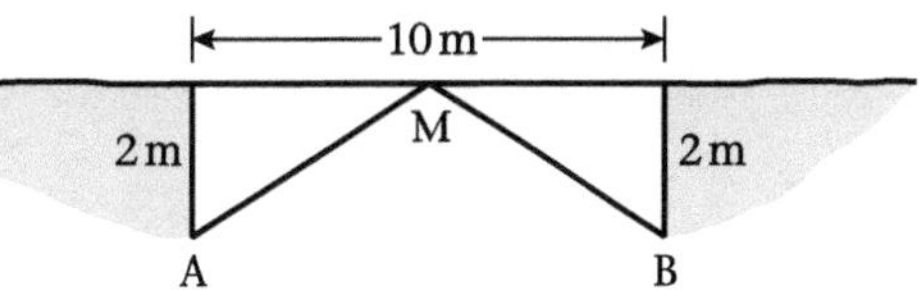

Fig. R26

17 Fig. R27 is part of a conversion graph. It is used to find the car insurance premiums to be paid according to the value of the cars being insured.

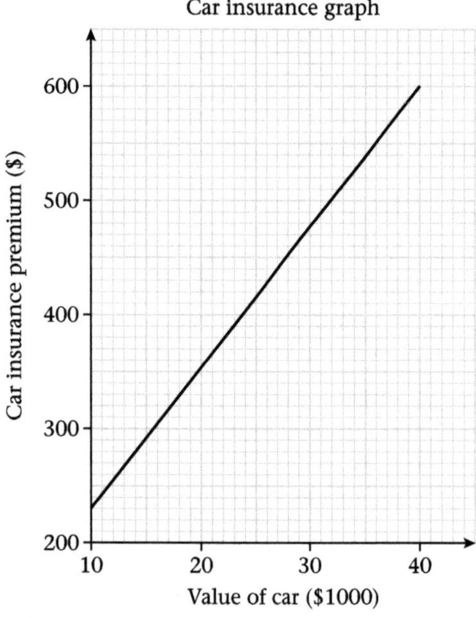

Fig. R27

(a) What is the insurance premium for a car of value \$27 500?
(b) If a woman pays an insurance premium of \$530, what is the value of her car?
(c) What is the difference in premiums paid by a man who insures his car for \$15 000 and a man who insures his car for \$35 000?

18 Find the solution sets for the following inequalities, given that x is an integer.

(a) $4x + 4 > 7$ (b) $30 - 5x < 2x + 9$

(c) $\frac{3}{2}x - \frac{7}{6} \leqslant \frac{5}{2} - \frac{1}{3}x$ (d) $-\frac{x}{5} > -\frac{13}{10}$

19 Fig. R28 shows a basic shape and a grid.

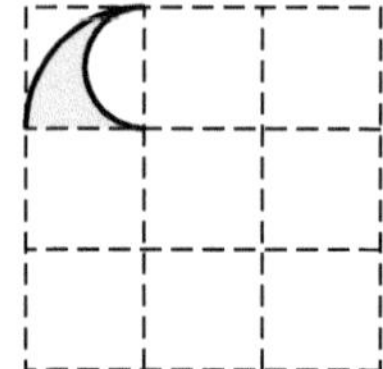

Fig. R28

Continue the pattern of Fig. R28 by reflecting the basic shape into the adjacent squares.

20 In $\triangle ABC$, AB = 10 cm, AC = 11 cm and $\widehat{B} = 90°$.

(a) Calculate BC.
(b) Calculate $\widehat{A}$ and $\widehat{C}$.

Give all answers correct to 2 s.f.

Chapter 17

Measurement (2)

Compound shapes, surface area of solids

Pre-requisites

- properties of solids and of shapes; units of measurement

Area of basic shapes (revision)

The formulae for the areas of some basic shapes are given in Fig. 17.1. These were previously found in Book 1.

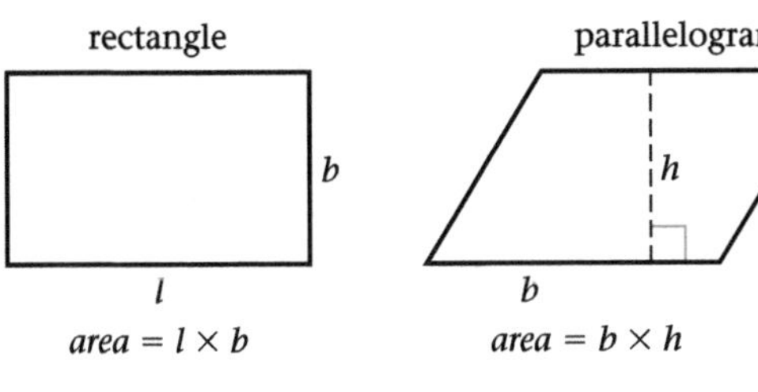

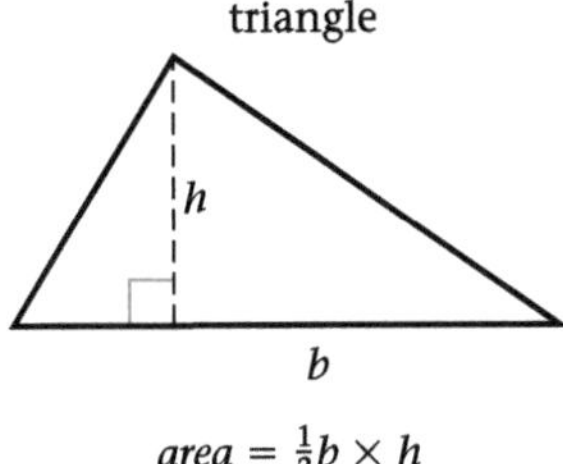

Fig. 17.1

Exercise 17a (Revision)

Calculate the areas of the shapes in Fig. 17.2.

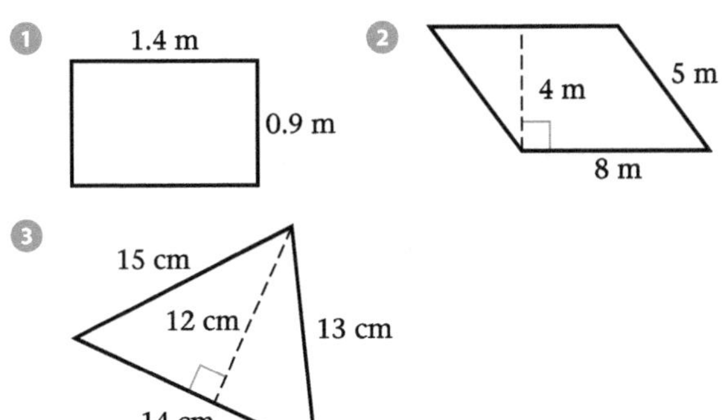

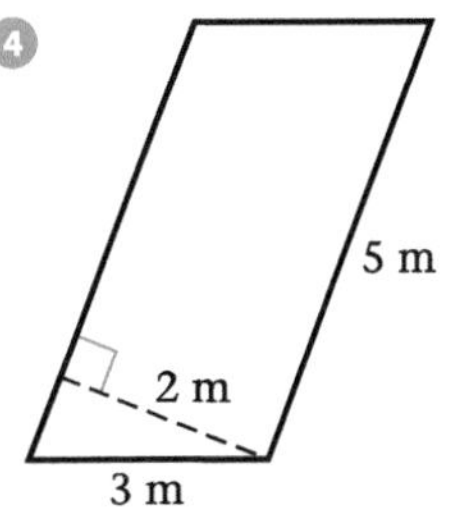

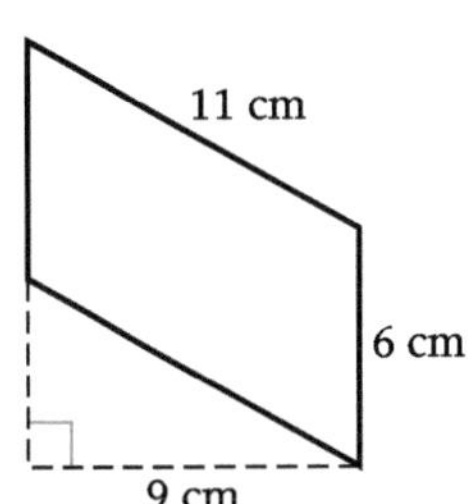

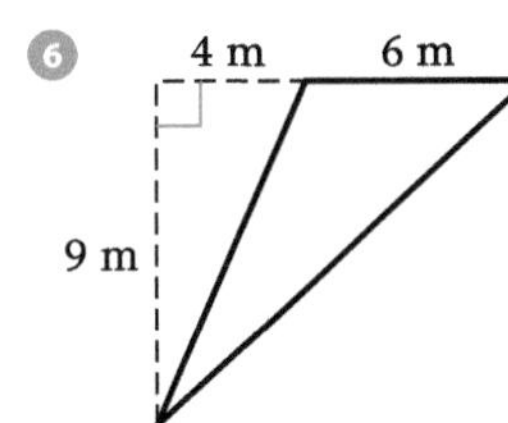

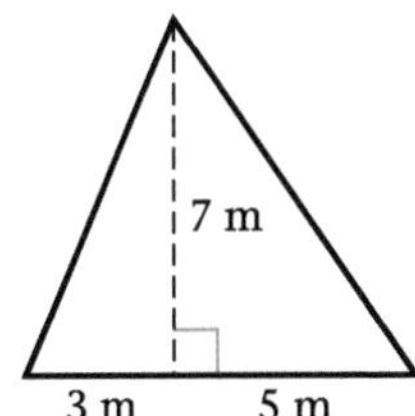

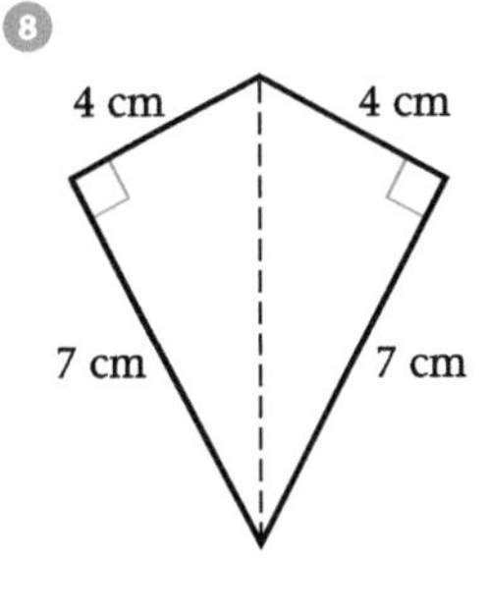

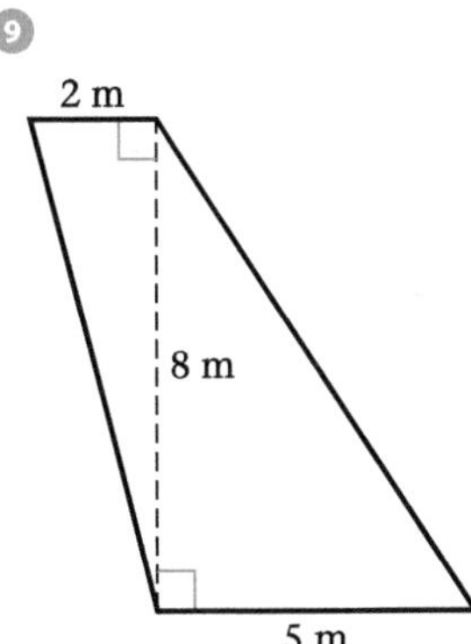

Fig. 17.2

10 Calculate the shaded area in Fig. 17.3. All dimensions are in cm.

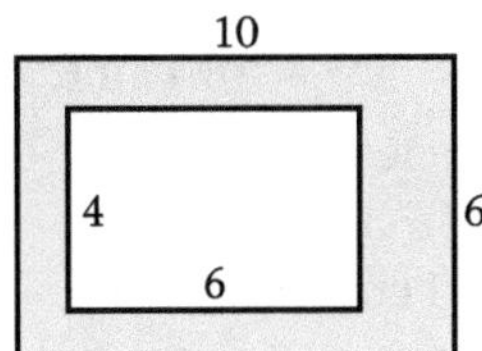

Fig. 17.3

Circles, rings, sectors

Example 1

What is the diameter of a circle of area $3850\,m^2$?

Area $= \pi r^2 = 3850 m^2$

Thus $\frac{22}{7} r^2 = 3850$

$r^2 = 3850 \times \frac{7}{22}$

$= 175 \times 7$

$= 25 \times 7 \times 7$

$= 5^2 \times 7^2$

$r = 5 \times 7 = 35$ (taking square roots)

diameter $= 2 \times 35\,m = 70\,m$

Example 2

What is the area of a flat washer 4.8 cm in outside diameter, the hole being of diameter 2.2 cm?

The required area is shaded in Fig. 17.4.

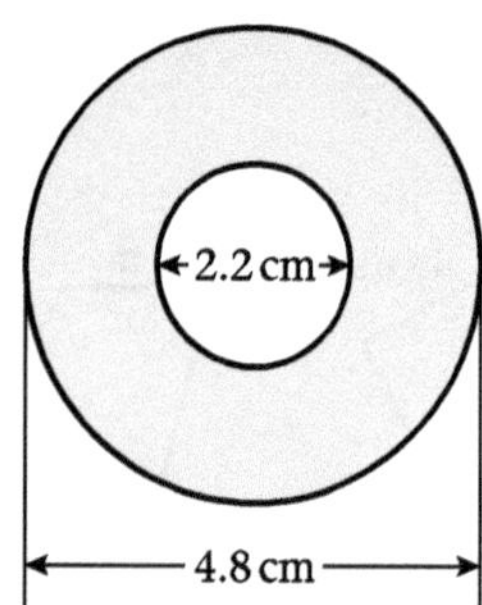

Fig. 17.4

Area $= \pi(2.4)^2 - \pi(1.1)^2\,cm^2$

$= \pi(5.76 - 1.21)\,cm^2$

$= \frac{22}{7} \times 4.55\,cm^2$

$= 14.3\,cm^2$

Alternatively

Area of outer circle of washer

$= \pi \times (2.4)^2\,cm^2$

$= \frac{22}{7} \times 5.76\,cm^2 = 18.1\,cm^2$

Area of inner circle of washer

$= \pi \times (1.1)^2\,cm^2$

$= \frac{22}{7} \times 1.21\,cm^2 = 3.8\,cm^2$

Required area

$= (18.1 - 3.8)\,cm^2$

$= 14.3\,cm^2$

Example 3

The sector of a circle of radius 7 cm has an angle of 108° at its centre. Calculate (a) the length of the arc of the sector, (b) the area of the sector.

Fig. 17.5 shows the sector of the circle.

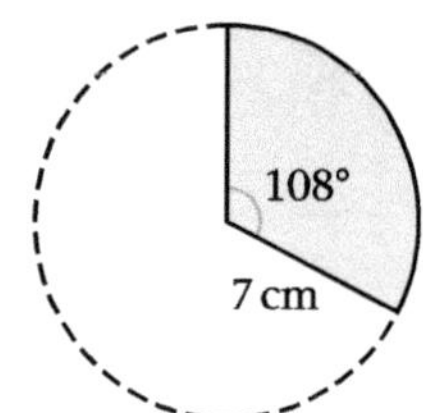

Fig. 17.5

(a) Length of arc $= \frac{108}{360}$ of $(2\pi \times 7)\,cm$

$= \frac{108}{360} \times 2 \times \frac{22}{7} \times 7\,cm$

$= \frac{3}{10} \times 44\,cm$

$= 13.2\,cm$

(b) Area of sector $= \frac{108}{360}$ of $(\pi \times 7^2)\,cm^2$

$= \frac{108}{360} \times \frac{22}{7} \times 7 \times 7\,cm^2$

$= \frac{3}{10} \times 22 \times 7\,cm^2$

$= 46.2\,cm^2$

Example 4

Find the area of the shaded segment for a quadrant of radius 14 cm.

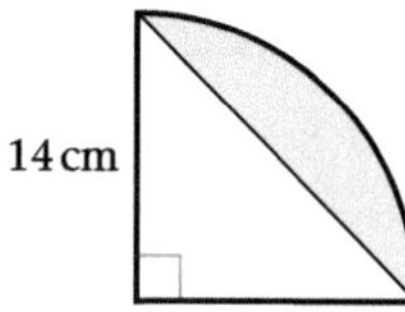

Fig. 17.6

A **quadrant** is a sector of a circle with an angle of 90°.

Area of quadrant $= \frac{1}{4} \times \pi \times 14^2\,cm^2$

$= \frac{1}{4} \times \frac{22}{7} \times 14 \times 14\,cm^2$

$= 11 \times 14 = 154\,cm^2$

Area of triangle $= \frac{1}{2} \times 14 \times 14\,cm^2$

$= 98\,cm^2$

Area of segment $= 154\,cm^2 - 98\,cm^2$

$= 56\,cm^2$

Exercise 17b

Throughout this exercise, take π to be $\frac{22}{7}$.

1. Find the area of each of the rings whose outside and inside diameters are as follows.
   (a) 8 m and 6 m
   (b) 22 cm and 20 cm
   (c) 15 m and 6 m
   (d) 8.6 cm and 8.2 cm

2. Complete Table 17.1 for sectors of circles. Make a rough sketch in each case.

Table 17.1

	Radius	Angle at centre	Length of arc	Area of sector
(a)	7 cm	90°		
(b)	35 m	72°		
(c)	4.2 cm	120°		
(d)	6.6 cm	135°		
(e)	14 m	300°		

3. Find the area of the shaded sections in Fig. 17.7. All dimensions are in cm.

(a) 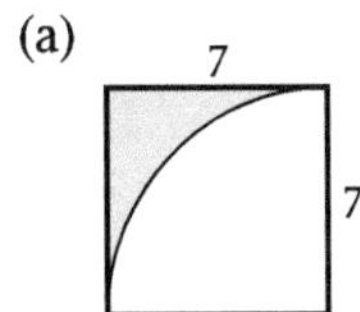

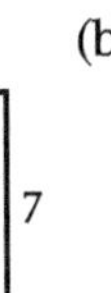

(b) 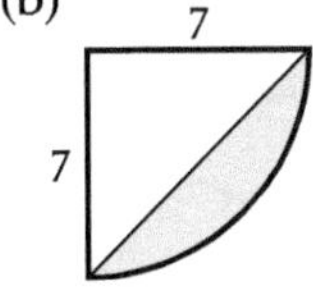

(c) 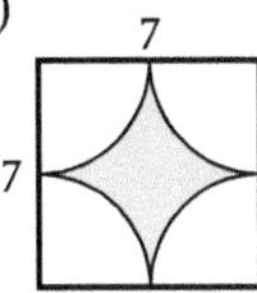

Fig. 17.7

4. Find the radii of circles with the following areas: (a) 154 cm^2, (b) 1386 cm^2, (c) $86\frac{5}{8}$ m^2, (d) 6.16 m^2.

5. Two circular bronze discs of radii 3 cm and 4 cm are melted down and cast into a single disc of the same thickness as before. What is the radius of the new disc?

6. The disc brake in a car is a flat metal ring 22 cm in diameter with a 6 cm diameter hole in the middle. Calculate the area of the metal.

7. The friction pad in a motorcycle shock absorber is a flat ring of fibre 10 cm in diameter with a 3 cm diameter hole in the middle. What is the area of the fibre?

8. The material for a pattern is cut in the form of a 210° sector of a circle of radius 3 m. What is the area of the material used?

9. Find the cross-sectional area of a round metal pipe if its outside diameter is 13.5 cm and the metal is 0.25 cm thick.

10. The windscreen wiper of a car sweeps through an angle of 150°. The blade of the wiper is 21 cm long and the radius of the unswept sector is 6 cm. See Fig. 17.8.

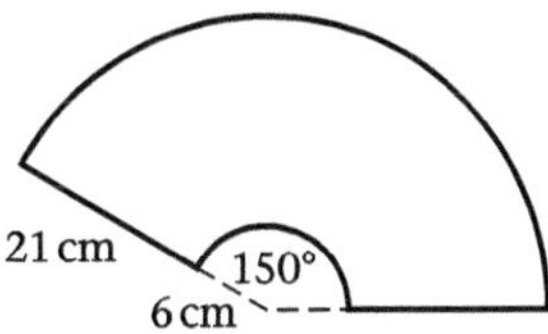

Fig. 17.8

What area of the windscreen is swept clean?

Surface area of a cylinder

Consider a cylinder of height h which has circular faces of radius r.

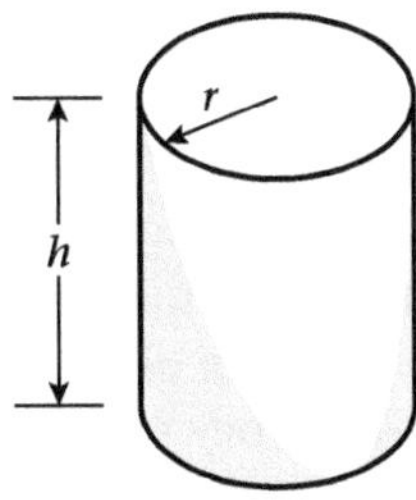

Fig. 17.9

Area of the two circular faces $= \pi r^2 + \pi r^2 = 2\pi r^2$

The curved surface is equivalent to a plane rectangle.

Length of rectangle = circumference of circular face $= 2\pi r$

breadth of rectangle $= h$

area of rectangle $= 2\pi r \times h = 2\pi rh$

Thus, area of curved surface $= 2\pi rh$

Total surface area of closed cylinder $= \mathbf{2\pi r^2 + 2\pi rh}$

Example 5

A cylinder of height 12 cm and radius 5 cm is made of cardboard. Use the value of 3.1 for π to calculate the total area of cardboard needed to make (a) a closed cylinder, (b) a cylinder open at one end.

(a) Notice that 'radius of a cylinder' is short for 'radius of the circular face of a cylinder'.

Surface area of closed cylinder
$= 2\pi r^2 + 2\pi rh$

Area of cardboard
$= (2 \times 3.1 \times 25) + (2 \times 3.1 \times 5 \times 12)\,\text{cm}^2$
$= (50 \times 3.1) + (120 \times 3.1)\,\text{cm}^2$
$= 155 + 372\,\text{cm}^2$
$= 527\,\text{cm}^2 = 530\,\text{cm}^2$ to 2 s.f.

(b) Surface area of open cylinder
$= \pi r^2 + 2\pi rh$

Area of cardboard
$= (3.1 \times 25) + (2 \times 3.1 \times 5 \times 12)\,\text{cm}^2$
$= 77.5 + 372\,\text{cm}^2$
$= 449.5\,\text{cm}^2 = 450\,\text{cm}^2$ to 2 s.f.

Note: The data in the question is given to 2 significant figures. The final answers can be rounded to 2 significant figures.

Exercise 17c

In this exercise, round the final answers to 2 significant figures.

1. A strip of thin paper is wound 8 times round a cylindrical pencil of diameter 7 mm (Fig. 17.10). Use the value $\frac{22}{7}$ for π to find the length of the paper. (Neglect the thickness of the paper.)

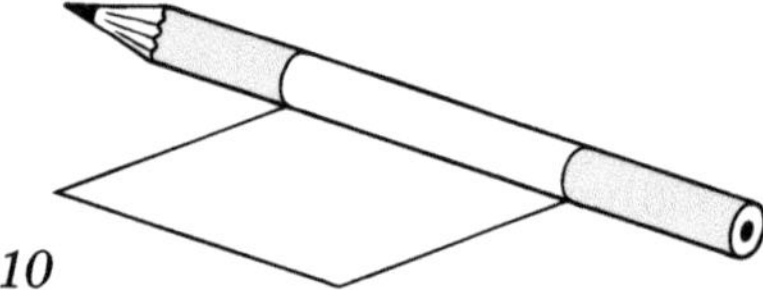

Fig. 17.10

2. A newspaper is rolled into a cylindrical shape of approximate diameter 4 cm. It is wrapped for posting with a strip of paper which goes about $2\frac{1}{2}$ times round the newspaper.
Use the value 3 for π to find the approximate length of the wrapping paper.

3. In Fig. 17.11 all dimensions are in cm. Use the value $\frac{22}{7}$ for π to calculate the surface areas of the closed cylinders.

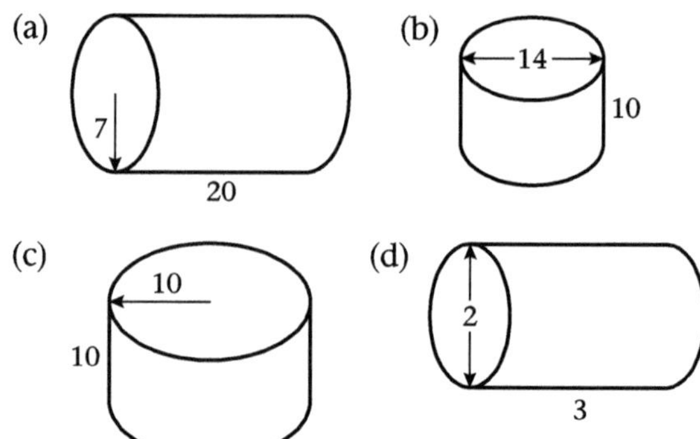

Fig. 17.11

4. A closed tin is in the shape of a cylinder of diameter 10 cm and height 15 cm. Use the value 3.14 for π to find
   (a) the total surface area of the tin,
   (b) the value of the tin to the nearest cent if tin plate costs \$1.50 per m^2.

5. A plastic container is in the shape of an open cylinder. It has a lid which is also an open cylinder. The dimensions of the container and lid are given in Fig. 17.12.

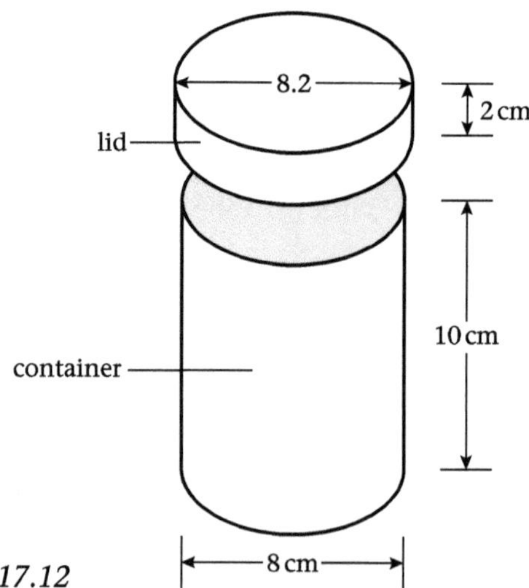

Fig. 17.12

Use the value 3.1 for π to find
(a) the surface area of the container,
(b) the surface area of the lid,
(c) the total area of plastic needed to make both.

Summary

Area of a sector of angle $\theta° = \frac{\theta}{360} \times \pi r^2$,
where r is the radius of the circle.

Curved surface area of a cylinder = $2\pi rh$ where r is the radius of the base and h is the height of the cylinder. Then
total surface area of a closed cylinder
$= 2\pi r^2 + 2\pi rh$

Practice Exercise P17.1

1. The diameter of the circle at the bottom of a closed cylindrical tin is 7.5 cm and the height of the tin is 8 cm.
Using $\pi = 3.14$, calculate
   (a) the surface area of the top and the bottom of the tin
   (b) the surface area of the curved side of the tin
   (c) the total surface area of the tin.

2. The top of a circular table has a diameter of 80 cm. The top of the table is covered with a circular plate of glass of diameter 75 cm. Calculate
   (a) the area of the top of the table
   (b) the area of the plate of glass
   (c) the area of the table top that is not covered by the plate of glass.

3. A flat metal ring 23.5 cm in diameter has a hole 8 cm in diameter in the middle. Calculate the area of the metal.

4. Calculate the area of the shaded sections in Fig.17.13. Use $\pi = 3.14$

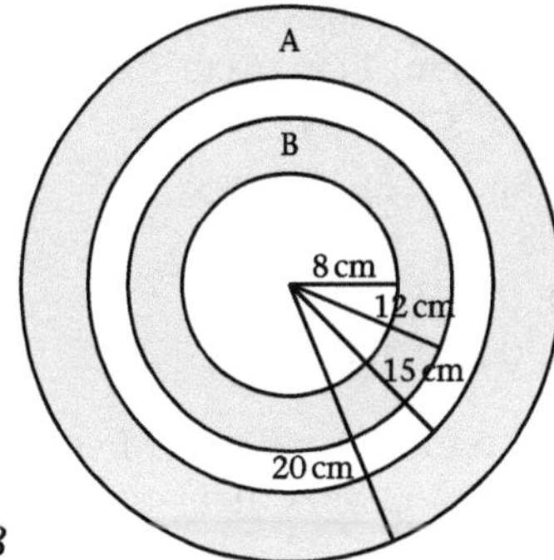

Fig. 17.13

5. A drinking mug is in the shape of an open cylinder. The curved surface of the mug is painted in different patterns as shown in Fig. 17.14.

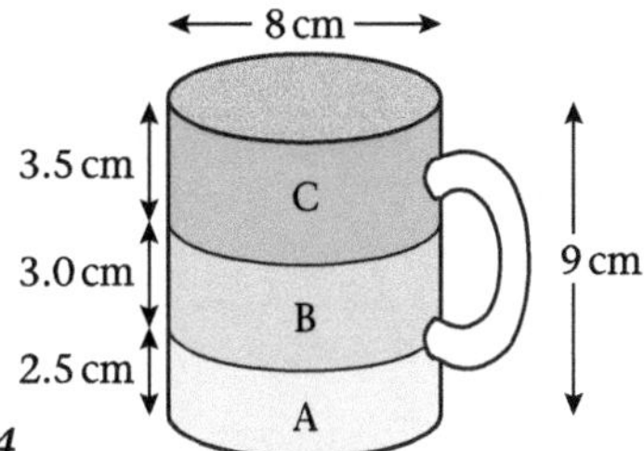

Fig. 17.14

Calculate the surface area of each of the three sections.

6. A signpost is built in the shape of a revolving closed cylinder on a fixed cylindrical base. Fig.17.15. Calculate:

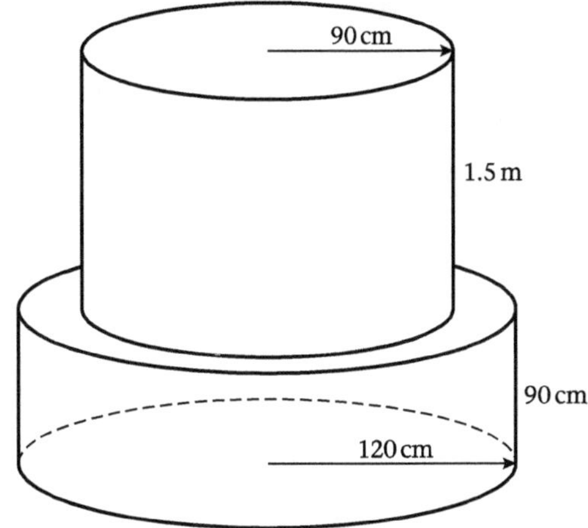

Fig. 17.15

   (a) the area of the curved surface of the cylindrical base.
   (b) the area of the surface available for signs on the revolving section.

7. A heavy roller is of diameter 84 cm and length 1.5 m. Calculate the area covered by the roller when it has been rolled through 24 complete revolutions.

Chapter 18

Statistics (2)

Mean, median, mode

Pre-requisites
- ordering of numbers

Averages

The **average** of a set of numbers is a very important statistic. The average is typical of the set of numbers and, therefore, provides information about them. For example:

(a) If a football team's **average score** is 5.2 goals, we know that the team is good at scoring goals.

(b) If two classes have **average ages** of 8.7 years and 16.9 years, we expect that the first is a Primary School class and the second is a Secondary School class.

(c) If the **average life** of a battery is 20 hours, we expect a new battery to last about 20 hours, maybe a little more or a little less.

Arithmetic mean

The **arithmetic mean**, or just **mean** for short, is the most common average. The averages in (a), (b) and (c) above are all examples of arithmetic means.

If there are n numbers in a set, then,

$$\text{arithmetic mean} = \frac{\text{sum of the numbers in the set}}{n}$$

Example 1

In 5 tests a student's marks were 13, 17, 18, 8 and 10. What is her average mark?

Average (mean) mark

$$= \frac{13 + 17 + 18 + 8 + 10}{5}$$

$$= \frac{66}{5} = 13.2$$

Example 2

A hockey team has played 8 games and has a mean score of 3.5 goals per game. How many goals has the team scored?

$$\text{Mean score} = \frac{\text{total number of goals}}{\text{number of games}}$$

$$3.5 = \frac{\text{total number of goals}}{8}$$

Multiply both sides by 8

$3.5 \times 8 = \text{total number of goals}$

Total number of goals scored = 28

Exercise 18a

1. Calculate the mean of the following sets of numbers.
   (a) 9, 11, 13
   (b) 7, 8, 12
   (c) 1, 9, 4, 6
   (d) 15, 3, 5, 9
   (e) 1, 8, 6, 8, 7
   (f) 4, 6, 2, 1, 7
   (g) 5, 12, 3, 9, 10, 3
   (h) 8, 9, 11, 12, 15, 17
   (i) 3, 1, 9, 8, 2, 3, 0, 7, 2, 5
   (j) 8, 2, 3, 1, 7, 8, 8, 4, 1, 1

2. Calculate the mean of the following.
   (a) 4 cm, 7 cm, 1 cm, 6 cm
   (b) \$6, \$7, \$7, \$9, \$12
   (c) 3.9 kg, 5.2 kg, 5.3 kg
   (d) $1\frac{1}{4}$, $3\frac{3}{4}$, $4\frac{3}{4}$
   (e) 0.9, 0.8, 0.6, 0.4, 0.9, 1.1, 0.2, 0.3, 0.5, 0.6

3. A market trader's profit after five days of trading was \$63.85. Calculate his mean profit per day.

4 On six working days a garage mended 6, 5, 2, 0, 3, 2 punctures. Calculate the mean number of punctures per day mended by the garage.

5 In four successive days a trader sold 24, 48, 12 and 60 oranges. Calculate her mean daily sale of oranges.

6 The temperatures at midday during a week in Roseau were 23 °C, 25 °C, 24 °C, 26 °C, 25 °C, 26 °C, 26 °C. Find, to the nearest degree, the average midday temperature for the week.

7 In the first six days of the month of June, the rainfall was 39 mm, 21 mm, 17 mm, 11 mm, 0 mm, 2 mm. It did not rain on any of the other days of the month. Calculate (a) the mean daily rainfall for the first six days, (b) the mean daily rainfall for the whole month. (June has 30 days.)

8 After 15 matches a football team's goal average was 1.8. How many goals has the team scored?

9 The average age of a mother and her three children is 10 years. If the ages of the children are 1, 4 and 7 years, how old is the mother?

10 In a test out of 40, the marks of 15 students were 31, 18, 6, 26, 36, 24, 23, 14, 29, 28, 32, 9, 11, 22, 21.

(a) Calculate the mean mark for the test.

(b) Express the mean mark as a percentage of the total mark.

11 Ten *Atlas* batteries were tested to find their average life. The times, in hours, that the batteries lasted were as follows: 10.8, 10.6, 11.4, 8.9, 10.1, 10.6, 9.9, 12.6, 10.5, 11.9.

(a) Find, to the nearest tenth of an hour, the average life of the ten batteries.

(b) Which of the following advertisements is more accurate?

(i) *Atlas batteries* are guaranteed to last 10 hours.

(ii) *Atlas batteries* have an average life of over 10 hours.

Example 3

For 3 days a trader made a profit at the rate of \$5.30 per day. For the next 4 days her profit averaged \$6.70 per day. What was her average daily profit for the week?

For the first 3 days:

profit = 3 × \$5.30 = \$15.90

For the next 4 days:

profit = 4 × \$6.70 = \$26.80

For all 7 days:

total profit = \$15.90 + \$26.80

= \$42.70

$$\text{Average daily profit} = \frac{\$42.70}{7} = \$6.10$$

Example 4

A man walked 12 km at 3 km/h and cycled 18 km at 9 km/h. What was his average speed for the whole journey?

$$\text{Average speed} = \frac{\text{total distance}}{\text{total time}}$$

Total distance = 12 km + 18 km = 30 km

Time for 1st part of journey

$$= \frac{12}{3}\text{ h} = 4\text{ h}$$

Time for 2nd part of journey

$$= \frac{18}{9}\text{ h} = 2\text{ h}$$

Total time = 4 h + 2 h = 6 h

$$\text{Average speed} = \frac{30}{6}\text{ km/h} = 5\text{ km/h}$$

Exercise 18b

1 On a journey, a motorist travels the first 40 km in $\frac{1}{2}$ hour, the next 34 km in 25 min and the last 7 km in 5 min. What is the average speed for the whole journey?

2 Diana lives 4 km from school. She walks 1 km at 6 km/h and travels the rest of the way by bus at 30 km/h.

(a) How many minutes does the whole journey take?

(b) What is her average speed in km/h?

3 Bob lives 5 km from school. He walks 1 km at 4 km/h and travels the rest of the way by bus at 16 km/h. What is his average speed for the whole distance?

4 A motorist averages 48 km/h for the first 30 km of a journey and 64 km/h for the next 120 km. What is the average speed for the whole journey?

5 For 4 weeks a man's average wage was $41 per week. For the next 6 weeks his average wage was $38 per week. What was his average weekly wage for the 10 weeks?

6 A factory employs 50 workers. 40 earn $30/ hour and 10 earn $40/hour. What is the average hourly rate of pay?

7 A 3rd grade contains three classes of 36, 33 and 31 students. In an examination the average marks for the classes were 65, 56 and 52 respectively. What was the average mark for the 3rd grade altogether?

8 The road from A to B is 10 km uphill followed by 20 km downhill. A motorcyclist averages 36 km/h uphill and 90 km/h downhill. Calculate the average speed (a) from A to B, (b) from B to A.

9 A class contains 10 girls and 20 boys. The average height of the girls is 1.58 m and the average height of the boys is 1.67 m. Calculate the average height of the students in the class.

10 A lorry driver travelled 84 km between two towns. The first 60 km of road was untarred and the average speed over this part was 30 km/h. If the average speed for the whole journey was 36 km/h, calculate the average speed over the good part of the road.

The arithmetic mean, very often referred to as average, is a statistic used as a representative of the set of data. In all of the examples above the arithmetic mean was used. Other representative values are sometimes used.

The median

The **median** of a set of numbers is the middle number when the numbers are arranged in order of size.

Example 5

Find the median of 17, 34, 13, 22, 27, 44, 8, 31, 13.

Arrange the numbers in order of increasing size:
8, 13, 13, 17, 22, 27, 31, 34, 44
There are 9 numbers. The 5th number is in the middle. The 5th number is 22. Median = 22.

Note that the result would be the same if the numbers were arranged in order of decreasing size (i.e. rank order). Also notice that every number is written down even if some numbers appear more than once. In Example 5, there are two 13s; each is written down and counted.

If there is an even number of terms in the set, find the mean of the middle two terms. Take this calculated value to be the median.

Example 6

Find the median of 8.3, 11.3, 9.4, 13.8, 12.9, 10.5.

Arrange the set of numbers in order of size:
8.3, 9.4, 10.5, 11.3, 12.9, 13.8
There are 6 numbers. The median is the mean of the 3rd and 4th numbers:

$$\text{median} = \frac{10.5 + 11.3}{2} = \frac{21.8}{2} = 10.9$$

Example 7

Find the median of the following percentages:
43%, 76%, 64%, 37%, 76%, 54%.

First, arrange the percentages in rank order.
76%, 76%, 64%, 54%, 43%, 37%
The median is the middle value. However, there is no single middle value in the list. The 3rd and 4th percentages are in the middle. Take the median as the mean of these values.

$$\text{Median} = \frac{64 + 54}{2}\% = \frac{118}{2}\% = 59\%$$

The mode

In many examples of statistical data, some numbers appear more than once. The **mode** is the number that appears most often. In Example 5, the number 13 appears twice. 13 is the mode of this data.

Example 8

The following are the number of days absent during a term for a class of 21 students: 7, 5, 0, 5, 0, 3, 0, 15, 0, 2, 2, 0, 1, 3, 5, 32, 1, 0, 0, 1, 2. Find the mode, median and mean days absent.

Arrange the number of days absent in order:
0, 0, 0, 0, 0, 0, 0, 1, 1, 1, 2, 2, 2, 3, 3, 5, 5, 5, 7, 15, 32
0 occurs most often. Mode = 0 days.
The median is the 11th number.
Median = 2 days

$$\text{Mean} = \frac{\text{total number of days absent}}{\text{total number of students}}$$

$$= \frac{0 + 0 + \ldots + 15 + 32}{21}$$

$$= \frac{84}{21} = 4 \text{ days}$$

The frequency is the number of times that a piece of data occurs. Hence the mode is the piece of data with the highest frequency.

Example 9

The distribution of ages of a group of 30 Teacher Training College students is given in Table 18.1.

Table 18.1

Age in years	20	21	22	23
Frequency	4	11	9	6

Find the mode, median and mean ages of the students.

(a) The mode is the piece of data that occurs most often. The highest frequency is 11. The mode is 21 years. Or we may say the modal age is 21 years.

(b) There are 30 students. The median age is the mean of the ages of the 15th and 16th students.

In this case it is not necessary to make an ordered list of all the ages. Since there are 4 students aged 20 years and 11 students aged 21 years, the 15th student is aged 21 years (4 + 11 = 15). The 16th student is the first of the 22 year age group.

$$\text{Median age} = \frac{21 + 22}{2} = 21.5 \text{ years}$$

(c) $\text{Mean age} = \dfrac{\text{total ages of all the students}}{\text{number of students}}$

Using the frequency table,

4 students are aged 20
  sum of their ages = $20 \times 4 = 80$ years

11 students are aged 21
  sum of their ages = $21 \times 11 = 231$ years

9 students are aged 22
  sum of their ages = $22 \times 9 = 198$ years

6 students are aged 23
  sum of their ages = $23 \times 6 = 138$ years

$$\text{Mean age} = \frac{80 + 231 + 198 + 138}{30}$$

$$= \frac{647}{30} = 21.57 \text{ years}$$

Notice, in Example 9, that the mode, median and mean are quite close in value to each other. This usually happens when the frequency figures rise and fall fairly smoothly. It is possible for some of these averages to be equal to each other.

Exercise 18c

1. Find the mode, median and mean of the following sets of numbers.
   (a) 7, 7, 9, 12, 15
   (b) 4, 5, 5, 7, 8, 10
   (c) 4, 8, 11, 11, 12, 12, 12
   (d) 2, 3, 6, 6, 7, 7, 7, 8, 8, 9
   (e) 15, 13, 13, 12, 11, 11, 10, 10, 10, 10, 9, 9, 8, 8, 7

2. Arrange the following numbers in order of size. Find the mode, median and mean of each set.
   (a) 2, 4, 3, 4
   (b) 7, 5, 2, 9, 5, 8
   (c) 1, 0, 14, 0, 5, 10
   (d) 7, 5, 11, 7, 12, 8, 6, 9, 7
   (e) 6, 5, 3, 6, 3, 2, 4, 6, 4, 5, 6, 4

3 In a test, the grades go from A (best) to E (poorest). The bar chart in Fig. 18.1 shows the number of students getting each of these grades.
Find (a) the number of students who took the test; (b) the mode for the test; (c) the median grade for the test.

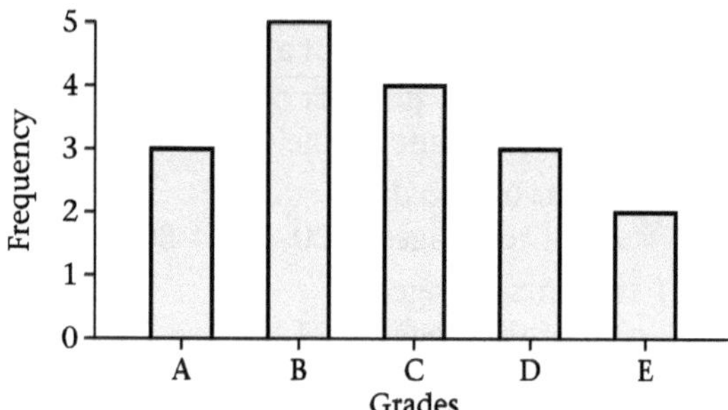

Fig. 18.1

4 16 people were asked which size of shoe they wear. Their answers are shown in the frequency table (Table 18.2).

Table 18.2

Shoe size	5	6	7	8	9	10
Frequency	1	2	5	4	3	1

Find (a) the modal shoe size, (b) the median shoe size.

5 Ten students walk to school each day. The distances they walk, to the nearest kilometre, are given in the frequency table (Table 18.3).

Table 18.3

Distance (km)	1	2	3	4	5
Frequency	4	2	2	1	1

Find the mode, median and mean distances walked.

6 Table 18.4 gives the frequencies of the ages of students in a choir.

Table 18.4

Age in years	14	15	16	17
Frequency	3	4	5	4

Find
(a) the number of students in the choir;
(b) the modal and median ages of the choir;
(c) the mean age of the choir.

Example 10

Nine boys take a test. Their marks are: Ian 4, Ben 8, Dan 7, Rex 9, Joe 4, Ron 3, Bob 6, Tom 4, Sam 7. Place the marks in rank order. Hence find the mean, median and mode for the test.

Table 18.5 gives the names, marks and positions of the boys in rank order, i.e. from highest mark to lowest mark.

Table 18.5

Name	Mark	Position
Rex	9	1
Ben	8	2
Dan	7	3
Sam	7	3
Bob	6	5
Ian	4	6
Joe	4	6
Tom	4	6
Ron	3	9

$$\text{mean} = \frac{9 + 8 + 7 + 7 + 6 + 4 + 4 + 4 + 3}{9}$$
$$= \frac{52}{9} = 5\tfrac{7}{9}$$

Bob is in the middle position. His mark is 6.
Median = 6.
4 marks is the most common score.
Mode = 4.

Example 11

A dressmaker buys 3 metres of fabric at \$12.50 per metre. The following week she buys 2 metres of the same fabric at \$13.00 per metre. Find the average cost per metre of the fabric.

3 metres of fabric at \$12.50 = \$37.50
2 metres of fabric at \$13.00 = \$26.00
Total cost of 5 metres = \$63.50
Average cost of 1 metre $= \frac{\$63.50}{5}$
= \$12.70

Example 12

A motorist travelled 96 km at an average speed of 60 km/h. He returned at an average speed of 48 km/h. What was his average speed for the whole journey?

96 km at 60 km/h takes $\frac{96}{60}$ hours = 1.6 h

96 km at 48 km/h takes $\frac{96}{48}$ hours = 2.0 h

Altogether he travelled 192 km in 3.6 h.

$$\text{Average speed} = \frac{192}{3.6}\text{ km/h} = \frac{1920}{36}\text{ km/h}$$
$$= \frac{160}{3}\text{ km/h} = 53\tfrac{1}{3}\text{ km/h}$$

Always remember that

$$\textbf{average speed} = \frac{\textbf{total distance travelled}}{\textbf{total time taken}}$$

Execise 18d

1. Arrange the following sets of numbers in order of size. Find the mean, median and mode of each set.
   (a) 7, 10, 7, 9, 7 (b) 5, 3, 0, 7, 3, 6
   (c) 1, 9, 5, 6, 1, 4
   (d) 8, 3, 1, 7, 3, 4, 8, 3, 4, 9, 5
   (e) 0.1, 0, 1.5, 0, 0.6, 1.1
   (f) 159.5, 155.8, 153.7, 157.2, 155.8

2. The weekly wages of 5 Local Government trainees are \$120.83, \$123.09, \$118.71, \$129.19, \$126.83. Find (a) the mean, (b) the median wage.

3. Six men together weigh $\frac{1}{2}$ tonne. What is the average weight of the men in kg?

4. A fisherman's catch on four different days was 28 kg, 16.5 kg, 34 kg, 19.9 kg of fish. What was his average catch per day?

5. The mean age of 4 women is 19 yr 11 mo. When a fifth woman joins them, the mean age of all 5 is 20 yr 7 mo. How old is the fifth woman?

6. There are 8 men and 1 woman in a boat. The average mass of the 9 people is 79 kg. Without the woman, the average is 81.5 kg. What is the mass of the woman?

7. A trader mixes 3 kg of sugar at 86c/kg with 2 kg of sugar at 76c/kg. What is the cost/kg of the mixture?

8. The mean daily rainfall for a week was 5.5 mm. For the first 6 days the mean rainfall was 1 mm. How much rain fell on the 7th day?

9. A bridge $1\frac{1}{4}$ km long was built at a cost of \$$7\frac{1}{2}$ million. What was the average cost per metre of building the bridge?

10. Some students were asked how many brothers and sisters they had. The bar chart in Fig. 18.2 shows the number of students who had 0, 1, 2, 3, 4, 5, 6, 7 or 8 brothers and sisters.

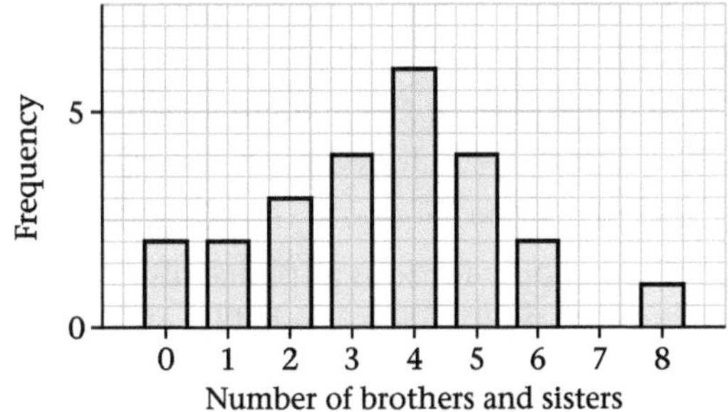

Fig. 18.2

Find (a) the number of students in the survey, (b) the modal number of brothers and sisters, (c) the median, and (d) the mean number of brothers and sisters.

11. A man travelled 30 km in a car at an average speed of 40 km/h. He returned at an average speed of 60 km/h. Find his average speed for the whole journey.

12. A woman travelled for $\frac{1}{2}$ hour in a car at an average speed of 40 km/h. For the next $\frac{1}{2}$ hour her average speed was 60 km/h. What was her average speed for the whole time?

13. In a test the average marks for three classes were 74, 58 and 51. If the classes contained 25, 22 and 23 students respectively, what was the average mark for the three classes together?

14 Five students took a test in four subjects. Their results are given in Table 18.6.

Table 18.6

	English	History	Maths	Science
Rudy	51	69	54	57
Sonya	68	60	67	73
Tinga	80	73	49	42
Ural	26	14	37	35
Vera	34	44	38	48

Find (a) the mean mark of each student, (b) the mean mark in each subject, (c) the median mark in each subject.

Mean, median, mode are all used as representative values of a set of data. Which one is used depends on the particular situation.

Exercise 18e (Group Discussion)

1 A motorist drives 80 km at an average speed of 63 km/h. Which average is this, mean, median or mode?

2 20 men apply for jobs as police officers. There are only 10 jobs available, and these are given to the 10 men who are above the average height of those who applied. Which average is this, mean, median or mode?

3 A trader sells shoes. He wants to know the average size of shoe that people buy. Which average is most useful?

Summary

The **mean**, **median** and **mode** are three measures of **statistical averages**.

The arithmetic mean

$$= \frac{\text{sum of numbers in the set}}{n}$$

where there are n numbers in the set.

The median is the middle number or the mean of the two middle numbers, when the numbers are arranged in ascending or descending order.

The mode is the number that occurs most often in a set of numbers.

Numbers here may refer to observations in data.

For a given set of data, one of the three averages may be more appropriate than the other two to describe the distribution.

Practice Exercise P18.1

1 The heights of twelve children are shown below:

4.1 m 4.4 m 4.8 m 4.0 m 5.1 m 3.9 m
4.2 m 3.7 m 4.0 m 4.3 m 5.2 m 4.7 m

(a) Arrange the heights of the children in order from shortest to tallest.
(b) What is the median height of the children?
(c) Calculate the mean height of the children, giving your answer to one decimal place.

2 The following is the score of 10 teams in a 20/20 cricket tournament:

Table 18.7

Team	Runs scored	Wickets lost
1	86	6
2	124	7
3	98	5
4	75	3
5	82	4
6	88	3
7	109	9
8	74	4
9	78	3
10	86	6

(a) What is the total number of runs scored in the tournament?
(b) What is the mean score per team?
(c) What is the mean number of wickets lost per team?
(d) Which team do you think made the best score? Why?

3 The mean age of five boys is 10 years 6 months.

(a) What is the total age of the five boys?

A sixth boy who is 10 years old joins the group of boys.

(b) What is the total age of the six boys?

(c) What is the mean age of the six boys?

4 A student counted the number of vowels on five pages of a history book. The results are shown in Table 18.8

Table 18.8

vowels	a	e	i	o	u
1st page	60	121	54	82	32
2nd page	50	82	46	73	25
3rd page	54	95	48	62	26
4th page	62	119	63	79	21
5th page	54	98	50	64	23
Total					

(a) Complete Table 18.8 to show the total number of each vowel used on the five pages.

(b) Which vowel is the mode?

(c) What is the total number of vowels used on the five pages?

(d) What is the mean number of vowels used on the pages?

(e) Draw a bar chart to show the number of vowels used on each of the five pages.

5 Five students were asked to keep a record for a week, of the amount of time spent waiting for a bus to get to school. The information is shown in Table 18.9.

Table 18.9

Students / Waiting time (mins)	A	B	C	D	E
Monday	30	30	15	24	30
Tuesday	15	27	6	15	30
Wednesday	12	10	8	10	15
Thursday	25	20	15	20	15
Friday	40	45	24	30	45

(a) Show in a table, the total amount of waiting time for the five days, for each student.

(b) Draw a bar chart to show this information.

(c) Which student spent most time waiting for a bus?

(d) What is the mean waiting time for this student?

(e) Which student spent least time waiting for a bus?

(f) On which day did they spend the most time waiting on a bus?

(g) On which day would one of the students most likely be late for school?

6 The final mark for the students in a class in mathematics is the total of the marks gained in three tests. The marks for five students in the class are shown in Table 18.10

Table 18.10

	Test 1	Test 2	Test 3	Total %
Adeff	18	25	27	
Brandon	21	22	31	74
Jermaine	26	28	38	
Kyle	22	18	36	
Rayan	20	24	28	

(a) Complete the table to show the final mark for the students.

(b) Calculate the mean mark for the five students. Give your answer to the nearest whole number.

(c) What is the median mark for the students?

(d) Draw a bar chart to show the marks for each student.

7 A football team scored the following number of goals in ten matches:

3, 2, 5, 3, 3, 0, 3, 2, 1, 3.

Find the mode and the mean score.

8 The mean age of a group of 7 boys is 9 years. What is the sum of the ages of the 7 boys? Three boys of mean age 8 years join the group. What is the mean age of the 10 boys?

Chapter 19

Solving triangles (2)

The right-angled triangle – the tangent ratio, angle of elevation, angle of depression

Pre-requisites

- properties of right-angled triangles; ratio

In Chapter 14 Pythagoras' theorem was used to solve for the sides of right-angled triangles. It is often necessary to solve triangles to find the sizes of the angles.

Tangent of an angle

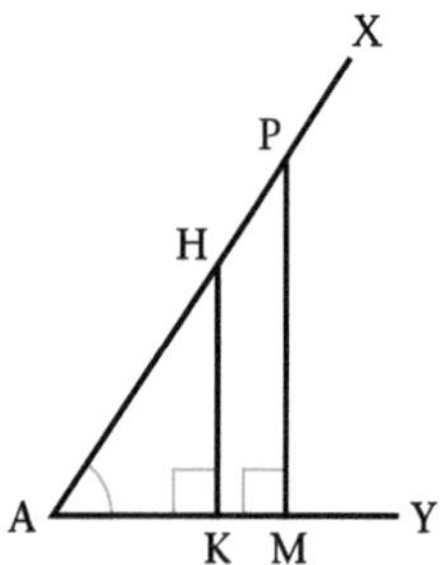

Fig. 19.1

Fig. 19.1 shows an angle A with arms AX and AY. H and P are any two points on AX. HK and PM meet AY perpendicularly at K and M. Thus △AHK is equiangular and, therefore, similar to △APM.

Thus $\frac{HK}{KA} = \frac{PM}{MA}$

If any number of points are taken on AX and the perpendiculars drawn, the ratio

$\frac{\text{length of perpendicular}}{\text{length of base-line}}$ is the same for each.

Hence the value of the ratio $\frac{HK}{KA}$ depends only on the size of $\hat{A}$. When HK is perpendicular to KA, the ratio $\frac{HK}{KA}$ is called the **tangent** of the angle A. This is usually shortened to **tan A**.

Fig. 19.2 shows △AHK in two positions.

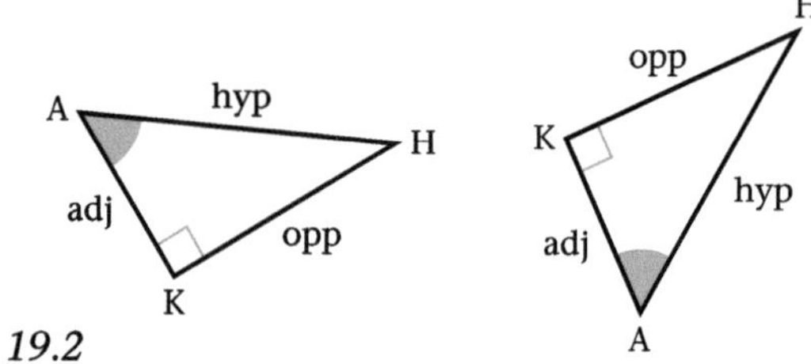

Fig. 19.2

Since △AHK is a right-angled triangle, the sides of the triangle are defined as follows:

AH the hypotenuse,
HK the side opposite to $\hat{A}$,
KA the side adjacent to $\hat{A}$.

These are abbreviated to **hyp**, **opp**, **adj** respectively, so that

$$\textbf{tan A} = \frac{\textbf{opp}}{\textbf{adj}}$$

Finding the tangent of an angle by measurement

In Fig. 19.3, $X\hat{A}Y = 41°$. Perpendiculars BP, CQ, DR, ES have been drawn so that AP = 3 cm, AQ = 4 cm, AR = 5 cm and AS = 6 cm.

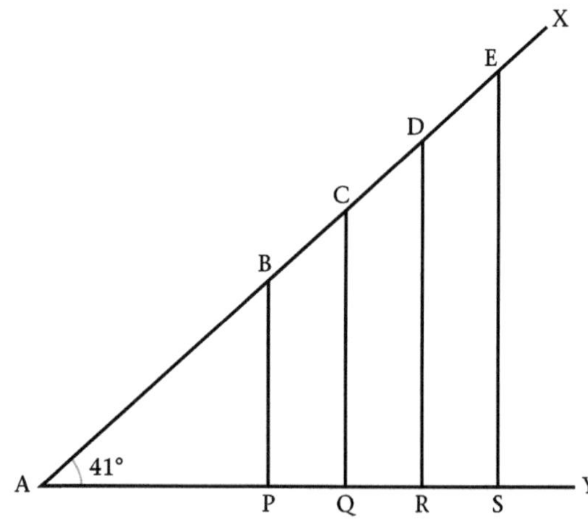

Fig. 19.3

By measurement, BP = 2.6 cm, CQ = 3.5 cm, DR = 4.3 cm and ES = 5.2 cm. Use a ruler to check these measurements.

Hence $\frac{BP}{PA} = \frac{2.6}{3} = 0.87$

Notice that the tangent of the angle is the ratio of the lengths of the two sides of the triangle which are opposite and adjacent to the angle.

$$\frac{CQ}{QA} = \frac{3.5}{4} = 0.87$$

$$\frac{DR}{RA} = \frac{4.3}{5} = 0.86$$

$$\frac{ES}{SA} = \frac{5.2}{6} = 0.87$$

The value of the ratio is roughly the same each time, i.e. tan 41° ≃ 0.87.

The working is made easier if the base-line (adj) is a convenient length such as 10 cm.

Exercise 19a (Group activity)

1. Copy Fig. 19.3, making $X\widehat{A}Y = 30°$, AP = 7 cm, PQ = QR = RS = 1 cm. Measure BP, CQ, DR, ES. Hence calculate the value of tan 30° by the above method.
2. Use the method of question 1 with $X\widehat{A}Y = 51°$. Hence find four values from which tan 51° may be calculated.

Example 1

Find the value of tan 57° by drawing and measurement.

Fig. 19.4 is a scale drawing of the method.

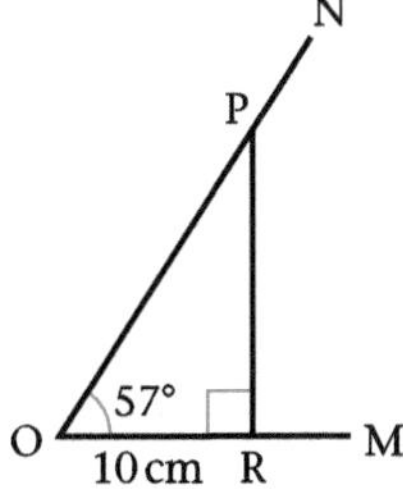

Fig. 19.4

Draw an angle MON of 57°. On OM, mark off OR equal to 10 cm. From R, draw a line perpendicular to OM to meet ON at P. Measure RP.

It is found that RP = 15.4 cm (approx.).

tan 57° $= \frac{PR}{RO} = \frac{15.4}{10} = 1.54$ (approx.)

Exercise 19b

Find the tangents of the following angles by drawing and measurement.

1. 42°
2. 62°
3. 38°
4. 71°
5. 45°
6. 27°
7. 77°
8. 14°
9. 33°

Example 2

Find by drawing and measurement the angle whose tangent is $\frac{3}{7}$.

The lengths of the opposite and adjacent sides are to be in the ratio 3 : 7. Thus the lengths could be 3 cm and 7 cm, or 6 cm and 14 cm, and so on. The bigger the drawing, the better the chance of accurate measurement. Fig. 19.5 is a scale drawing of the required triangle.

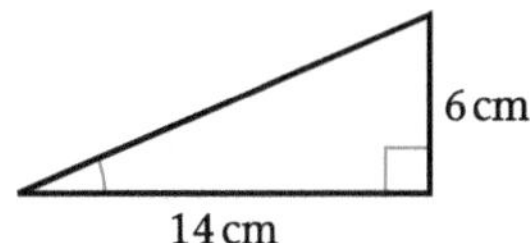

Fig. 19.5

By measurement, the angle whose tangent is $\frac{3}{7}$ is 23° (approx.).

Exercise 19c

Find by drawing and measurement the angles whose tangents are as follows.

1. $\frac{5}{9}$
2. $\frac{2}{7}$
3. $\frac{4}{3}$
4. $\frac{8}{5}$
5. $\frac{11}{4}$
6. $\frac{5}{8}$
7. $\frac{9}{10}$
8. $\frac{10}{9}$
9. $\frac{5}{7}$

Angle of elevation

In Fig. 19.6(b) (page 170) the boy B is looking at the top of the tree T. To do this he has to raise his line of sight through an angle $e°$ from the horizontal. The angle $e°$ in Fig. 19.6 is called the **angle of elevation** of T from B.

(a)

(b)

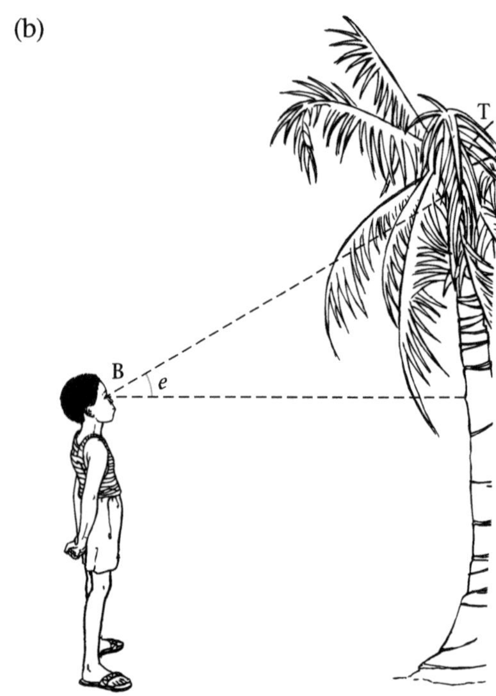

Fig. 19.6

Angle of depression

In Fig. 19.7, the girl in the window at G is looking down at her friend at F. To do this she has to lower her line of sight from the horizontal through an angle $d°$. The angle $d°$ in Fig. 19.7 is called the **angle of depression** of F from G.

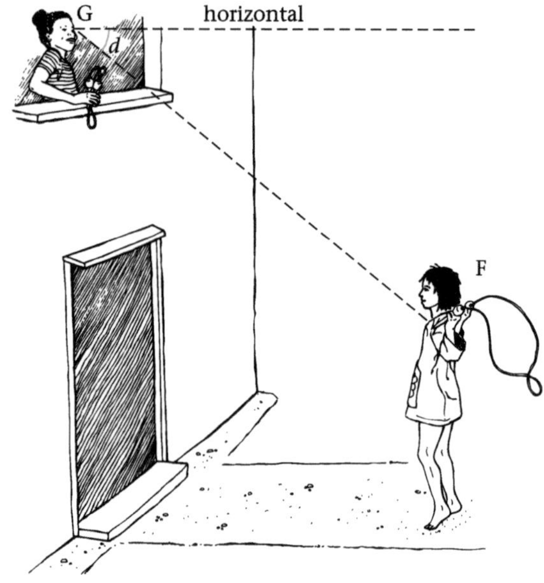

Fig. 19.7

Fig. 19.8 shows that there is a connection between angles of elevation and depression.

The angle of elevation of the cat, C, from the dog, D, is equal in size to the angle of depression of D from C. They are alternate angles.

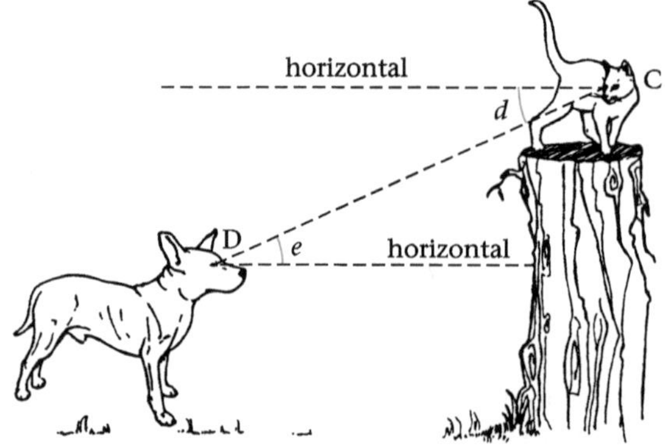

Fig. 19.8

Exercise 19d

You will need a protractor for this exercise.

1. Assume that Figs. 19.6, 19.7 and 19.8 are all scale drawings. Measure the following.
   (a) The angle of elevation, $e°$, in Fig. 19.6.
   (b) The angle of depression, $d°$, in Fig. 19.7.
   (c) The angle of elevation of C from D in Fig. 19.8.
   (d) The angle of depression of D from C in Fig. 19.8.

2. (a) Measure the angle of elevation of A from C in Fig. 19.9.
   (b) Hence state the angle of depression of C from A.

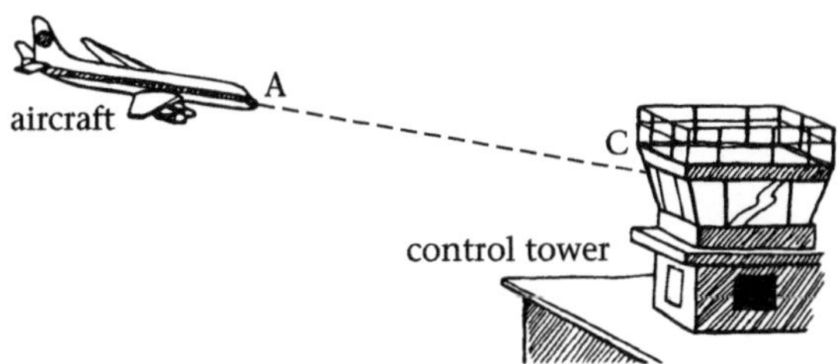

Fig. 19.9

3. (a) Measure the angle of depression of man B from man A in Fig. 19.10.
   (b) Hence state the angle of elevation of A from B.

Fig. 19.10

4. (a) Measure the angle of elevation of the light bulb from student P in Fig. 19.11.
   (b) Measure the angle of elevation of the light bulb from student Q.

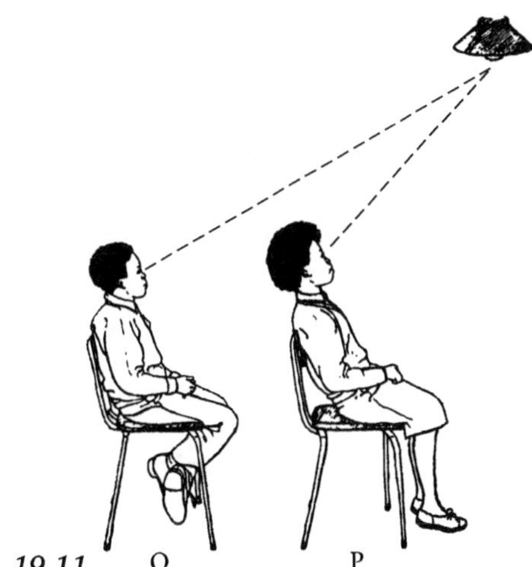

Fig. 19.11

5. In Fig. 19.12, measure the angle of depression of the coin from the man.

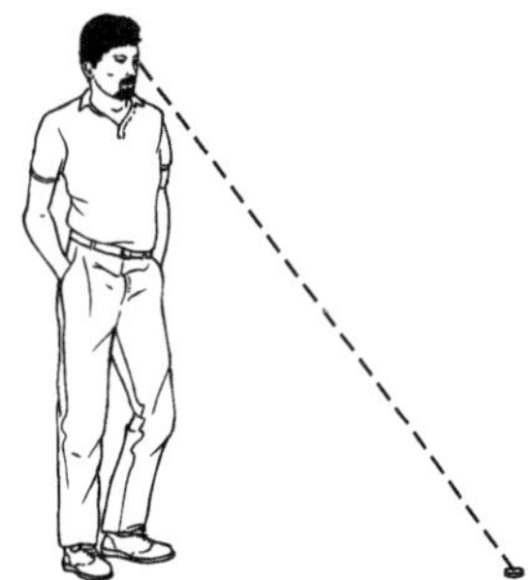

Fig. 19.12

Use of tangent of an angle

The use of tables for finding the tangent of an angle will be explained later in this chapter. Meanwhile, Table 19.1 gives the tangents of some chosen angles.

The values in Table 19.1 are given correct to 3 significant figures.

Table 19.1

Angle A	Tan A
25°	0.466
30°	0.577
35°	0.700
40°	0.839
45°	1.00
50°	1.19
55°	1.43
60°	1.73
65°	2.14
70°	2.75

Example 3

The angle of elevation of the top of a building is 25° from a point 70 m away on level ground. Calculate the height of the building.

It is important to draw a diagram to show the information given.

In Fig. 19.13, HK represents the height of the building, AK is on level ground.

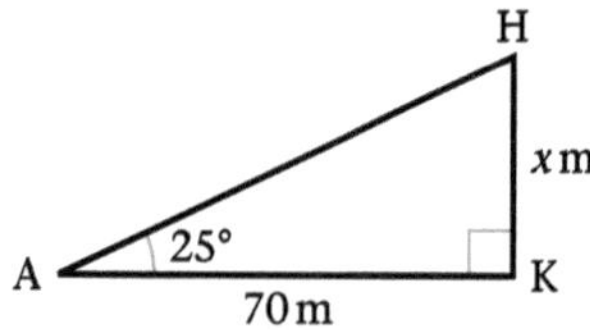

Fig. 19.13

$$\frac{HK}{KA} = \tan 25°.$$

Let HK be x m. KA = 70 m and, from Table 19.1, $\tan 25° = 0.466$.

$$\text{Hence, } \frac{x}{70} = 0.466$$

$$x = 0.466 \times 70$$

$$= 4.66 \times 7$$

$$= 32.62$$

However, the answer cannot be given to this **degree of accuracy**. The working depends on a value taken from three-figure tables. The third significant figure is only *approximate*, so that when 4.66 is multiplied by 7, the best that can be obtained is accuracy to 2 significant figures.

The height of the building is 33 m to 2 s.f.

Exercise 19e

Use the values in Table 19.1 in this exercise. Give all answers correct to 2 s.f.

Note: Remember to draw a figure and put in the known lengths and angles before attempting to solve a problem.

1. Find the value of x in each of the triangles in Fig. 19.14.

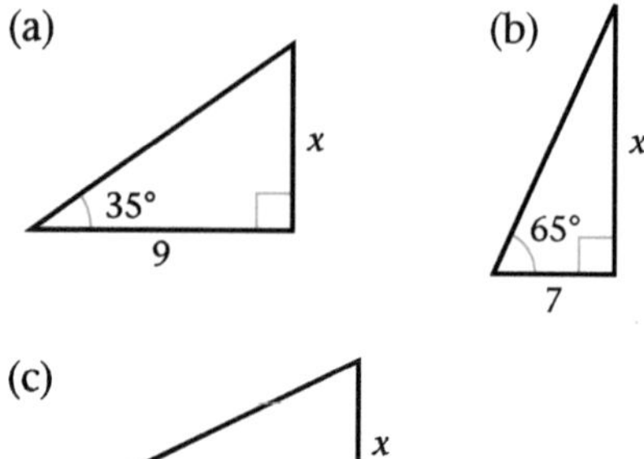

Fig. 19.14

2. Find the value of y in each of the triangles in Fig. 19.15.

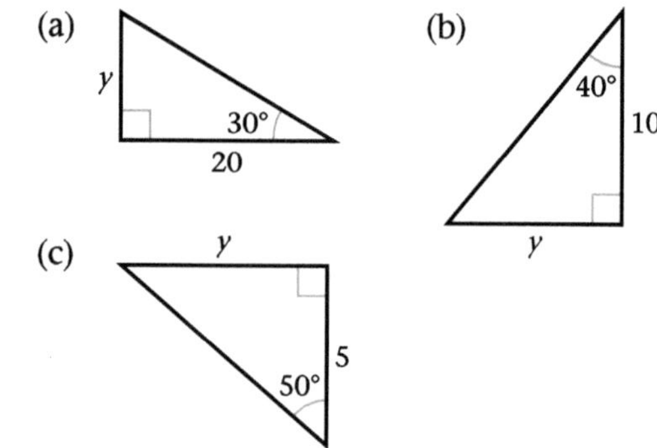

Fig. 19.15

3. Find the value of z in each of the triangles in Fig. 19.16.

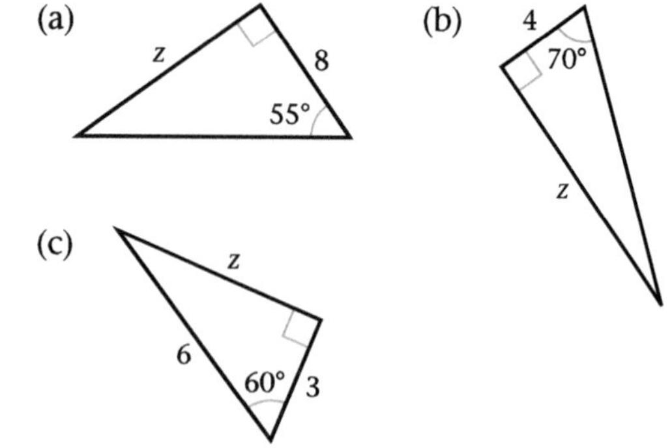

Fig. 19.16

4. When the angle of elevation of the sun is 45°, a boy's shadow on level ground is 1.6 m long. Find the height of the boy.

5. An aerial mast has a shadow 40 m long on level ground when the elevation of the sun is 70°. Calculate the height of the mast.

6. The angle of elevation of the top of a building from a point 80 m away on level ground is 25°. Calculate the height of the building.

Tangent tables

The tangent tables on page 220 can be used to find the tangents of angles from 0° to 90°. Table 19.2 gives three lines taken from the tangent table.

Table 19.2

θ	.0	.1	.2	.3	.4	.5	.6	.7	.8	.9
32	0.625	.627	.630	.632	.635	.637	.640	.642	.644	.647
59	1.664	1.671	1.678	1.684	1.691	1.698	1.704	1.711	1.718	1.725
71	2.904	2.921	2.937	2.954	2.971	2.989	3.006	3.024	3.092	3.060

Notice the following:

1. The table gives the tangent of any angle from 0° to 90° in intervals of 0.1°.
2. Each tangent is given correct to 3 or 4 significant figures.
3. As angles increase towards 90°, the sizes of their tangents increase rapidly.

Exercise 19f (Oral or written)

Use the table on page 221 to find the tangents of the following.

1. 13°
2. 64°
3. 35°
4. 56°
5. 74°
6. 88°
7. 23.1°
8. 36° 6′
9. 45.1°
10. $32\frac{1}{2}$°
11. 42.5°
12. 19° 30′
13. 56° 12′
14. 63.8°
15. 18.3°

Use of the calculator

The tan function on your calculator can be used to find the tangent of angles. The difference is that the values are expressed to 3 or 4 s.f. in the tables.

Example 4

Use the calculator to find the tangents of
(a) 29° (b) 36.8° (c) 85.1°

Give all answers correct to 3 s.f.

The number of digits in the window of the calculator will not be the same on all calculators. Remember you are required to give the answers correct to 3 s.f.

(a) tan 29° = 0.5543091
= 0.554 to 3 s.f.

(b) tan 36.8° = 0.7480956
= 0.748 to 3 s.f.

(c) tan 85.1° = 11.664495
= 11.7 to 3 s.f.

Example 5

Use the calculator to find the tangents of (a) 32° 12′, (b) 42° 48′.

Give values correct to 3 s.f.

(a) tan 32° 12′ = tan 32.2°
= 0.6297336
= 0.630 to 3 s.f.

(b) tan 42° 48′ = tan 42.8°
= 0.9260102
= 0.926 to 3 s.f.

Exercise 19g (Oral or written)

Use the calculator to find the tangents of the following angles. Give values correct to 3 s.f.

1. 27.7°
2. 48.6°
3. 67° 24′
4. 78.6°
5. 78.8°
6. 25.9°
7. 87° 6′
8. 87° 12′
9. 87° 18′
10. 68° 12′
11. 39.4°
12. 11.9°
13. 71.9°
14. 80° 24′
15. 55.7°

Example 6

A cone is 6 cm high and its vertical angle is 54°. Calculate the radius of its base.

In Fig. 19.17, the **vertical angle** is the angle between opposite slant heights VA and VB.

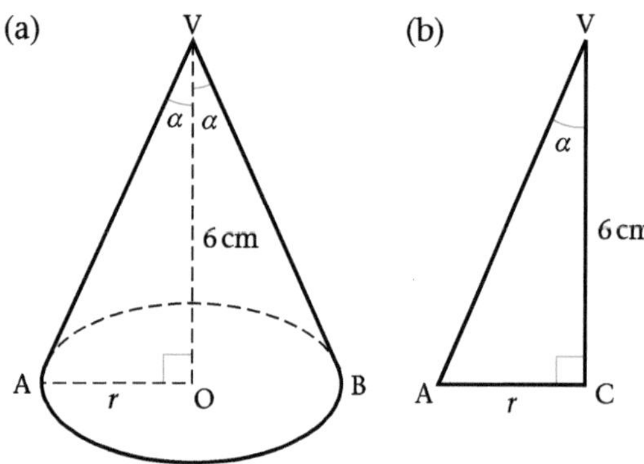

Fig. 19.17

Thus, with the lettering of the diagram, the vertical angle is 2α.

$$2\alpha = 54°$$

thus $\alpha = 27°$

In $\triangle$AVO, $\tan \alpha = \frac{r}{6}$

$$\begin{aligned} r &= 6 \tan 27° \\ &= 6 \times 0.51 \\ &= 3.06 \\ &= 3.1 \text{ to 2 s.f.} \end{aligned}$$

The radius of the base of the cone is 3.1 cm.

Exercise 19h

Give all calculated lengths correct to 2 significant figures. Give all calculated angles correct to 3 s.f.

1. Calculate the lengths marked x in the triangles shown in Fig. 19.18, all lengths being in metres.

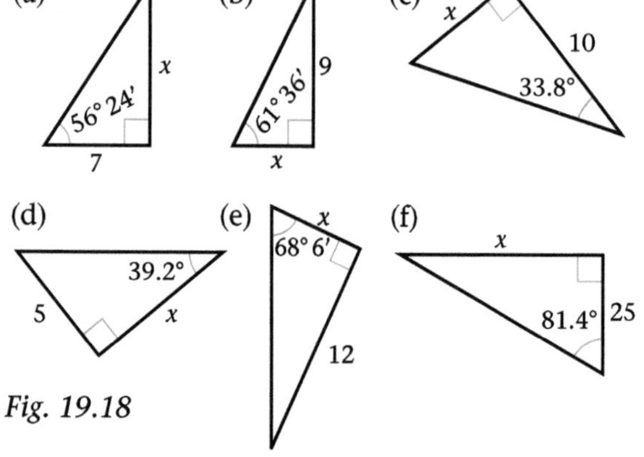

Fig. 19.18

2. A cone is 8 cm high and its vertical angle is 62°. Find the diameter of its base.
3. An isosceles triangle has a vertical angle of 116°, and its base is 8 cm long. Calculate its height.
4. From a point on level ground 40 m away, the angle of elevation of the top of a tree is $32\frac{1}{2}°$. Calculate the height of the tree.
5. The cone-shaped roof of a cylindrical tank 3.6 m in diameter rises symmetrically to a vertex. If the roof slopes at 48° to the horizontal, calculate the height of the vertex above the top of the tank.
6. In Fig. 19.19, O is the centre of the circle.

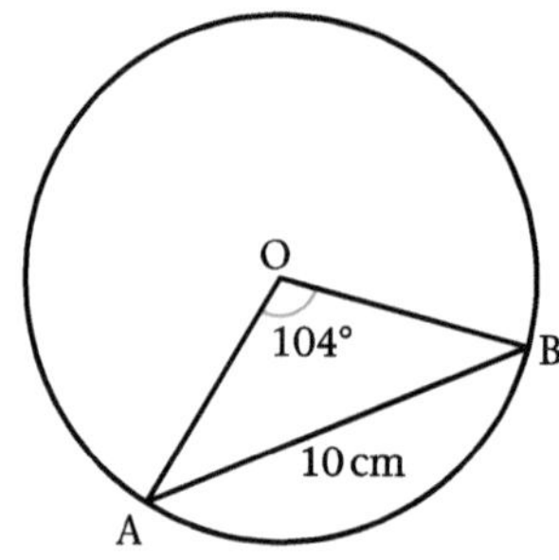

Fig. 19.19

Calculate the perpendicular distance of O from AB.

7. An aeroplane, coming in to land, passes over a point 1 km away from its landing place on level ground. If its angle of elevation is 15°, calculate the height of the plane in metres.
8. From a point 100 m from the foot of a building, the angle of elevation of the top of the building is 18° 42′. Find the height of the building.
9. Fig. 19.20 shows how a man finds the width of a river.

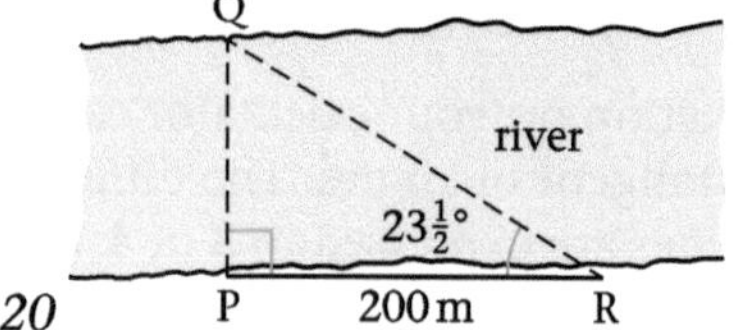

Fig. 19.20

He places a stone at P on one bank directly opposite a post Q on the other bank. From P he walks 200 m along the bank to R. He finds that $P\widehat{R}Q = 23\frac{1}{2}°$. Calculate the width of the river.

10

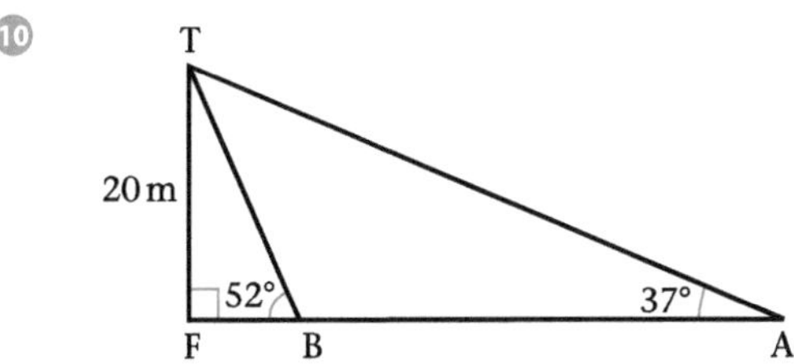

Fig. 19.21

In Fig. 19.21, TF is a flag-pole 20 m high. The angle of elevation of its top from a point A on level ground is 37°. From another point B, in line with A and F, the foot of the pole, the angle of elevation is 52°. Calculate the distance AB. (*Hint*: First find FA, then FB and subract)

Summary

In a right-angled triangle, if α is one of the acute angles, then

$$\tan \alpha = \frac{\text{side opposite } \alpha}{\text{side adjacent to } \alpha}$$

where **tan** is the **tangent ratio**. This ratio is used in finding the angles and sides of triangles.

If P is a point lying above the horizontal through a point O, then the **angle of elevation** of the point P from the point O is the angle that OP makes with the horizontal through O.

If P lies below the horizontal through O then the angle is called the **angle of depression**.

Practice Exercise P19.1

You may use your calculator in this exercise.

1. Triangle LMN is a right-angled triangle with a right angle at M. MN = 8 cm and $L\widehat{N}M = 26°$.

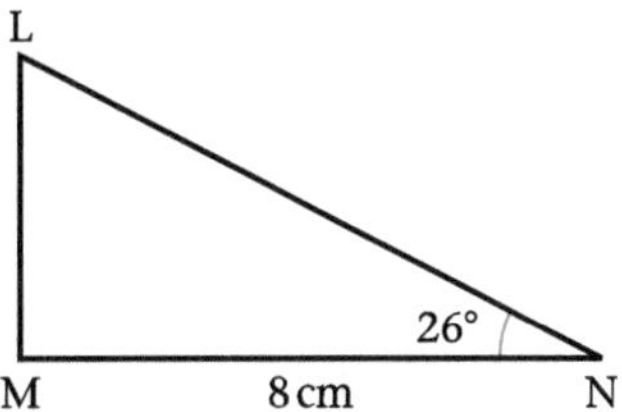

Fig. 19.22

Calculate the length of LM, giving your answer to one decimal place.

2. A player is standing at the centre of a circular field of radius 3 m. The angle of elevation of the top of a pole at a point on the circumference of the field is 35°.
   (a) Draw a diagram to show this information.
   (b) Calculate the height of the pole giving your answer to the nearest whole number of measure.

3. The angle of elevation of the top of a building from a point on the ground 42 m from the foot of the building is 40°. Calculate the height of the building, giving your answer to the nearest metre.

4. A ladder is leaning against a wall at a point 2.4 m from the foot of the building. A painter at the top of the ladder estimates the angle of depression of a paint pan on the ground, to be 32°. Calculate the estimate for the distance of the paint pan from the foot of the building.

5. The angle of elevation of the top of a building is 35° from a point on the ground. If the height of the building is 40 m, calculate the distance of the point on the ground from the foot of the building.

6 Two points P and Q are in a straight line on the ground. The angle of elevation of the top of a building is 19° from P and 27° from Q, and the height of the building is 32 m.

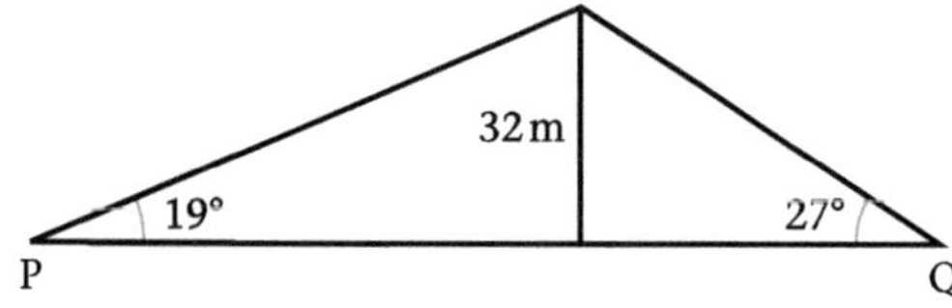

Fig. 19.23

Calculate

(a) The distance of P from the foot of the building.

(b) The distance of Q from the foot of the building

(c) The distance between P and Q on the ground.

7 AB is the diameter of the circle, Fig.19.24. BC = 4. 2 cm, $A\hat{C}B$ is a right angle and $B\hat{A}C = 27°$.

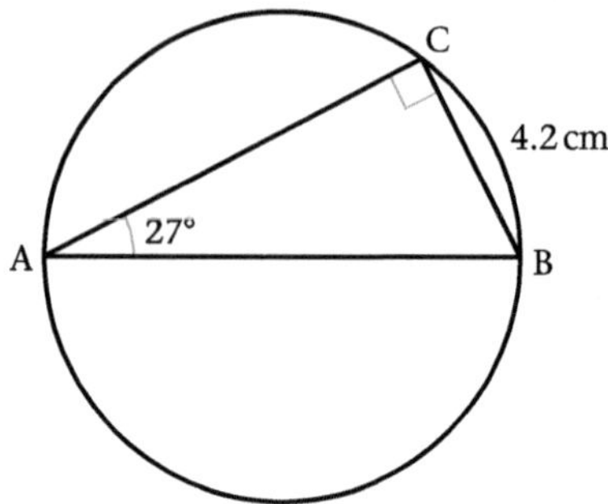

Fig. 19.24

Calculate the length of the chord AC.

Factorisation (1)

Factors and fractions

Pre-requisites

- common factors; lowest common multiple; grouping, multiplying and dividing algebraic terms

Removing brackets (revision)

The expression $3(2x - y)$ means 3 times $(2x - y)$.

Example 1

Remove brackets from (a) $3(2x - y)$, (b) $(3a + 8b)5a$, (c) $-2n(7y - 4z)$.

(a) $3(2x - y) = 3 \times 2x - 3 \times y$
$= 6x - 3y$

(b) $(3a + 8b)5a = 3a \times 5a + 8b \times 5a$
$= 15a^2 + 40ab$

(c) $-2n(7y - 4z) = (-2n) \times 7y - (-2n) \times 4z$
$= -14ny - (-8nz)$
$= -14ny + 8nz$
$= 8nz - 14ny$

Exercise 20a (Oral revision)

Remove brackets from the following.

1. $2(x + y)$
2. $5(7 - a)$
3. $(n + 9)3$
4. $8(2a - b)$
5. $5(-x - 3y)$
6. $(-3p + q)4$
7. $-2(m + n)$
8. $-3(a - b)$
9. $(p + q)(-4)$
10. $-7(3d - 2)$
11. $-9(-2k - 3r)$
12. $(-7s + t)(-6)$
13. $x(x + 2)$
14. $y(y - 1)$
15. $(a + b)a$
16. $n(3n - 2)$
17. $p(2s + 3t)$
18. $(5 - 3n)m$
19. $2a(5a - 8b)$
20. $3x(x + 9)$
21. $5p(9r - 8s)$
22. $-6a(2a - 7b)$
23. $(3a - 4b)3b$
24. $2\pi r(r + h)$

Common factors

Example 2

Find the HCF of $6xy$ and $18x^2$.

$6xy = 6 \times x \times y$
$18x^2 = 3 \times 6 \times x \times x$

The HCF of $6xy$ and $18x^2$ is $6 \times x = 6x$.

Exercise 20b (Oral revision)

Find the HCF of the following.

1. $5a$ and $5z$
2. $6x$ and $15y$
3. $7mnp$ and mp
4. $5xy$ and $15x$
5. $12a$ and $8a^2$
6. $13ab$ and $26b$
7. ab^2 and a^2b
8. $6d^2e$ and $3de^2$
9. $8pq$ and $24p^2$
10. $10ax^2$ and $14a^2x$
11. $9xy$ and $24pq$
12. $30ad$ and $28ax$

Factorisation by taking out common factors

To **factorise** an expression is to write it as a product of its factors. In order to do this, find the HCF of the terms in the expression.

Example 3

Factorise the following. (a) $9a - 3z$ (b) $5x^2 + 15x$ (c) $2mh - 8m^2h$

(a) The HCF of $9a$ and $3z$ is 3.
Hence,

$$9a - 3z = 3\left(\frac{9a}{3} - \frac{3z}{3}\right)$$
$$= 3(3a - z)$$

(b) The HCF of $5x^2$ and $15x$ is $5x$.
Hence,

$$5x^2 + 15x = 5x\left(\frac{5x^2}{5x} + \frac{15x}{5x}\right)$$
$$= 5x(x + 3)$$

(c) The HCF of $2mh$ and $8m^2h$ is $2mh$.
Hence,

$$2mh - 8m^2h = 2mh\left(\frac{2mh}{2mh} - \frac{8m^2h}{2mh}\right)$$
$$= 2mh(1 - 4m)$$

The above examples show that factorisation is the opposite of removing brackets. *Note:* It is not necessary to write the first line of working as above; this has been included to show the method.

Exercise 20c (Oral revision)

Factorise the following. Questions 1–12 correspond to questions 1–12 in Exercise 20b.

1. $5a + 5z$
2. $6x - 15y$
3. $7mnp - mp$
4. $5xy + 15x$
5. $12a + 8a^2$
6. $13ab - 26b$
7. $ab^2 - a^2b$
8. $6d^2e - 3de^2$
9. $8pq + 24p^2$
10. $10ax^2 + 14a^2x$
11. $9xy + 24pq$
12. $30ad - 28ax$
13. $5am - 20bm$
14. $5a^3 - 3a^2b$
15. $\pi r^2 + \pi rs$
16. $7d^2 - d$
17. $33bd - 3de$
18. $9pq + 12t$
19. $ab - 2b$
20. $3dh + 15dk$
21. $x^2 + 9xy$
22. $2a^2 + 10a$
23. $am + a$
24. $24x^2y - 6xy$

Simplifying calculations by factorisation

Example 4

By factorising, simplify $79 \times 37 + 21 \times 37$.

37 is a common factor of 79×37 and 21×37

$$79 \times 37 + 21 \times 37 = 37(79 + 21)$$
$$= 37 \times 100$$
$$= 3\,700$$

Example 5

Factorise the expression $\pi r^2 + 2\pi rh$. Hence find the value of $\pi r^2 + 2\pi rh$ when $\pi = \frac{22}{7}$, $r = 14$ and $h = 43$.

$\pi r^2 + 2\pi rh = \pi r(r + 2h)$

When $\pi = \frac{22}{7}$, $r = 14$ and $h = 43$,

$$\pi r^2 + 2\pi rh = \pi r(r + 2h)$$
$$= \tfrac{22}{7} \times 14(14 + 2 \times 43)$$
$$= 22 \times 2(14 + 86)$$
$$= 44 \times 100$$
$$= 4400$$

Exercise 20d

Simplify questions 1–15 by factorising.

1. $34 \times 48 + 34 \times 52$
2. $61 \times 87 - 61 \times 85$
3. $128 \times 27 - 28 \times 27$
4. $693 \times 7 + 693 \times 3$
5. $\frac{8}{13} \times 125 + \frac{5}{13} \times 125$
6. $\frac{22}{7} \times 10 + \frac{22}{7} \times 4$
7. $121 \times 67 + 79 \times 67$
8. $67 \times 23 - 67 \times 13$
9. $\frac{22}{7} \times 3\frac{1}{4} - \frac{22}{7} \times 2\frac{1}{4}$
10. $53 \times 49 - 53 \times 39$
11. $\frac{3}{4} \times 133 - \frac{3}{4} \times 93$
12. $35 \times 29 + 35 \times 11$
13. $27 \times 354 + 27 \times 646$
14. $\frac{22}{7} \times 1\frac{1}{4} + \frac{22}{7} \times 2\frac{3}{4}$
15. $762 \times 87 - 562 \times 87$
16. Factorise the expression $\pi R^2 - \pi r^2$. Hence find the value of the expression when $\pi = \frac{22}{7}$, $R = 9$ and $r = 5$.
17. Factorise the expression $2\pi r^2 + 2\pi rh$. Hence find the value of the expression when $\pi = \frac{22}{7}$, $r = 5$ and $h = 16$.
18. Factorise the expression $\pi r^2h + \frac{1}{3}\pi r^2H$. Hence find the value of the expression when $\pi = \frac{22}{7}$, $r = 3$, $h = 10$ and $H = 12$..

Factorisation of larger expressions

Example 6

Factorise $2x(5a + 2) - 3y(5a + 2)$.

This expression is of the same kind as $2xm - 3ym$, in which m is common to both terms, so that

$$2xm - 3ym = m(2x - 3y)$$

In the given expression,

$$2x(5a + 2) = 2x \text{ times } (5a + 2)$$

and $3y(5a + 2) = 3y$ times $(5a + 2)$

Hence the products $2x(5a + 2)$ and $3y(5a + 2)$ have the factor $(5a + 2)$ in common. Thus,
$2x(5a + 2) - 3y(5a + 2) = (5a + 2)(2x - 3y)$

Example 7

Factorise $2d^3 + d^2(3d - 1)$.

$2d^3 + d^2(3d - 1)$ have the factor d^2 in common. Thus,

$$\begin{aligned} 2d^3 + d^2(3d - 1) &= d^2[2d + (3d - 1)] \\ &= d^2(2d + 3d - 1) \\ &= d^2(5d - 1) \end{aligned}$$

Example 8

Factorise $(a + m)(2a - 5m) - (a + m)^2$.

The two parts of the expression have the factor $(a + m)$ in common. Thus,

$$\begin{aligned} &(a + m)(2a - 5m) - (a + m)^2 \\ &\quad = (a + m)[(2a - 5m) - (a + m)] \\ &\quad = (a + m)(2a - 5m - a - m) \\ &\quad = (a + m)(a - 6m) \end{aligned}$$

Example 9

Factorise $(x - 2y)(z + 3) - x + 2y$.

Notice that -1 is a factor of the last two terms. The given expression may be written as follows.

$$\begin{aligned} &(x - 2y)(z + 3) - x + 2y \\ &\quad = (x - 2y)(z + 3) - 1(x - 2y) \end{aligned}$$

The two parts of the expression now have $(x - 2y)$ as a common factor. Hence,

$$\begin{aligned} &(x - 2y)(z + 3) - 1(x - 2y) \\ &\quad = (x - 2y)[(z + 3) - 1] \\ &\quad = (x - 2y)(z + 3 - 1) \\ &\quad = (x - 2y)(z + 2) \end{aligned}$$

Exercise 20e

Factorise the following.

1. $3m + m(u - v)$
2. $2a - a(3x + y)$
3. $x(3 - a) + bx$
4. $(4m - 3n)p - 5p$
5. $a(m + 1) + b(m + 1)$
6. $a(n + 2) - b(n + 2)$
7. $ax - x(b - 4c)$
8. $5x(a - b) - 2y(a - b)$
9. $3h(5u - v) + 2k(5u - v)$
10. $m(u - v) + m^2$
11. $d(3h + k) - 4d^2$
12. $5a^2 + a(b - c)$
13. $4x^2 - x(3y + 2z)$
14. $3d^3 - d^2(e - 4f)$
15. $a(3u - v) + a(u + 2v)$
16. $(5x - y)a - (3x + 5y)a$
17. $3(3u + 2v) - a(3u + 2v)$
18. $(4a - b)3x + (4a - b)2y$
19. $h(2a - 7b) - 3k(2a - 7b)$
20. $m(3m - 2) + 2m^2$
21. $a^2(5a - 3b) - 3a^3$
22. $5x^2 - x(x + 4)$
23. $2d(3m - 4n) - 3e(3m - 4n)$
24. $(a + 2b)(x - y) - 3(x - y)$
25. $p(2m + n) + (q - r)(2m + n)$
26. $(h + k)(r + s) + (h + k)(r - 2s)$
27. $(3x - y)(u + v) + (x + 2y)(u + v)$
28. $(b - c)(3d + e) - (b - c)(d - 2e)$
29. $(a + 2b)^2 - 3(a + 2b)$
30. $(3m - 2n)^2 + 5p(3m - 2n)$
31. $(2u - 3v)(3m - 4n) - (2u - 3v)(m + 2n)$
32. $a(x + 2y) + (x + 2y)^2$
33. $3u(2x + y) - (2x + y)^2$

34 $(f-g)4e-(f-g)^2$

35 $(a-3b)(2u-v)+(a-3b)(u+7v)$

36 $(5m+2n)(6a+b)-(5m+2n)(a-4b)$

37 $(x+3y)(m-n)+x+3y$

38 $(2a-3b)(c+d)-2a+3b$

39 $7u-2v+(7u-2v)^2$

40 $(2u-7v)^2+7v-2u$

Harder fractions

Example 10

Solve the equation $\dfrac{5}{x-3}=2$.

$$\frac{5}{x-3}=2$$

There is *one* denominator, $x-3$. Notice that the *whole* of $x-3$ is the denominator; it cannot be split into parts. Multiply both sides by $(x-3)$.

$$(x-3)\times\frac{5}{x-3}=2(x-3)$$

On the LHS, the $(x-3)$s cancel, leaving 5; clear brackets on the RHS.

$$5=2x-6$$

Add 6 to both sides

$$11=2x$$

Divide both sides by 2

$$5\tfrac{1}{2}=x$$

Check: When $x=5\frac{1}{2}$,

$$\text{LHS}=\frac{5}{5\frac{1}{2}-3}=\frac{5}{2\frac{1}{2}}=2=\text{RHS}$$

Example 11

Solve $\dfrac{5}{7a-1}-\dfrac{4}{9}=0$.

$$\frac{5}{7a-1}-\frac{4}{9}=0$$

The denominators are $(7a-1)$ and 9. Their LCM is $9(7a-1)$. Multiply each term by $9(7a-1)$.

$$9(7a-1)\times\frac{5}{7a-1}-9(7a-1)\times\frac{4}{9}=9(7a-1)\times 0$$

$$9\times5-4(7a-1)=0$$

Clear brackets and collect terms.

$$45-28a+4=0$$
$$49-28a=0$$
$$49=28a$$
$$a=\frac{49}{28}=\frac{7}{4}$$
$$a=1\tfrac{3}{4}$$

The solution is $a=1\frac{3}{4}$. The check is left as an exercise.

In a fraction like $\dfrac{7}{2r+3}$, the division line acts like a bracket on the terms in the denominator: $\dfrac{7}{(2r+3)}$. Examples 10 and 11 show that the bracket must be kept in the working until it can be cleared properly.

Exercise 20f

Solve the following equations.

1 $\dfrac{6m-3}{7}=\dfrac{2m+1}{7}$

2 $\dfrac{3(2a+1)}{4}=\dfrac{5(a+5)}{6}$

3 $\dfrac{2a-1}{3}-\dfrac{a+5}{5}=\dfrac{1}{2}$

4 $\dfrac{4x-3}{2}=\dfrac{9x-6}{8}+2\frac{3}{4}$

5 $\dfrac{3}{h}=\dfrac{1}{5}+\dfrac{8}{35}$

6 $\dfrac{5}{2x}-\dfrac{1}{x}=\dfrac{1}{6}$

7 $\dfrac{2}{x}+\dfrac{3}{2x}=\dfrac{7}{8}$

8 $\dfrac{9}{4x}-\dfrac{5}{x}+\dfrac{11}{3}=0$

Exercise 20g

1 I add 45 to a certain number and then divide the sum by 2. The final result is 5 times the original number. Find the original number.

2 $\frac{1}{5}$ of an even number added to $\frac{1}{6}$ of the next even number makes a total of 15. Find the two numbers.
(*Hint*: let the numbers be x and $x+2$).

3 A trader buys x watches (all alike) for \$67.20.
(a) Write down the cost of one watch in terms of x.
(b) If the watches cost \$9.60 each, find the number of watches bought.

4 A student walks 3 km at a speed of v km/h.
(a) Write down the time taken, in hours, in terms of v.
(b) If the journey takes 35 min, find the value of v.

5 A trader sells a number of books and takes in \$615 altogether. If the average selling price of a book is \$7.50, find the number of books sold.

6 A bag of mangoes has a total mass of 56 kg. If the average mass of a mango is $1\frac{3}{5}$ kg, find the number of mangoes in the bag. (Ignore the mass of the bag.)

7 A mother is 24 years older than her daughter. If the daughter's age is x years,
(a) express the mother's age in terms of x;
(b) find x when the daughter's age is $\frac{1}{3}$ of her mother's age.

Exercise 20h

Solve the following equations.

1 $\frac{12}{x-1} = 3$ 2 $\frac{4}{1+x} = 1$

3 $2 = \frac{7}{y+2}$ 4 $\frac{4}{t-2} = 3$

5 $\frac{1}{z+4} = 1$ 6 $5 = \frac{15}{1-r}$

7 $\frac{6}{x+7} + 3 = 0$ 8 $2 - \frac{1}{k-2} = 0$

9 $2 = \frac{2}{4-3a}$ 10 $\frac{13}{2x+1} = 5$

11 $\frac{1}{x+3} = \frac{1}{5}$ 12 $\frac{1}{7} = \frac{1}{a-3}$

13 $\frac{1}{2} - \frac{1}{y-5} = 0$ 14 $\frac{1}{b+5} + \frac{1}{4} = 0$

15 $\frac{3}{2-e} = \frac{3}{8}$ 16 $\frac{4}{9} = \frac{2}{c-8}$

17 $\frac{5}{2n-5} = \frac{3}{2}$ 18 $2\frac{1}{2} = \frac{10}{3d+7}$

19 $\frac{2}{5} + \frac{3}{a-8} = 0$ 20 $\frac{5}{3x+2} - \frac{1}{4} = 0$

Summary

To **factorise** an expression is to write the expression as a product of its factors. To do this find the HCF of the terms in the expression.

Calculations may be made much simpler by first factorising the expression. In simplifying fractions, first find the LCM of the terms in the denominators.

Practice Exercise P20.1

Factorise each of the following expressions

1 $5d + 10$ 2 $4m^2 + 12m$

3 $10m + 5n$ 4 $5xy - 7xz$

5 $16vw + 6uv$ 6 $3apr - art$

7 $6d + 3cd$ 8 $3vw - uv$

9 $2d^2 + d$ 10 $hk - 2k^2$

11 $2x + 4y - 8w$ 12 $2r^4h + 3r^2h^3$

Practice Exercise P20.2

Factorise the following, first simplifying the expressions if necessary

1 $2(2w + v) + 3w + 5v$

2 $(b-4)(b-3) + b(b+5) - 12$

3 $5x + 4 - 2x - 10$

4 $b - 4 + 3b - 2$

5 $3(4m - 1) - (m + 8)$

6 $2(n-3) + 5(n-3)$

7 $3a + 2(2a + 11) - 1$

8 $3(2y - 2) - 4(y + 7)$

9 $4(m-2) + 2(2m-3) - m$

Practice Exercise P20.3

Simplify the following

1 $7(a+2) - a(a+2)$

2 $7y - 3(y-2) + 2(2-y) + 4$

3 $2(1-2y) - 3y(2y-1)$

4 $4d(2d+1) + 3(2d+1)$

5 $10cd + 14d^2 - 30c^2d - 42cd^2$

Chapter 21

Consumer arithmetic (2)

Simple and compound interest, currency exchange rates

Pre-requisites

- percentages; rates

Interest

Bankers want people to save money. They give extra payments to encourage saving. The extra money is called **interest**.

For example, a man saves \$100 in a bank for a year. If he receives \$8 interest at the end of the year so that he gets a total of \$108, then the **interest rate** is 8% per annum, i.e. 8% of his money saved for that one year is interest. He still has his \$100 plus \$8 interest from the bank. Interest which is paid like this is called **simple interest.**

If the interest rate does not change, he gets $\$(8 \times 2)$ for 2 years, $\$(8 \times 3)$ for 3 years, $\$(8 \times \frac{6}{12})$ for 6 months, $\$(8 \times \frac{5}{2})$ for $2\frac{1}{2}$ years, and so on.

Example 1

Find the simple interest on \$600 for 5 years at 9% per annum.

$$\text{Yearly interest} = 9\% \text{ of } \$600 = \frac{9}{100} \times \$600 = \$54$$

$$\text{Interest for 5 years} = \$54 \times 5 = \$270$$

Exercise 21a

Find the simple interest on the following.

1. \$400 for 1 year at 5% per annum
2. \$700 for 1 year at 4% per annum
3. \$100 for 3 years at 6% per annum
4. \$100 for 2 years at 4% per annum
5. \$100 for 4 years at $4\frac{1}{2}$% per annum
6. \$100 for $3\frac{1}{2}$ years at 4% per annum
7. \$300 for 2 years at 5% per annum
8. \$200 for 4 years at 6% per annum
9. \$7200 for 1 month at 5% per annum
10. \$600 for $2\frac{1}{2}$ years at 5% per annum
11. \$350 for 4 years at 3% per annum
12. \$250 for 2 years at 6% per annum
13. \$1500 for 3 months at 4% per annum
14. \$2550 for 4 months at 6% per annum
15. \$250 for 6 months at 5% per annum

Sometimes people have to borrow money. When someone borrows money that person has to pay interest to the lender.

The original sum of money put in the bank or the loan is called the **principal**. The sum of the principal and the interest is called the **amount**.

Exercise 21b

1. For each question in Exercise 21a, calculate the amount.

We see that
Interest for 1 year = principal × rate per annum
Interest for t years = principal × rate per annum × t
Hence, we get the formula

$I = \frac{P \times r \times t}{100}$ where I is the interest
P is the principal
$\frac{r}{100}$ is the rate per annum
and t is the time in years

and
$A = P + I$ where A is the amount.
Using the formula for Example 1,

$$\text{Interest} = \$\frac{600 \times 9 \times 5}{100} = \$270$$

Example 2

A man borrows \$160 000 to buy a house. He is charged interest at a rate of 11% per annum. In the first year he paid the interest on the loan. He also paid back \$10 000 of the money he borrowed. How much did he pay back altogether? If he paid this money by monthly instalments, how much did he pay per month?

Interest on \$160 000

$$\begin{aligned} \text{for 1 year} &= 11\% \text{ of } \$160\,000 \\ &= \tfrac{11}{100} \times \$160\,000 \\ &= 11 \times \$1600 \\ &= \$17\,600 \end{aligned}$$

Total money paid in

$$\begin{aligned} \text{1st year} &= \$10\,000 + 17\,600 \\ &= \$27\,600 \end{aligned}$$

Monthly payments = \$27 600 ÷ 12 = \$2300

(Notice that the man now owes \$150 000. Interest will be paid on this new principal in the second year.)

Exercise 21c

1. Find the total amount to be paid back (i.e. loan + interest) on the following loans.
   (a) \$5 for 2 weeks at \$1 interest per week
   (b) \$20 for 3 weeks at 10 cents in the dollar interest per week
   (c) \$1000 for 1 year at 9% simple interest per annum
   (d) \$10 000 for 15 years at 8% simple interest per annum
   (e) \$6000 for 3 years at $7\frac{1}{2}$% simple interest per annum
   (f) \$860 for $2\frac{1}{2}$ years at $8\frac{1}{2}$% simple interest per annum

2. A man borrows \$400 on a short term loan. He is charged interest of \$1 on each \$10 per week. How much does he pay back altogether if he borrows the money (a) for 1 week, (b) for 3 weeks, (c) for 10 weeks?

3. A girl borrows \$20 for 4 weeks. She agrees to pay \$25 back at the end of the 4 weeks.
   (a) How much interest does she pay over the 4 weeks?
   (b) How much interest does she pay per week?
   (c) Find the percentage rate of interest per week, that she pays.

4. A man gets a \$180 000 loan to buy a house. He pays interest at a rate of 9% per annum. In the first year he paid the interest on the loan. He also paid back \$14 000 of the money he borrowed.
   (a) How much did he pay in the first year altogether?
   (b) If he paid this money in monthly instalments, how much did he pay each month?

5. A woman borrows \$3000 to help to pay for a car. She agrees to pay the money back over 2 years, paying simple interest at 9% per annum.
   (a) Calculate the simple interest on \$3000 at 9% per annum for 2 years.
   (b) Hence find the total amount she must pay back.
   (c) If the total money is paid back in monthly instalments over 2 years, how much will she pay each month?

Sometimes we know the amount we need, that is, including the interest that we want to get on savings, or that we can pay on a loan. Then, we may have to calculate the principal, or the time, or the rate of interest.

It is important in solving such a problem to identify correctly which sum of money is the principal, the interest or the amount.

Example 3

Calculate the sum of money that gets \$276 interest in 2 years when the interest rate is 6% per annum.

In this problem, we have been given the interest and we have to find the principal, P.

Interest for 1 year = $\$\frac{276}{2}$ = \$138

Then
$$\begin{aligned} 138 &= \frac{6}{100} \times P \\ P &= \frac{138 \times 100}{6} \\ &= 2300 \end{aligned}$$

Sum of money = \$2300

Using the formula,

$$276 = \frac{P \times 6 \times 2}{100}$$
$$P \times 6 \times 2 = 276 \times 100$$
$$P = \frac{276 \times 100}{6 \times 2}$$
$$= 2300$$

Sum of money = \$2300

Example 4

Calculate the time for \$1000 to amount to \$1300 at a rate of 5% per annum.

Principal = \$1000
Amount = \$1300
Interest = \$(1300 − 1000)
= \$300

Using the formula,

$$300 = \frac{1000 \times 5 \times n}{100}$$
$$300 = 50n$$
$$n = 6$$

Time required = 6 years

Exercise 21d

Calculate the time when the interest on

1. \$100 at 6% is \$12
2. \$300 at 5% is \$45
3. \$2100 at 12% is \$126
4. \$15 000 at 4.5% is \$675
5. \$1200 at 3% is \$108

Calculate the interest rate per annum when the interest on

6. \$100 for 3 years is \$12
7. \$7500 for 5 years is \$750
8. \$25 000 for 3 months is \$250
9. \$1000 for 9 months is \$60
10. \$100 000 for 6 months is \$1750

Example 5

Find the sum that amounts to \$2040 in 6 months at 4% per annum.

Let \$$P$ be the principal
Interest = \$(2040 − P)

$$\text{Then } 2040 - P = \frac{P \times 4 \times 6}{100 \times 12}$$
$$2040 - P = \frac{P}{50}$$
$$2040 = P + \frac{P}{50}$$
$$2040 = \frac{51P}{50}$$
$$P = 2000$$

Original sum = \$2000

Exercise 21e

1. Find the original sum of money saved when the interest paid after 4 years at 5% per annum is \$350.
2. Calculate the rate of interest per annum when \$5000 amounts to \$5200 after 6 months.
3. Find the time required for \$10 000 to amount to \$12 200 when the interest rate is $5\frac{1}{2}$% per annum.
4. A bank pays interest at the rate of 3% per annum on the money deposited. What sum of money will provide an income of \$400 per month?
5. If \$24 300 is the amount of a loan after 3 months at 5% per annum, calculate the interest paid.

Compound interest

Interest, I, on a principal, P, is calculated by using the formula

$$I = \frac{PRT}{100}$$

In practice, the amount, A, at the end of a given interval of time, is invested as the principal for the next interval. Interest is then calculated using this new value of P. Since this new value of P is greater, more interest is earned although the rate per cent and the time interval remain the same. Interest that is calculated by this method is called **compound interest**. The interest is said to be compounded, that is, added to the principal.

In some savings plans, the interest may be compounded daily, monthly or semi-annually. Ready reckoner tables may be used to read off the values of the interest for different rates. The interest may be rounded off to the nearest dollar or to the nearest cent before calculating the value of the new principal.

Example 6

Calculate the total interest on \$10 000 for 3 years at 4% per annum if interest is compounded annually.

$$\begin{aligned} \text{Principal for 1st year} &= \$10\,000 \\ \text{Interest for 1st year at 4\%} &= \$10\,000 \times \frac{4}{100} \times 1 \\ &= \$400 \\ \text{Principal for 2nd year} &= \$10\,000 + 400 \\ &= \$10\,400 \\ \text{Interest for 2nd year at 4\%} &= \$10\,400 \times \frac{4}{100} \times 1 \\ &= \$416 \\ \text{Principal for 3rd year} &= \$10\,400 + 416 \\ &= \$10\,816 \\ \text{Interest for 3rd year at 4\%} &= \$10\,816 \times \frac{4}{100} \times 1 \\ &= \$432.64 \end{aligned}$$

Total interest earned for 3 years

$= \$400 + 416 + 432.64$

$= \$1248.64$

Note: If interest was being paid at simple interest,

then total interest $= \$\left(10\,000 \times \frac{4}{100} \times 3\right)$

$= \$1200$

Sometimes it may be arranged that the interest on the repayment of a loan is calculated using the balance of the loan remaining at that time as the principal. By this method, payment at stated intervals consists partly of repayment of part of the loan and partly of payment of interest calculated on the decreasing balance.

Example 7

Interest is charged on a loan of \$5000 at 18% per annum. At the end of each month the interest due plus \$500 of the loan are paid. Calculate the total sum of money paid at the end of three months.

Month 1: Interest on \$5000 at 18% p.a.

$$\begin{aligned} &= \$\left(5000 \times \frac{18}{100} \times \frac{1}{12}\right) \\ &= \$75 \\ \text{Payment} &= \$500 + 75 \\ &= \$575 \\ \text{Balance of loan} &= \$4500 \end{aligned}$$

Month 2: Interest on \$4500 at 18% p.a.

$$\begin{aligned} &= \$\left(4500 \times \frac{18}{100} \times \frac{1}{12}\right) \\ &= \$67.50 \\ \text{Payment} &= \$500 + 67.50 \\ &= \$567.50 \\ \text{Balance of loan} &= \$4000 \end{aligned}$$

Month 3: Interest on \$4000 at 18% p.a.

$$\begin{aligned} &= \$\left(4000 \times \frac{18}{100} \times \frac{1}{12}\right) \\ &= \$60 \\ \text{Payment} &= \$500 + 60 \\ &= \$560 \end{aligned}$$

$$\begin{aligned} \text{Total sum paid} &= \$575 + 567.50 + 560 \\ &= \$1702.50 \end{aligned}$$

Note:

1. If interest was charged at simple interest, interest for 3 months at 18% p.a.

   $= \$\left(5000 \times \frac{18}{100} \times \frac{3}{12}\right)$

   $= \$225$

   Total repayment for 3 months

   $= \$(500 \times 3) + 225$

   $= \$1725$

2. Outstanding balance of loan $= \$5000 - 1500$

   $= \$3500$

3. The outstanding balance of the loan is the same but the amount paid when interest was charged at simple interest will be \$22.50 more (that is, \$1725 − 1702.50).

Housing mortgage payments are usually calculated by a method similar to that in Example 7.

Exercise 21f

1. Calculate the interest compounded annually on
   (a) \$300 for 2 years at 5% per annum,
   (b) \$800 for 3 years at 2% per annum,
   (c) \$6400 for 2 years at 4% per annum,
   (d) \$5250 for 4 years at 3% per annum.

2. Calculate the amount in each example in question 1.

3. Calculate the total amount received if compound interest is paid on
   (a) \$4000 at 5% per annum for 2 years.
   (b) \$3000 at 6% per annum for 3 years.

4. Calculate the total interest on \$20 000 at 6% per annum for 3 years when
   (a) the interest is not added to the principal, that is, at simple interest;
   (b) the interest is added at the end of each year, that is, interest is compounded annually.

5. A man takes a loan of \$100 000 at $12\frac{1}{2}$% per annum. He pays \$5000 and the interest on the decreasing balance at the end of every six-month period. Calculate the amount he owes at the end of 2 years.

6. Calculate the sum of money invested at 6% per annum if interest of \$45 is paid at the end of each year.

7. A woman saves \$10 000 for 3 years. Interest is compounded annually at 5% per annum for the first year and at 6% per annum for the second and third years. Calculate the total amount at the end of
   (a) the first year
   (b) the third year.

Currency exchange rates

A member of any country expects to pay and be paid in the currency of his or her own country.

However, when money is moved from one country to another, it is necessary to exchange the currency of the first country for that of the second. The various currencies of the world are linked together in agreed ratios, or **exchange rates**. This enables transfer of money and payment for goods between countries to take place.

Table 21.1 gives the currency units in use in some countries and their approximate exchange rate for US\$1.

Exchange rates vary from day to day. Table 21.1 may be taken as a rough guide only. In Examples 8, 9, 10 and in Exercise 21g, values are taken from Table 21.1, unless other rates are given.

Table 21.1

Country	Currency unit	Units to US\$1
Barbados	Bds\$	2.00
Curaçao	NAf	1.79
Eastern Caribbean	EC\$	2.70
Guyana	G\$	100.00
Haiti	Gourde (H\$)	16.50
Jamaica	Ja\$	35.00
Trinidad & Tobago	TT\$	6.30
Canada	Can\$	1.40
France	Euro (€)	1.50
Italy	Euro (€)	1.50
Japan	Yen	120
Kenya	Shilling (Sh)	60.75
Nigeria	Naira (N)	82.65
UK	Pound (£)	0.60
Germany	Euro (€)	1.50
Mexico	Peso	7.80

Note: Rates vary for buying or selling transactions. However this difference is small and only becomes significant for very large sums of money.

Example 8

An American traveller takes US\$500 to St Kitts. How many EC\$ can this be exchanged for?

$$\begin{aligned} \text{US\$1} &= \text{EC\$2.70} \\ \text{US\$500} &= 500 \times \text{EC\$2.70} \\ &= \text{EC\$1350} \end{aligned}$$

Example 9

What is the US$ value of a Ja$100 note?

$$\begin{aligned} \text{Ja\$35} &= \text{US\$1} \\ \text{Ja\$1} &= \text{US\$}\frac{1}{35} \\ \text{Ja\$100} &= \text{US\$}\left(100 \times \frac{1}{35}\right) \\ &= \text{US\$}\frac{100}{35} \\ &= \text{US\$2.86 (correct to 2 d.p.)} \end{aligned}$$

Notice that calculations should be rounded to 2 decimal places.

Example 10

How many Bds$ can be bought for TT$2000?

$$\begin{aligned} \text{TT\$6.30} &= \text{Bds\$2.00 (= US\$1)} \\ \text{TT\$1} &= \text{Bds\$}\frac{6}{2.3} \\ \text{TT\$2000} &= \text{Bds\$2000} \times \frac{6}{2.3} \\ &= \text{Bds\$}\frac{40\,000}{63} \\ &= \text{Bds\$635 (to the nearest dollar)} \end{aligned}$$

Exercise 21g

1. Exchange US$20 into the following currencies.
   (a) euro (b) TT$ (c) yen
   (d) G$ (e) naira (f) Bds$
2. Exchange the following amounts into US$. Give all answers correct to 2 d.p.
   (a) Bds$610 (b) EC$32 (c) Ja$392
   (d) 50€ (e) 300€ (f) £6000
3. Find the US$ value of a TT dollar to the nearest cent.
4. What is the US dollar value of a Guyana dollar to the nearest cent?
5. How many euros can a Jamaican buy with Ja$476?
6. How many EC$ can be bought with US$2500?
7. How many US$ can be bought for 5000 yen?
8. How many Bds$ can an Italian tourist buy with 500 euros?
9. How many £ sterling can a Japanese visitor buy with 10 000 yen?
10. A bank exchanges Curaçao currency for dollars at the rate of 1.8NAf to the $.
    (a) Calculate, in NAf, the amount received in exchange for $52.50.
    (b) Calculate, correct to the nearest cent, the amount received in exchange for 60 NAf.
11. How much does the £ sterling value of an American $100 note rise or fall, if the exchange rate changes from $1.57 to $1.55 to the £? (Answer to the nearest penny.)
12. In 1998 the Xe$ exchange rate with the US$ was Xe$1 = US$1.6; in 2006, the exchange rate was Xe$1 = U$0.24. In terms of US$ by what percentage had the value of the Xe$ fallen since 1998?

Summary

Interest is extra money paid on savings or loans. The interest paid depends on the money saved (or loaned), the time period of the savings (or loan), and the interest rate expressed as a percentage.

Compound interest is found using the same basic formula for calculating interest as for **simple interest**, that is

$$I = \frac{PRT}{100}$$

However, in calculating compound interest, the interest is added to the principal at the end of an agreed time interval to give a new value for P.

An everyday application of rates is the use of **currency exchange rates** to enable transfer of money and payment for goods between different countries. These rates are not constant but may change daily.

Practice Exercise P21.1

1. (a) Calculate the simple interest paid on a loan of $8000 with a rate of interest of 10% over 5 years.
   (b) Suppose that at the end of each year the interest is added to the principal. How much interest is paid?
   (c) How much more interest was paid in (b) than (a)?

2. Four friends take loans from different banks that charge simple interest per annum. Copy and complete Table 21.2

Table 21.2

	Principal	Rate of interest	Repayment period	Interest	Amount
Clive	$400	7.2%	2 yrs		
Rose	$2250		8.5 yrs		$3588.75
Evan		5.8%		$504.60	$1954.60
Reese			3.5 yrs	$157.50	$907.50

Practice Exercise P21.2

1. Copy and complete Table 21.3

Table 21.3

Compound interest	At the end of the 1st year		At the end of the 2nd year		At the end of the 3rd year	
	Interest	amt	Interest	amt	Interest	amt
$700 at 5%						
$450 at12%						

2. Find the final amount and total compound interest paid on:
   (a) $340 at 7% p.a. for two years
   (b) $190 at 9% p.a. for two years
   (c) $1200 at 15% p.a. for three years.

3. What is the total compound interest paid on the following?
   (a) $490 at 7% for 2 years
   (b) $2550 at 15% for 3 years.

Practice Exercise P21.3

Table of Exchange rates

Table 21.4

	Bds$	EC$	Ja$	TT$	Pesos	US$	UK£
Bds$1.00	1	1.35	30.87	3.10	5.72	0.50	0.28
EC$1.00	0.74	1	22.85	2.30	4.23	0.37	0.20
Ja$1.00	0.03	0.04	1	0.10	0.19	0.02	0.01
TT$1.00	0.32	0.44	9.96	1	1.85	0.16	0.09
Pesos 1.00	5.72	0.24	5.40	0.54	1	0.09	0.05
US$1.00	2.00	2.70	61.59	6.19	11.41	1	0.55
UK£1.00	3.63	4.91	112.20	11.27	20.79	1.82	1

Use the exchange rates table above to calculate the following conversions.

1. Convert 150 United States dollars (US$150) into the following currencies.
   (a) Barbados dollars (Bds$)
   (b) Jamaican dollar (Ja$)
   (c) Mexican Pesos
   (d) United Kingdom pounds (UK£)

2. Clara works in Trinidad. She needs to go to the United States on business so she exchanges TT$800 into US currency. The bank buys her dollars at a rate of US$0.15 for each TT$.
   (a) How many US dollars does she get?
   She has to cancel her trip. The bank buys the US$ back from her at a rate of TT$6.17 for each US$.
   (b) How much money has she lost in the two transactions?

3. Millennium Motors plans to import three SL cars from Europe. These cars cost 12 000 euros each. Shipping and handling expenses are 20% of the cost of the cars. Customs duties are imposed at the rate of 40%. Local expenses amount to $2000 per car.
   (a) What is the amount of shipping and handling costs
      (i) in euros?
      (ii) in dollars, given that 1 euro is equivalent to $1.50?

(b) How much money is paid in customs duties?
(c) What is the total cost of each car to Millennium Motors?
(d) What must Millennium Motors charge for each car in order to
(i) not lose any money?
(ii) make a profit of 25%?

Refer to the currency exchange table below to answer the following questions.

Table 21.5

	Bds $	EC $	Ja $	TT $	UK £	US $
Bds$1.00	1.00	1.34	19.22	3.23	0.36	0.50
EC$1.00	0.74	1.00	14.34	2.41	0.27	0.37
Ja$1.00	0.05	0.07	1.00	0.17	0.02	0.03
TT$1.00	0.30	0.41	5.95	1.00	0.11	0.16
UK£1.00	2.80	3.70	52.6	8.94	1.00	1.40
US$1.00	2.00	2.67	36.4	6.18	0.71	1.00

4. Use a scale of 1 cm = TT$10 on the 'horizontal' axis, and 1 cm = Ja$50 on the 'vertical axis' and plot the graph showing the conversion from TT$ to Ja$.
Use your graph to estimate the value, in Ja$, of
(a) TT$4.00 (b) TT$25.00 (c) TT$45.00

5. Use the graph drawn to estimate the value, in TT$, of the following amounts in Ja$
(a) 60 (b) 150 (c) 200 (d) 750.

6. Use the currency exchange table to draw a currency conversion graph for
(a) Bds$ and EC$
(b) US$ and UK£.

7. Desmond, a tourist from the UK, travels to Barbados where he converts UK £750 to Bds$. After spending $1200 in local currency, he changes the remaining money into US$. How much money, in US$ does he receive?

8. Aunt Agatha lives in Grenada. One day she receives a money order from her nephew in Atlanta, Georgia, US for US$300. When she goes to the bank to cash the money order, she has to pay 2.5% bank charges. How much money does she actually receive in EC currency?

9. Ken, who lives in Jamaica, has to go to UK on a training course. His expenses, in £ sterling, are £2400. His ticket, which is priced in US currency costs $875. How much money, in Jamaican currency, does Ken need to pay for his ticket and meet his expenses in UK?

Chapter 22

Geometrical constructions (2)

Constructing special angles, bisecting lines and angles

Pre-requisites

- use of ruler, set square and compasses

Drawing accurately

Remember that in geometry, to construct means to draw accurately. In Chapter 2 we looked at points to note when doing constructions. Below are some other points:

1 Where possible, arrange that the angles of intersection between lines and arcs are about 90°.

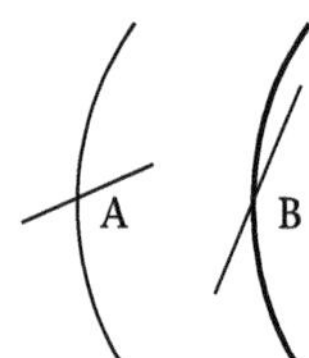

Fig. 22.1

In Fig. 22.1, there is a clear point of intersection at A. At B there is a large 'area of intersection'; this is because the lines are too thick and the angle between them is too small.

2 It is often helpful to draw lines which are longer than the required segments, using compasses to mark off the required points as in Fig. 22.2.

Fig. 22.2

Similarly, when drawing an angle, extend its arms beyond its vertex as in Fig. 22.3. This will improve the accuracy of the drawing.

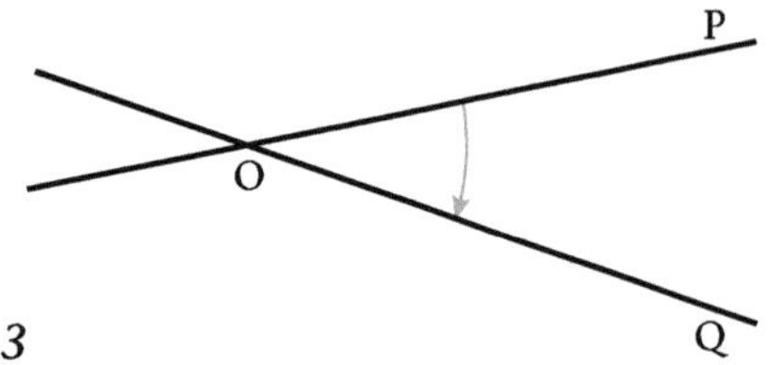

Fig. 22.3

To bisect a straight line segment

In Fig. 22.4, the **line segment** AB is the part of the line between A and B, including the points A and B.

Fig. 22.4

To **bisect** the line segment AB means to divide it into two equal parts.

(a) Open a pair of compasses so that the radius is about $\frac{3}{4}$ of the length of AB.

(b) Place the sharp point of the compasses on A. Draw two arcs, one above, the other below the middle of AB, as in Fig. 22.5.

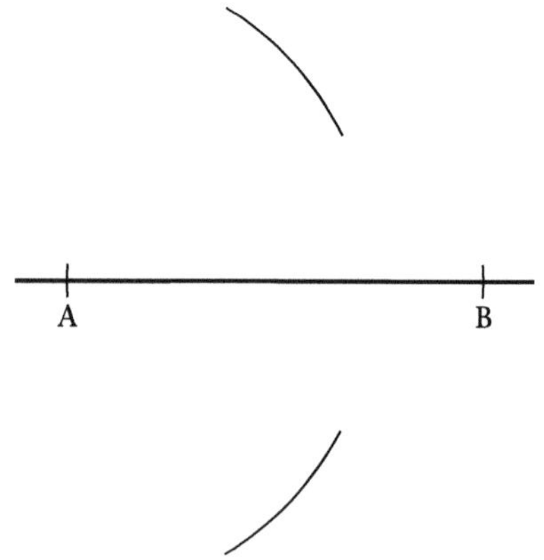

Fig. 22.5

(c) *Keep the same radius* and place the sharp point of the compasses on B. Draw two arcs so that they cut the first arcs at P and Q as in Fig. 22.6.

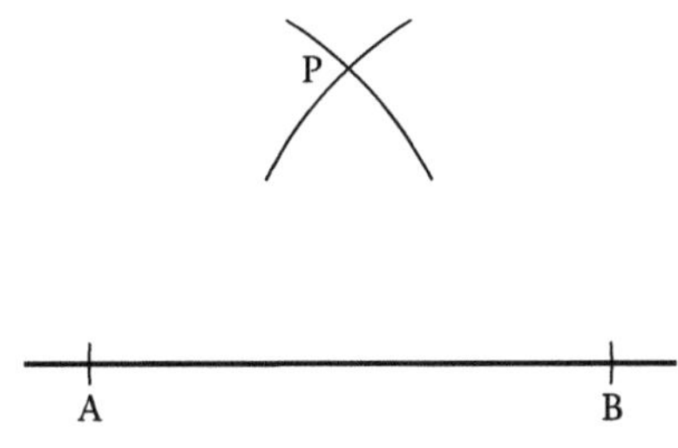

Fig. 22.6

(d) Draw a straight line through P and Q so that it cuts AB at M

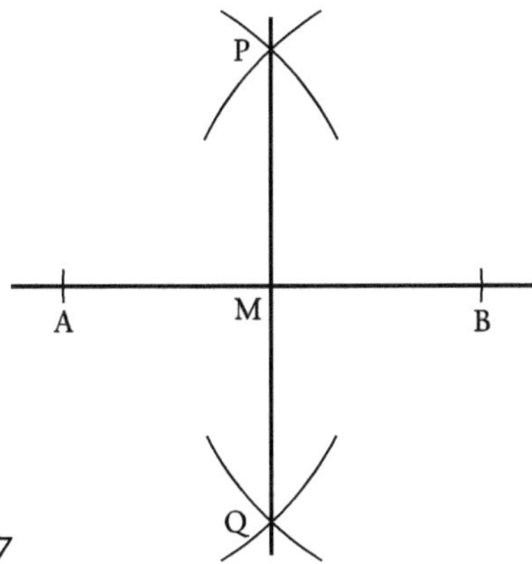

Fig. 22.7

M is the mid-point of AB. PQ meets AB perpendicularly. PQ is the **perpendicular bisector** of AB. Use a ruler and protractor to check that AM = MB and $A\widehat{M}P = B\widehat{M}P = 90°$ in Fig. 22.7.

Exercise 22a

Use ruler and compasses **only** in this exercise.

1. Draw any line segment AB. Use the above method to find the mid-point of AB. Check by measurement that your answer is correct.
2. Fig. 22.8 represents a paper triangle, ABC.

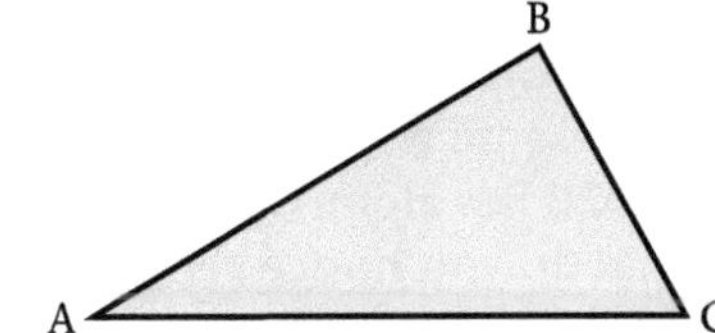

Fig. 22.8

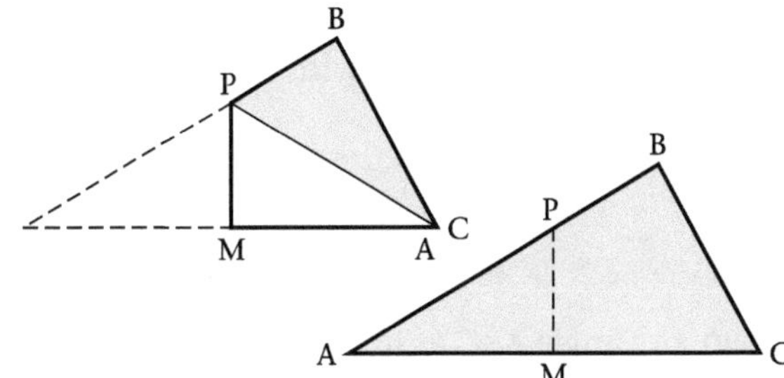

Fig. 22.9

△ABC is folded so that A meets C. This gives a fold line PM as shown in Fig. 22.9.

(a) What can be said about the point M?
(b) What can be said about line PM?

3. (a) Cut out a large paper triangle, ABC (as in Fig. 22.8). △ABC should be scalene and acute-angled.
   (b) Fold △ABC so that A meets C.
   (c) Open out △ABC, then make a second fold so that B meets C.
   (d) In the same way, make a third fold so that A meets B.
   (e) What do you notice about the three folds?
   (f) What can be said about each fold?
4. Draw any triangle ABC.
   (a) Construct the perpendicular bisector of each side.
   (b) What do you notice?
5. Draw any circle and any two chords AB and XY. (Neither chord should be a diameter.)
   (a) Construct the perpendicular bisectors of AB and XY.
   (b) What do you notice?
6. Draw any circle and any diameter AB.
   (a) Construct the perpendicular bisector of AB and extend it if necessary to cut the circumference at P and Q.
   (b) What kind of chord is PQ?
   (c) Join AP, PB, BQ, QA. What kind of quadrilateral is APBQ?
7. Draw any triangle ABC.
   (a) Construct the mid-point, M, of AB.
   (b) Construct the mid-point, N, of BC.
   (c) Measure MN and AC.
   (d) What do you notice?

8 (a) Draw a line 10 cm long.
(b) Construct a square with this line as diagonal.
(c) Measure a side of the square.

To bisect a given angle

Given any angle ABC as in Fig. 22.10:

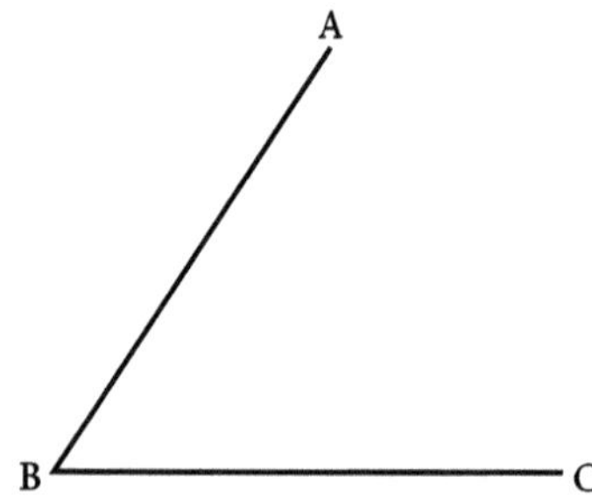

Fig. 22.10

(a) With centre B and any radius (i.e. open a pair of compasses to any radius and place the sharp point at B) draw an arc to cut the arms BA, BC at P, Q. See Fig. 22.11.

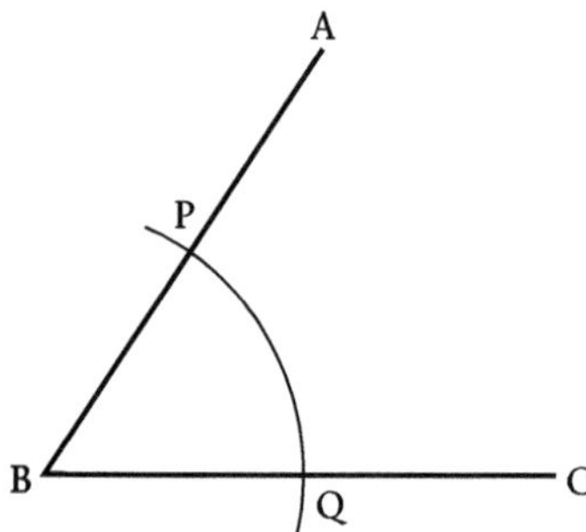

Fig. 22.11

(b) With centres P, Q and equal radii, draw arcs to cut each other at R.

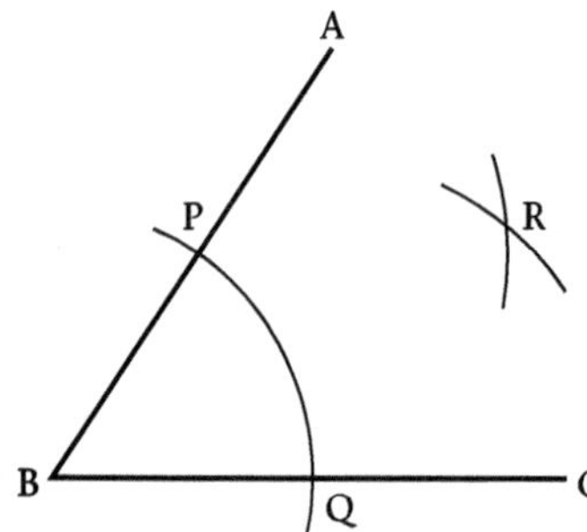

Fig. 22.12

(c) Join BR.

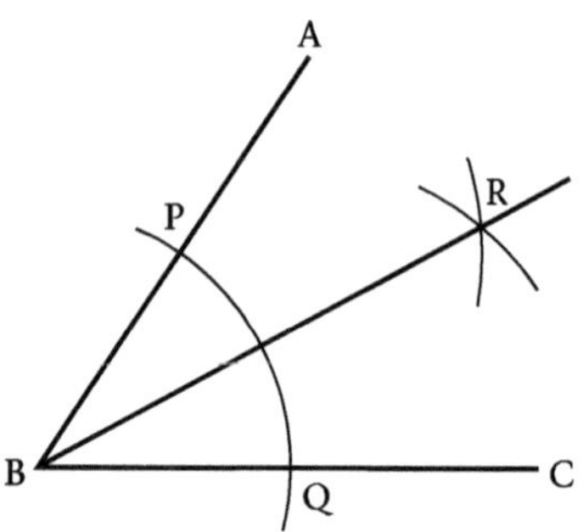

Fig. 22.13

BR bisects $\hat{ABC}$. BR is the bisector of $\hat{ABC}$. Use a protractor to check that $A\hat{B}R = C\hat{B}R$ in Fig. 22.13.

Exercise 22b

Use ruler and compasses *only* in this exercise.

1 Draw any angle ABC. Use the above method to construct the bisector of $A\hat{B}C$. Use a protractor to check your result.

2 A paper triangle like that of Fig. 22.8 is folded so that AB lies along AC. This gives a fold line AR as shown in Fig. 22.14.

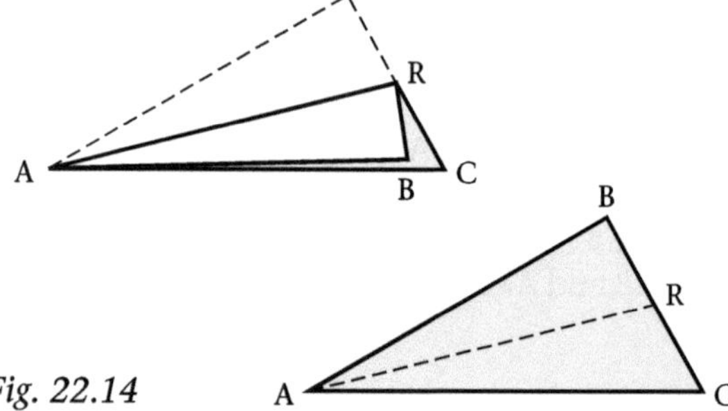

Fig. 22.14

(a) What can be said about angles $B\hat{A}R$ and $C\hat{A}R$?
(b) What is the correct name for the line AR?

3 (a) Cut out a large paper triangle, ABC, such as that of Fig. 22.8.
(b) Fold △ABC so that AB lies along AC.
(c) Open out △ABC, then make a second fold so that BC lies along BA.
(d) In the same way, make a third fold so that CB lies along CA.
(e) What do you notice about the three folds?

4 (a) Draw a scalene triangle PQR such that $\hat{Q}$ is obtuse.
(b) Construct the bisectors of $\hat{P}$, $\hat{Q}$ and $\hat{R}$.
(c) If necessary, produce each bisector so that it cuts the other two.
(d) What do you notice about the three bisectors?

5 (a) Construct an isosceles triangle XYZ such that XY = YZ = 8 cm.
(b) Construct the bisector of $\hat{Y}$.
(c) Construct the perpendicular bisector of side XZ.
(d) What do you notice?

6 (a) Draw any obtuse angle ABC.
(b) RB is the bisector of $A\hat{B}C$. Construct RB.
(c) SB is the bisector of $R\hat{B}C$. Construct SB.
(d) TB is the bisector of $S\hat{B}C$. Construct TB.
(e) What fraction of $A\hat{B}C$ is $T\hat{B}C$?

7 (a) Draw a triangle with sides 6, 8, 10 cm.
(b) Bisect the smallest angle.
(c) The bisector cuts the opposite side into two parts. Measure the lengths of the two parts.

8 (a) Draw a circle of radius 75 mm.
(b) Construct two diameters at right angles to each other.
(c) Construct two more diameters bisecting the angles between those drawn first.
(d) Join the ends of the diameters to form a regular polygon. What sort of polygon is it?
(e) Measure the length of one of the sides of the polygon.

Angle constructions

To construct an angle of 90°

Given a point B on a straight line AC:

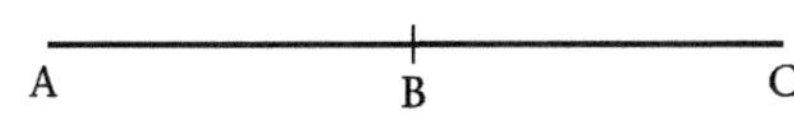

Fig. 22.15

It is required to construct a line BR through B such that $R\hat{B}A = R\hat{B}C = 90°$.

(a) With centre B and any radius draw arcs to cut AC at P and Q. See Fig. 22.16.

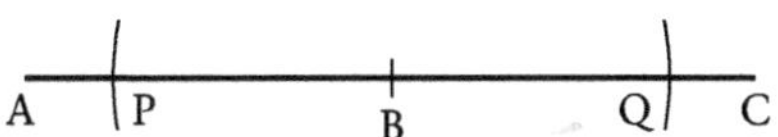

Fig. 22.16

(b) With centres P, Q and equal radii, draw arcs to cut each other at R. See Fig. 22.17.

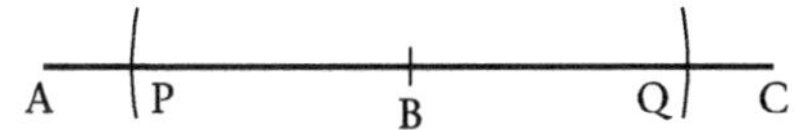

Fig. 22.17

(c) Join BR.

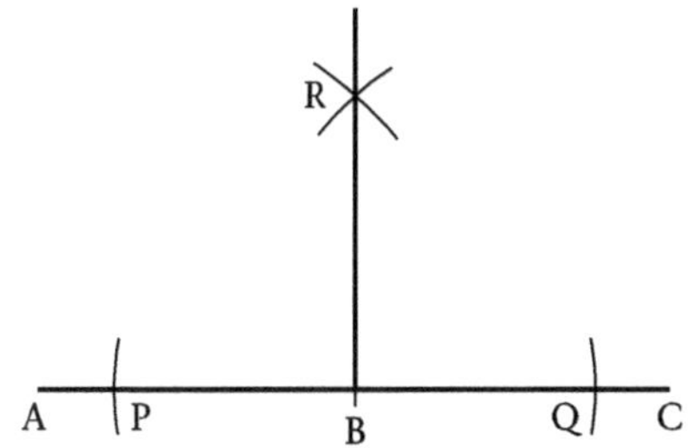

Fig. 22.18

BR is perpendicular to AC. Hence $R\hat{B}A = R\hat{B}C = 90°$. Use a protractor to check this result in Fig. 22.18. Notice that this method is equivalent to bisecting an angle of 180°.

The following method is useful if B is near the edge of the paper.

Fig. 22.19

(a) Draw any circle to pass through B and cut AC at Q. Mark the centre of the circle, O.
(b) Join QO. Produce QO to cut the circle again at R. See Fig. 22.20 (a).
(c) Join BR. $R\hat{B}C = 90°$.

Use a protractor to check that $R\hat{B}C = 90°$ in Fig. 22.20 (b).

(a)

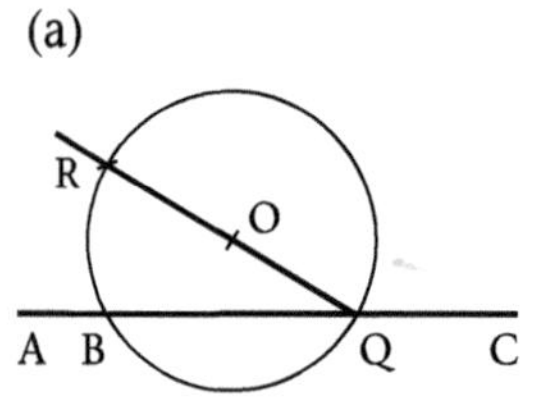

(b)

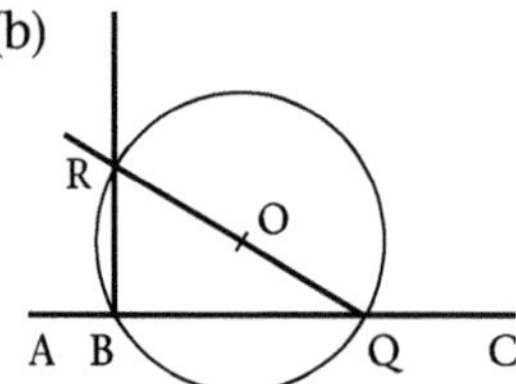

Fig. 22.20

To construct an angle of 45°

$45° = \frac{1}{2}$ of 90°. To construct an angle of 45°, first construct an angle of 90° and then bisect it. This is shown in Fig. 22.21.

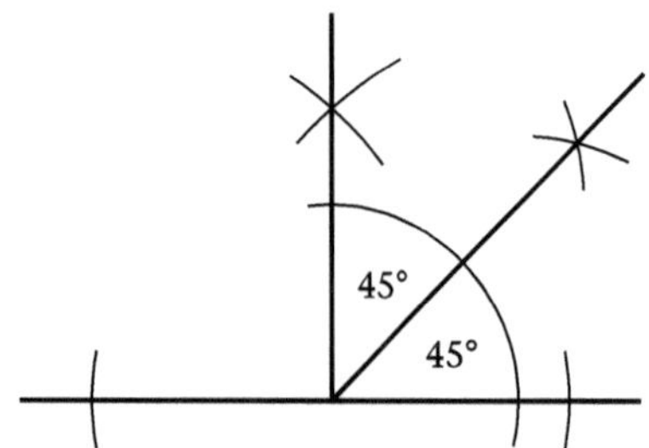

Fig. 22.21

Use a protractor to check the data in Fig. 22.21.

Exercise 22c

Use ruler and compass *only* in this exercise.

1. Construct angles of 90° and 45°.
2. Construct an angle of 135°.
3. (a) Construct a square with sides each 83 mm long.
   (b) Measure the length of a diagonal.
4. (a) Construct a rectangle measuring 7.4 cm by 10.3 cm
   (b) Measure the length of a diagonal.
5. (a) Construct △PQR such that $\widehat{Q} = 90°$, PQ = 5 cm and QR = 6 cm.
   (b) Measure the length of its hypotenuse PR.
6. (a) Draw any circle and any chord AB. (AB should *not* be a diameter.)
   (b) Construct another chord BC such that $A\widehat{B}C = 90°$.
   (c) Join AC. What do you notice about the line AC?
7. (a) Construct △XYZ such that $\widehat{X} = \widehat{Z} = 45°$ and XZ = 8 cm.
   (b) Measure either of the sides XY or YZ.
8. (a) Construct an isosceles triangle with the equal sides 9 cm long and the angle between them 45°.
   (b) Measure the third side.

To construct an angle of 60°

Given a straight line BC as shown in Fig. 22.22.

Fig. 22.22

To construct a point such that $A\widehat{B}C = 60°$,

(a) With centre B and any radius, draw an arc to cut BC at X. Notice in Fig. 22.23, the arc is extended well above BC.

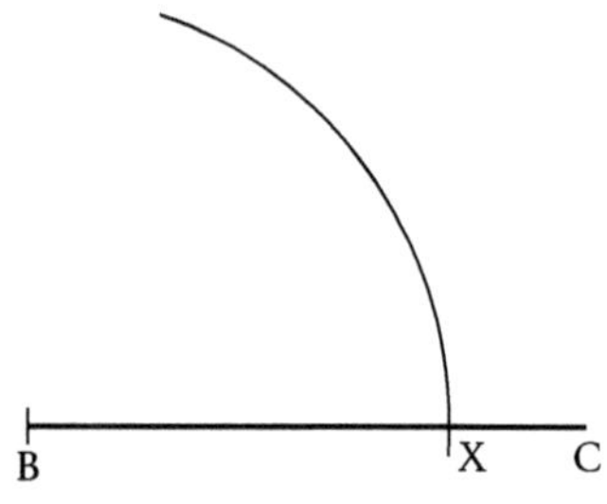

Fig. 22.23

(b) With centre X and the *same* radius, draw an arc to cut the first arc at A. See Fig. 22.24.

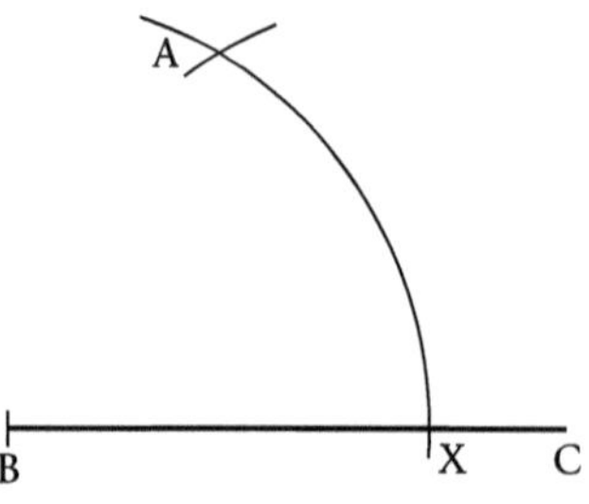

Fig. 22.24

(c) Join AB.

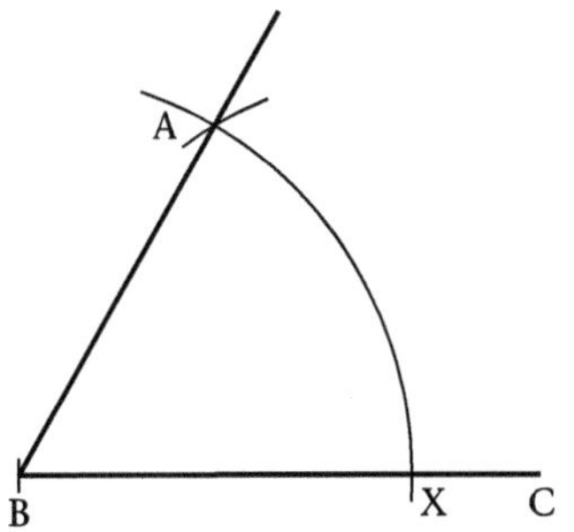

Fig. 22.25

$A\hat{B}C = 60°$. Use a protractor to check that $A\hat{B}C = 60°$ in Fig. 22.25. Notice that the points A, B and X are the vertices of an equilateral triangle.

To construct an angle of 30°

$30° = \frac{1}{2}$ of 60°. To construct an angle of 30°, first construct an angle of 60° and then bisect it. This is shown in Fig. 22.26.

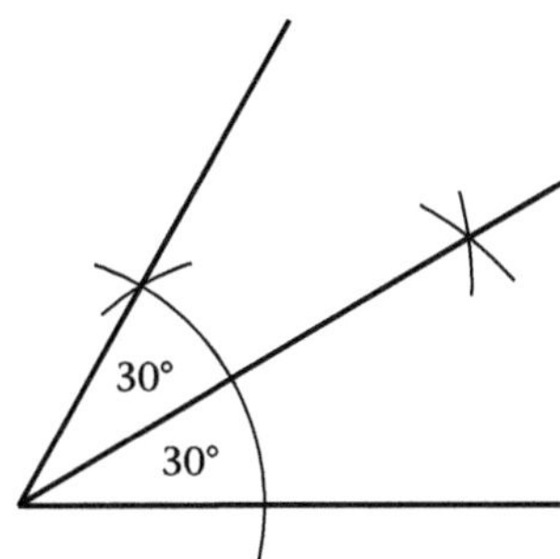

Fig. 22.26

Use a protractor to check the data in Fig. 22.26.

Exercise 22d

Use ruler and compasses *only* in this exercise.

1. Construct angles of 60°, 30°, 15°.
2. Construct angles of 120°, 105°.
3. Construct an equilateral triangle with sides of length 7.2 cm.
4. (a) Draw a circle of radius 5 cm.
   (b) Construct radii at 60° intervals in the circle.
   (c) Hence construct a regular hexagon.
   (d) How long are the sides of the hexagon?
5. Construct a parallelogram with sides 6 cm and 9 cm, the angle between these sides being 60°. Measure the diagonals.
6. (a) Construct a rhombus with sides 6 cm such that one of its acute angles is 75°.
   (b) Measure the diagonals of the rhombus.
7. (a) Construct $\triangle LMN$ in which LM = 105 mm, $\hat{M} = 60°$ and $\hat{N} = 90°$.
   (b) Measure LN and MN.
8. (a) Construct a kite ABCD in which AB = 5 cm, AD = 8 cm and $\hat{A} = \hat{C} = 105°$.
   (b) Measure the diagonal AC.

To copy an angle

Given any angle $A\hat{B}C$ as in Fig. 22.27.

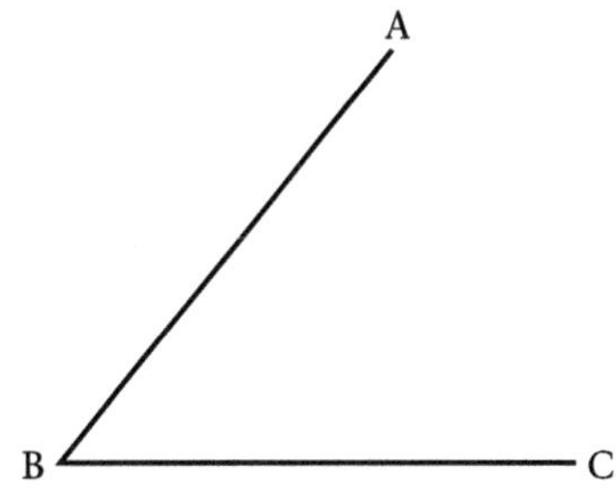

Fig. 22.27

To make a copy of $A\hat{B}C$,

(a) Draw any line XY. Mark a point B′ on XY. With centre B and any radius, draw an arc to cut BA, BC at P, Q respectively. See Fig. 22.28(a). Then with centre B′ and the *same* radius, draw an arc to cut XY at Q′. See Fig. 22.28(b).

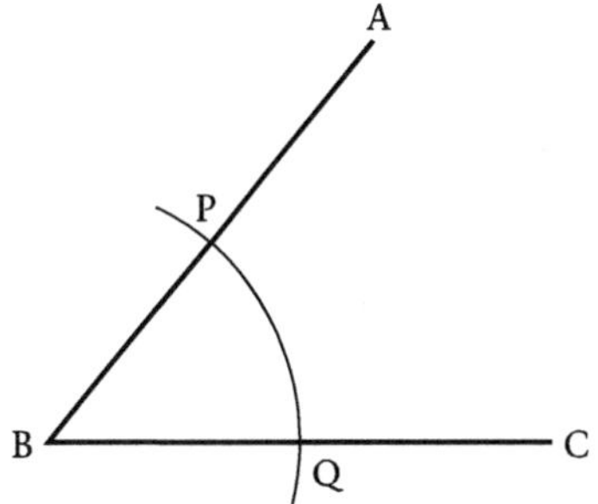

Fig. 22.28(a)

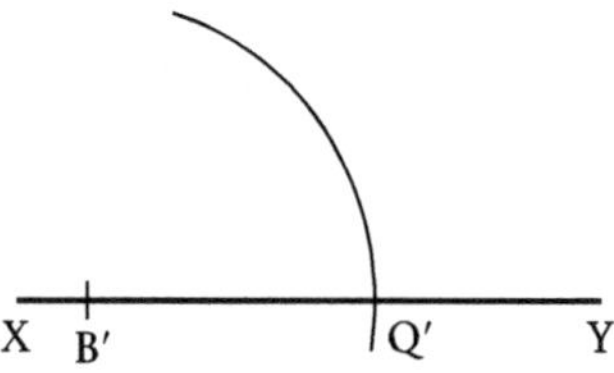

Fig. 22.28(b)

(b) With centre Q, open the compasses until the radius = QP. Make an arc at P as a check. See Fig. 22.29(a). Then with centre Q′ and the *same* radius, draw an arc to cut the arc through Q′ at P′. See Fig. 22.29(b).

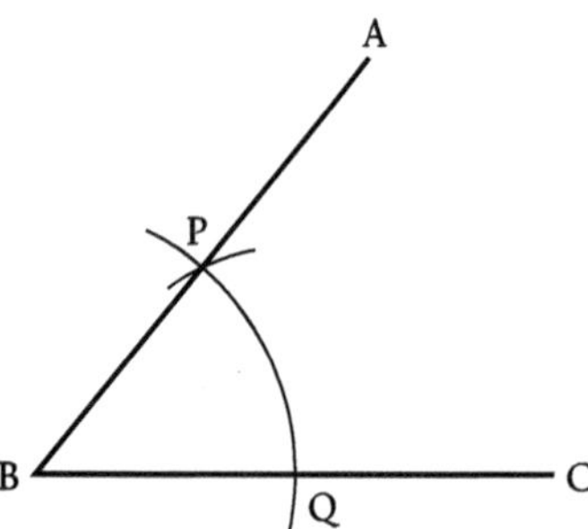

Fig. 22.29(a)

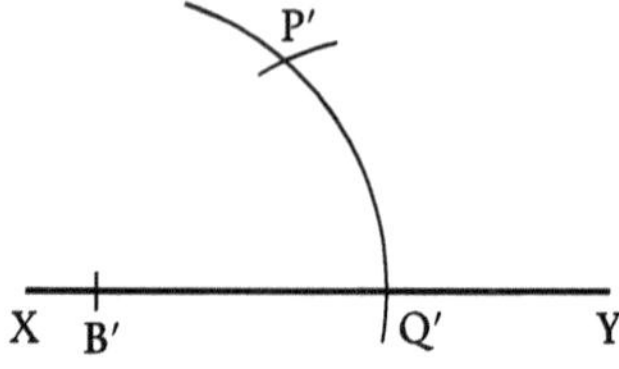

Fig. 22.29(b)

(c) Draw a line through B′ and P′.

In Fig. 22.30, $A'\widehat{B}'Y = A\widehat{B}C$ of Fig. 22.27.

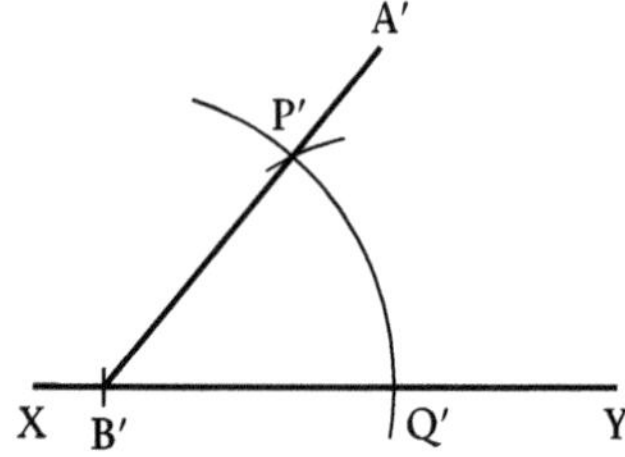

Fig. 22.30

Note: The lengths of the arms of the angles may be different; however, the sizes of the angles are the same.

Exercise 22e

Copy all angles using ruler and compasses *only*. Check your work using a protractor.

1. Use the method given above to copy the angles in Fig. 22.31 into your exercise book. Use a protractor to check your accuracy.

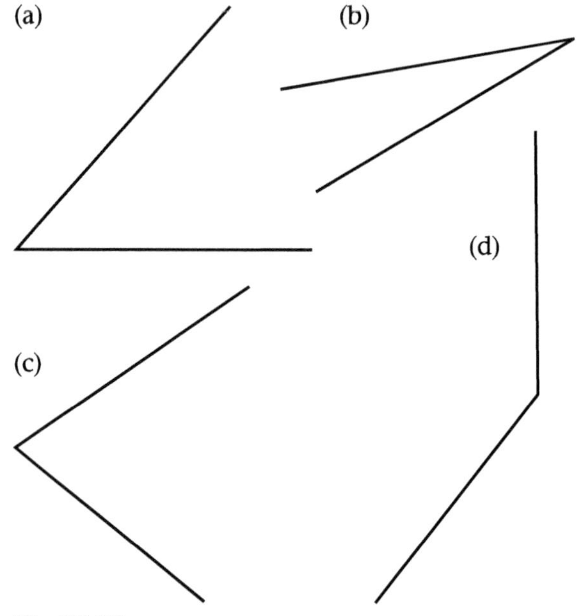

Fig. 22.31

2. (a) Exchange exercise books with a friend. Draw 3 angles of any size in your friend's book. (Your friend will draw 3 angles in your book.)
   (b) Get your own book back. Copy the 3 angles your friend has drawn.
   (c) Use a protractor to check the accuracy of your work.

Exercise 22f (Further practice)

In this exercise draw a rough first. This will give you an idea of the shape of the final drawing. All construction lines should be left in your work.

1. (a) Construct an isosceles △ABC so that AB = AC, BC = 75 mm and the length of the perpendicular from A to BC is 60 mm.
   (b) Measure AB.

2 (a) Construct $\triangle ABC$ with sides 7 cm, 8 cm, 9 cm.
(b) Draw the perpendicular bisectors of all three sides. These should meet at one point O.
(c) With centre O and radius OA, draw a circle.
(d) What is the radius of the circle? Does the circle also pass through B and C?

3 (a) Draw a triangle PQR with sides 69 mm, 102 mm, 135 mm.
(b) Use the method of question 2 to construct the circle passing through P, Q and R.
(c) Measure the radius of the circle.

4 (a) Construct a triangle with sides 6 cm, 8 cm, 9 cm.
(b) Use ruler and compasses to find the mid-point of each side.
(c) Join each vertex to the mid-point of the opposite side. (These three lines are called **medians.**)
(d) Do the three medians meet at a point?
(e) By careful measurement, find the ratio in which this point divides the length of each median.

5 (a) Use the method of question 4 with a triangle of any size.
(b) Do the three medians behave in the same way as before?

6 (a) Construct $\triangle XYZ$ in which XY = 8.3 cm, YZ = 11.9 cm and $X\hat{Y}Z = 60°$.
(b) Construct M, the mid-point of XZ.
(c) Measure YM.

7 (a) Construct $\triangle ABC$ with AB = 68 mm, AC = 102 mm and $\hat{B} = 120°$.
(b) Construct the perpendicular bisectors of AB and BD and let them meet at O.
(c) Measure OA, OB, and OC.

8 (a) Construct $\triangle ABC$ in which AB = 9 cm, BC = 12 cm, $A\hat{B}C = 60°$.
(b) Construct the bisector of $\hat{A}$ and let it meet BC at D.
(c) Measure DC.

9 (a) Construct $\triangle ABC$ in which AB = 99 mm, BC = 114 mm, CA = 126 mm.
(b) Use ruler and compasses to find the position of M, the mid-point of BC.
(c) Through M, construct lines parallel to AC, AB to meet AB, AC in H, K respectively.
(d) Measure HK.

10 (a) Construct a parallelogram ABCD with BD = 104 mm, DC = 48 mm and $B\hat{D}C = 30°$.
(b) Measure AC.

11 (a) Construct a trapezium PQRS in which PQ is parallel to SR, PQ = 6 cm, PS = 5 cm, SR = 11 cm and QS = 9 cm.
(b) Measure QR.

12 (a) The diagonals of a parallelogram bisect each other. Construct a parallelogram with one side 10 cm long, and diagonals 15 cm and 10 cm long. (Draw a sketch first.)
(b) Measure the side of the parallelogram which is not given.

Summary

To **construct** is to draw accurately.

Construction lines must be 'thin', fine lines; points should be 'pinpoints' or marked by a cross (×).

All construction lines must be clearly shown on a final diagram.

Practice Exercise P22.1

Remember not to erase construction lines.

1. Using ruler and compasses only,
   (a) Construct a triangle ABC with AB = 7 cm, BC = 6.5 cm and AC = 6 cm.
   (b) Construct the perpendicular bisector of AB, BC and AC.
   (c) What do you notice?

2. Construct a triangle PQR with PQ = 6 cm, $P\hat{Q}R = 60°$ and $R\hat{P}Q = 45°$.
   Measure and state the length of PR.

3. Construct an isosceles triangle ABC with AB = 8.4 cm, and $A\hat{B}C = C\hat{A}B = 45°$.

4. (a) Construct a parallelogram PQRS with PQ = 7 cm, $Q\hat{P}S = 60°$, PS = 4.4 cm.
   (b) Then draw SR || PQ and = 7 cm.
   Complete the parallelogram by joining QR.

5. Using ruler and compasses only
   (a) Draw a line AB = 6.5 cm.
   (b) Construct an angle of 60° at P.
   (c) Mark a point S, 4.4 cm from P.
   (d) Through S draw a line SR = 4.4 cm and parallel to PQ.
   (e) Join QR.
   What shape is PQRS?

6. Draw a line PQ = 5 cm. Construct an an angle of 105° at P and Q. Mark a point S, 6.2 cm from P and a point R 6.2 cm from Q. Join RS.
   What shape is PQRS?

7. (a) Construct triangle LMN such that $L\hat{M}N = 90°$, LM = 4.8 cm and $M\hat{L}N = 50°$
   (b) Measure the length of LN.

8. (a) Draw a circle centre O and radius 5 cm and draw a diameter XY.
   (b) With centre Y and radius 6 cm mark a point Z on the circumference of the circle.
   (c) Join YZ.
   (d) Measure XZ and $Y\hat{Z}X$.
   (e) What kind of triangle is triangle ABC?

9. (a) Draw an isosceles triangle ABD with $A\hat{B}D = A\hat{D}B = 51\,°$ and BD = 5.2 cm.
   (b) With centre B and radius 7.5 cm strike an arc below BD.
   (c) With centre D and radius 7.5 cm strike an arc to cut the first at a point C.
   (d) Join BC, DC.
   What shape is ABCD?
   (e) Construct the perpendicular bisector of BD to pass through A and C.

10. (a) Construct a square PQRS of side 6 cm.
    (b) Construct the perpendicular bisector of PQ and QR.
    What do you notice?
    (c) Mark the point at which the perpendiculars meet, O.
    (d) With centre O and radius OP, draw a circle.
    What do you notice?

Chapter 23

Solving triangles (3)

The right-angled triangle – sine and cosine ratios

Pre-requisites

- properties of right-angled triangles; ratio

Sine and cosine

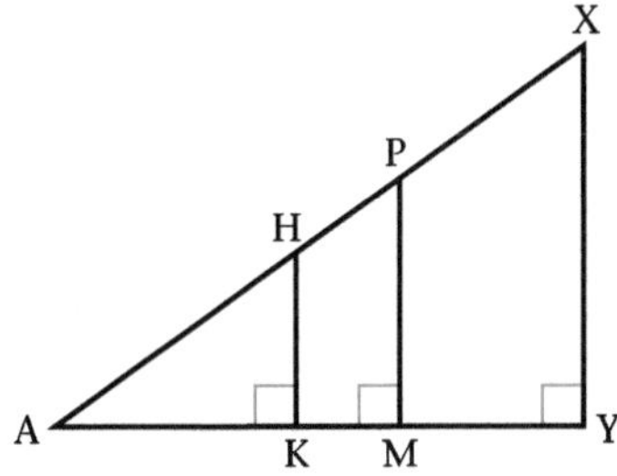

Fig. 23.1

In Fig. 23.1, △s HAK, PAM, XAY are similar.

Thus, $\frac{KH}{AH} = \frac{MP}{AP} = \frac{YX}{AX}$

The value of this ratio depends only on the size of $\hat{A}$. The ratio is called the **sine of $\hat{A}$**. This is usually shortened to **sin A**.

Similarly, $\frac{AK}{AH} = \frac{AM}{AP} = \frac{AY}{AX}$ is a ratio whose size depends only on the size of $\hat{A}$. This ratio is called the **cosine of $\hat{A}$**, usually shortened to **cos A**.

Note: It is only possible to define these ratios as sine and cosine because the △s HAK, PAM and XAY are right-angled triangles. However, the value of these ratios remains the same for the same value of A when the angle A is in any other triangle or polygon.

Fig. 23.2 shows △AHK in various positions. The sides of the triangle are as follows:

AH, the hypotenuse,
KH, the side opposite to $\hat{A}$,
AK, the side adjacent to $\hat{A}$.

These are abbreviated to **hyp**, **opp**, **adj**, respectively, so that

$$\sin A = \frac{\textbf{opp}}{\textbf{hyp}} \qquad \cos A = \frac{\textbf{adj}}{\textbf{hyp}}$$

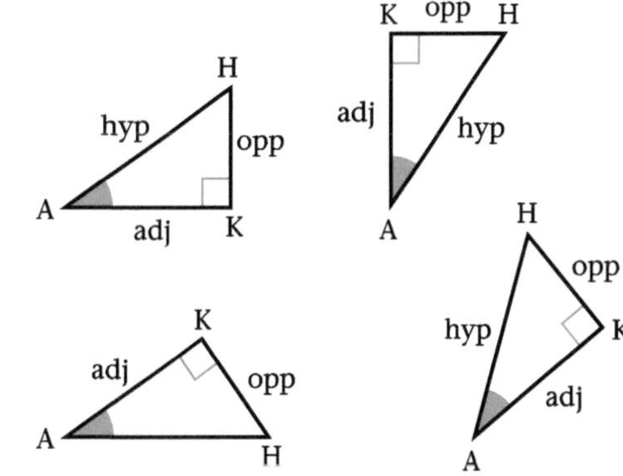

Fig. 23.2

Finding the sine and cosine of angles by measurement

Example 1

Triangles ABC and PQR are drawn accurately.

(a) Measure angles ABC, ACB, PQR, PRQ

(b) State the sine and cosine of each angle

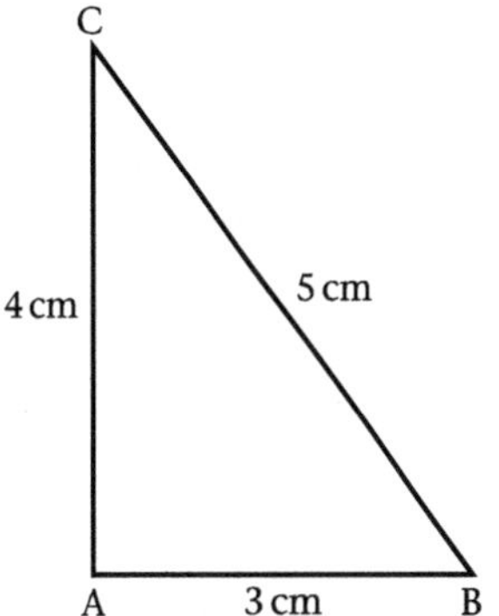

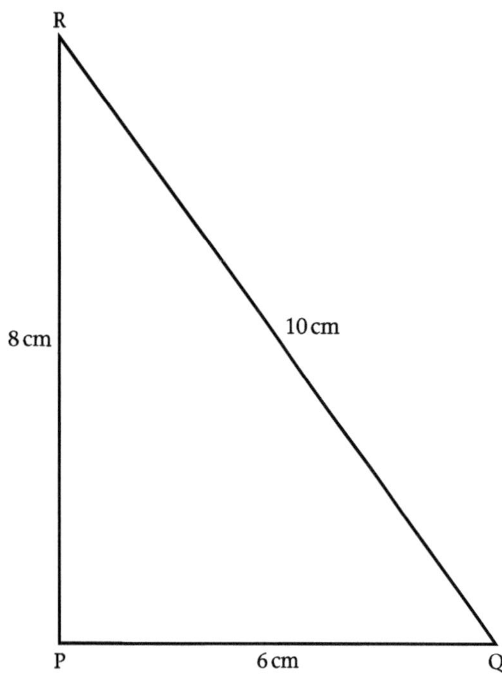

Fig. 23.2

(a) $A\hat{B}C = 53°$, $A\hat{C}B = 37°$,
$P\hat{Q}R = 53°$, $P\hat{R}Q = 37°$

(b) The sine of an angle is the ratio of the lengths of the opposite side to the hypotenuse.

$$\text{sine } A\hat{B}C = \text{sine } 53° = \frac{4\text{ cm}}{5\text{ cm}} = \frac{4}{5}$$

$$\text{sine } P\hat{Q}R = \text{sine } 53° = \frac{8\text{ cm}}{10\text{ cm}} = \frac{4}{5}$$

The cosine of an angle is the ratio of the lengths of the adjacent side to the hypotenuse.

$$\text{cosine } A\hat{B}C = \text{cosine } 53° = \frac{3\text{ cm}}{5\text{ cm}} = \frac{3}{5}$$

$$\text{cosine } P\hat{Q}R = \text{cosine } 53° = \frac{6\text{ cm}}{10\text{ cm}} = \frac{3}{5}$$

Similarly

$$\text{cosine } A\hat{C}B = \text{cosine } 37° = \frac{4\text{ cm}}{5\text{ cm}} = \frac{4}{5}$$

$$\text{cosine } P\hat{R}Q = \text{cosine } 37° = \frac{8\text{ cm}}{10\text{ cm}} = \frac{4}{5}$$

It is very important to identify the sides that are opposite or adjacent to particular angles.

Example 2

Find the value of sine 60° by drawing and measurement.

Fig. 23.4 is a scale drawing of the method.

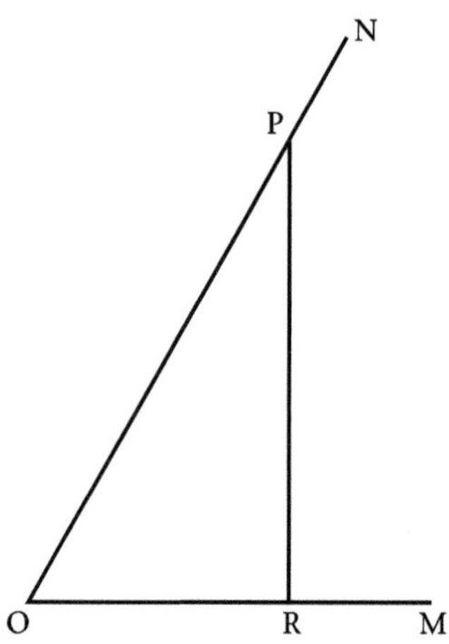

Fig. 23.4

Draw an angle MON = 60°. On ON mark off OP = 10 cm. From P, draw a line perpendicular to OM to meet OM at R. Measure PR. It is found that PR = 8.7 cm (approx.)

$$\text{sine } 60° = \frac{PR}{OP} = \frac{8.7}{10} = 0.87 \text{ (approx.)}$$

Example 3

Using the diagram in Example 2 find the value of cos 60°.

$$\cos 60° = \frac{OR}{OP} = \frac{5}{10} = 0.5$$

Exercise 23a (Group Activity)

1. Find by drawing and measurement, as in Example 2, approximate values for
   (a) sin 20°, cos 20°, (b) sin 40°, cos 40°,
   (c) sin 65°, cos 65°.
   Check the results with members of other groups.
2. Using the calculator find the approximate values for all the ratios in question 1. Give your answers correct to 2 s.f. Compare the results for question 1.
3. What do you notice about the sines and cosines of angles as the angles increase from 0° to 90°?

Solving triangles

Sines and cosines of angles are used to find the lengths of unknown sides in triangles. Table 23.1 gives the sines and cosines of some chosen angles.

Table 23.1

Angle A	sin A	cos A
30°	0.500	0.866
35°	0.574	0.819
40°	0.643	0.766
45°	0.707	0.707
50°	0.766	0.643
55°	0.819	0.574
60°	0.866	0.500

The values in Table 23.1 are given to 3 significant figures.

Example 4

Calculate the value of x in Fig. 23.5.

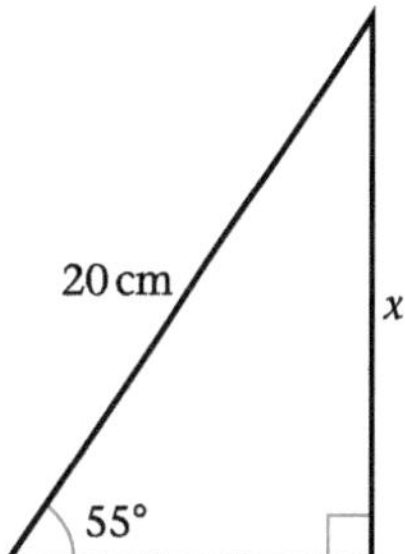

Fig. 23.5

In Fig. 23.5, the hypotenuse is given and x is *opposite* the given angle. Use the *sine* of the given angle.

$$\sin 55° = \frac{x}{20}$$

$$x = 20 \times \sin 55° \text{ cm}$$

$$= 20 \times 0.819 \text{ cm}$$

$$= 16.38 \text{ cm}$$

$$= 16 \text{ cm to 2 s.f.}$$

Example 5

A pole 8 m long leans against a vertical wall making an angle of 40° with the wall. Calculate the height of the top of the pole from the ground.

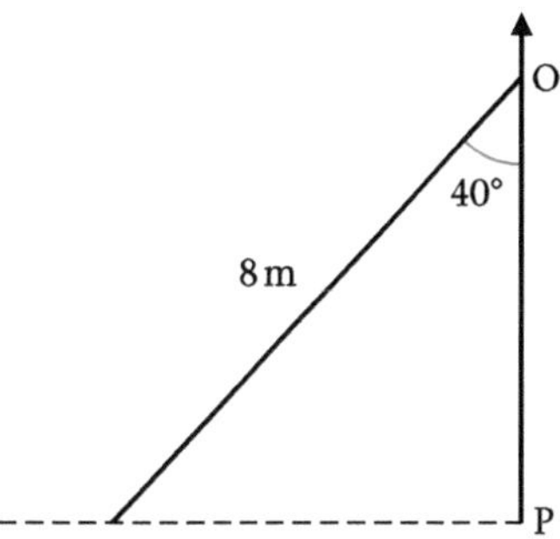

Fig. 23.6

It is required to find the length of OP. In Fig. 23.6 OP is *adjacent* to the known angle. Use the *cosine* of 40°.

$$\cos 40° = \frac{OP}{8}$$

$$OP = 8 \times \cos 40° \text{ m}$$

$$= 8 \times 0.839 \text{ m}$$

$$= 6.712 \text{ m}$$

$$= 6.7 \text{ m to 2 s.f.}$$

The height of the top of the pole from the ground is 6.7 m.

Notice that if the unknown side is *opposite* the given angle, use the *sine* of the angle; if the unknown side is *adjacent* to the given angle, use the *cosine* of the angle.

Exercise 23b

Use the values in Table 23.1 or a calculator in this exercise. Give all answers correct to 2 s.f.

1 Find the value of x in each of the triangles in Fig. 23.7.

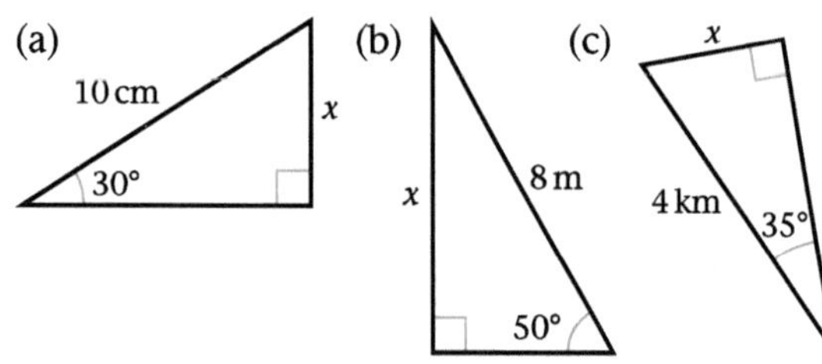

Fig. 23.7

2 Find the value of y in each of the triangles in Fig. 23.8.

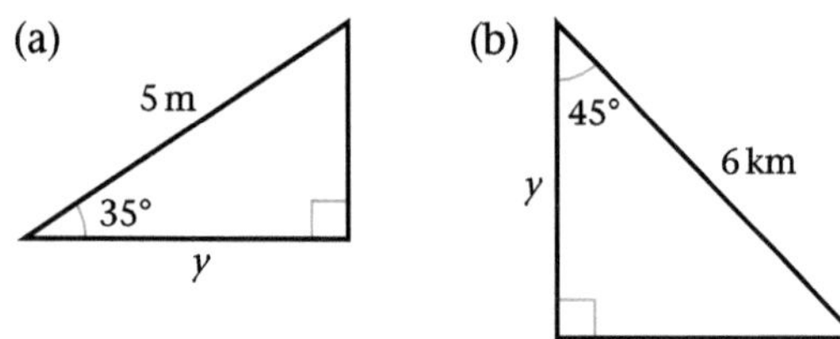

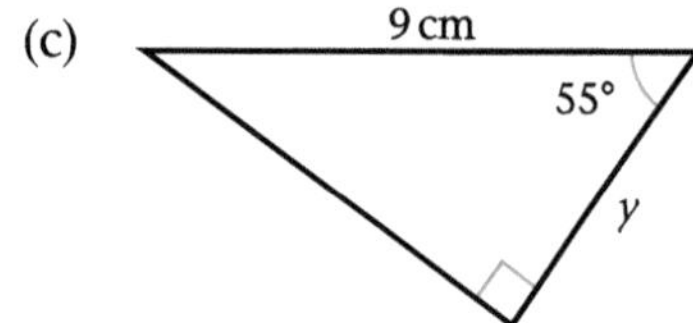

Fig. 23.8

3 Find the value of z in each of the triangles in Fig. 23.9.

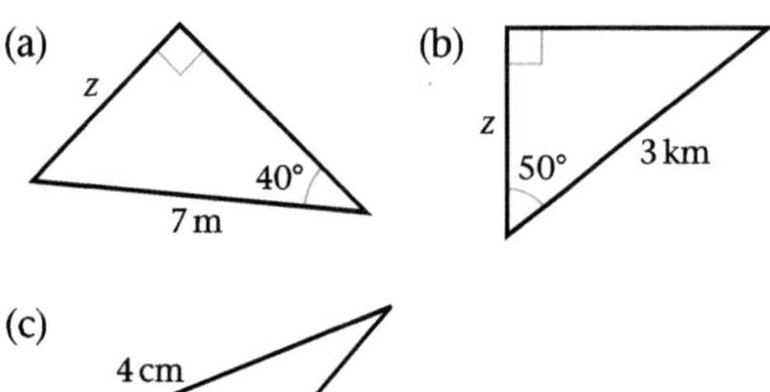

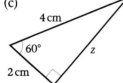

Fig. 23.9

4 A ladder, 5 m long, leans against a wall so that it makes an angle of 60° with the horizontal ground. Calculate how far up the wall the ladder reaches.

5 A diagonal of a square is 20 cm long. How long is each side?

6 The vertical angle of a cone is 70° and its slant height is 11 cm. Calculate the height of the cone.

7 A rhombus of side 10 cm has obtuse angles of 110°. Sketch the rhombus, showing its diagonals and as many angles as possible. Hence calculate the lengths of the diagonals of the rhombus.

Using sine and cosine tables

Three-figure sine and cosine tables are given on pages 222 and 223. These are used in much the same way as tangent tables.

Notice the following:

1 In the sine table, as angles increase from 0° to 90°, their sines increase from 0 to 1.

2 In the cosine table, as angles increase from 0° to 90°, their cosines *decrease* from 1 to 0.

Exercise 23c (Oral or written)

Use the tables on pages 222 and 223 or the calculator to find the value of the following. Give calculator values to 3 s.f.

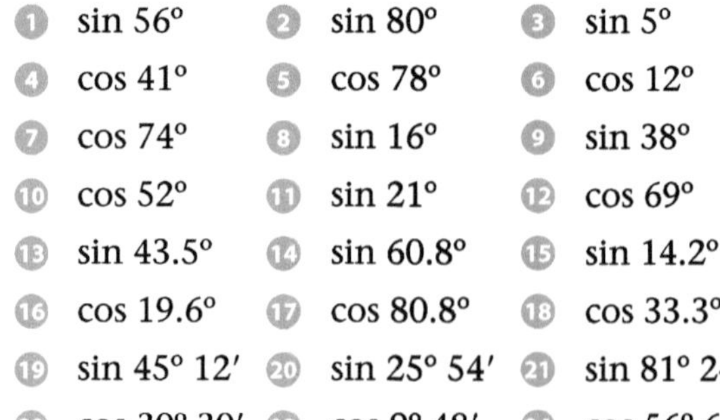

1 sin 56°	2 sin 80°	3 sin 5°
4 cos 41°	5 cos 78°	6 cos 12°
7 cos 74°	8 sin 16°	9 sin 38°
10 cos 52°	11 sin 21°	12 cos 69°
13 sin 43.5°	14 sin 60.8°	15 sin 14.2°
16 cos 19.6°	17 cos 80.8°	18 cos 33.3°
19 sin 45° 12′	20 sin 25° 54′	21 sin 81° 24′
22 cos 30° 30′	23 cos 9° 48′	24 cos 56° 6′
25 cos 54.7°	26 sin 35.3°	27 sin 28.6°
28 cos 61.4°	29 cos 66° 24′	30 sin 23° 36′

Remember where measurements of angles are given in degrees and minutes, the minutes should be written as decimal fractions of degrees.

Example 6

Calculate the length of the hypotenuse of the triangle in Fig. 23.10.

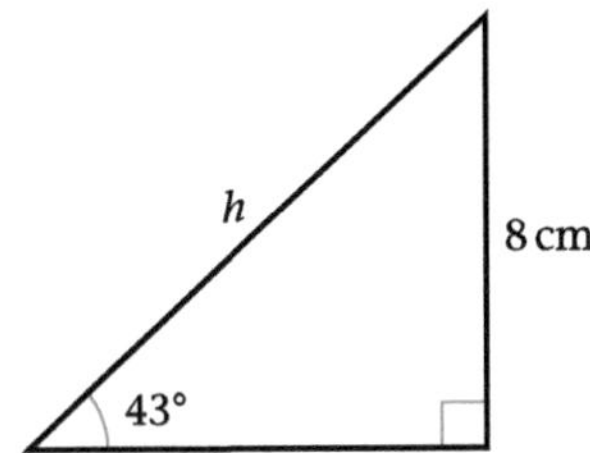

Fig. 23.10

In Fig. 23.10

$$\sin 43° = \frac{8}{h}$$

$$h \times \sin 43° = 8$$

$$h = \frac{8}{\sin 43°} \text{ cm}$$

$$= \frac{8}{0.682} \text{ cm}$$

$= 11.73$ cm
(calculator value)

$= 12$ cm to 2 s.f.

Example 7

A car travels 120 m along a straight road which is inclined at 8° to the horizontal. Calculate the vertical distance through which the car rises.

Fig. 23.11 is a sketch of the road.

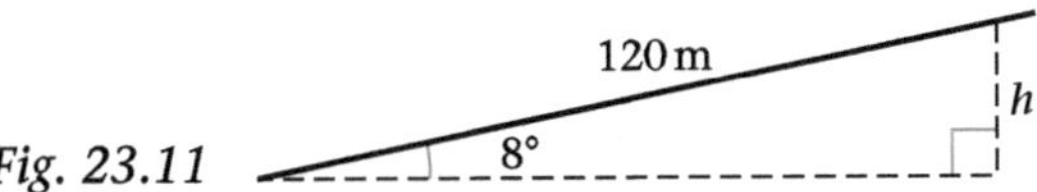

Fig. 23.11

In Fig. 23.11, h is the vertical distance.

$$\sin 8° = \frac{h}{120}$$

$$h = 120 \times \sin 8° \text{ m}$$
$$= 120 \times 0.139 \text{ m}$$
$$= 16.68 \text{ m}$$
$$= 17 \text{ m to 2 s.f.}$$

Exercise 23d

Give all calculated lengths correct to 2 significant figures. Give all calculated angles correct to the nearest 0.1°.

1. Calculate the lengths a, b, c, d, e, f, g, h in Fig. 23.12, all lengths being in cm.

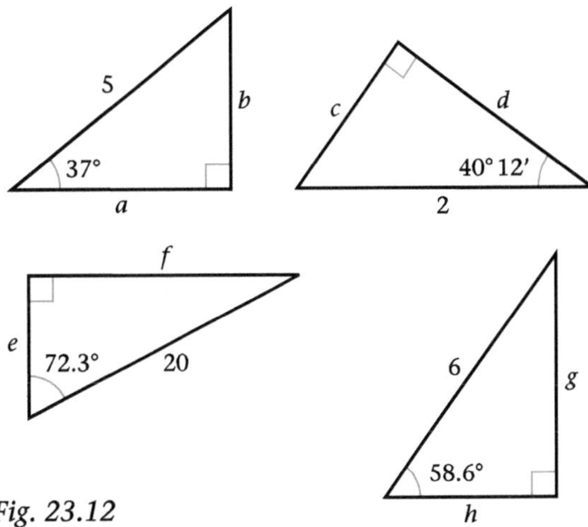

Fig. 23.12

2. Calculate the lengths of AB and BC in Fig. 23.13.

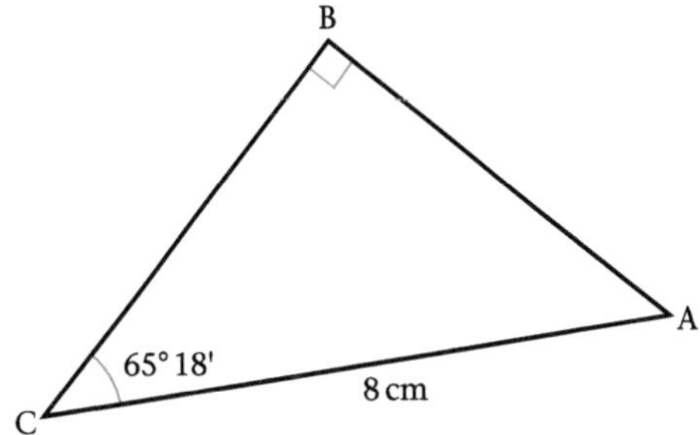

Fig. 23.13

3. Calculate the length of the hypotenuse in each of the triangles in Fig. 23.14, all lengths being in cm.

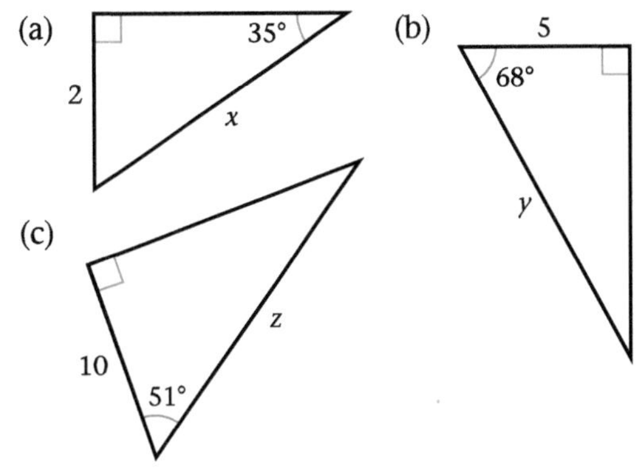

Fig. 23.14

4. Make suitable construction lines, then calculate the lengths BC, XY and PQ in Fig. 23.15.

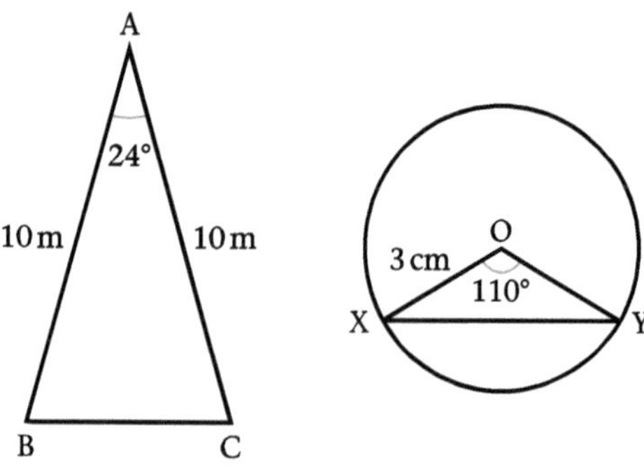

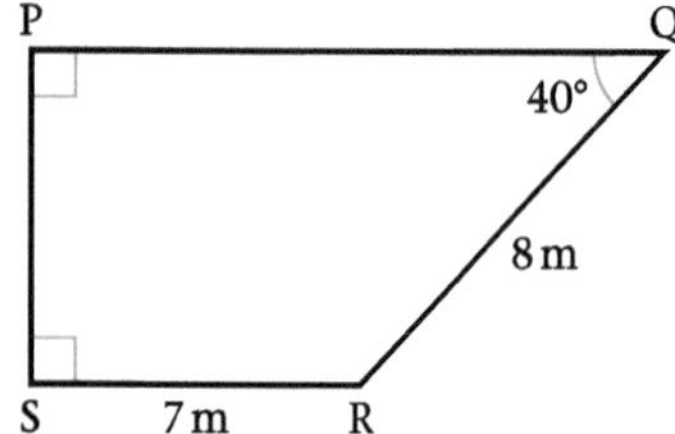

Fig. 23.15

5. The roof of a hut is made from sheets of corrugated iron of length 2 m, inclined at 18° to the horizontal. (See Fig. 23.16.)

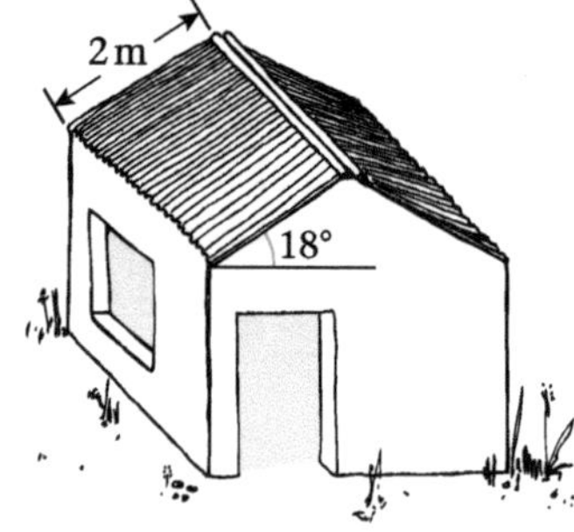

Fig. 23.16

Calculate the width of the hut.

6. A stone is suspended from a point P by a piece of string 50 cm long. It swings backwards and forwards.
Calculate the angle the string makes with the vertical when the stone is in the position shown in Fig. 23.17.

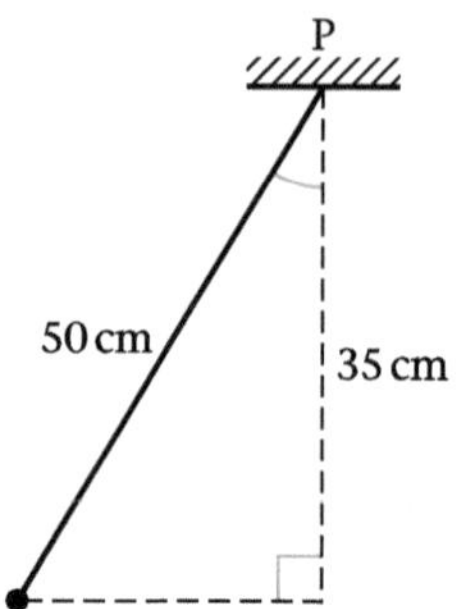

Fig. 23.17

7. An aeroplane is flying at a height of 200 m. Its angle of elevation to an observer on the ground is 23°. (See Fig. 23.18.)

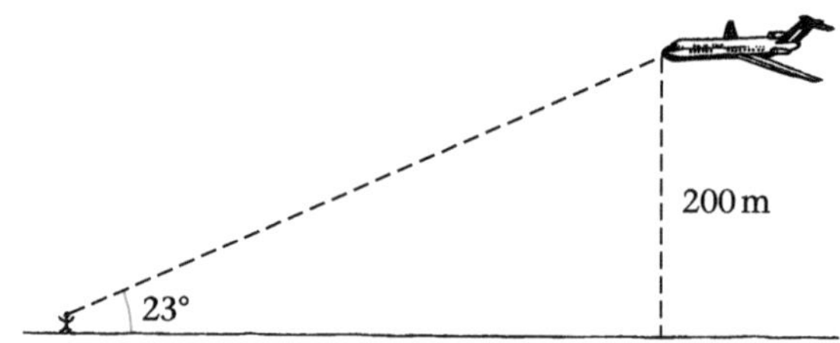

Fig. 23.18

Calculate the distance of the aeroplane from the observer.

8. Fig. 23.19 shows some men using a board to slide loads from a platform on to a lorry. The platform is 1.5 m higher than the lorry. It is found that the best position for the board is when it is inclined at 20° to the horizontal. Calculate the length of the board.

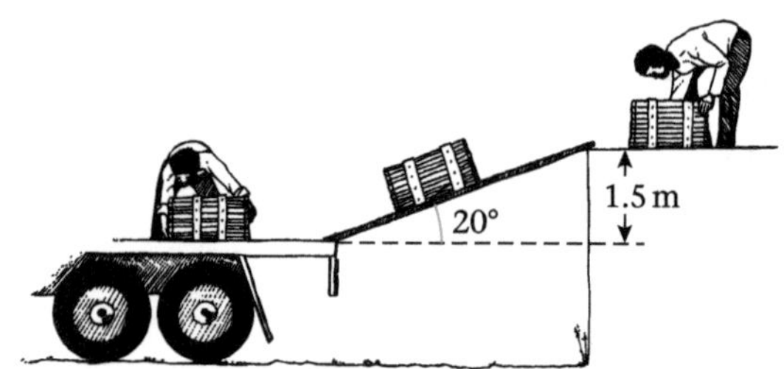

Fig. 23.19

9. The arms of a pair of compasses are 10 cm long and the angle between them is 35°. Calculate the radius of the circle that the compasses will draw.

Summary

In addition to the tangent ratio, two further trigonometric ratios, **sine (sin)** and **cosine (cos)**, can be used to solve triangles.

In a right-angled triangle, where θ is an acute angle

$$\sin\theta = \frac{\text{side opposite }\theta}{\text{hypotenuse}}$$

$$\cos\theta = \frac{\text{side adjacent to }\theta}{\text{hypotenuse}}$$

Practice Exercise P23.1

Give all calculated lengths to one decimal place and all calculated angles to one decimal place.

1. A ladder 5 m long, leans against a vertical wall. If the foot of the ladder makes an angle of 64° with the horizontal ground, calculate the distance of the foot of the ladder from the wall.

2. From the top of a building the angle of depression of a point on the ground is 32°. If the height of the building is 24 m, calculate the distance from the top of the building to the point on the ground.

3. A helicopter is flying at a height of 120 m. An observer on the ground notes the angle of elevation of the helicopter to be 24°. (See Fig.23.20.)

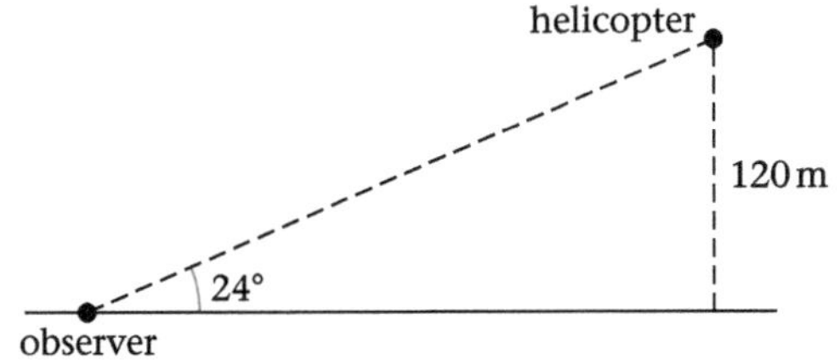

Fig. 23.20

Calculate the distance of the observer from the helicopter.

4. AC and BC are chords of the circle of radius 7.5 cm. $B\hat{C}A$ is a right angle and $B\hat{A}C = 28°$. AB is the diameter of the circle. (See Fig. 23.21.)

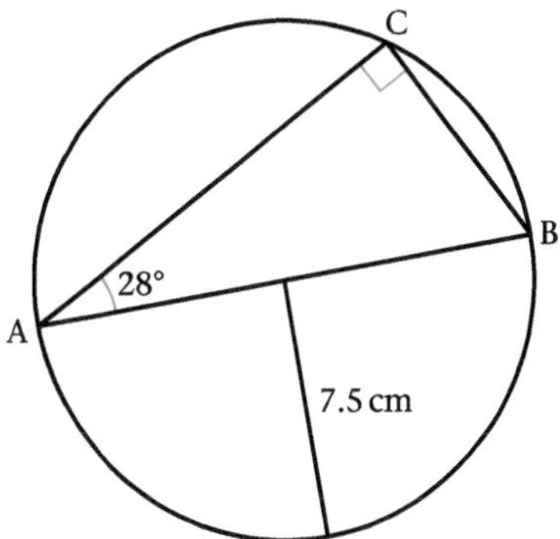

Fig. 23.21

Calculate the lengths of AC and BC.

5. ABCD is a rectangle 5 cm wide by 12 cm long. Calculate
   (a) The length of the diagonal
   (b) The angle between the diagonal AC and a long side of the rectangle.

6. In triangle PQR, PQ = PR = 24 cm and QR = 20 cm. Calculate the height of the triangle and the size of $Q\hat{P}R$.

7. A see-saw tilts at an angle of 16° to the horizontal ground. If the length of the see-saw is 1.8 m, calculate the height of the up-end above the ground.

8. A ladder 2 m long, leans against a wall. If the top of the ladder is 1.5 m above the foot of the wall, calculate the angle between the foot of the ladder and the horizontal ground.

Revision exercises and tests

Chapters 17–23

Revision exercise 9 (Chapters 17, 20)

1 Calculate the area of a trapezium in which the parallel sides are 7 cm and 13 cm long and 9 cm apart.

2 Find the area of a path 2 m wide which surrounds a circular plot 12 m in diameter. (Make a sketch; use the value $3\frac{1}{7}$ for π.)

3 Use the value $\frac{22}{7}$ for π to find the surface area of a closed cylinder of radius 7 cm and height 30 cm.

4 Factorise the following.

(a) $7x - 28$ (b) $5m + 8mn$
(c) $27ab + 36b^2$ (d) $35p^2q - 14pq^2$

5 Simplify

(a) $\frac{k+5}{4} + \frac{2k}{3}$ (b) $\frac{4h-3}{3} - \frac{h+2}{5}$
(c) $\frac{n-4}{2b} + \frac{3}{b}$ (d) $\frac{2x}{x+1} - \frac{6x-1}{3(x+1)}$
(e) $\frac{3}{y} - \frac{2}{y-1}$

6 Factorise and simplify, if possible

(a) $h^2 + 7h$
(b) $m^2 + m(m - 1)$
(c) $3(n - 4) + (n - 4)^2$
(d) $4x^2 + 4x(2 - x)$
(e) $3(y + 1)^2 + 5(y + 1)$

7 If $n = 37^2 + 37 \times 63$ use factorisation to find the value of n.

8 Solve the following equations.

(a) $5h - 3(h - 2) = 0$
(b) $\frac{3}{x} + \frac{2}{x+1} = \frac{8}{x(x+1)}$
(c) $\frac{3}{y+3} = \frac{2}{5}$

9 Calculate the shaded areas in the shapes in Fig. R29. Use the value $\frac{22}{7}$ for π.

(a) 10, 10, 7

(b)
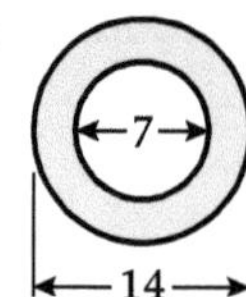

(c)
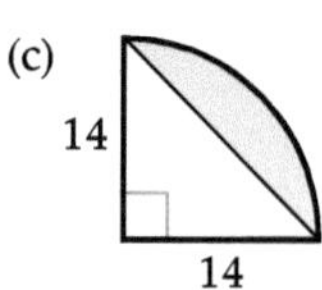

(d) 7, 7

(e)
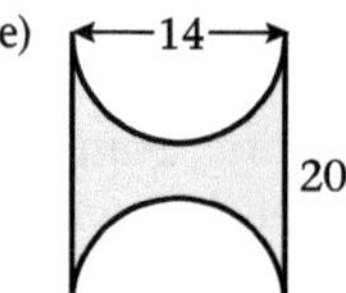

(f)
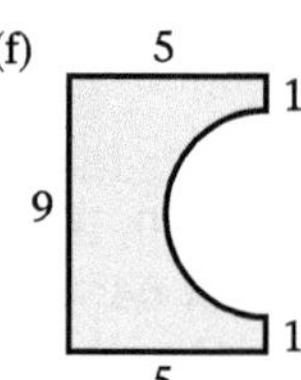

Fig. R29

10 Discs of diameter 6 cm are cut from a sheet 130 cm long and 70 cm wide as shown in Fig. R30.

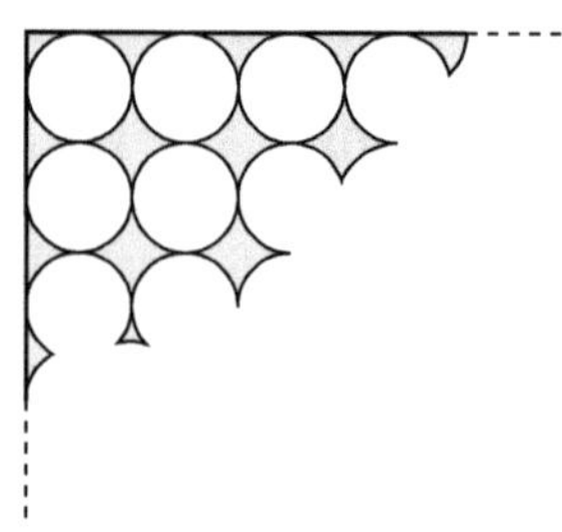

Fig. R30

(a) How many discs can be cut in this way?
(b) What area of the sheet is wasted?

Revision test 9 (Chapters 17, 20)

1 The number of cm² in 1 m² is

A 100 B 1000
C 10 000 D 100 000

2 A square has the same perimeter as a 5 cm by 7 cm rectangle. The area of the square is

A 25 cm^2 B 35 cm^2
C 36 cm^2 D 49 cm^2

3 A sector of a circle of radius 12 cm has an angle of 160°. In terms of π, the area of the sector is

A $\frac{16\pi}{36}\text{ cm}^2$ B $\frac{32\pi}{3}\text{ cm}^2$
C $16\pi\text{ cm}^2$ D $64\pi\text{ cm}^2$

4 The highest common factor of $10a^2b$ and $8ab$ is

A $80a^2b$ B $2ab$ C $5a$ D $1\frac{1}{4}a$

5 $\frac{1}{2a} - \frac{1}{2b} =$

A $\frac{1}{a - b}$ B $\frac{1}{2a - 2b}$
C $\frac{b - a}{2ab}$ D $\frac{b - a}{4ab}$

6 Find the circumference and area of a circle of diameter 56 cm. (Use $\frac{22}{7}$ for π.)

7 Calculate the area of the shape shown in Fig. R31. Use $\frac{22}{7}$ for π.

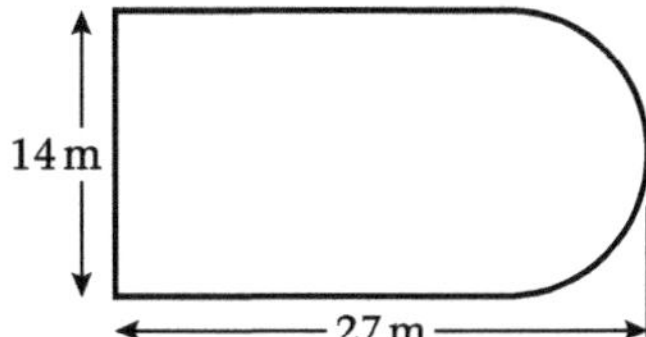

Fig. R31

8 Four discs, each of radius 1 cm, are cut from a 5 cm by 5 cm cardboard square. Use the value 3.14 for π to find the area of cardboard left over.

9 Factorise the following, simplifying where possible.
(a) $3a^2 + a(2a + b)$
(b) $(5x - 2y)(a - b) - (2x - y)(a - b)$

10 (a) Factorise $\pi r^2 + 2\pi rh$.
(b) Hence find the value of the expression when $\pi = \frac{22}{7}$, $r = 4$ and $h = 5$.

Revision exercise 10 (Chapters 18, 21)

Use the following data in questions 1, 2, 3, 4. Table R4 shows the distribution of ages of a group of police cadets.

Table R4

Age in years	21	22	23	24
Frequency	5	10	6	9

1 Which age is the mode of the above data?

2 (a) How many cadets were in the group?
(b) Find the median age of the cadets.

3 Calculate the mean age of the cadets.

4 Draw a bar chart to show the data in Table R4.

5 The mean of eight numbers is 9. The mean of seven of the numbers is 10. What is the eighth number?

6 In a test the marks of four boys were 23, 18, 24, 27 and the marks of three girls were 21, 16, 29. Find the mean mark of (a) the boys, (b) the girls, (c) all seven students.

7 A man borrowed \$125 at simple interest. After 8 months he paid back \$130 as repayment of his debt plus interest on it. Calculate the percentage rate of interest per annum.

8 Calculate the difference between the interest at $5\frac{1}{2}$% per annum on \$15 000 for 3 years
(a) at simple interest
(b) when compounded annually.

9 Use Table 21.1 on page 186 to change the following amounts of money to US\$.
(a) Ja\$1000 (b) £10 000
(c) EC\$1350 (d) Can\$850
(e) Bds\$720 (f) TT\$11 500

10 An article was bought for £8.50 in London. An import duty of 20% was paid on it in Xanica. Given that £1 = Xa\$1.35 what is the total cost of the article in Xa dollars?

Revision test 10 (Chapters 18, 21)

Use the following set of numbers in questions 1, 2 and 3: 2, 2, 2, 5, 5, 8, 9, 10, 11

1. The mode of the above set of numbers is
   A 2 B 3 C 5 D 6
2. The median of the above set of numbers is
   A 2 B 3 C 5 D 9
3. The mean of the above set of numbers is
   A 3 B 5 C 6 D 9
4. How much interest does \$400 make in 5 years at 7% per annum?
   A \$20 B \$28 C \$35 D \$140
5. If EC\$1 = \$1.26, change \$500 to Eastern Caribbean dollars to the nearest EC\$1.
   A EC\$390 B EC\$397
   C EC\$414 D EC\$630
6. For a six-week period a salesman makes the following profits (in \$): 440, 340, 350, 230, 510, 380. Draw a bar chart to represent the information.
7. Oranges are packed for sale in plastic bags. The distribution of a sample is given in Table R5.

Table R5

Oranges/bag	10	11	12	13	14	15
No. of bags	1	2	6	7	3	1

   (a) State the modal quantity.
   (b) Calculate the mean number of oranges per bag.

8. Calculate the compound interest on \$12 000 at $6\frac{1}{4}$% per annum for 2 years.
9. Convert G\$720 into Ja\$ if US\$1 = G\$90.00 and US\$1 = Ja\$35.00.
10. I bought a radio for \$145 in Barbados. I paid an import duty of 50% on entering Trinidad. Given that Bds\$1 = TT\$2.90, calculate the total cost of the radio in TT\$.

Revision exercise 11 (Chapters 19, 22, 23)

1. Find the tangent of 60° by construction and measurement.
2. Make accurate scale drawings to find the values of x and z in Fig. R32.

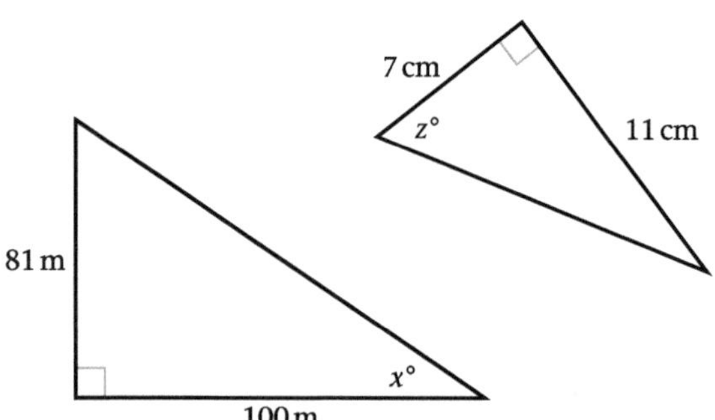

Fig. R32

3. When the angle of elevation of the sun is 64°, a boy's shadow on level ground is 80 cm long. Calculate the height of the boy to the nearest 5 cm.
4. Draw any quadrilateral so that its 4 sides are of different lengths. Use ruler and compasses to find the mid-point of each side. Join the mid-points to form a new quadrilateral. What kind of quadrilateral is it?
5. (a) Construct $\triangle XYZ$ such that $\hat{Y} = 90°$, XY = 8 cm and YZ = 5 cm.
   (b) Measure the length of the hypotenuse XZ.
   (c) Check your result by calculation.
6. By drawing and measuring find approximately (a) the value of cos 44°, (b) the size of the angle whose sine is $\frac{4}{5}$.
7. In $\triangle ABC$, $\hat{A} = 38°$, $\hat{B} = 90°$ and AC = 9 cm. Calculate (a) $\hat{C}$, (b) AB, (c) BC.
8. The diagonals of a rectangle are 10 cm long and intersect at an angle of 120°. Make a sketch of the rectangle. Hence use trigonometry to calculate the length and width of the rectangle.

9 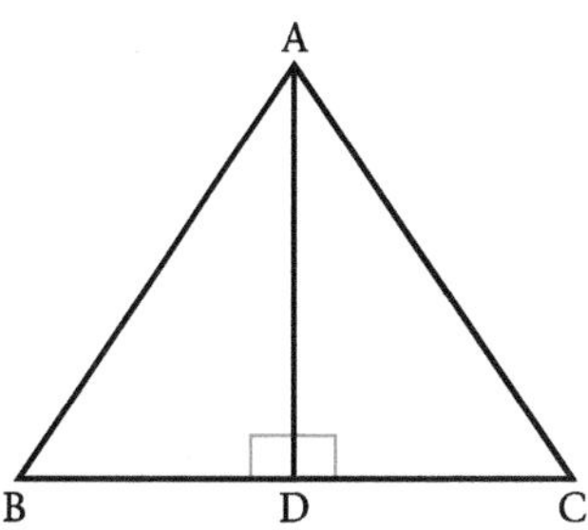

Fig. R33

In Fig. R33, ABC is an isosceles triangle with AB = AC = 15 cm and AD = 8 cm. AD is perpendicular to BC. Calculate
(a) $A\widehat{B}C$ (b) BD (c) BC.

10 From the top of a tree, the angle of depression of a stone on horizontal ground is 55°. If the stone is 8 m from the foot of the tree, find, by calculation, the height of the tree. Give your answer to 3 significant figures.

Revision test 11 (Chapters 19, 22, 23)

1 In Fig. R34 the tangent of $\widehat{X}$ is given by the ratio.
A $\frac{XZ}{XY}$ B $\frac{XZ}{YZ}$ C $\frac{YZ}{XZ}$ D $\frac{YZ}{XY}$

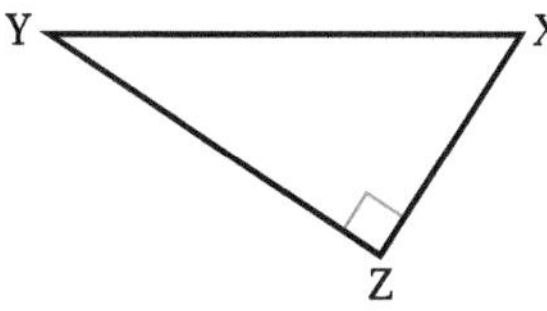

Fig. R34

2 15° 44′ to the nearest 0.1° is
A 15.4° B 15.6° C 15.7° D 15.8°

3 Which one of the following angles can be constructed using ruler and compasses only?
A 115° B 125° C 135° D 145°

4 A vertical fence post has a shadow 1 m long when the angle of elevation of the sun is 45°. The height of the fence post is
A 0.5 m B 1 m C 1.4 m D 2 m

5 In Fig. R35, $\sin \widehat{P} =$
A $\frac{p}{q}$ B $\frac{q}{p}$ C $\frac{r}{q}$ D $\frac{p}{r}$

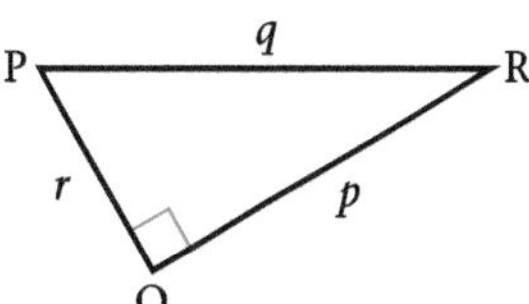

Fig. R35

6 In △ABC, $A\widehat{B}C$ = 23°, $B\widehat{C}A$ = 90° and AB = 7.5 cm. Using two different ratios, calculate (a) $C\widehat{A}B$, (b) AC correct to 2 decimal places.

7 Draw a line AB 6 cm long. Construct the perpendicular bisector of AB. Hence construct an isosceles triangle ACB such that CA = CB = 8 cm. Measure $\widehat{A}$.

8 By drawing and measurement, find approximately (a) the value of sin 37°, (b) the angle whose cosine is $\frac{7}{10}$.

9 The angle of elevation of the top of a tower from a point 23 m from its base on level ground is 50°. Calculate the height of the tower to the nearest metre.

10 A ladder 6 m long leans against a vertical wall and makes an angle of 80° with the horizontal ground. Calculate, to 2 s.f., how far up the wall the ladder reaches.

General revision test C (Chapters 17–23)

1 The mean of 3, 5, 4, 8, 6, 4, 6, 2, 3, 6 is
A 4.5 B 4.7 C 6 D 10

2 $-6y + 8 =$
A $-2(3y + 4)$ B $-2(3y + 8)$
C $2(-3y + 4)$ D $2(-3y + 8)$

3 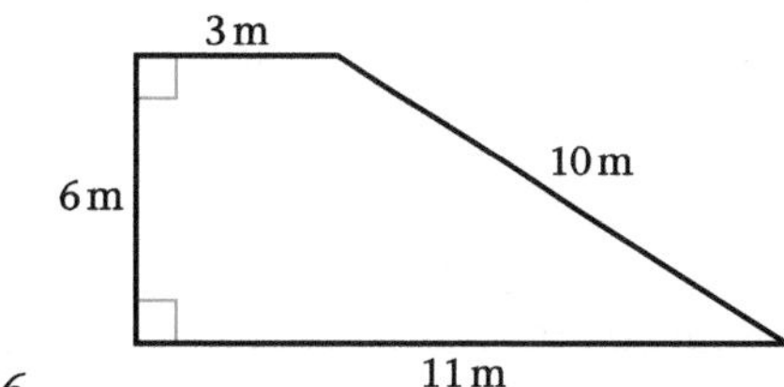

Fig. R36

The area of the trapezium in Fig. R36 is
A 33 m² B 42 m² C 56 m² D 66 m²

4 If TT\$1.00 = Ja\$5.50, then Ja\$14.30 = TT\$$x$, where x =

A 2.60 B 8.80 C 19.80 D 78.65

5 The mean of three numbers is 6. The mode of the numbers is 7. The lowest of the three numbers is

A 2 B 3 C 4 D 6

6 $\frac{3}{a+3} - 2 =$

A $-\frac{1}{a}$ B $\frac{1}{a+3}$

C $\frac{1-2a}{a}$ D $\frac{-3-2a}{a+3}$

7 A cylinder is of height 5 cm and base radius 2 cm. Use the value 3.14 for π to calculate the area of its curved surface to the nearest cm^2.

A 25 cm^2 B 31 cm^2
C 63 cm^2 D 126 cm^2

8 The average mass of 6 people is 58 kg. If the lightest person has a body mass of 43 kg, what is the average mass of the other 5 people?

A 58 kg B 59 kg
C 61 kg D 68 kg

9

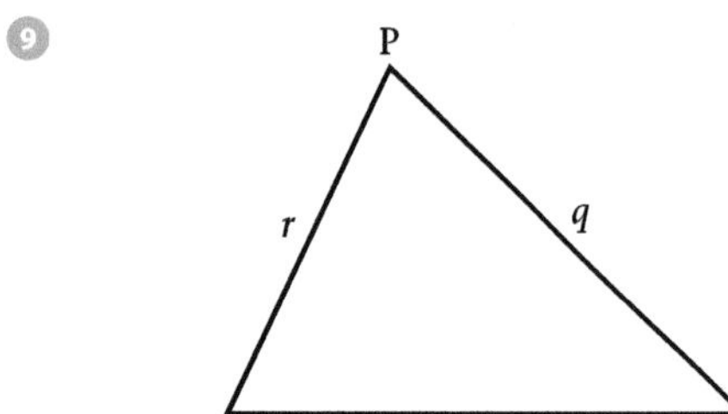

Fig. R37

In Fig. R37, which of the following give(s) the perpendicular distance of P from QR?

I $r \sin \widehat{Q}$, II $q \sin \widehat{R}$, III $p \sin \widehat{R}$.

A I only B II only
C III only D I and II only

10 The interest on a loan for 2 years at 6% per annum simple interest was \$480. The sum of money loaned was

A \$5760 B \$4000
C \$1600 D \$1440

11

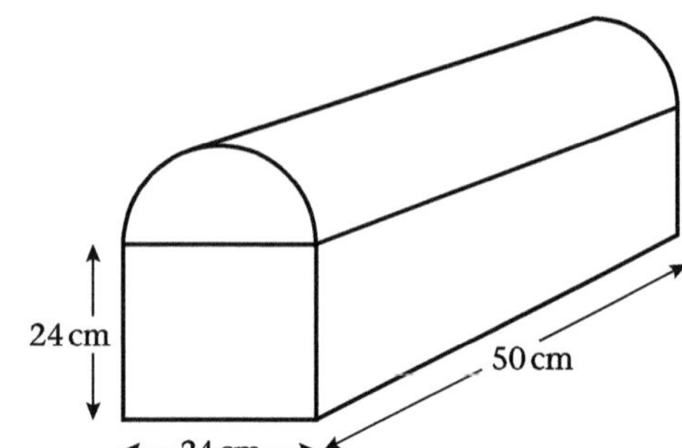

Fig. R38

The cross-section of a chest is a square joined to a semi-circle as in Fig. R38. Calculate to the nearest cm^2 the surface area of the chest.

12 The acute angles of a rhombus are 76°. If the shorter diagonal is of length 8 cm, calculate the length of the sides of the rhombus to the nearest mm.

13 In Fig. R39 AB = 8 cm, BC = 16.5 cm.

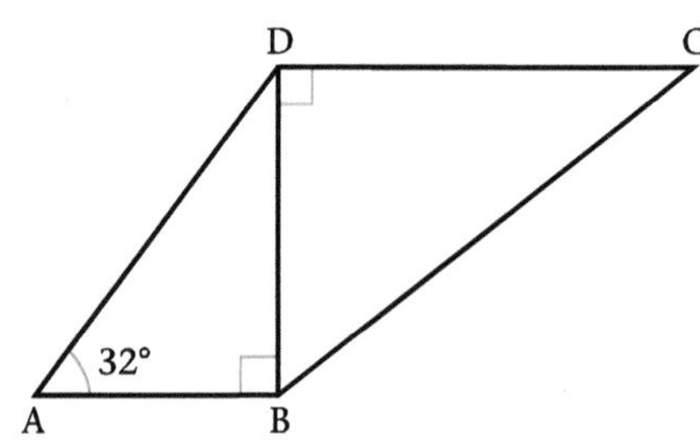

Fig. R39

Calculate (a) BD to the nearest whole number
(b) $B\widehat{C}D$ to 1 decimal place.

14 In Fig. R40, PQR is a right-angled isosceles triangle.

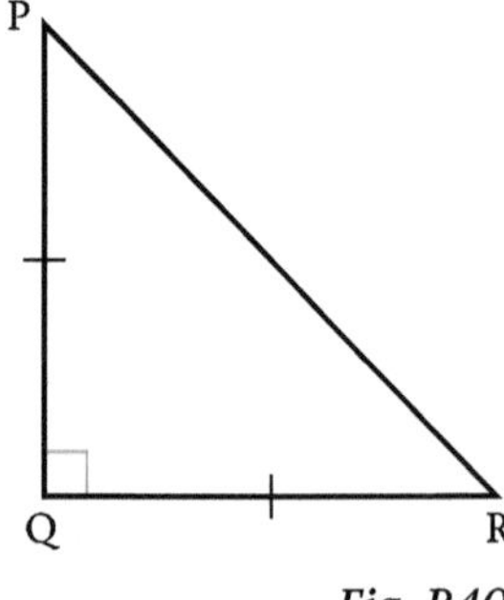

Fig. R40

Show that
(a) $\sin \widehat{R} = \sin \widehat{P}$
(b) $\cos \widehat{R} = \cos \widehat{P}$
(c) $\tan \widehat{R} = \tan \widehat{P} = 1$

15 A small factory employs 10 workmen, 2 supervisors and 1 manager. The workmen get \$1350 per week, the supervisors get \$1550 per week and the manager gets \$1860 per week.

(a) Calculate the mean weekly wage at the factory.
(b) Compare this result with the modal wage.

Which average is most representative of the weekly wages?

16 Solve the following equations.

(a) $3x + 5 = 4x + 2$
(b) $\frac{2 + 5y}{4} = 3$
(c) $\frac{3n - 4}{n} = \frac{1}{3}$
(d) $\frac{x + 1}{2} - \frac{x - 3}{3} = 4$

17 Calculate the amount when \$20 000 is invested for 2 years

(a) at 15% per annum simple interest
(b) at $14\frac{1}{2}$% per annum compounded annually.

18 From a point on level ground 60 m away, the angle of elevation of the top of a tree is $24\frac{1}{2}°$. Calculate the height of the tree to the nearest metre.

19 Using ruler and compasses only, construct $\triangle PQR$ such that $PQ = 8.4$ cm, $\widehat{Q} = 60°$ and $QR = 4.2$ cm. Measure $\widehat{P}$ and find the length of PR.

20 Fig. R41 shows notes that a boy made of a survey of a tree.

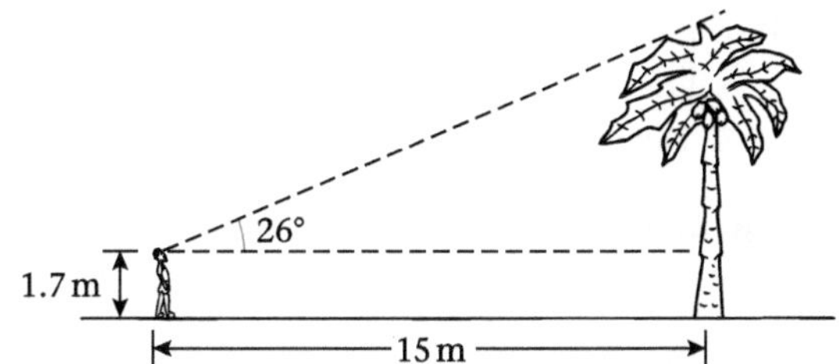

Fig. R41

Make a scale drawing and hence find the height of the tree to the nearest $\frac{1}{2}$ metre.

Practice examination

Paper 1 ($1\frac{1}{2}$ hours)

Section A (20 marks)

Answer **all** *questions.*
In each question, choose **one** *of the letters A, B, C, D which corresponds to the correct answer.*

1. 486 kg expressed in tonnes is
   A 48.6 t B 4.86 t
   C 0.486 t D 0.048 6 t

2. The highest common factor of 36, 72, 90 is
   A 9 B 18 C 90 D 360

3. Solve $\frac{2x - 1}{5} + 2x = 19$. $x =$
   A $1\frac{2}{3}$ B 5 C 8 D 24

4. In Fig. P1, PQRS is a parallelogram and STUP is a square.

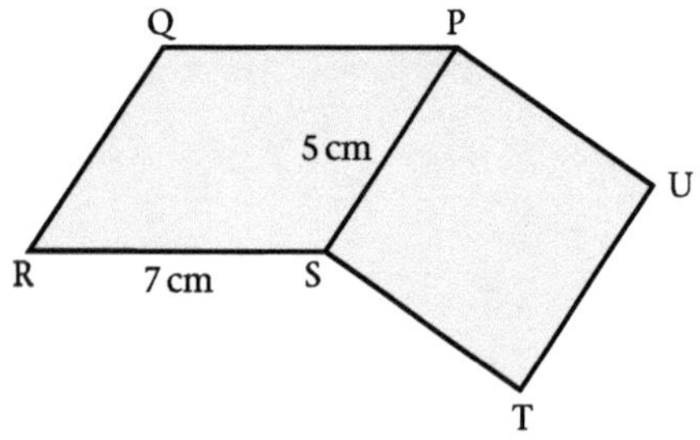

Fig. P1

   If PS = 5 cm and RS = 7 cm the perimeter of PQRSTU is
   A 29 cm B 34 cm C 39 cm D 44 cm

5. The value of $\frac{a - 3b}{a}$ when $a = 2$ and $b = -8$ is
   A −24 B −11 C 13 D 24

6. How much simple interest does $600 make in 3 years at 7% per annum?
   A $18 B $21 C $42 D $126

7. X and Y are subsets of U as shown in Fig. P2.

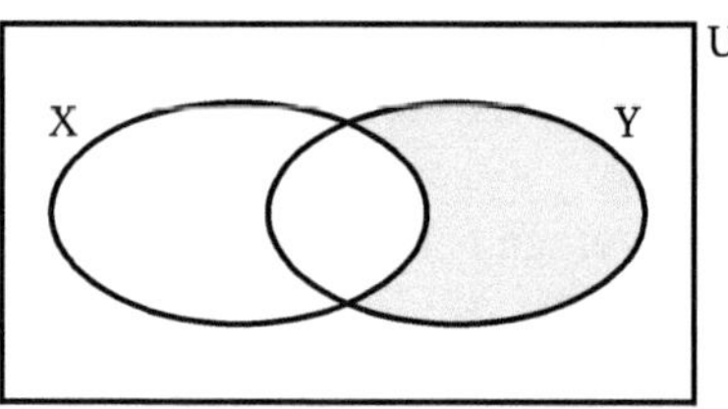

Fig. P2

   The shaded region represents the set
   A X′ ∩ Y B X ∩ Y′
   C X′ ∪ Y′ D X ∪ Y′

8. Change $100\,100_{two}$ to base ten.
   A 14 B 36 C 44 D 900

9. When 27 people share a sack of meal they each get 4 kg. When 12 people share the same sack of meal they each get
   A 3 kg B 8 kg C 9 kg D 12 kg

10. Express $\frac{a}{3} + \frac{4}{b}$ as a single fraction.
    A $\frac{a + 4}{3 + b}$ B $\frac{ab + 12}{3b}$
    C $\frac{ab + 12}{3 + b}$ D $\frac{a + 4}{3b}$

11. Which of the points in Fig. P3 has coordinates (−4, 1)?

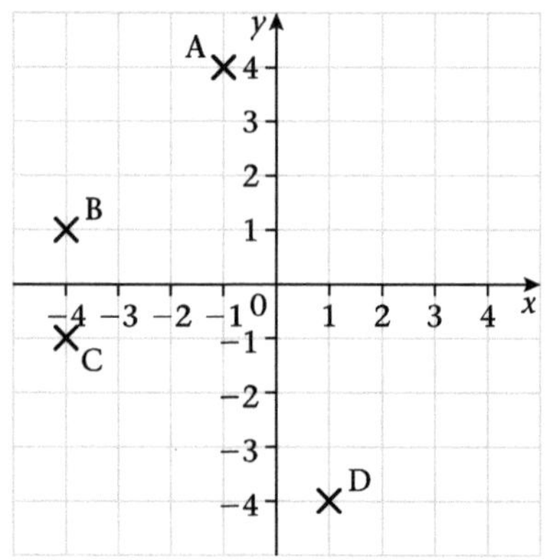

Fig. P3

12 In Fig. P4, which one of the following is an expression for h in terms of d and α?

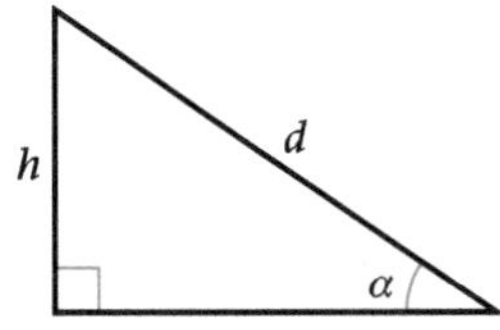

Fig. P4

A $d \sin \alpha$ B $\dfrac{d}{\sin \alpha}$

C $d \cos \alpha$ D $\dfrac{d}{\cos \alpha}$

13 What is the mode of the following numbers?

8, 9, 5, 6, 2, 4, 8, 0

A 4 B 5 C 6 D 8

14

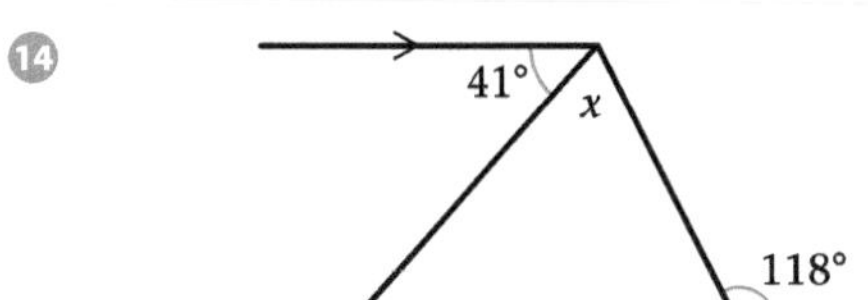

Fig. P5

In Fig. P5, $x =$

A 41° B 49° C 62° D 77°

15 In the Venn diagram shown in Fig. P6,

Q = {rational numbers}

Z = {integers}

W = {whole numbers}

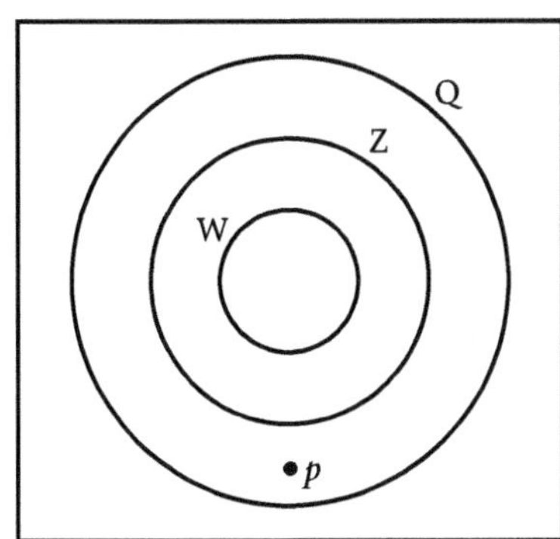

Fig. P6

A possible value of p is

A 0 B −1 C 3 D $\frac{2}{5}$

16 The sum of the angles of a polygon is 900°. The number of sides the polygon has is

A 6 B 7 C 10 D 14

17 Calculate $(4 \times 10^3) \times (8 \times 10^2)$, giving the answer in standard form.

A 3.2×10^4 B 3.2×10^5

C 3.2×10^6 D 3.2×10^7

18 Fig. P7 is the graph of a set of numbers, S.

Fig. P7

Which one of the following defines S?

A $\{x: -1 < x < 2\}$

B $\{x: -1 \leqslant x \leqslant 2\}$

C $\{x: -1 < x \leqslant 2\}$

D $\{x: -1 \leqslant x < 2\}$

19 $(-3)^2 - (-5) + 2 =$

A −12 B −2 C 2 D 16

20 In a map, 1 cm represents 50 km. The scale used is

A 1:50 B 1:5000

C 1:500 000 D 1:5 000 000

Section B (60 marks)

Answer **all** *questions.*
Show all necessary working.

21 Arrange the following fractions in order from lowest to highest.

$\frac{4}{5}, \frac{9}{10}, \frac{3}{4}, \frac{17}{20}$ [2]

22 Calculate the exact value of $\dfrac{0.01 \times 0.4}{0.0002}$ [3]

23 Find the value of $(13_{\text{five}})^2$ in base five. [3]

24 Find the value of $\dfrac{x^2 + y^2}{2}$ when $x = -7$ and $y = 3$. [3]

25 Express 0.009 238 correct to

(a) 2 decimal places,

(b) 2 significant figures. [2]

26 Express the mean of $2x$, $x - 2$ and 17 as simply as possible. [2]

27

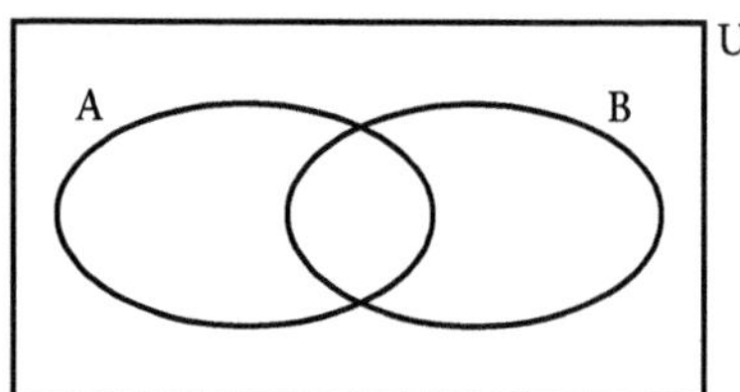

Fig. P8

In Fig. P8, n(A) = 15, n(B) = 12, n(A ∩ B) = 8 and n(A ∪ B)′ = 18. Find n(U). [3]

28 Simplify (a) $4h + 8k - 3k - 10h$,
(b) $x(x - 3) - 5(x - 3)$. [3]

29 (a) Calculate the rate of interest if the amount at the end of 2 years on $5000 is $5600.
(b) Calculate the amount if the interest is compounded annually. [4]

30 Solve $2(x - 2) + 5(x - 9) = 0$. [3]

31 A triangle PQR with P(−2, 0), Q(4, 2), R(1, 6) is translated by a vector $\begin{pmatrix} 2 \\ -3 \end{pmatrix}$ to triangle STU. State the coordinates of S, T and U. [3]

32 If the angles of a quadrilateral are $4x°$, $(95 - x)°$, $(95 + x)°$ and $6x°$, find the value of x. [3]

33 In Fig. P9, AB = AC and $\widehat{A} = 64°$.

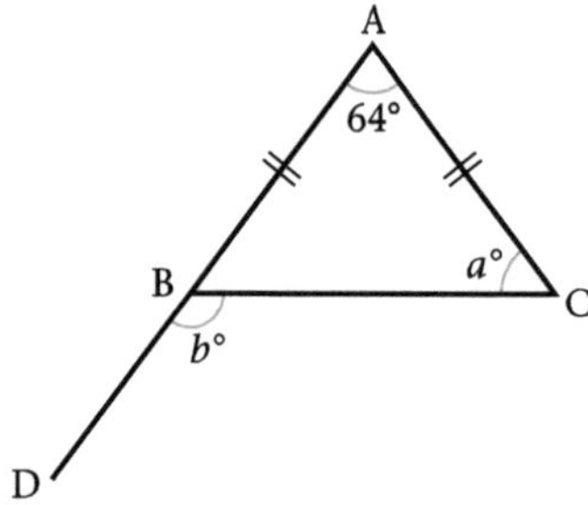

Fig. P9

Find the sizes of the angles marked $a°$ and $b°$. [3]

34 Express $\frac{d-2}{6} + \frac{d+3}{3}$ as a single fraction in its lowest terms. [3]

35 The exchange rate is Bds$2 = US$1 and Ja$35 = US$1. How many Bds$ can be bought for Ja$1540? [3]

36 Find the value of $2\pi\sqrt{\frac{l}{g}}$ when $\pi = 3\frac{1}{7}$, $l = 98$ and $g = 32$. [3]

37 Using ruler and compasses only, construct *any* right-angled triangle. Leave all construction lines on your drawing [2]

38 In Fig. P10, M is the mid-point of one of the sides of the equilateral triangle.

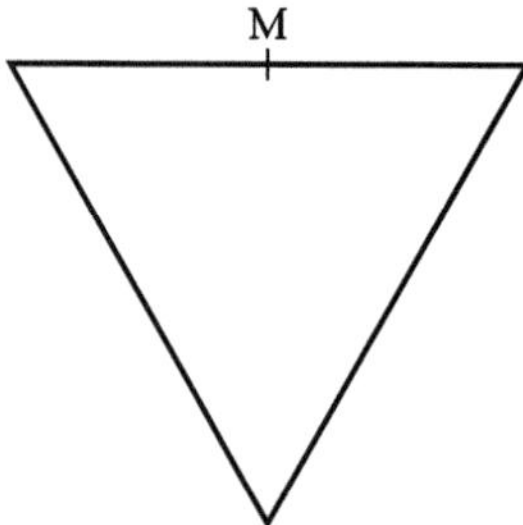

Fig. P10

Make a copy of Fig. P10. Draw the image of the triangle after a rotation of 60° clockwise about M. Draw any line(s) of symmetry of the combined figure. [4]

39 A refrigerator costs $2850. In a sale, its price is reduced by 15c in the $ and then rounded to the nearest $10. Find the sale price. [4]

40 A brother is 5 years older than his sister. The ratio of their ages is 4 : 3. Find their ages. [4]

Paper 2 (2 hours)

Answer all *questions in* Section A *and any* four *questions from* Section B.
Show all necessary working.

Section A (60 marks)

1 (a) Evaluate $14.56 \div 0.52$. [3]
(b) Express 165 g as a percentage of 1 kg. [2]

2 (a) In a cafe a cup of tea costs t cents. How many cups can I buy with n dollars? [2]
(b) Find y if $10y + 8 = y - 19$. [3]

3 (a) If $3014_{five} = 2112_{five} + x_{five}$, find x. [3]
(b) If $53_y + 62_y = 135_y$, what is the value of the base y? [2]

4 (a) State the coordinates of the vertices of the triangle ABC (Fig. P11) [3]
(b) State the coordinates of the points which remain fixed after a reflection in
(i) the x-axis, (ii) the y-axis. [2]

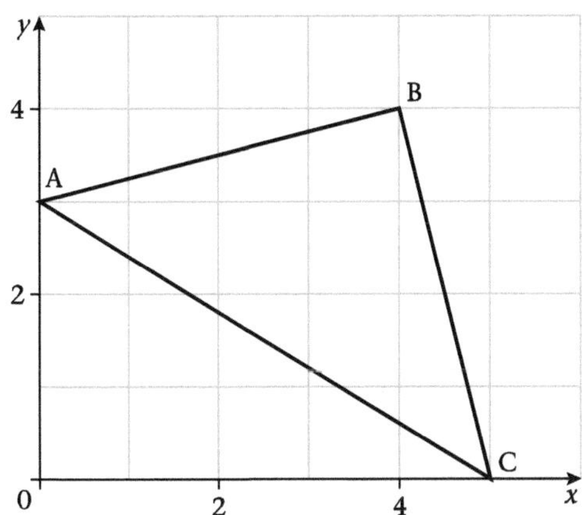

Fig. P11

5 In Fig. P12, PQRS is a field with dimensions as shown.
Calculate the area of the field. [5]

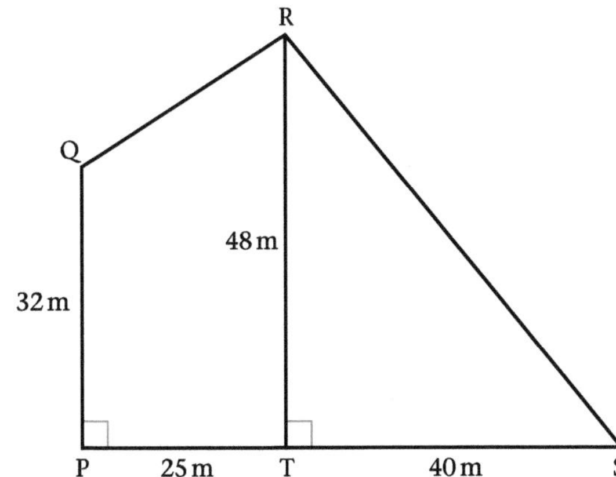

Fig. P12

6 A metal is made of copper and zinc in the ratio 3 : 2 by volume. The mass of copper and zinc are 8.9 g/cm³ and 7.1 g/cm³ respectively. Find the mass of 100 cm³ of the metal. [5]

7 Using ruler and compasses only, construct
(a) a triangle PQR with PQ = 7 cm, angle $P\hat{Q}R = 60°$, QR = 6.4 cm, [3]
(b) the bisector of angle PQR to meet PQ at S. [1]
Measure PS. [1]

8 If U = {1, 2, 3, 4, 6, 9, 12, 18, 36}, S = {perfect squares}, E = {even numbers}, list the members of the following sets:
(a) E′ [1]
(b) E′ ∪ S [2]
(c) E ∩ S′ [2]

9 LMN is a triangle in which LM = 16 cm, MN = 32 cm and angle $L\hat{M}N = 90°$. Calculate
(a) the length of LN giving your answer correct to 1 decimal place; [2]
(b) the area of the square on side LN giving your answer correct to 3 significant figures. [2]
What difference would the length of LN to 2 decimal places, make to the answer to (b)?

10 (a) Solve the inequality $1 - 3x \leqslant 13$ and sketch a line graph of the solution set. [3]
(b) In Fig. P13 use set-builder notation to describe the set of points represented by the *unshaded* region. [2]

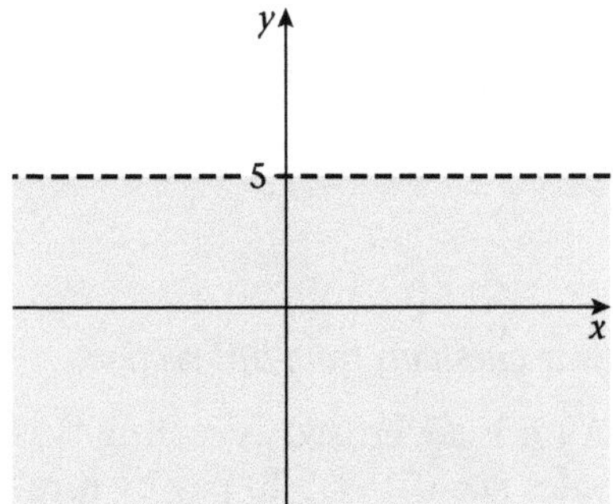

Fig. P13

11 In Fig. P14 the regular pentagon ABCDE and the square ABPQ have a common side AB as shown.

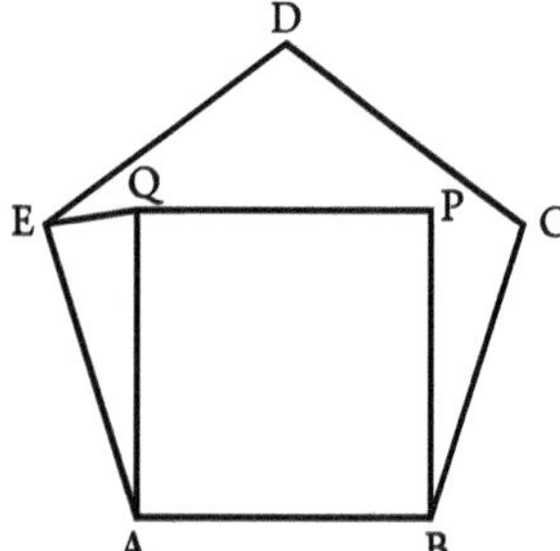

Fig. P14

Calculate the angles of $\triangle AEQ$. [5]

12

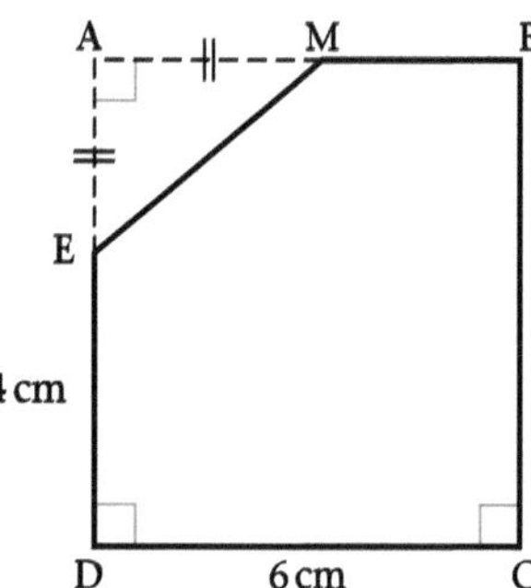

Fig. P15

The figure MBCDE (not drawn to scale) represents a scale drawing of the cross-section of a container (Fig. P15). M is the mid-point of AB and AM = AE. The scale used is 1 : 40.

(a) Calculate the actual width, DC, in metres, of the container. [1]

(b) The length of the container is 3.0 m. Calculate its capacity. [4]

Section B (60 marks)

Answer **four** *questions from this section.*

13 (a) A car is advertised as costing '\$8950 plus sales tax of 18%'. To pay by hire purchase requires a deposit of 40% of the cost price and 36 monthly payments of \$208. Calculate

(i) the cost price (including sales tax); [3]

(ii) the total cost when paying by hire purchase. [6]

(b) Village A has 200 families with an average of 3.9 children each. Village B has 300 families with an average of 4.4 children each. Find

(i) the total number of children in the two villages; [3]

(ii) the average number of children per family for the two villages. [3]

14 A vertical pole stands on level ground. A boy whose eye, E, is 1.5 m vertically above his feet, F, stands 8 m from the pole. The angle of elevation of the top of the pole from E is 33°.

(a) Make a rough sketch to show this information. [3]

(b) *Either* by calculation *or* by scale drawing, find

(i) the distance of E from the top of the pole to the nearest 0.1 m; [4]

(ii) the height of the pole to the nearest 0.1 m; [4]

(iii) the angle of depression of F from the top of the pole, to the nearest degree. [4]

15 The coordinates of quadrilateral ABCD are A(2, 0), B(0, 4), C(4, 6), D(6, 2).

(a) Using a scale of 1 cm to 1 unit on both axes, draw ABCD on graph paper. [4]

(b) What kind of quadrilateral is ABCD? [1]

(c) State the order of rotational symmetry of ABCD about (i) point (3, 3), (ii) the origin. [3]

(d) Reflect the quadrilateral ABCD in the line $x = 0$. State the coordinates of the vertices of the reflected shape.

(e) The coordinates of a quadrilateral PQRS are P(−2, 0), Q(0, −4), R(−4, −6) and S(−6, −2). Describe a single transformation which would map PQRS onto ABCD. [3]

16 (a) Solve the equations

(i) $\frac{9}{h+1} = 3$, (ii) $\frac{3}{y-1} = \frac{4}{y}$.

(b) Find the solution set of

$\{x : 3x \leqslant 15 - 2x\} \cap \{x : 4x + 1 = 9\}$ [5]

(c) A knife has a mass of m grams. It is 20 g heavier than a spoon.
  (i) Write down an expression in m for the mass of the spoon. [1]
  (ii) The total mass of 4 of the spoons and a knife is 330 g. Make an equation in m and solve it to find the mass of a knife. [4]

17 (a) Table P1 gives the ages of a group of students.

Table P1

Age (yr)	12	13	14	15	16	17	18
Frequency	2	0	1	3	4	3	2

  (i) How many students are in the group? [1]
  (ii) What is the modal age of the group? [1]
  (iii) What is the median age of the group? [2]
  (iv) Find the mean age of the group. [3]

(b) At 5 p.m. Diane leaves her office to cycle home. She rides steadily for 10 min and covers 2 km. She then stops and talks to a friend for 20 min. She then cycles the remaining 3 km at a steady rate and arrives home at 5.40 p.m.
  (i) Represent the above information on a distance–time graph. Use scales of 2 cm to 1 km and 2 cm to 10 min. [3]
  (ii) How far was Diane from home at 5.35 p.m? [2]
  (iii) Find Diane's average speed for the whole journey. [3]

18 Fig. P16 is a sketch (not to scale) of a solid.
  (a) The solid has 1 curved face and n plane faces. What is the value of n? [1]
  (b) How many planes of symmetry has the solid? [1]
  (c) Find the radius of the semicircular groove. [2]
  (d) Use a value $\frac{22}{7}$ for π to calculate the area in cm^2 of the shaded front face of the solid. [4]
  (e) Hence or otherwise calculate its volume in cm^3. [2]
  (f) Calculate the total surface area of the solid. [5]

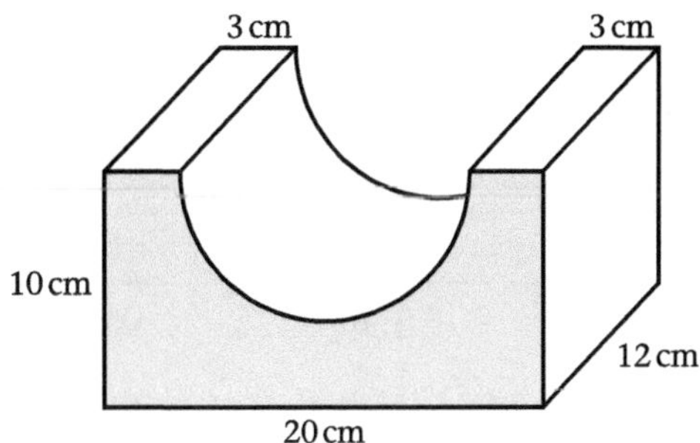

Fig. P16

Mensuration tables and formulae, three-figure tables

SI units

Length

The **metre** is the basic unit of length.

unit	abbreviation	basic unit
1 kilometre	1 km	1000 m
1 hectometre	1 hm	100 m
1 decametre	1 dam	10 m
1 metre	1 m	1 m
1 decimetre	1 dm	0.1 m
1 centimetre	1 cm	0.01 m
1 millimetre	1 mm	0.001 m

The most common measures are the millimetre, the metre and the kilometre.

1 m = 1000 mm
1 km = 1000 m = 1 000 000 mm

Mass

The **gram** is the basic unit of mass.

unit	abbreviation	basic unit
1 kilogram	1 kg	1000 g
1 hectogram	1 hg	100 g
1 decagram	1 dag	10 g
1 gram	1 g	1 g
1 decigram	1 dg	0.1 g
1 centigram	1 cg	0.01 g
1 milligram	1 mg	0.001 g

The **tonne** (t) is used for large masses. The most common measures of mass are the milligram, the gram, the kilogram and the tonne.

1 g = 1000 mg
1 kg = 1000 g = 1 000 000 mg
1 t = 1000 kg = 1 000 000 g

Time

The **second** is the basic unit of time.

unit	abbreviation	basic unit
1 second	1 s	1 s
1 minute	1 min	60 s
1 hour	1 h	3600 s

Area

The **square metre** is the basic unit of area. Units of area are derived from units of length.

unit	abbreviation	relation to other units of area
square millimetre	mm^2	
square centimetre	cm^2	$1\ cm^2 = 100\ mm^2$
square metre	m^2	$1\ m^2 = 10\,000\ cm^2$
square kilometre	km^2	$1\ km^2 = 1\,000\,000\ m^2$
hectare (for land measure)	ha	$1\ ha = 10\,000\ m^2$

Volume

The **cubic metre** is the basic unit of volume. Units of volume are derived from units of length.

unit	abbreviation	relation to other units of volume
cubic millimetre	mm^3	
cubic centimetre	cm^3	$1\ cm^3 = 1000\ mm^3$
cubic metre	m^3	$1\ m^3 = 1\,000\,000\ cm^3$

Capacity

The **litre** is the basic unit of capacity. 1 litre takes up the same space as 1000 cm^3.

Unit	Abbreviation	Relation to other units of capacity	Relation to units of volume
millilitre	$m\ell$		$1\,m\ell = 1\,cm^3$
litre	ℓ	$1\,\ell = 1000\,m\ell$	$1\,\ell = 1000\,cm^3$
kilolitre	$k\ell$	$1\,k\ell = 1000\,\ell$	$1\,k\ell = 1\,m^3$

The calendar

Remember this poem:

> Thirty days have September,
> April, June and November,
> All the rest have thirty-one,
> Excepting February alone;
> This has twenty-eight days clear,
> And twenty-nine in each Leap Year.

For a Leap Year, the year date must be divisible by 4.
Thus 2004 was a Leap Year.
Century year dates, such as 1900 and 2000, are Leap Years only if they are divisible by 400. Thus 1900 was not a Leap Year but 2000 was a Leap Year.

Mensuration formulae

	Perimeter	Area
square side s	$4s$	s^2
rectangle length l, breadth b	$2(l + b)$	lb
circle radius r	$2\pi r$	πr^2
trapezium height h, parallels of length a and b		$\frac{1}{2}(a + b)\,h$
triangle base b, height h		$\frac{1}{2}bh$
parallelogram base b, height h		bh

	Surface area	Volume
cube edge s	$6s^2$	s^3
cuboid length l, breadth b, height h	$2(lb + bh + lh)$	lbh
right-triangular prism length l, breadth b, height h		$\frac{1}{2}lbh$
cylinder base radius r, height h	$2\pi rh + 2\pi r^2$	πr^2h

Multiplication table

×	1	2	3	4	5	6	7	8	9	10
1	1	2	3	4	5	6	7	8	9	10
2	2	4	6	8	10	12	14	16	18	20
3	3	6	9	12	15	18	21	24	27	30
4	4	8	12	16	20	24	28	32	35	40
5	5	10	15	20	25	30	35	40	45	50
6	6	12	18	24	30	36	42	48	54	60
7	7	14	21	28	35	42	49	56	63	70
8	8	16	24	32	40	48	56	64	72	80
9	9	18	27	36	45	54	63	72	81	90
10	10	20	30	40	50	60	70	80	90	100

Divisibility tests

Any whole number is exactly divisible by

2	if its last digit is even
3	if the sum of its digits is divisible by 3
4	if its last two digits form a number divisible by 4
5	if its last digit is 5 or 0
6	if its last digit is even and the sum of its digits is divisible by 3
8	if its last three digits form a number divisible by 8
9	if the sum of its digits is divisible by 9
10	if its last digit is 0

Angle and length

In an n-sided polygon,
sum of angles = $(n - 2) \times 180°$

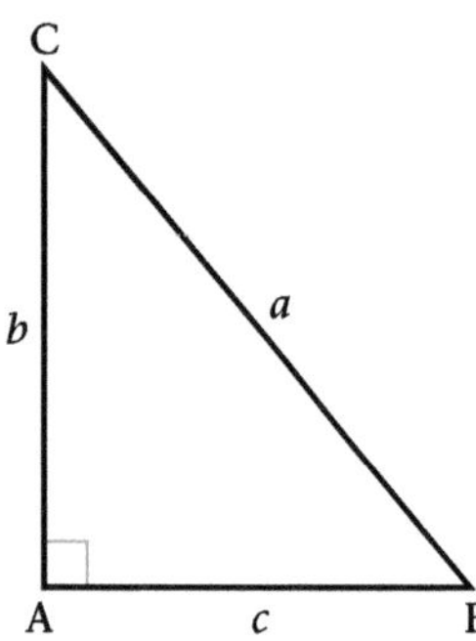

Fig. T1

In the right-angled triangle shown in Fig. T1,

$a^2 = b^2 + c^2$ *(Pythagoras' theorem)*

$\tan B = \frac{b}{c}$ $\quad\quad$ $\tan C = \frac{c}{b}$

$\sin B = \frac{b}{a}$ $\quad\quad$ $\sin C = \frac{c}{a}$

$\cos B = \frac{c}{a}$ $\quad\quad$ $\cos C = \frac{b}{a}$

Symbols

Symbol	Meaning
$=$	is equal to
$\neq$	is not equal to
$\simeq$	is approximately equal to
$>$	is greater than
$<$	is less than
$\geqslant$	is greater than or equal to
$\leqslant$	is less than or equal to
°	degree (angle)
°C	degrees Celsius (temperature)
A, B, C ...	points
AB	the line through point A and the point B, *or* the distance between points A and B
$\parallel$	lines parallel to
$\triangle$ABC	triangle ABC
$A\hat{B}C$	the angle ABC
∟	lines meeting at right angles
π	pi (3.14 ...)
%	per cent
A = $\{p, q, r\}$	A is the set p, q, r
B = {1, 2, 3, ...}	B is the infinite set 1, 2, 3 and so on
C = $\{x : x$ is an integer$\}$	Set builder notation. C is the set of numbers x such that x is an integer
n (A)	number of elements in set A
$\in$	is an element of
$\notin$	is not an element of
A′	complement of A
{ } or $\varnothing$	the empty set
U	the universal set
A $\subset$ B	A is a subset of B
A $\supset$ B	A contains B
$\not\subset$, $\not\supset$	negations of $\subset$ and $\supset$
A $\cup$ B	union of A and B
A $\cap$ B	intersection of A and B

Tangents of angles

$\theta \rightarrow \tan \theta$

θ	.0	.1	.2	.3	.4	.5	.6	.7	.8	.9
0	0.000	002	003	005	007	009	010	012	014	016
1	.017	019	021	023	024	026	028	030	031	033
2	.035	037	038	040	042	044	045	047	049	051
3	.052	054	056	058	059	061	063	065	066	068
4	.070	072	073	075	077	079	080	082	084	086
5	0.087	089	091	093	095	096	098	100	102	103
6	.105	107	109	110	112	114	116	117	119	121
7	.123	125	126	128	130	132	133	135	137	139
8	.141	142	144	146	148	149	151	153	155	157
9	.158	160	162	164	166	167	169	171	173	175
10	0.176	178	180	182	184	185	187	189	191	193
11	.194	196	198	200	202	203	205	207	209	211
12	.213	214	216	218	220	222	224	225	227	229
13	.231	233	235	236	238	240	242	244	246	247
14	.249	251	253	255	257	259	260	262	264	266
15	0.268	270	272	274	275	277	279	281	283	285
16	.287	289	291	292	294	296	298	300	302	304
17	.306	308	310	311	313	315	317	319	321	323
18	.325	327	329	331	333	335	337	338	340	342
19	.344	346	348	350	352	354	356	358	360	362
20	0.364	366	368	370	372	374	376	378	380	382
21	.384	386	388	390	392	394	396	398	400	402
22	.404	406	408	410	412	414	416	418	420	422
23	.424	427	429	431	433	435	437	439	441	443
24	.445	447	449	452	454	456	458	460	462	464
25	0.466	468	471	473	475	477	479	481	483	486
26	.488	490	492	494	496	499	501	503	505	507
27	.510	512	514	516	518	521	523	525	527	529
28	.532	534	536	538	541	543	545	547	550	552
29	.554	557	559	561	563	566	568	570	573	575
30	0.577	580	582	584	587	589	591	594	596	598
31	.601	603	606	608	610	613	615	618	620	622
32	.625	627	630	632	635	637	640	642	644	647
33	.649	652	654	657	659	662	664	667	669	672
34	.675	677	680	682	685	687	690	692	695	698
35	0.700	703	705	708	711	713	716	719	721	724
36	.727	729	732	735	737	740	743	745	748	751
37	.754	756	759	762	765	767	770	773	776	778
38	.781	784	787	790	793	795	798	801	804	807
39	.810	813	816	818	821	824	827	830	833	836
40	0.839	842	845	848	851	854	857	860	863	866
41	.869	872	875	879	882	885	888	891	894	897
42	.900	904	907	910	913	916	920	923	926	929
43	.933	936	939	942	946	949	952	956	959	962
44	.966	969	972	976	979	983	986	990	993	997

θ	.0	.1	.2	.3	.4	.5	.6	.7	.8	.9
45	1.000	003	007	011	014	018	021	025	028	032
46	.036	039	043	046	050	054	057	061	065	069
47	.072	076	080	084	087	091	095	099	103	107
48	.111	115	118	122	126	130	134	138	142	146
49	.150	154	159	163	167	171	175	179	183	188
50	1.192	196	200	205	209	213	217	222	226	230
51	.235	239	244	248	253	257	262	266	271	275
52	.280	285	289	294	299	303	308	313	317	322
53	.327	332	337	342	347	351	356	361	366	371
54	.376	381	387	392	397	402	407	412	418	423
55	1.428	433	439	444	450	455	460	466	471	477
56	.483	488	494	499	505	511	517	522	528	534
57	.540	546	552	558	564	570	576	582	588	594
58	.600	607	613	619	625	632	638	645	651	658
59	.664	671	678	684	691	698	704	711	718	725
60	1.732	739	746	753	760	767	775	782	789	797
61	.804	811	819	827	834	842	849	857	865	873
62	.881	889	897	905	913	921	929	937	946	954
63	1.963	971	980	988	997	2.006	2.014	2.023	2.032	2.041
64	2.050	059	069	078	087	097	106	116	125	135
65	2.145	154	164	174	184	194	204	215	225	236
66	.246	257	267	278	289	300	311	322	333	344
67	.356	367	379	391	402	414	426	438	450	463
68	.475	488	500	513	526	539	552	565	578	592
69	.605	619	633	646	660	675	689	703	718	733
70	2.747	762	778	793	808	824	840	856	872	888
71	2.904	921	937	954	971	989	3.006	3.024	3.042	3.060
72	3.078	096	115	133	152	172	191	211	230	251
73	.271	291	312	333	354	376	398	420	442	465
74	.487	511	534	558	582	606	630	655	681	706
75	3.732	758	785	812	839	867	895	923	952	981
76	4.011	041	071	102	134	165	198	230	264	297
77	4.331	366	402	437	474	511	548	586	625	665
78	4.705	745	787	829	872	915	959	5.005	5.050	5.097
79	5.145	193	242	292	343	396	449	503	558	614
80	5.671	730	789	850	912	976	6.041	6.107	6.174	6.243
81	6.314	386	460	535	612	691	772	855	940	7.026
82	7.115	207	300	396	495	596	700	806	916	8.028
83	8.144	264	386	513	643	777	915	9.058	9.205	9.357
84	9.514	9.677	9.845	10.02	10.20	10.39	10.58	10.78	10.99	11.20
85	11.43	11.66	11.91	12.16	12.43	12.71	13.00	13.30	13.62	13.95
86	14.30	14.67	15.06	15.46	15.89	16.35	16.83	17.34	17.89	18.46
87	19.08	19.74	20.45	21.20	22.02	22.90	23.86	24.90	26.03	27.27
88	28.64	30.14	31.82	33.69	35.80	38.19	40.92	44.07	47.74	52.08
89	57.29	63.66	71.62	81.85	95.49	114.6	143.2	191.0	286.5	573.0

Sines of angles

$\theta \rightarrow \sin\theta$

θ	.0	.1	.2	.3	.4	.5	.6	.7	.8	.9
0	0.000	002	003	005	007	009	010	012	014	016
1	.017	019	021	023	024	026	028	030	031	033
2	.035	037	038	040	042	044	045	047	049	051
3	.052	054	056	058	059	061	063	065	066	068
4	.070	071	073	075	077	078	080	082	084	085
5	0.087	089	091	092	094	096	098	099	101	103
6	.105	106	108	110	111	113	115	117	118	120
7	.122	124	125	127	129	131	132	134	136	137
8	.139	141	143	144	146	148	150	151	153	155
9	.156	158	160	162	163	165	167	168	170	172
10	0.174	175	177	179	181	182	184	186	187	189
11	.191	193	194	196	198	199	201	203	204	206
12	.208	210	211	213	215	216	218	220	222	223
13	.225	227	228	230	232	233	235	237	239	240
14	.242	244	245	247	249	250	252	254	255	257
15	0.259	261	262	264	266	267	269	271	272	274
16	.276	277	279	281	282	284	286	287	289	291
17	.292	294	296	297	299	301	302	304	306	307
18	.309	311	312	314	316	317	319	321	322	324
19	.326	327	329	331	332	334	335	337	339	340
20	0.342	344	345	347	349	350	352	353	355	357
21	.358	360	362	363	365	367	368	370	371	373
22	.375	376	378	379	381	383	384	386	388	389
23	.391	392	394	396	397	399	400	402	404	405
24	.407	408	410	412	413	415	416	418	419	421
25	0.423	424	426	427	429	431	432	434	435	437
26	.438	440	442	443	445	446	448	449	451	452
27	.454	456	457	459	460	462	463	465	466	468
28	.469	471	473	474	476	477	479	480	482	483
29	.485	486	488	489	491	492	494	495	497	498
30	0.500	502	503	505	506	508	509	511	512	514
31	.515	517	518	520	521	522	524	525	527	528
32	.530	531	533	534	536	537	539	540	542	543
33	.545	546	548	549	550	552	553	555	556	558
34	.559	561	562	564	565	566	568	569	571	572
35	0.574	575	576	578	579	581	582	584	585	586
36	.588	589	591	592	593	595	596	598	599	600
37	.602	603	605	606	607	609	610	612	613	614
38	.616	617	618	620	621	623	624	625	627	628
39	.629	631	632	633	635	636	637	639	640	641
40	0.643	644	645	647	648	649	651	652	653	655
41	.656	657	659	660	661	663	664	665	667	668
42	.669	670	672	673	674	676	677	678	679	681
43	.682	683	685	686	687	688	699	691	692	693
44	.695	696	697	698	700	701	702	703	705	706

θ	.0	.1	.2	.3	.4	.5	.6	.7	.8	.9
45	0.707	708	710	711	712	713	714	716	717	718
46	.719	721	722	723	724	725	727	728	729	730
47	.731	733	734	735	736	737	738	740	741	742
48	.743	744	745	747	748	749	750	751	752	754
49	.755	756	757	758	759	760	762	763	764	765
50	0.766	767	768	769	771	772	773	774	775	776
51	.777	778	779	780	782	783	784	785	786	787
52	.788	789	790	791	792	793	794	795	797	798
53	.799	800	801	802	803	804	805	806	807	808
54	.809	810	811	812	813	814	815	816	817	818
55	0.819	820	821	822	823	824	825	826	827	828
56	.829	830	831	832	833	834	835	836	837	838
57	.839	840	841	842	842	843	844	845	846	847
58	.848	849	850	851	852	853	854	854	855	856
59	.857	858	859	860	861	862	863	863	864	865
60	0.866	867	868	869	869	870	871	872	873	874
61	.875	875	876	877	878	879	880	880	881	882
62	.883	884	885	885	886	887	888	889	889	890
63	.891	892	893	893	894	895	896	896	897	898
64	.899	900	900	901	902	903	903	904	905	906
65	0.906	907	908	909	909	910	911	911	912	913
66	.914	914	915	916	916	917	918	918	919	920
67	.921	921	922	923	923	924	925	925	926	927
68	.927	928	928	929	930	930	931	932	932	933
69	.934	934	935	935	936	937	937	938	938	939
70	0.940	940	941	941	942	943	943	944	944	945
71	.946	946	947	947	948	948	949	949	950	951
72	.951	952	952	953	953	954	954	955	955	956
73	.956	957	957	958	958	959	959	960	960	961
74	.961	962	962	963	963	964	964	965	965	965
75	0.966	966	967	967	968	968	969	969	969	970
76	.970	971	971	972	972	972	973	973	974	974
77	.974	975	975	976	976	976	977	977	977	978
78	.978	979	979	979	980	980	980	981	981	981
79	.982	982	982	983	983	983	984	984	984	985
80	0.985	985	985	986	986	986	987	987	987	987
81	.988	988	988	988	989	989	989	990	990	990
82	.990	991	991	991	991	991	992	992	992	992
83	.993	993	993	993	993	994	994	994	994	994
84	.995	995	995	995	995	995	996	996	996	996
85	0.996	996	996	997	997	997	997	997	997	997
86	.998	998	998	998	998	998	998	998	998	999
87	.999	999	999	999	999	999	999	999	999	999
88	0.999	999	1.000	1.000	1.000	1.000	1.000	1.000	1.000	1.000
89	1.000	1.000	1.000	1.000	1.000	1.000	1.000	1.000	1.000	1.000
90	1.000									

 Mensuration tables and formulae, three-figure tables

Cosines of angles

$\theta \rightarrow \cos\theta$

θ	.0	.1	.2	.3	.4	.5	.6	.7	.8	.9
0	1.000	000	000	000	000	000	000	000	000	000
1	1.000	000	000	000	000	000	000	000	000	0.999
2	0.999	999	999	999	999	999	999	999	999	999
3	.999	999	998	998	998	998	998	998	998	998
4	.998	997	997	997	997	997	997	997	996	996
5	0.996	996	996	996	996	995	995	995	995	995
6	.995	994	994	994	994	994	993	993	993	993
7	.993	992	992	992	992	991	991	991	991	991
8	.990	990	990	990	989	989	989	988	988	988
9	.988	987	987	987	987	986	986	986	985	985
10	0.985	985	984	984	984	983	983	983	982	982
11	.982	981	981	981	980	980	980	979	979	979
12	.978	978	977	977	977	976	976	976	975	975
13	.974	974	974	973	973	972	972	972	971	971
14	.970	970	969	969	969	968	968	967	967	966
15	0.966	965	965	965	964	964	963	963	962	962
16	.961	961	960	960	959	959	958	958	957	957
17	.956	956	955	955	954	954	953	953	952	952
18	.951	951	950	949	949	948	948	947	947	946
19	.946	945	944	944	943	943	942	941	941	940
20	0.940	939	938	938	937	937	936	935	935	934
21	.934	933	932	932	931	930	930	929	928	928
22	.927	927	926	925	925	924	923	923	922	921
23	.921	920	919	918	918	917	916	916	915	914
24	.914	913	912	911	911	910	909	909	908	907
25	0.906	906	905	904	903	903	902	901	900	900
26	.899	898	897	896	896	895	894	893	893	892
27	.891	890	889	889	888	887	886	885	885	884
28	.883	882	881	880	880	879	878	877	876	875
29	.875	874	873	872	871	870	869	869	868	867
30	0.866	865	864	863	863	862	861	860	859	858
31	.857	856	855	854	854	853	852	851	850	849
32	.848	847	846	845	844	843	842	842	841	840
33	.839	838	837	836	835	834	833	832	831	830
34	.829	828	827	826	825	824	823	822	821	820
35	0.819	818	817	816	815	814	813	812	811	810
36	.809	808	807	806	805	804	803	802	801	800
37	.799	798	797	795	794	793	792	791	790	789
38	.788	787	786	785	784	783	782	780	779	778
39	.777	776	775	774	773	772	771	769	768	767
40	0.766	765	764	763	762	760	759	758	757	756
41	.755	754	752	751	750	749	748	747	745	744
42	.743	742	741	740	738	737	736	735	734	733
43	.731	730	729	728	727	725	724	723	722	721
44	.719	718	717	716	714	713	712	711	710	708

θ	.0	.1	.2	.3	.4	.5	.6	.7	.8	.9
45	0.707	706	705	703	702	701	700	698	697	696
46	.695	693	692	691	690	688	687	686	685	683
47	.682	681	679	678	677	676	674	673	672	670
48	.669	668	667	665	664	663	661	660	659	657
49	.656	655	653	652	651	649	648	647	645	644
50	0.643	641	640	639	637	636	635	633	632	631
51	.629	628	627	625	624	623	621	620	618	617
52	.616	614	613	612	610	609	607	606	605	603
53	.602	600	599	598	596	595	593	592	591	589
54	.588	586	585	584	582	581	579	578	576	575
55	0.574	572	571	569	568	566	565	564	562	561
56	.559	558	556	555	553	552	550	549	548	546
57	.545	543	542	540	539	537	536	534	533	531
58	.530	528	527	525	524	522	521	520	518	517
59	.515	514	512	511	509	508	506	505	503	502
60	0.500	498	497	495	494	492	491	489	488	486
61	.485	483	482	480	479	477	476	474	473	471
62	.469	468	466	465	463	462	460	459	457	456
63	.454	452	451	449	448	446	445	443	442	440
64	.438	437	435	434	432	431	429	427	426	424
65	0.423	421	419	418	416	415	413	412	410	408
66	.407	405	404	402	400	399	397	396	394	392
67	.391	389	388	386	384	383	381	379	378	376
68	.375	373	371	370	368	367	365	363	362	360
69	.358	357	355	353	352	350	349	347	345	344
70	0.342	340	339	337	335	334	332	331	329	327
71	.326	324	322	321	319	317	316	314	312	311
72	.309	307	306	304	302	301	299	297	296	294
73	.292	291	289	287	286	284	282	281	279	277
74	.276	274	272	271	269	267	266	264	262	261
75	0.259	257	255	254	252	250	249	247	245	244
76	.242	240	239	237	235	233	232	230	228	227
77	.225	223	222	220	218	216	215	213	211	210
78	.208	206	204	203	201	199	198	196	194	193
79	.191	189	187	186	184	182	181	179	177	175
80	0.174	172	170	168	167	165	163	162	160	158
81	.156	155	153	151	150	148	146	144	143	141
82	.139	137	136	134	132	131	129	127	125	124
83	.122	120	118	117	115	113	111	110	108	106
84	.105	103	101	099	098	096	094	092	091	089
85	0.087	085	084	082	080	078	077	075	073	071
86	.070	068	066	065	063	061	059	058	056	054
87	.052	051	049	047	045	044	042	040	038	037
88	.035	033	031	030	028	026	024	023	021	019
89	.017	016	014	012	010	009	007	005	003	002
90	0.000									

Squares

$x \rightarrow x^2$

x	0	1	2	3	4	5	6	7	8	9
1.0	1.00	1.02	1.04	1.06	1.08	1.10	1.12	1.14	1.17	1.19
1.1	1.21	1.23	1.25	1.28	1.30	1.32	1.35	1.37	1.39	1.42
1.2	1.44	1.46	1.49	1.51	1.54	1.56	1.59	1.61	1.64	1.66
1.3	1.69	1.72	1.74	1.77	1.80	1.82	1.85	1.88	1.90	1.93
1.4	1.96	1.99	2.02	2.04	2.07	2.10	2.13	2.16	2.19	2.22
1.5	2.25	2.28	2.31	2.34	2.37	2.40	2.43	2.46	2.50	2.53
1.6	2.56	2.59	2.62	2.66	2.69	2.72	2.76	2.79	2.82	2.86
1.7	2.89	2.92	2.96	2.99	3.03	3.06	3.10	3.13	3.17	3.20
1.8	3.24	3.28	3.31	3.35	3.39	3.42	3.46	3.50	3.53	3.57
1.9	3.61	3.65	3.69	3.72	3.76	3.80	3.84	3.88	3.92	3.96
2.0	4.00	4.04	4.08	4.12	4.16	4.20	4.24	4.28	4.33	4.37
2.1	4.41	4.45	4.49	4.54	4.58	4.62	4.67	4.71	4.75	4.80
2.2	4.84	4.88	4.93	4.97	5.02	5.06	5.11	5.15	5.20	5.24
2.3	5.29	5.34	5.38	5.43	5.48	5.52	5.57	5.62	5.66	5.71
2.4	5.76	5.81	5.86	5.90	5.95	6.00	6.05	6.10	6.15	6.20
2.5	6.25	6.30	6.35	6.40	6.45	6.50	6.55	6.60	6.66	6.71
2.6	6.76	6.81	6.86	6.92	6.97	7.02	7.08	7.13	7.18	7.24
2.7	7.29	7.34	7.40	7.45	7.51	7.56	7.62	7.67	7.73	7.78
2.8	7.84	7.90	7.95	8.01	8.07	8.12	8.18	8.24	8.29	8.35
2.9	8.41	8.47	8.53	8.58	8.64	8.70	8.76	8.82	8.88	8.94
3.0	9.00	9.06	9.12	9.18	9.24	9.30	9.36	9.42	9.49	9.55
3.1	9.61	9.67	9.73	9.80	9.86	9.92	9.99	10.05	10.11	10.18
3.2	10.24	10.30	10.37	10.43	10.50	10.56	10.63	10.69	10.76	10.82
3.3	10.89	10.96	11.02	11.09	11.16	11.22	11.29	11.36	11.42	11.49
3.4	11.56	11.63	11.70	11.76	11.83	11.90	11.97	12.04	12.11	12.18
3.5	12.25	12.32	12.39	12.46	12.53	12.60	12.67	12.74	12.82	12.89
3.6	12.96	13.03	13.10	13.18	13.25	13.32	13.40	13.47	13.54	13.62
3.7	13.69	13.76	13.84	13.91	13.99	14.06	14.14	14.21	14.29	14.36
3.8	14.44	14.52	14.59	14.67	14.75	14.82	14.90	14.98	15.05	15.13
3.9	15.21	15.29	15.37	15.44	15.52	15.60	15.68	15.76	15.84	15.92
4.0	16.00	16.08	16.16	16.24	16.32	16.40	16.48	16.56	16.65	16.73
4.1	16.81	16.89	16.97	17.06	17.14	17.22	17.31	17.39	17.47	17.56
4.2	17.64	17.72	17.81	17.89	17.98	18.06	18.15	18.23	18.32	18.40
4.3	18.49	18.58	18.66	18.75	18.84	18.92	19.01	19.10	19.18	19.27
4.4	19.36	19.45	19.54	19.62	19.71	19.80	19.89	19.98	20.07	20.16
4.5	20.25	20.34	20.43	20.52	20.61	20.70	20.79	20.88	20.98	21.07
4.6	21.16	21.25	21.34	21.44	21.53	21.62	21.72	21.81	21.90	22.00
4.7	22.09	22.18	22.28	22.37	22.47	22.56	22.66	22.75	22.85	22.94
4.8	23.04	23.14	23.23	23.33	23.43	23.52	23.62	23.72	23.81	23.91
4.9	24.01	24.11	24.21	24.30	24.40	24.50	24.60	24.70	24.80	24.90
5.0	25.00	25.10	25.20	25.30	25.40	25.50	25.60	25.70	25.81	25.91
5.1	26.01	26.11	26.21	26.32	26.42	26.52	26.63	26.73	26.83	26.94
5.2	27.04	27.14	27.25	27.35	27.46	27.56	27.67	27.77	27.88	27.98
5.3	28.09	28.20	28.30	28.41	28.52	28.62	28.73	28.84	28.94	29.05
5.4	29.16	29.27	29.38	29.48	29.59	29.70	29.81	29.92	30.03	30.14

x	0	1	2	3	4	5	6	7	8	9
5.5	30.25	30.36	30.47	30.58	30.69	30.80	30.91	31.02	31.14	31.25
5.6	31.36	31.47	31.58	31.70	31.81	31.92	32.04	32.15	32.26	32.38
5.7	32.49	32.60	32.72	32.83	32.95	33.06	33.18	33.29	33.41	33.52
5.8	33.64	33.76	33.87	33.99	34.11	34.22	34.34	34.46	34.57	34.69
5.9	34.81	34.93	35.05	35.16	35.28	35.40	35.52	35.64	35.76	35.88
6.0	36.00	36.12	36.24	36.36	36.48	36.60	36.72	36.84	36.97	37.09
6.1	37.21	37.33	37.45	37.58	37.70	37.82	37.95	38.07	38.19	38.32
6.2	38.44	38.56	38.69	38.81	38.94	39.06	39.19	39.31	39.44	39.56
6.3	39.69	39.82	39.94	40.07	40.20	40.32	40.45	40.58	40.70	40.83
6.4	40.96	41.09	41.22	41.34	41.47	41.60	41.73	41.86	41.99	42.12
6.5	42.25	42.38	42.51	42.64	42.77	42.90	43.03	43.16	43.30	43.43
6.6	43.56	43.69	43.82	43.96	44.09	44.22	44.36	44.49	44.62	44.76
6.7	44.89	45.02	45.16	45.29	45.43	45.56	45.70	45.83	45.97	46.10
6.8	46.24	46.38	46.51	46.65	46.79	46.92	47.06	47.20	47.33	47.47
6.9	47.61	47.75	47.89	48.02	48.16	48.30	48.44	48.58	48.72	48.86
7.0	49.00	49.14	49.28	49.42	49.56	49.70	49.84	49.98	50.13	50.27
7.1	50.41	50.55	50.69	50.84	50.98	51.12	51.27	51.41	51.55	51.70
7.2	51.84	51.98	52.13	52.27	52.42	52.56	52.71	52.85	53.00	53.14
7.3	53.29	53.44	53.58	53.73	53.88	54.02	54.17	54.32	54.46	54.61
7.4	54.76	54.91	55.06	55.20	55.35	55.50	55.65	55.80	55.95	56.10
7.5	56.25	56.40	56.55	56.70	56.85	57.00	57.15	57.30	57.46	57.61
7.6	57.76	57.91	58.06	58.22	58.37	58.52	58.68	58.83	58.98	59.14
7.7	59.29	59.44	59.60	59.75	59.91	60.06	60.22	60.37	60.53	60.68
7.8	60.84	61.00	61.15	61.31	61.47	61.62	61.78	61.94	62.09	62.25
7.9	62.41	62.57	62.73	62.88	63.04	63.20	63.36	63.52	63.68	63.84
8.0	64.00	64.16	64.32	64.48	64.64	64.80	64.96	65.12	65.29	65.45
8.1	65.61	65.77	65.93	66.10	66.26	66.42	66.59	66.75	66.91	67.08
8.2	67.24	67.40	67.57	67.73	67.90	68.06	68.23	68.39	68.56	68.72
8.3	68.89	69.06	69.22	69.39	69.56	69.72	69.89	70.06	70.22	70.39
8.4	70.56	70.73	70.90	71.06	71.23	71.40	71.57	71.74	71.91	72.08
8.5	72.25	72.42	72.59	72.76	72.93	73.10	73.27	73.44	73.62	73.79
8.6	73.96	74.13	74.30	74.48	74.65	74.82	75.00	75.17	75.34	75.52
8.7	75.69	75.86	76.04	76.21	76.39	76.56	76.74	76.91	77.09	77.26
8.8	77.44	77.62	77.79	77.97	78.15	78.32	78.50	78.68	78.85	79.03
8.9	79.21	79.39	79.57	79.74	79.92	80.10	80.28	80.46	80.64	80.82
9.0	81.00	81.18	81.36	81.54	81.72	81.90	82.08	82.26	82.45	82.83
9.1	82.81	82.99	83.17	83.36	83.54	83.72	83.91	84.09	84.27	84.46
9.2	84.64	84.82	85.01	85.19	85.38	85.56	85.75	85.93	86.12	86.30
9.3	86.49	86.68	86.86	87.05	87.24	87.42	87.61	87.80	87.98	88.17
9.4	88.36	88.55	88.74	88.92	89.11	89.30	89.49	89.68	89.87	90.06
9.5	90.25	90.44	90.63	90.82	91.01	92.10	91.39	91.58	91.78	91.97
9.6	92.16	92.35	92.54	92.74	92.93	93.12	93.32	93.51	93.70	93.90
9.7	94.09	94.28	94.48	94.67	94.87	95.06	95.26	95.45	95.65	95.84
9.8	96.04	96.24	96.43	96.63	96.83	97.02	97.22	97.42	97.61	97.81
9.9	98.01	98.21	98.41	98.60	98.80	99.00	99.20	99.40	99.60	99.80

Square roots from 1 to 9.99

$x \rightarrow \sqrt{x}$

x	0	1	2	3	4	5	6	7	8	9
1.0	1.00	1.00	1.01	1.01	1.02	1.02	1.03	1.03	1.04	1.04
1.1	1.05	1.05	1.06	1.06	1.07	1.07	1.08	1.08	1.09	1.09
1.2	1.10	1.10	1.10	1.11	1.11	1.12	1.12	1.13	1.13	1.14
1.3	1.14	1.14	1.15	1.15	1.16	1.16	1.17	1.17	1.17	1.18
1.4	1.18	1.19	1.19	1.20	1.20	1.20	1.21	1.21	1.22	1.22
1.5	1.22	1.23	1.23	1.24	1.24	1.24	1.25	1.25	1.26	1.26
1.6	1.26	1.27	1.27	1.28	1.28	1.28	1.29	1.29	1.30	1.30
1.7	1.30	1.31	1.31	1.32	1.32	1.32	1.33	1.33	1.33	1.34
1.8	1.34	1.35	1.35	1.35	1.36	1.36	1.36	1.37	1.37	1.37
1.9	1.38	1.38	1.39	1.39	1.39	1.40	1.40	1.40	1.41	1.41
2.0	1.41	1.42	1.42	1.42	1.43	1.43	1.44	1.44	1.44	1.45
2.1	1.45	1.45	1.46	1.46	1.46	1.47	1.47	1.47	1.48	1.48
2.2	1.48	1.49	1.49	1.49	1.50	1.50	1.50	1.51	1.51	1.51
2.3	1.52	1.52	1.52	1.53	1.53	1.53	1.54	1.54	1.54	1.55
2.4	1.55	1.55	1.56	1.56	1.56	1.57	1.57	1.57	1.57	1.58
2.5	1.58	1.58	1.59	1.59	1.59	1.60	1.60	1.60	1.61	1.61
2.6	1.61	1.62	1.62	1.62	1.62	1.63	1.63	1.63	1.64	1.64
2.7	1.64	1.65	1.65	1.65	1.66	1.66	1.66	1.66	1.67	1.67
2.8	1.67	1.68	1.68	1.68	1.69	1.69	1.69	1.69	1.70	1.70
2.9	1.70	1.71	1.71	1.71	1.71	1.72	1.72	1.72	1.73	1.73
3.0	1.73	1.73	1.74	1.74	1.74	1.75	1.75	1.75	1.75	1.76
3.1	1.76	1.76	1.77	1.77	1.77	1.77	1.78	1.78	1.78	1.79
3.2	1.79	1.79	1.79	1.80	1.80	1.80	1.81	1.81	1.81	1.81
3.3	1.82	1.82	1.82	1.82	1.83	1.83	1.83	1.84	1.84	1.84
3.4	1.84	1.85	1.85	1.85	1.85	1.86	1.86	1.86	1.87	1.87
3.5	1.87	1.87	1.88	1.88	1.88	1.88	1.89	1.89	1.89	1.89
3.6	1.90	1.90	1.90	1.91	1.91	1.91	1.91	1.92	1.92	1.92
3.7	1.92	1.93	1.93	1.93	1.93	1.94	1.94	1.94	1.94	1.95
3.8	1.95	1.95	1.95	1.96	1.96	1.96	1.96	1.97	1.97	1.97
3.9	1.97	1.98	1.98	1.98	1.98	1.99	1.99	1.99	1.99	2.00
4.0	2.00	2.00	2.00	2.01	2.01	2.01	2.01	2.02	2.02	2.02
4.1	2.02	2.03	2.03	2.03	2.03	2.04	2.04	2.04	2.04	2.05
4.2	2.05	2.05	2.05	2.06	2.06	2.06	2.06	2.07	2.07	2.07
4.3	2.07	2.08	2.08	2.08	2.08	2.09	2.09	2.09	2.09	2.10
4.4	2.10	2.10	2.10	2.10	2.11	2.11	2.11	2.11	2.12	2.12
4.5	2.12	2.12	2.13	2.13	2.13	2.13	2.14	2.14	2.14	2.14
4.6	2.14	2.15	2.15	2.15	2.15	2.16	2.16	2.16	2.16	2.17
4.7	2.17	2.17	2.17	2.17	2.18	2.18	2.18	2.18	2.19	2.19
4.8	2.19	2.19	2.20	2.20	2.20	2.20	2.20	2.21	2.21	2.21
4.9	2.21	2.22	2.22	2.22	2.22	2.22	2.23	2.23	2.23	2.23
5.0	2.24	2.24	2.24	2.24	2.24	2.25	2.25	2.25	2.25	2.26
5.1	2.26	2.26	2.26	2.26	2.27	2.27	2.27	2.27	2.28	2.28
5.2	2.28	2.28	2.28	2.29	2.29	2.29	2.29	2.30	2.30	2.30
5.3	2.30	2.30	2.31	2.31	2.31	2.31	2.32	2.32	2.32	2.32
5.4	2.32	2.33	2.33	2.33	2.33	2.33	2.34	2.34	2.34	2.34

x	0	1	2	3	4	5	6	7	8	9
5.5	2.35	2.35	2.35	2.35	2.35	2.36	2.36	2.36	2.36	2.36
5.6	2.37	2.37	2.37	2.37	2.37	2.38	2.38	2.38	2.38	2.39
5.7	2.39	2.39	2.39	2.39	2.40	2.40	2.40	2.40	2.40	2.41
5.8	2.41	2. 41	2.41	2.41	2.42	2.42	2.42	2.42	2.42	2.43
5.9	2.43	2.43	2.43	2.44	2.44	2.44	2.44	2.44	2.45	2.45
6.0	2.45	2.45	2.45	2.46	2.46	2.46	2.46	2.46	2.47	2.47
6.1	2.47	2.47	2.47	2.48	2.48	2.48	2.48	2.48	2.49	2.49
6.2	2.49	2.49	2.49	2.50	2.50	2.50	2.50	2.50	2.51	2.51
6.3	2.51	2.51	2.51	2.52	2.52	2.52	2.52	2.52	2.53	2.53
6.4	2.53	2.53	2.53	2.54	2.54	2.54	2.54	2.54	2.55	2.55
6.5	2.55	2.55	2.55	2.56	2.56	2.56	2.56	2.56	2.57	2.57
6.6	2.57	2.57	2.57	2.57	2.58	2.58	2.58	2.58	2.58	2.59
6.7	2.59	2.59	2.59	2.59	2.60	2.60	2.60	2.60	2.60	2.61
6.8	2.61	2.61	2.61	2.61	2.62	2.62	2.62	2.62	2.62	2.62
6.9	2.63	2.63	2.63	2.63	2.63	2.64	2.64	2.64	2.64	2.64
7.0	2.65	2.65	2.65	2.65	2.65	2.66	2.66	2.66	2.66	2.66
7.1	2.66	2.67	2.67	2.67	2.67	2.67	2.68	2.68	2.68	2.68
7.2	2.68	2.69	2.69	2.69	2.69	2.69	2.69	2.70	2.70	2.70
7.3	2.70	2.70	2.71	2.71	2.71	2.71	2.71	2.71	2.72	2.72
7.4	2.72	2.72	2.72	2.73	2.73	2.73	2.73	2.73	2.73	2.74
7.5	2.74	2.74	2.74	2.74	2.75	2.75	2.75	2.75	2.75	2.75
7.6	2.76	2.76	2.76	2.76	2.76	2.77	2.77	2.77	2.77	2.77
7.7	2.77	2.78	2.78	2.78	2.78	2.78	2.79	2.79	2.79	2.79
7.8	2.79	2.79	2.80	2.80	2.80	2.80	2.80	2.81	2.81	2.81
7.9	2.81	2.81	2.81	2.82	2.82	2.82	2.82	2.82	2.82	2.83
8.0	2.83	2.83	2.83	2.83	2.84	2.84	2.84	2.84	2.84	2.84
8.1	2.85	2.85	2.85	2.85	2.85	2.85	2.86	2.86	2.86	2.86
8.2	2.86	2.87	2.87	2.87	2.87	2.87	2.87	2.88	2.88	2.88
8.3	2.88	2.88	2.88	2.89	2.89	2.89	2.89	2.89	2.89	2.90
8.4	2.90	2.90	2.90	2.90	2.91	2.91	2.91	2.91	2.91	2.91
8.5	2.92	2.92	2.92	2.92	2.92	2.92	2.93	2.93	2.93	2.93
8.6	2.93	2.93	2.94	2.94	2.94	2.94	2.94	2.94	2.95	2.95
8.7	2.95	2.95	2.95	2.95	2.96	2.96	2.96	2.96	2.96	2.96
8.8	2.97	2.97	2.97	2.97	2.97	2.97	2.98	2.98	2.98	2.98
8.9	2.98	2.98	2.99	2.99	2.99	2.99	2.99	2.99	3.00	3.00
9.0	3.00	3.00	3.00	3.00	3.01	3.01	3.01	3.01	3.01	3.01
9.1	3.02	3.02	3.02	3.02	3.02	3.02	3.03	3.03	3.03	3.03
9.2	3.03	3.03	3.04	3.04	3.04	3.04	3.04	3.04	3.05	3.05
9.3	3.05	3.05	3.05	3.05	3.06	3.06	3.06	3.06	3.06	3.06
9.4	3.07	3.07	3.07	3.07	3.07	3.07	3.08	3.08	3.08	3.08
9.5	3.08	3.08	3.09	3.09	3.09	3.09	3.09	3.09	3.10	3.10
9.6	3.10	3.10	3.10	3.10	3.10	3.11	3.11	3.11	3.11	3.11
9.7	3.11	3.12	3.12	3.12	3.12	3.12	3.12	3.13	3.13	3.13
9.8	3.13	3.13	3.13	3.14	3.14	3.14	3.14	3.14	3.14	3.14
9.9	3.15	3.15	3.15	3.15	3.15	3.15	3.16	3.16	3.16	3.16

Square roots from 10 to 99.9

$x \to \sqrt{x}$

x	.0	.1	.2	.3	.4	.5	.6	.7	.8	.9
10	3.16	3.18	3.19	3.21	3.22	3.24	3.26	3.27	3.29	3.30
11	3.32	3.33	3.35	3.36	3.38	3.39	3.41	3.42	3.44	3.45
12	3.46	3.48	3.49	3.51	3.52	3.54	3.55	3.56	3.58	3.59
13	3.61	3.62	3.63	3.65	3.66	3.67	3.69	3.70	3.71	3.73
14	3.74	3.75	3.77	3.78	3.79	3.81	3.82	3.83	3.85	3.86
15	3.87	3.89	3.90	3.91	3.92	3.94	3.95	3.96	3.97	3.99
16	4.00	4.01	4.02	4.04	4.05	4.06	4.07	4.09	4.10	4.11
17	4.12	4.14	4.15	4.16	4.17	4.18	4.20	4.21	4.22	4.23
18	4.24	4.25	4.27	4.28	4.29	4.30	4.31	4.32	4.34	4.35
19	4.36	4.37	4.38	4.39	4.40	4.42	4.43	4.44	4.45	4.46
20	4.47	4.48	4.49	4.51	4.52	4.53	4.54	4.55	4.56	4.57
21	4.58	4.59	4.60	4.62	4.63	4.64	4.65	4.66	4.67	4.68
22	4.69	4.70	4.71	4.72	4.73	4.74	4.75	4.76	4.77	4.79
23	4.80	4.81	4.82	4.83	4.84	4.85	4.86	4.87	4.88	4.89
24	4.90	4.91	4.92	4.93	4.94	4.95	4.96	4.97	4.98	4.99
25	5.00	5.01	5.02	5.03	5.04	5.05	5.06	5.07	5.08	5.09
26	5.10	5.11	5.12	5.13	5.14	5.15	5.16	5.17	5.18	5.19
27	5.20	5.21	5.22	5.22	5.23	5.24	5.25	5.26	5.27	5.28
28	5.29	5.30	5.31	5.32	5.33	5.34	5.35	5.36	5.37	5.38
29	5.39	5.39	5.40	5.41	5.42	5.43	5.44	5.45	5.46	5.47
30	5.48	5.49	5.50	5.50	5.51	5.52	5.53	5.54	5.55	5.56
31	5.57	5.58	5.59	5.59	5.60	5.61	5.62	5.63	5.64	5.65
32	5.66	5.67	5.67	5.68	5.69	5.70	5.71	5.72	5.73	5.74
33	5.74	5.75	5.76	5.77	5.78	5.79	5.80	5.81	5.81	5.82
34	5.83	5.84	5.85	5.86	5.87	5.87	5.88	5.89	5.90	5.91
35	5.92	5.92	5.93	5.94	5.95	5.96	5.97	5.97	5.98	5.99
36	6.00	6.01	6.02	6.02	6.03	6.04	6.05	6.06	6.07	6.07
37	6.08	6.09	6.10	6.11	6.12	6.12	6.13	6.14	6.15	6.16
38	6.16	6.17	6.18	6.19	6.20	6.20	6.21	6.22	6.23	6.24
39	6.24	6.25	6.26	6.27	6.28	6.28	6.29	6.30	6.31	6.32
40	6.32	6.33	6.34	6.35	6.36	6.36	6.37	6.38	6.39	6.40
41	6.40	6.41	6.42	6.43	6.43	6.44	6.45	6.46	6.47	6.47
42	6.48	6.49	6.50	6.50	6.51	6.52	6.53	6.53	6.54	6.55
43	6.56	6.57	6.57	6.58	6.59	6.60	6.60	6.61	6.62	6.63
44	6.63	6.64	6.65	6.66	6.66	6.67	6.68	6.69	6.69	6.70
45	6.71	6.72	6.72	6.73	6.74	6.75	6.75	6.76	6.77	6.77
46	6.78	6.79	6.80	6.80	6.81	6.82	6.83	6.83	6.84	6.85
47	6.86	6.86	6.87	6.88	6.88	6.89	6.90	6.91	6.91	6.92
48	6.93	6.94	6.94	6.95	6.96	6.96	6.97	6.98	6.99	6.99
49	7.00	7.01	7.01	7.02	7.03	7.04	7.04	7.05	7.06	7.06
50	7.07	7.08	7.09	7.09	7.10	7.11	7.11	7.12	7.13	7.13
51	7.14	7.15	7.16	7.16	7.17	7.18	7.18	7.19	7.20	7.20
52	7.21	7.22	7.22	7.23	7.24	7.25	7.25	7.26	7.27	7.27
53	7.28	7.29	7.29	7.30	7.31	7.31	7.32	7.33	7.33	7.34
54	7.35	7.36	7.36	7.37	7.38	7.38	7.39	7.40	7.40	7.41
55	7.42	7.42	7.42	7.44	7.44	7.45	7.46	7.46	7.47	7.48
56	7.48	7.49	7.50	7.50	7.51	7.52	7.52	7.53	7.54	7.54
57	7.55	7.56	7.56	7.57	7.58	7.58	7.59	7.60	7.60	7.61
58	7.62	7.62	7.63	7.64	7.64	7.65	7.66	7.66	7.67	7.67
59	7.68	7.69	7.69	7.70	7.71	7.71	7.72	7.73	7.73	7.74
60	7.75	7.75	7.76	7.77	7.77	7.78	7.78	7.79	7.80	7.80
61	7.81	7.82	7.82	7.83	7.84	7.84	7.85	7.85	7.86	7.87
62	7.87	7.88	7.89	7.89	7.90	7.91	7.91	7.92	7.92	7.93
63	7.94	7.94	7.95	7.96	7.96	7.97	7.97	7.98	7.99	7.99
64	8.00	8.01	8.01	8.02	8.02	8.03	8.04	8.04	8.05	8.06
65	8.06	8.07	8.07	8.08	8.09	8.09	8.10	8.11	8.11	8.12
66	8.12	8.13	8.14	8.14	8.15	8.15	8.16	8.17	8.17	8.18
67	8.19	8.19	8.20	8.20	8.21	8.22	8.22	8.23	8.23	8.24
68	8.25	8.25	8.26	8.26	8.27	8.28	8.28	8.29	8.29	8.30
69	8.31	8.31	8.32	8.32	8.33	8.34	8.34	8.35	8.35	8.36
70	8.37	8.37	8.38	8.38	8.39	8.40	8.40	8.41	8.41	8.42
71	8.43	8.43	8.44	8.44	8.45	8.46	8.46	8.47	8.47	8.48
72	8.49	8.49	8.50	8.50	8.51	8.51	8.52	8.53	8.53	8.54
73	8.54	8.55	8.56	8.56	8.57	8.57	8.58	8.58	8.59	8.60
74	8.60	8.61	8.61	8.62	8.63	8.63	8.64	8.64	8.65	8.65
75	8.66	8.67	8.67	8.68	8.68	8.69	8.69	8.70	8.71	8.71
76	8.72	8.72	8.73	8.73	8.74	8.75	8.75	8.76	8.76	8.77
77	8.77	8.78	8.79	8.79	8.80	8.80	8.81	8.81	8.82	8.83
78	8.83	8.84	8.84	8.85	8.85	8.86	8.87	8.87	8.88	8.88
79	8.89	8.89	8.90	8.91	8.91	8.92	8.92	8.93	8.93	8.94
80	8.94	8.95	8.96	8.96	8.97	8.97	8.98	8.98	8.99	8.99
81	9.00	9.01	9.01	9.02	9.02	9.03	9.03	9.04	9.04	9.05
82	9.06	9.06	9.07	9.07	9.08	9.08	9.09	9.09	9.10	9.10
83	9.11	9.12	9.12	9.13	9.13	9.14	9.14	9.15	9.15	9.16
84	9.17	9.17	9.18	9.18	9.19	9.19	9.20	9.20	9.21	9.21
85	9.22	9.22	9.23	9.24	9.24	9.25	9.25	9.26	9.26	9.27
86	9.27	9.28	9.28	9.29	9.30	9.30	9.31	9.31	9.32	9.32
87	9.33	9.33	9.34	9.34	9.35	9.35	9.36	9.36	9.37	9.38
88	9.38	9.39	9.39	9.40	9.40	9.41	9.41	9.42	9.42	9.43
89	9.43	9.44	9.44	9.45	9.46	9.46	9.47	9.47	9.48	9.48
90	9.49	9.49	9.50	9.50	9.51	9.51	9.52	9.52	9.53	9.53
91	9.54	9.54	9.55	9.56	9.56	9.57	9.57	9.58	9.58	9.59
92	9.59	9.60	9.60	9.61	9.61	9.62	9.62	9.63	9.63	9.64
93	9.64	9.65	9.65	9.66	9.66	9.67	9.67	9.68	9.69	9.69
94	9.70	9.70	9.71	9.71	9.72	9.72	9.73	9.73	9.74	9.74
95	9.75	9.75	9.76	9.76	9.77	9.77	9.78	9.78	9.79	9.79
96	9.80	9.80	9.81	9.81	9.82	9.82	9.83	9.83	9.84	9.84
97	9.85	9.85	9.86	9.86	9.87	9.87	9.88	9.88	9.89	9.89
98	9.90	9.90	9.91	9.91	9.92	9.92	9.93	9.93	9.94	9.94
99	9.95	9.95	9.96	9.96	9.97	9.97	9.98.	9.98	9.99	9.99

Reciprocals

$x \to \frac{1}{x}$

x	0	1	2	3	4	5	6	7	8	9
1.0	1.000	0.990	980	971	962	952	943	935	926	917
1.1	0.909	901	893	885	877	870	862	855	847	840
1.2	.833	826	820	813	806	800	794	787	781	775
1.3	.769	763	758	752	746	741	735	730	725	719
1.4	.714	709	704	699	694	690	685	680	676	671
1.5	0.667	662	658	654	649	645	641	637	633	629
1.6	.625	621	617	613	610	606	602	599	595	592
1.7	.588	585	581	578	575	571	568	565	562	559
1.8	.556	552	549	546	543	541	538	535	532	529
1.9	.526	524	521	518	515	513	510	508	505	503
2.0	0.500	498	495	493	490	488	485	483	481	478
2.1	.476	474	472	469	467	465	463	461	459	457
2.2	.455	452	450	448	446	444	442	441	439	437
2.3	.435	433	431	429	427	426	424	422	420	418
2.4	.417	415	413	412	410	408	407	405	403	402
2.5	0.400	398	397	395	394	392	391	389	388	386
2.6	.385	383	382	380	379	377	376	375	373	372
2.7	.370	369	368	366	365	364	362	361	360	358
2.8	.357	356	355	353	352	351	350	348	347	346
2.9	.345	344	342	341	340	339	338	337	336	334
3.0	0.333	332	331	330	329	328	327	326	325	324
3.1	.323	322	321	319	318	317	316	315	314	313
3.2	.313	312	311	310	309	308	307	306	305	304
3.3	.303	302	301	300	299	299	298	297	296	295
3.4	.294	293	292	292	291	290	289	288	287	287
3.5	0.286	285	284	283	282	282	281	280	279	279
3.6	.278	277	276	275	275	274	273	272	272	271
3.7	.270	270	269	268	267	267	266	265	265	264
3.8	.263	262	262	261	260	260	259	258	258	257
3.9	.256	256	255	254	254	253	253	252	251	251
4.0	0.250	249	249	248	248	247	246	246	245	244
4.1	.244	243	243	242	242	241	240	240	239	239
4.2	.238	238	237	236	236	235	235	234	234	233
4.3	.233	232	231	231	230	230	229	229	228	228
4.4	.227	227	226	226	225	225	224	224	223	223
4.5	0.222	222	221	221	220	220	219	219	218	218
4.6	.217	217	216	216	216	215	215	214	214	213
4.7	.213	212	212	211	211	211	210	210	209	209
4.8	.208	208	207	207	207	206	206	205	205	204
4.9	.204	204	203	203	202	202	202	201	201	200
5.0	0.200	200	199	199	198	198	198	197	197	196
5.1	.196	196	195	195	195	194	194	193	193	193
5.2	.192	192	192	191	191	190	190	190	189	189
5.3	.189	188	188	188	187	187	187	186	186	186
5.4	.185	185	185	184	184	183	183	183	182	182

x	0	1	2	3	4	5	6	7	8	9
5.5	0.182	181	181	181	181	180	180	180	179	179
5.6	.179	178	178	178	177	177	177	176	176	176
5.7	.175	175	175	175	174	174	174	173	173	173
5.8	.172	172	172	172	171	171	171	170	170	170
5.9	.169	169	169	169	168	168	168	168	167	167
6.0	0.167	166	166	166	166	165	165	165	164	164
6.1	.164	164	163	163	163	163	162	162	162	162
6.2	.161	161	161	161	160	160	160	159	159	159
6.3	.159	158	158	158	158	157	157	157	157	156
6.4	.156	156	156	156	155	155	155	155	154	154
6.5	0.154	154	153	153	153	153	152	152	152	152
6.6	.152	151	151	151	151	150	150	150	150	149
6.7	.149	149	149	149	148	148	148	148	147	147
6.8	.147	147	147	146	146	146	146	146	145	145
6.9	.145	145	145	144	144	144	144	143	143	143
7.0	0.143	143	142	142	142	142	142	141	141	141
7.1	.141	141	140	140	140	140	140	139	139	139
7.2	.139	139	139	138	138	138	138	138	137	137
7.3	.137	137	137	136	136	136	136	136	136	135
7.4	.135	135	135	135	134	134	134	134	134	134
7.5	0.133	133	133	133	133	132	132	132	132	132
7.6	.132	131	131	131	131	131	131	130	130	130
7.7	.130	130	130	129	129	129	129	129	129	128
7.8	.128	128	128	128	128	127	127	127	127	127
7.9	.127	126	126	126	126	126	126	125	125	125
8.0	0.125	125	125	125	124	124	124	124	124	124
8.1	.123	123	123	123	123	123	123	122	122	122
8.2	.122	122	122	122	121	121	121	121	121	121
8.3	.120	120	120	120	120	120	120	119	119	119
8.4	.119	119	119	119	118	118	118	118	118	118
8.5	0.118	118	117	117	117	117	117	117	117	116
8.6	.116	116	116	116	116	116	115	115	115	115
8.7	.115	115	115	115	114	114	114	114	114	114
8.8	.114	114	113	113	113	113	113	113	113	112
8.9	.112	112	112	112	112	112	112	111	111	111
9.0	0.111	111	111	111	111	110	110	110	110	110
9.1	.110	110	110	110	109	109	109	109	109	109
9.2	.109	109	108	108	108	108	108	108	108	108
9.3	.108	107	107	107	107	107	107	107	107	106
9.4	.106	106	106	106	106	106	106	106	105	105
9.5	0.105	105	105	105	105	105	105	104	104	104
9.6	.104	104	104	104	104	104	104	103	103	103
9.7	.103	103	103	103	103	103	102	102	102	102
9.8	.102	102	102	102	102	102	101	101	101	101
9.9	.101	101	101	101	101	101	100	100	100	100

Answers

Exercise 1a (page 1)

1. (a) $\{-5, 0, 3\}$ [*or* any positive or negative non-fractional numbers]
   (b) $\{0, 2, 8\}$ [*or* any number $\geqslant 0$]
   (c) $\{\frac{-2}{3}, \frac{3}{1}, \frac{21}{5}\}$ [or $\frac{p}{q}$, where p and q are integers and $q \neq 0$]
   (d) $\{2, 5, 12\}$ [*or* any number $\geqslant 1$]
2. (a) 0 (b) -3 [*or* 0 or any negative number]
3. (a) $\in$ (b) $\notin$ (c) $\notin$ (d) $\in$ (e) $\in$

Exercise 1b (page 2)

3, 5, 7, 9, 10

Exercise 1c (page 2)

1. 0.23
2. $0.8\dot{1}\dot{8}$, recurring decimal; Q
4. 7.81; irrational number

Exercise 1d (page 3)

1. commutative
2. associative
3. associative
4. distributive
5. distributive
7. (a) $pq + pr$

Exercise 1e (page 3)

1. (a) 20, 24, 28, 32, 36, 40, 44, 48, 52, 56, 60, 64, 68, 72, 76, 80, 84, 88, 92, 96
   (b) 30, 36, 42, 48, 54, 60, 66, 72, 78, 84, 90
   (c) 40, 48, 56, 64, 72, 80, 88
   (d) 45, 54, 63, 72, 81, 90
3. (a) 17, 20, 23, 26 (b) 26, 31, 36, 41
   (c) 56, 67, 78, 89 (d) 5, 4, 3, 2
   (e) 15, 21, 28, 36 (f) 32, 64, 128, 256
   (g) 31, 43, 57, 73 (h) 26, 37, 50, 65
   (i) 36, 49, 64, 81 (j) 34, 55, 89, 144
4.

Number of tins in bottom row	1	2	3	4	5	6	7	8
Total number of tins in stack	1	3	6	10	15	21	28	36

5. (a)

1^2	2^2	3^2	4^2	5^2	6^2	7^2	8^2	9^2	10^2
1	4	9	16	25	36	49	64	81	100

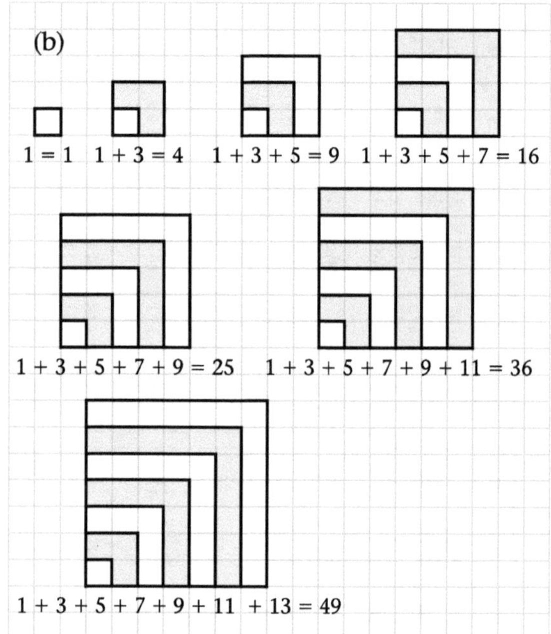
(b)
1 = 1
1 + 3 = 4
1 + 3 + 5 = 9
1 + 3 + 5 + 7 = 16
1 + 3 + 5 + 7 + 9 = 25
1 + 3 + 5 + 7 + 9 + 11 = 36
1 + 3 + 5 + 7 + 9 + 11 + 13 = 49

Yes, if n is any whole number, n^2 = sum of the first n odd numbers

(c)

Number	Pattern	Total
1	1	1
2	1 + 2 + 1	4
3	1 + 2 + 3 + 2 + 1	9
4	1 + 2 + 3 + 4 + 3 + 2 + 1	16
5	1 + 2 + 3 + 4 + 5 + 4 + 3 + 2 + 1	25
6	1 + 2 + 3 + 4 + 5 + 6 + 5 + 4 + 3 + 2 + 1	36
7	1 + 2 + 3 + 4 + 5 + 6 + 7 + 6 + 5 + 4 + 3 + 2 + 1	49

1, 4, 9, 16, 25, 36, 49
the total column contains square numbers

6 $100 = 10^2 = (1 + 2 + 3 + 4)^2$
$225 = 15^2 = (1 + 2 + 3 + 4 + 5)^2$
7 28, 36, 45
8 36, 49, 64, 81, 100
9 1, 1, 2, 3, 5, 8, 13, 21, 34, 55

Practice Exercise P1.1 (page 5)

1 (a) {2, 4, 6, 8, 10, 12, 14, 16, 18, 20}
(b) {1, 3, 5, 7, 9, 11, 13, 15, 17, 19}
(c) {2, 3, 5, 7, 11, 13, 17, 19, 23, 29}
(d) {4, 6, 8, 9, 10, 12, 14, 15, 16, 18}
(e) {1, 4, 9, 16, 25, 36, 49, 64, 81, 100}
2 *answers will vary*
3 (a) {32, 64, 128, 256}
(b) {31, 43, 57, 73}
(c) {41, 48, 55, 62}
(d) {13, 21, 34, 55}
(e) {28, 36, 45, 55}
(f) {12, 15, 18, 21}
(g) {17, 20, 23, 26}
(h) {15, 21, 28, 36}
(i) {36, 49, 64, 81}
(j) {26, 37, 50, 65}

Practice Exercise P1.2 (page 5)

1 (a) even numbers (b) square numbers
(c) odd numbers (d) multiples of 3
(e) powers of 2 (f) multiples of 4
2 (a) 2, 4, 6 , 8 ,10 (b) multiples of 2
(c) 3, 6, 9, 12, 15 (d) 4, 8, 12, 16, 20
3 (a) 8, 9, 10, 11, 12 (b) 20, 25, 30, 35, 40
(c) 18, 15, 12, 9, 6 (d) 4, 4.1, 4.2, 4.3, 4.4
(e) 5, 10, 20, 40, 80, (f) 4, 2, 1, $\frac{1}{2}$, $\frac{1}{4}$
4 (a) +2 (b) +4 (c) −10
(d) +7 (e) −0.1 (f) ×10
5 (a) +2, +4, +6 (b) 23, 33 (c) 26, 34

Practice Exercise P1.3 (page 6)

1 (a) 330 (b) 270 (c) 2
(d) 338 (e) 4.88 (f) 40
2 (a) multiplication before addition
(b) calculate brackets first
(c) calculate brackets first
(d) multiplication before subtraction
(e) division before addition
(f) calculate brackets first
3 (i) (a) $3\frac{6}{7}$ (b) Q, R
(ii) (a) $\sqrt{3}$ (b) R
(iii) (a) 0 (b) W, Z, Q, R
(iv) (a) −16 (b) Z, Q, R

Practice Exercise P1.4 (page 6)

1 $60 - 15 = 45$
2 $16 + 55 = 71$
3 $52 + 16 = 68$
4 $45 - 32 = 13$
5 $15 + 45 = 60$
6 $73 + 6 = 79$
7 $75 - 20 = 55$
8 $4 \times 5 = 20$
9 $312 \div 13 = 24$
10 $630 \div 6 = 105$

Practice Exercise P1.5 (page 6)

1 (a) False (b) LHS = − (RHS)
2 (a) True
(b) Multiplication before addition
3 (a) True
(b) 0 does not change the value in addition
4 (a) True (b) distributive law
5 (a) False (b) division is not associative
6 (a) True
(b) 1 does not change the value in multiplication
7 (a) True (b) multiplication is associative

Practice Exercise P1.6 (page 7)

1 (a) 4 (b) 20 (c) 39 (d) 24
(e) 400 (f) 20 (g) 21 (h) 11
(i) 4 (j) 30 (k) −12 (l) 11
2 (a) 560 (b) 21 (c) 3.1
(d) 7.9 (e) 0.7 (f) 31

Practice Exercise P1.7 (page 7)

1 $E[(1300 + 2800 + 500) × 1.42]; $E6532
2 $5[(10.85)^2 - (8.32)^2]\text{cm}^2$; $242.50\,\text{cm}^2$
3 (a) $[96 + 3 {2(22 + 16.50)}] ÷ 4; $81.75
(b) $(200 × 2) − 327; $73

Practice Exercise P1.8 (page 7)

1 (a) (i) 36 (ii) 28
(b) (i) 15 (ii) 228
2 (a) * obeys the commutative law ;
does not obey the commutative law
(b) * obeys the associative law
does not obey the commutative law

Exercise 2a (page 9)

3 It is impossible to construct such a triangle since the two short sides will not meet; the missing word is *greater*.
4 AC = 9.6 cm (an accurate drawing will give between 9.5 cm and 9.7 cm)
5 CD = 7.6 cm (an accurate drawing will give between 7.5 cm and 7.7 cm)
6 PX = 2.7 cm (an accurate drawing will give between 2.6 cm and 2.8 cm)

Exercise 2b (page 10)

2 the third angle is (a) 65° (b) 43° (c) 70°
3 (not drawn to scale but construction lines indicated)

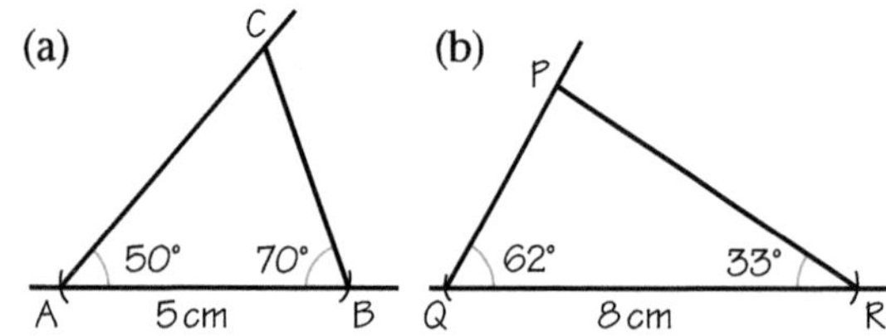

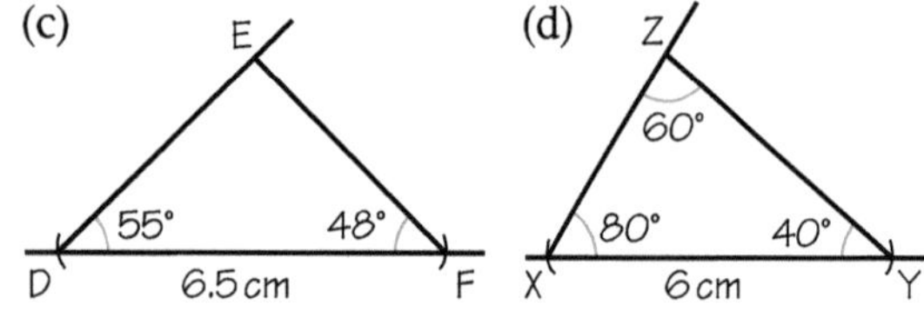

4 AM = 5.5 cm (between 5.4 and 5.6 would be accurate enough)

Exercise 2c (page 11)

Each measurement is given to the nearest mm; an acceptable range of measurements is given in the form $a \leqslant x \leqslant b$.

2 (a) 10 cm (9.9 cm $\leqslant x \leqslant$ 10.1 cm)
(b) 7 cm (6.9 cm $\leqslant x \leqslant$ 7.1 cm)
(c) 5.3 cm (5.2 cm $\leqslant x \leqslant$ 5.4 cm)
(d) 62 mm (61 mm $\leqslant x \leqslant$ 63 mm)
(e) 49 mm (48 mm $\leqslant x \leqslant$ 50 mm)
(f) 3.8 cm (3.7 cm $\leqslant x \leqslant$ 3.9 cm)
3 AD = 3.6 cm (3.5 cm $\leqslant x \leqslant$ 3.8 cm)

Exercise 2d (page 12)

4 AC = 4.9 cm; BD = 6.7 cm
5 XA = XC = 2.2 cm; XB = XD = 2.8 cm
6 4.3 cm; 6.7 or 6.8 cm
7 DC = 4.1 cm, $B\widehat{A}D = 110°$, $B\widehat{C}D = 66\frac{1}{2}°$, $A\widehat{D}C = 113\frac{1}{2}°$
8 BD = 5.7 cm

Exercise 2e (page 14)

5 The areas of the parallelograms are 14.1 cm^2 and 11.6 cm^2 approximately

Practice Exercise P2.1 (page 15)

1 PQ = 8.2 cm
2 BC = 6.3 cm
3 Student diagram
4 (e) $P\widehat{Q}S = 115°$
5 Diagonals are 8.8 cm and 4.6 cm
6 Student diagram
7 The perpendicular lines intersect each other at a single point within the triangle

Exercise 3a (page 16)

1 (a) +3, +4, +5, +6, +7 (b) −6, −5, −4, −3, −2 (c) −3, −2, −1, 0, +1
2 (a) 11 (b) −5 (c) 11
(d) 5 (e) 18 (f) 0
(g) 7 (h) 1 (i) −1
(j) 50 (k) 30 (l) 20
(m) −15 (n) −22 (o) 150
(p) 10 (q) 0

Exercise 3b (page 17)

1 (a) 4 (b) −4 (c) 3
(d) 0 (e) 2 (f) −7
2 −4 °C
3 $24
4 (a) +5 (b) +3 (c) −12
(d) −11 (e) +8 (f) −30
5 (a) +4 (b) +16 (c) +5
(d) −4 (e) −7 (f) −7
6 $16.04
7 13 BC
8 (a) +6 (b) 0 (c) +450
(d) −120 (e) $+2x$ (f) $+18y$
9 (a) $-\frac{1}{2}$ (b) +3.5 (c) +8.7
(d) $+2\frac{1}{6}$ (e) −2.4 °C (f) +3.2 °C

Exercise 3c (page 18)

1. (a) -6 (b) $+6$ (c) -6
   (d) $+9$ (e) $+5$ (f) $+8$
   (g) $+2$ (h) -3 (i) -11
   (j) $+\frac{1}{6}$ (k) $-\frac{1}{2}$ (l) $+\frac{1}{3}$
2. (a) $+4$ (b) -15 (c) $+2$
   (d) $+5$ (e) -60 (f) $+4$
   (g) $+1\frac{1}{2}$ (h) $-\frac{1}{2}$

Exercise 3d (page 18)

1. $-5a$ 2. $-4x$ 3. $-3x$
4. $-3c$ 5. $6x$ 6. $9y$
7. $-6a$ 8. $6a$ 9. $-6a$
10. $-36x$ 11. $40x$ 12. $-40x$
13. $24x^2$ 14. $-18d^2$ 15. $-3a$
16. $3x$ 17. $-2y$ 18. $-6z$
19. $-2x$ 20. $-7x$ 21. $-x$
22. $7x$ 23. $-9y$ 24. $-10z$

Exercise 3e (page 19)

1. (a) $-4, -6, -8, -10$
   (b) $2 \times (-2), 2 \times (-3), 2 \times (-4), 2 \times (-5)$
   (c) $2 \times (-2) = -4$
   $2 \times (-3) = -6$
   $2 \times (-4) = -8$
   $2 \times (-5) = -10$
   (d) $-4, -8, -12, -16$
   (e) $(-1) \times 4, (-2) \times 4, (-3) \times 4, (-4) \times 4$
   (f) $(-1) \times 4 = -4$
   $(-2) \times 4 = -8$
   $(-3) \times 4 = -12$
   $(-4) \times 4 = -16$
   (g) $+10, +15, +20, +25$
   (h) $(-2) \times (-5), (-3) \times (-5),$
   $(-4) \times (-5), (-5) \times (-5)$
   (i) $(-2) \times (-5) = +10$
   $(-3) \times (-5) = +15$
   $(-4) \times (-5) = +20$
   $(-5) \times (-5) = +25$
2. (a) -12 (b) $+28$ (c) -15
   (d) $+6$ (e) -48 (f) $+56$
   (g) $+1$ (h) 0 (i) -1
3. -90 m
4. (a) 8 (b) -7 (c) -6
   (d) 10 (e) -9 (f) -3
5. (a) -2 (b) -13 (c) $+1$
   (d) -6 (e) -9 (f) $+2$
   (g) 2 (h) -2 (i) -3
6. (a) 12 (b) -12 (c) -15
   (d) -24 (e) -22 (f) -90
7. (a) -42 (b) -1 (c) $+4$
   (d) $+4$ (e) -12 (f) $+\frac{2}{3}$
8. (a) $-5\frac{1}{2}$ (b) $+0.56$ (c) $-1\frac{1}{7}$ (d) -43.4
9. (a) $-\frac{7}{9}$ (b) $+0.8$ (c) $+\frac{3}{4}$ (d) -1.2
10. (a) $-42a$ (b) $-22y$ (c) $-16c$
    (d) $30y$ (e) $18xy$ (f) $-21ab$
11. (a) $-3x$ (b) $-5b$ (c) $6y$
    (d) $-3c$ (e) $2b$ (f) $-7x$
12. (a) $6t$ (b) $-9n$ (c) $3x$
    (d) $-11b$ (e) $-13x$ (f) $3m$

Practice Exercise P3.1 (page 20)

1. (a) 18, 14, 12, 7, 5, -17
   (b) 21, 18, 12, -6, -9, -15
   (c) 36, 23, -14, -17, -18, -34
   (d) 30, -11, -12, -18, -32, -43
2. (a) 5 (b) 6 (c) 6 (d) 6 (e) 2 (f) 5

Practice Exercise P3.2 (page 20)

1. (a) -4 (b) 1 (c) 8
   (d) -5 (e) -5 (f) -7
   (g) -14 (h) -8 (i) 8
   (j) -17 (k) -15 (l) -22
   (m) -4 (n) -7 (o) 33
2. (a) $-\frac{1}{2}$ (b) $-\frac{1}{2}$ (c) 3
   (d) $\frac{3}{4}$ (e) -4 (f) 2

Practice Exercise P3.3 (page 20)

1. 2 2. -4 3. -11 4. -20
5. 7 6. $\frac{1}{2}$ 7. -9 8. -5
9. 2 10. -1 11. 7 12. -40

Practice Exercise P3.4 (page 20)

1. -2 2. 2 3. -2 4. 4
5. 12 6. 3 7. -24 8. -5
9. 18 10. 24 11. -3 12. -7
13. -5 14. -2

Practice Exercise P3.5 (page 20)

1. (a) 6 °C (b) 7 °C (c) 5 °C
   (d) 5 °C (e) 2 °C (f) 10 °C
   (g) 8 °C (h) 8 °C (i) 14 °C
2. (a) 1 °C (b) -5 °C (c) 2 °C
   (d) 9 °C (e) 6 °C

3. (a) 2 °C (b) 9 °C (c) 6 °C
   (d) 13 °C
4. (a) −6 °C (b) 2 °C (c) 2 °C
   (d) −3 °C (e) −7 °C (f) −12 °C

Exercise 4a (page 21)

5. (b) Country → Capital
   Grenada → St Georges
   Jamaica → Kingston
   Trinidad & Tobago → Port of Spain
   Barbados → Bridgetown
   Antigua → St Johns
   (b) Place → Country
   Montego Bay, Ocho Rios → Jamaica
   Arima, Chaguaramas → Trinidad
   Roseau → Dominica
   Castries → St Lucia
   Georgetown → Guyana
   St Christopher → St Kitts

Exercise 4b (page 22)

1. 4(a), (b)
2. (a), (d), (b) possibly but not necessarily
3. (a)

Exercise 4c (page 23)

1.

Days	1	2	3	4	5	6	7	8
↓	↓	↓	↓	↓	↓	↓	↓	↓
Money ($)	2.50	5.00	7.50	10.00	12.50	15.00	17.50	20.00

2.

Number	10	11	12	13	14	15	16	17	18
↓	↓	↓	↓	↓	↓	↓	↓	↓	↓
Square	100	121	144	169	196	225	256	289	324

3. 30°

Hours	1	2	3	4	5	6
↓	↓	↓	↓	↓	↓	↓
Angle	30°	60°	90°	120°	150°	180°
Hours	7	8	9	10	11	12
↓	↓	↓	↓	↓	↓	↓
Angle	210°	240°	270°	300°	330°	360°

4.

Hours	1	2	3	4	5
↓	↓	↓	↓	↓	↓
Distance (km)	40	80	120	160	200

5.

Buses	1	2	3	4	5
↓	↓	↓	↓	↓	↓
Passengers	32	64	96	128	160

6.

Days	1	2	3	4	5	6	7
↓	↓	↓	↓	↓	↓	↓	↓
Electricity	12	24	36	48	60	72	84

7.

Distance (m)	1	2	3	4	5	6	7	8	9	10
↓	↓	↓	↓	↓	↓	↓	↓	↓	↓	↓
Rise (cm)	5	10	15	20	25	30	35	40	45	50

Exercise 4d (page 24)

1. {(1, 40), (2, 80), (3, 120), (4, 160), (5, 200)}
2. (a)

Hours	1	2	3	4	5	6	7	8
↓	↓	↓	↓	↓	↓	↓	↓	↓
Wages ($)	50	100	150	200	250	300	350	400

   (b) {(1, 50), (2, 100), (3, 150), (4, 200), (5, 250), (6, 300), (7, 350), (8, 400)}
3. {(1, 54), (2, 108), (3, 162), (4, 216), (5, 270), (6, 324), (7, 378), (8, 432)}
4.

Time (min)	1	2	3	4	5	6	7	8	9	10
No. of drops	16	32	48	64	80	96	112	128	144	160

5. (a)

0	1	2	3	4
↓	↓	↓	↓	↓
3	5	7	9	11

   (b) (0, 3), (1, 5), (2, 7), (3, 9), (4, 11)
6. 2 → 13 6 → 61
   4 → 33 8 → 97
7. (0, 3) (1, 4) (2, 5)
8.

Jim's share ($)	20	32	38	44	50
Joe's share ($)	10	16	19	22	25

9. 0 → 2 3 → 11
   1 → 5 4 → 14
   2 → 8 5 → 17
10. 'multiply by 3 and then add 2'

Practice Exercise P4.1 (page 25)

1. (i) (a)

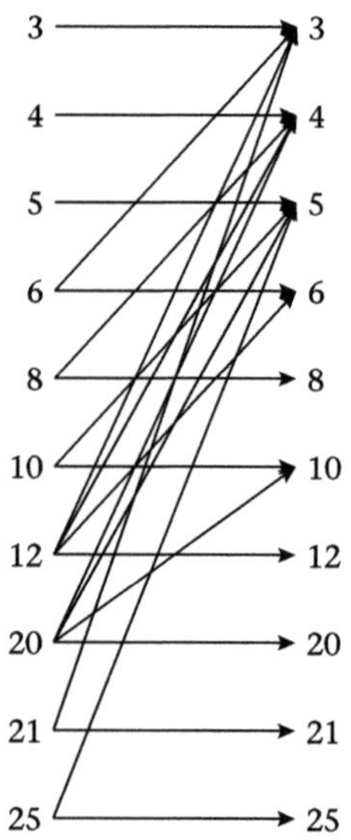

3, 4, 5, 6, 8, 10, 12, 20, 21, 25 → 3, 4, 5, 6, 8, 10, 12, 20, 21, 25

is a multiple of

(b) (3, 3), (4, 4), (5, 5), (6, 3), (6, 6), (8, 4), (8, 8), (10, 5), (10, 10), (12, 3), (12, 4), (12, 6), (12, 12), (20, 4), (20, 5), (20, 10), (20, 20), (21, 3), (21, 21), (25, 5), (25, 25)

(ii) (a)

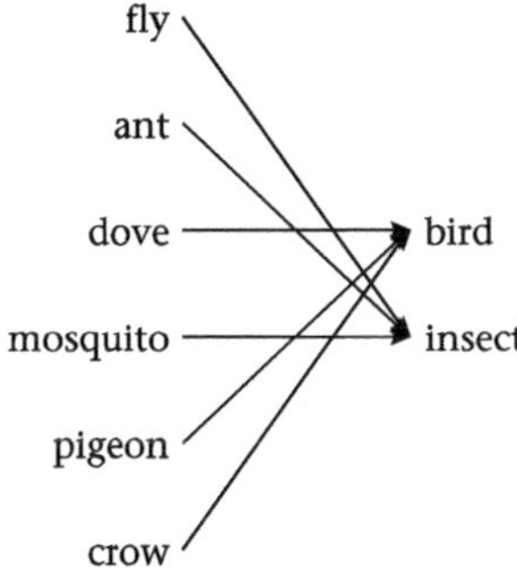

belongs to the species

(iii) (a)

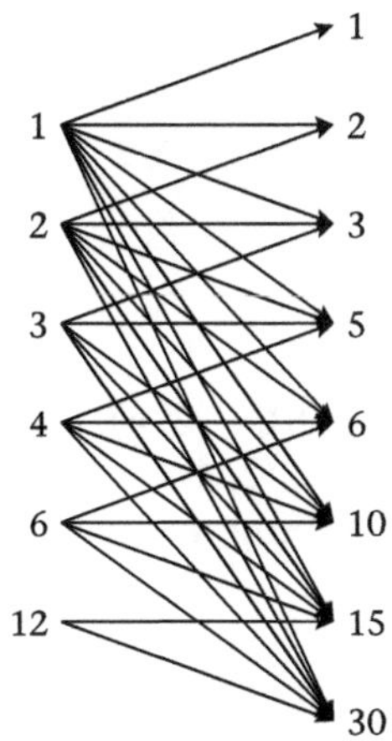

is less than or equal to

(b) (1, 1), (1, 2), (1, 3), (1, 5), (1, 6), (1, 10), (1, 15), (1, 30), (2, 2), (2, 3), (2, 5), (2, 6), (2, 10), (2, 15), (2, 30), (3, 3), (3, 5), (3, 6), (3, 10), (3, 15), (3, 30), (4, 5), (4, 6), (4, 10), (4, 15), (4, 30), (6, 6), (6, 10), (6, 15), (6, 30), (12, 15), (12, 30)

(iv) (a)

1 ⟶ 1

8 ⟶ 2

27 ⟶ 3

64 ⟶ 4

is the cube of

(b) (1, 1), (8, 2), (27, 3), (64, 4)

2 (i) (a)

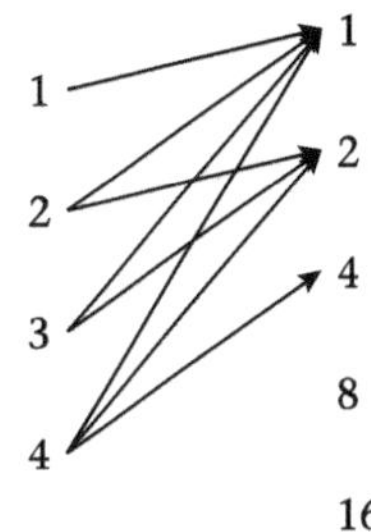

is greater than or equal to

(b) (1, 1), (2, 1), (2, 2), (3, 1), (3, 2), (4, 1), (4, 2), (4, 3)

(c) many-to-many

(ii) (a)

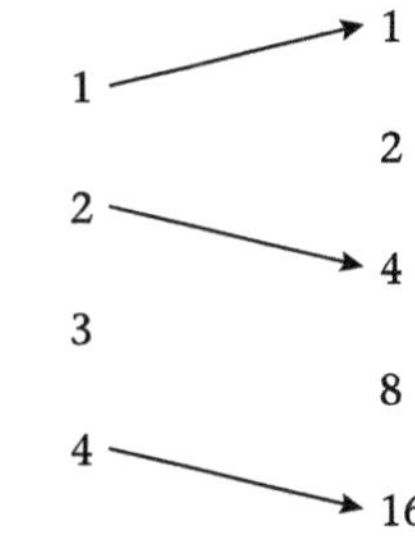

is the square root of

(b) (1, 1), (2, 4), (4, 16),

(c) one-to-one

(iii) (a)

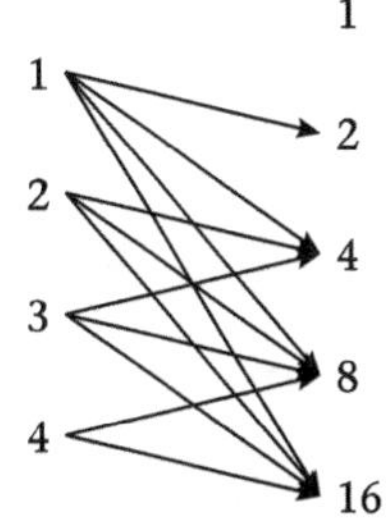

is less than

(b) (1, 2), (1, 4), (1, 8), (1, 16), (2, 4), (2, 8), (2, 16), (3, 4), (3, 8), (3, 16), (4, 8), 4, 16

(c) many-to-many

Practice Exercise P4.2 (page 25)

1 (a) $x \to 4$

(b) for all the values of x, x is mapped onto 4

(c) many-to-one

2 (a) 1 → 5, 1 → 6; 2 → 7, 2 → 8

(b) not a mapping, each member of domain mapped onto more than one member of the range

3 (a) $x \rightarrow 5x - 15$
(b) multiply x by 5 and subtract 15
(c) one-to-one

4 (a) $x \rightarrow 2x + 3$
(b) multiply x by 2 and add 3
(c) one-to-one

5 (a) 3 → 0; 4 → −1, 4 → 1; 7 → −2, 7 → 2

(b) not a mapping, members of domain mapped onto more than one member of the range

Practice Exercise P4.3 (page 26)

1 (a) (i) $\{-3, -1, 2, 5\}$ (ii) $\{-6, -2, 4, 10\}$
(b) $x \rightarrow 2x$
(c) a mapping , each and every member of the domain is mapped onto one member of the range
(d) one-to-one

2 (a) (i) $\{-3, -2, 0, 2, 3\}$ (ii) $\{2, 0, 4.5\}$
(b) $x \rightarrow \dfrac{x^2}{2}$
(c) a mapping, each and every member of the domain is mapped onto a member of the range
(d) many-to-one

3 (a) (i) $\{-1, 0, 1, 2, 3\}$ (ii) $\{-1, 0, 3, 8\}$
(b) $x \rightarrow x^2 - 1$
(c) a mapping, each and every member of the domain mapped onto a member of the range
(d) many-to-one

4 (a) (i) $\{10, 5, 2\}$ (ii) $\{3, 2, 1, -1, -2, -3\}$
(c) not a mapping, members of domain mapped onto more than one member of the range

5 (a) (i) $\{8, 6, 4, 2\}$ (ii) $\{2, 3, 4\}$
(c) not a mapping, one member of the domain not mapped onto any member of the range

Practice Exercise P4.4 (page 26)

1 (a) (i) $-9, -7, -5, -1$
(ii) $-53, -1, 3, 55$
(b) (i) $(-2, -9), (-1, -7), (0, -5), (3, -1)$
(ii) $(-3, -53), (-1, -1), (1, 3), (3, 55)$

2 (i) (a)

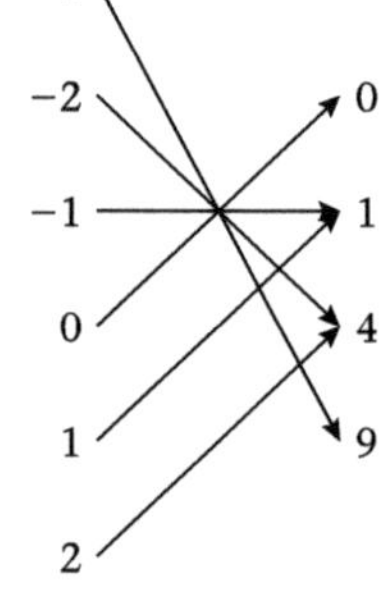

(b) $(-3, 9), (-2, 4), (-1, 1), (0, 0), (1, 1), (2, 4)$
(c) mapping, many-to-one

(ii) (a) 1 → 40; 2 → 80; 3 → 120; 4 → 160; 5 → 200

(b) $(1, 40), (2, 80), (3, 120), (4, 160), (5, 200)$
(c) mapping, one-to-one

3 (a) −2 → −7; −1 → −5; 0 → −3; 1 → −1; 2 → 1

(b) 3

4 (a) $b = \frac{1}{3}$ (b) $c = \pm 2$

Practice Exercise P4.5 (page 26)

1. $\{(-2, -1); (-1, -4); (0, -5); (1, -4); (3, 4)\}$
2. $\{(-1, 5); (0, 2); (2, -4); (3, -7); (5, -13)\}$
3. $\{(-1, -1); (1, -\frac{3}{5}); (4, 0); (10, \frac{6}{5})\}$

Practice Exercise P4.6 (page 26)

1. (a) F = {1, 2, 3, 4}
   (b) {(6, 1), (7, 2), (8, 3), (9, 4)}
   (c) not a mapping, every member of the domain is not mapped onto a member of the range
2. (a) M = {4, 5, 6, 7, 8, 9, 10}
   (b) (4, 4), (4, 8), (5, 5), (5, 10), (6, 6), (7, 7), (8, 8), (9, 9)}
   (c) not a mapping, members of the domain mapped onto more than one member of the range
3. (a) S = {2, 3}
   (b) {(4, 2), (9, 4)}
   (c) not a mapping, every member of the domain is not mapped onto a member of the range
4. (a) T = {1, 2}
   (b) {(4, 1), (8, 2)}
   (c) not a mapping, every member of the domain is not mapped onto a member of the range

Exercise 5a (page 27)

2. $3(7 + 2) = 3 \times 9 = 27$
   $3(7 + 2) = 3 \times 7 + 3 \times 2 = 21 + 6 = 27$
3. $10(6 - 4) = 10 \times 2 = 20$
   $10(6 - 4) = 10 \times 6 - 10 \times 4 = 60 - 40 = 20$
4. $9(4 + 3) = 9 \times 7 = 63$
   $9(4 + 3) = 9 \times 4 + 9 \times 3 = 36 + 27 = 63$
5. $6(4 - 2) = 6 \times 2 = 12$
   $6(4 - 2) = 6 \times 4 - 6 \times 2 = 24 - 12 = 12$
6. $7(1 + 9) = 7 \times 10 = 70$
   $7(1 + 9) = 7 \times 1 + 7 \times 9 = 7 + 63 = 70$
7. $3(8 - 2) = 3 \times 6 = 18$
   $3(8 - 2) = 3 \times 8 - 3 \times 2 = 24 - 6 = 18$
8. $3(8 + 2) = 3 \times 10 = 30$
   $3(8 + 2) = 3 \times 8 + 3 \times 2 = 24 + 6 = 30$
9. $4(11 - 3) = 4 \times 8 = 32$
   $4(11 - 3) = 4 \times 11 - 4 \times 3 = 44 - 12 = 32$
10. $5(10 + 2) = 5 \times 12 = 60$
    $5(10 + 2) = 5 \times 10 + 5 \times 2 = 50 + 10 = 60$

Exercise 5b (page 27)

1. $3x + 3y$
2. $36 - 4n$
3. $2a - 14$
4. $10p - 5q$
5. $18a + 24b$
6. $49x + 14y$
7. $8a + 14ac$
8. $24xy - 32y$
9. $21x - 24px$
10. $-3pq + 3q^2$
11. $-2a^2 - 2ab$
12. $-5bd - 10cd$
13. $2a + 4b$
14. $5u - 5v$
15. $9a + 3b$
16. $15a - 9ad$
17. $49c^2 - 14c$
18. $24s^2 + 6st$

Exercise 5c (page 28)

1. $-3m - 3n$
2. $-2u - 2v$
3. $-4a - 4b$
4. $-5a + 5b = 5b - 5a$
5. $-8x + 8y = 8y - 8x$
6. $-9p + 9q = 9q - 9p$
7. $-12mn - 8n$
8. $-21y + 35y^2 = 35y^2 - 21y$
9. $-4a^2 - 6a$
10. $2a^2 + 6ab$
11. $10xy - 55x^2$
12. $5pq - p^2$
13. $-4c - 16d$
14. $45m^2 - 18mn$
15. $-70t - 21t^2$
16. $11xy - 2x^2$
17. $15 - 36a$
18. $25ax + 35ay$

Exercise 5d (page 28)

1. $5a + 4b$
2. $3x + 6y$
3. $12p - 5q$
4. $11c + 3$
5. $18a - 5b$
6. $20x + 6y$
7. $4x - 10$
8. $5r - 12$
9. $-a - 6b$
10. $10 - a$
11. $28x - 32$
12. $18t - 24$
13. $9a + 14$
14. $16x + 13y$
15. $9x + 4y$
16. $5a - 9b$
17. $3x - 29y$
18. $-2x - 5y$
19. $9b - a$
20. $-15b$
21. $x^2 + x + 6$
22. $x^2 + 9x + 20$
23. $x^2 + 5x - 14$
24. $x^2 - 4x - 32$
25. $a^2 + 3a - 10$
26. $a^2 - 7a - 18$
27. $y^2 - 9y + 18$
28. $z^2 - 11z + 10$
29. $x^2 + 2ax + a^2$
30. $a^2 - b^2$

Exercise 5e (page 29)

1. $pr + qr + ps + qs$
2. $xy + 8y + 3x + 24$
3. $12b + 15ab + 4a + 5a^2$
4. $ac - bc + ad - bd$
5. $xy - 7y + x - 7$
6. $6p - 3p^2 + 2q - qp$
7. $wy + xy - wz - xz$
8. $ab + 6b - 9a - 54$
9. $2mp + 5np - 6mq - 15nq$
10. $ac - bc - ad + bd$
11. $xy - 4y - 5x + 20$
12. $5x^2 - xy - 15xy + 3y^2$

Exercise 5f (page 30)

1. $a^2 + 7a + 12$
2. $b^2 + 7b + 10$
3. $m^2 + m - 6$
4. $n^2 - 5n - 14$
5. $x^2 + 4x + 4$
6. $y^2 - 3y - 4$
7. $c^2 + 3c - 10$
8. $d^2 - 7d + 12$
9. $p^2 - 7p + 10$
10. $x^2 - 8x + 16$
11. $y^2 + 8y + 7$
12. $a^2 + 2a - 24$
13. $b^2 - 10b + 21$
14. $c^2 + 4c - 5$
15. $6 + 5d + d^2$
16. $10 + 3x - x^2$
17. $12 - 7y + y^2$
18. $m^2 + 5mn + 6n^2$
19. $a^2 - ab - 6b^2$
20. $x^2 - 7xy + 12y^2$
21. $p^2 + 4pq + 4q^2$
22. $m^2 + 2mn - 15n^2$
23. $2a^2 + 7a - 15$
24. $3x^2 - 2x - 8$
25. $6h^2 + hk - 2k^2$
26. $15x^2 + 26xy + 8y^2$
27. $9a^2 - 12ab + 4b^2$
28. $25h^2 + 10hk + k^2$
29. $10a^2 - 19ab + 6b^2$
30. $35m^2 - 4mn - 15n^2$

Exercise 5g (page 30)

1. (a) x^5 (b) a^7 (c) n^6 (d) $24a^6$ (e) $20x^{10}$ (f) $6c^7$
2. (a) m^8 (b) a^{10} (c) c^{14} (d) 10^9 (e) b^{15} (f) x^8 (g) $10e^{14}$ (h) 15×10^9 (i) $15y^8$

Exercise 5h (page 31)

1. (a) a^4 (b) c^3 (c) d (d) $3x^4$ (e) $2a^2$ (f) x^5
2. (a) x^2 (b) b^3 (c) c^6 (d) a^2 (e) 10^2 (f) x^6 (g) $2x$ (h) $4x^3$ (i) 2×10^3

Exercise 5i (page 32)

1. $\frac{1}{100}$
2. $\frac{1}{10000}$
3. $\frac{1}{1000000}$
4. x^3
5. $\frac{1}{a^2}$
6. 1
7. $\frac{1}{a^5}$
8. x^9
9. p^5
10. b^3
11. 1
12. a^6
13. 8
14. 9
15. 9
16. $6a$
17. $\frac{3a}{2}$
18. $18a$
19. $\frac{1}{c^2}$
20. $\frac{1}{b^6}$

Exercise 5j (page 32)

1. $1, 3, x, 3x$
2. $1, a, b, ab$
3. $1, x, x^2$
4. $1, a, b, ab, a^2, a^2b$
5. $1, 2, 3, 6, a, 2a, 3a, 6a$
6. $1, 5, a, 5a, b, 5b, ab, 5ab$
7. $1, x, y, z, xy, xz, yz, xyz$
8. $1, 2, 5, 10, m, 2m, 5m, 10m, m^2, 2m^2, 5m^2, 10m^2$
9. $1, 5, 25, p, 5p, 25p, q, 5q, 25q, pq, 5pq, 25pq$
10. $1, 2, 4, 8, d, 2d, 4d, 8d, e, 2e, 4e, 8e, de, 2de, 4de, 8de$
11. $1, 2, a, 2a, b, 2b, a^2, 2a^2, b^2, 2b^2, ab, 2ab, a^2b, 2a^2b, ab^2, 2ab^2, a^2b^2, 2a^2b^2$
12. $1, 2, 7, 14, a, 2a, 7a, 14a, b, 2b, 7b, 14b, ab, 2ab, 7ab, 14ab, b^2, 2b^2, 7b^2, 14b^2, ab^2, 2ab^2, 7ab^2, 14ab^2$

Exercise 5k (page 33)

1. a
2. 3
3. x
4. m
5. 2
6. g
7. x
8. 4
9. z
10. x
11. $4p$
12. ab
13. xy
14. pq
15. $2x$
16. ax
17. x
18. $2x$
19. $5x$
20. $3a$
21. mn
22. $2de$
23. $5m^2$
24. $3ab$
25. $6xy$
26. $7d$
27. $5a^2$
28. $4xy$
29. $9y$
30. 1

Exercise 5l (page 33)

1. (a) $(3x + y)$ (b) $(a - 3b)$ (c) $(x + y)$ (d) $(p + q)$ (e) $(a - b)$ (f) $(1 - y)$ (g) $(2 - r)$ (h) $(3b + 5x)$ (i) $(3b + 5d)$ (j) $(3y - z)$ (k) $(3c + 4d)$ (l) $(3x - 2)$ (m) $(2a + 1)$ (n) $(2m - 1)$ (o) $(p - 3q)$ (p) $(5a + 2x)$

2 (a) $6(2c + d)$ (b) $4(a - 2b)$
(c) $3(2z - 1)$ (d) $3(3x + 4y)$
(e) $x(y + z)$ (f) $c(b + d)$
(g) $x(9 + a)$ (h) $ab(c + d)$
(i) $yz(x - a)$ (j) $m(pq + ab)$
(k) $3a(b - 2c)$ (l) $4x(3a + 2b)$
(m) $x(3x - 1)$ (n) $2(3m^2 - 1)$
(o) $ax(2b + 7c)$ (p) $d^2(3e + 5)$
(q) $2am(2m - 3)$ (r) $-5(3x^2 + 2)$
(s) $-6g(3f + 2)$
(t) $-5y(x - 2)$ or $5y(2 - x)$

Exercise 5m (page 34)

1 ab 2 $5x$ 3 $6a$ 4 $12ab$
5 $10xy$ 6 $9a$ 7 $3x$ 7 abc
9 xyz 10 $6b$ 11 x^2 12 $3a^2$
13 $3m^2n$ 14 $18a^2b$ 15 $6x^2y^2$ 16 $30ab^2$

Exercise 5n (page 34)

1 $24b$ 2 bx 3 $4x$ 4 16
5 1 6 $3c$ 7 $2k$ 7 b
9 3 10 $12xz$ 11 $3ac$ 12 $6dm$

Exercise 5o (page 35)

1 $\frac{3a}{5}$ 2 $3b$ 3 $\frac{4x}{3}$
4 $\frac{5x}{3}$ 5 $\frac{a}{5}$ 6 $\frac{9x}{20}$
7 $\frac{7}{a}$ 8 $\frac{2}{5y}$ 9 $\frac{6}{x}$
10 $\frac{1}{a}$ 11 $\frac{2}{y}$ 12 $\frac{2}{x}$
13 $\frac{4}{3x}$ 14 $\frac{4}{5a}$ 15 $\frac{1}{2z}$
16 $\frac{1}{12a}$ 17 $\frac{8}{15x}$ 18 $\frac{5}{6a}$
19 $\frac{3 + x}{x}$ 20 $\frac{2a - b}{a}$ 21 $\frac{6d + 1}{2d}$
22 $\frac{9 + xy}{x}$ 23 $\frac{8b + 3}{4}$ 24 $\frac{2pq - 3}{2q}$
25 $\frac{b + a}{ab}$ 26 $\frac{y - x}{xy}$ 27 $\frac{3m + 4}{4m}$
28 $\frac{3d - 2c}{cd}$ 29 $\frac{15 - 8a}{12ab}$ 30 $\frac{6y + 5x}{15xy}$

Exercise 5p (page 36)

1 (a) $\frac{3a + 1}{2}$ (b) $\frac{5b - 1}{3}$
(c) $\frac{2c - 4}{5}$ (d) $\frac{2x + 7}{4}$
(e) $\frac{7z - 5}{4}$ (f) $\frac{10 - 7n}{12}$
(g) $\frac{11a - 2}{4}$ (h) $\frac{b + 1}{2}$
(i) $\frac{17u - 15}{12}$ (j) $\frac{m - 7}{24}$
(k) $\frac{13c - 12d}{30}$ (l) $\frac{7a - 6b}{18}$

2 (a) $\frac{x + 3}{2}$ (b) $\frac{x - 3}{2}$
(c) $\frac{2h + 3}{3}$ (d) $\frac{3b - a}{2}$

Practice Exercise P5.1 (page 37)

1 $3uv + 2uw - 3v^2 + 2u^2$
2 $6y^2 - xy - x^2$
3 $u(uv - v^2 + 4w^2)$
4 $y(y - 6x^2 - yx)$
5 $b(a - 6c)$
6 $a(3a - 4c)$

Practice Exercise P5.2 (page 37)

1 $3x + 3y$ 2 $ax + ay$ 3 $x^2 + xy$
4 $2x^2 + 2xy$ 5 $ax^2 + axy$ 6 $x^2y + xy^2$
7 $8x + 12y$ 8 $2ax + 3ay$ 9 $6xy + 9y^2$

Practice Exercise P5.3 (page 37)

1 $3a^2 - 3ab$ 2 $-c(c - \frac{1}{2})$
3 $(a + b)d$ 4 $-h(h + 4)$
5 $s^2 - 4st + 3s - t^2 - 2t$
6 $2mp - 2np + nm$
7 $-2k^2 + 3k - 2h^2 + h$
8 $-(jk + 2k - \frac{1}{3}j)$
9 $(g - f)f^2g$
10 $(m^2 + 2mn - n^2)mn$
11 $10u - 17$ 12 $3y + 10$
13 $-5(d - 1)$ 14 $2a - 1$
15 $x^2 + yx + 3y^2$ 16 $-d(2d - 3)$
17 $(5t^3 + 4t^2 - 3t - 6)t$
18 $3v^2 + 2w^2$
19 $-2k^2 + 3k - 2h^2 + h$
20 $n(-6u + 21v)$

Practice Exercise P5.4 (page 37)

1 $14x$ 2 $r + 23s$
3 $m(-23u + 16v)$ 4 0
5 $3a(x - 2y)$ 6 $-(7x + 3by)$

7 $8a - b$
8 $u^2 + 4uv - v^2$
9 $2h^2 + 2hk - 3k^2$
10 $6u(u + v)$

Practice Exercise P5.5 (page 37)

1 $3x^2 - xy$
2 $2x^2 + 3yx + y^2$
3 $2a^2 - 3ba - 2b^2$
4 $2u^2 + 3vu - 2v^2$
5 $2h^2 - 5hk - 3k^2$
6 $u^2 + 2vu + v^2$
7 $x^2 - y^2$
8 $4a^2 - b^2$
9 $-m^2 + 4mn - 4n^2$
10 $4h^2 + 4hk + k^2$

Practice Exercise P5.6 (page 38)

1 $6x^2$
2 $6x^3$
3 x^5
4 $6x^3y$
5 $2d^2$
6 $6d$
7 $2de$
8 $2d^2e^2$

Practice Exercise P5.7 (page 38)

1 $a^3(a + 4)$
2 $pq^2(p^2r - 2)$
3 $3d^2(1 - 2bd)$
4 $3y(xy + 4z)$
5 $3h^2(3k - j)$
6 $4b^3c^3(3ab^2c^2 - 1)$
7 $a^2f^3(8a^4 + 1)$
8 $3h^2k^2(hkn^6 + 3)$
9 $xy^2(a^9x^5y + 2)$
10 $6bc^5(2ab^2 - c)$

Practice Exercise P5.8 (page 38)

1 $\frac{1}{9}$
2 $\frac{1}{8}$
3 4
4 $8b^3$
5 $4k^6$
6 $\frac{v^2}{4}$
7 $\frac{1}{x^4}$
8 64
9 $\frac{6}{h^5}$
10 18
11 $6x^2$
12 $\frac{27x^2}{2}$

Practice Exercise P5.9 (page 38)

1 $\frac{7x}{6}$
2 $\frac{2x - 3}{x}$
3 $\frac{5u}{4}$
4 $\frac{13f}{3}$
5 $\frac{3k - 2}{k}$
6 $\frac{w}{12}$
7 $\frac{2 + 4h}{h}$
8 $\frac{4x - y}{6}$
9 $\frac{c}{3}$
10 $\frac{5}{3h}$
11 $\frac{k(9h - 2)}{3h}$
12 $\frac{2x^2 + y^2}{xy}$
13 $\frac{7x - 3}{6}$
14 $\frac{2a + 9}{12}$
15 $\frac{3c - 6 - d}{6}$
16 $\frac{3x - 2y - 1}{4}$
17 $\frac{8a - 5b}{6}$
18 $\frac{8x - 9y - 1}{12}$
19 $\frac{6a + 2b - 4c}{3}$
20 $\frac{13x - 14y}{15}$

Exercise 6a (page 40)

1 the missing values are:
(a) 30 m³ (b) 48 cm³ (c) 25 m³
(d) 27 cm³ (e) 2 cm (f) 3 m
(g) 6 cm (h) 34 m³
2 8 cm³
3 810 cm³
4 3 m
5 the cube (by 1 cm³)
6 20 m²
7 4 cm
8 150 m³; 10 people

Exercise 6b (page 41)

1 0.288 m³ (288 000 cm³)
2 750 000 cm³
3 20 000 cm³
4 2500 blocks
5 1000 cubes

Exercise 6c (page 42)

1 (a) 90 cm³ (b) 60 cm³ (c) 72 cm³
(d) $7\frac{1}{2}$ cm³ (e) 35 cm³ (f) 63 cm³
2 (a) 7500 cm³ (b) 27 000 cm³
(c) 8000 cm³ (d) 14 000 cm³
(e) 20 000 cm³ (f) 10 500 cm³
3 0.021 m³
4 9000 cm³
5 2140 cm³

Exercise 6d (page 44)

1 77 cm³
2 22 m³, 22 kℓ
3 52.5 kℓ
4 35 cm
5 3.85 ℓ
6 (a) $5028\frac{4}{7}$ cm³ (b) 3520 g
7 2.6 kg
8 (a) $\frac{22}{7}$ m³ (b) 6.6 t
9 9.8 cm

Practice Exercise P6.1 (page 45)

1 (a) 30 m³; 30 000 000 cm³
(b) 30 000
2 (a) 18 m³
(b) 14.4 m³
3 (a) 2464 cm³
(b) 8 cm
4 (a) 140 m³
(b) 947 containers

5. $565\,714.26\,cm^3$
6. $144\,cm^3$
7. (a) 36 cm
   (b) 6 cm
   (c) $216\,cm^3$
8. (a) $42\,240\,cm^3$ (b) $633\,600\,cm^3$
9. $14\,044.8\,cm^3$
10. (a) $924\,cm^3$
    (b) $92\,400\,cm^3$
    (c) 150 cm
11. 24
12. (a) $492\,000\,cm^3$
    (b) 6 m
    (c) $369\,600\,cm^3$
13. 6.3 kg
14. 3.2 cm

Exercise 7a (page 47)

1. (a) million (b) billion
2. (a) $40\,000\,000\,cm^2$ (b) 1 million
3. 1 million
4. just over $11\frac{1}{2}$ days
5. (d) (there are over $31\frac{1}{2}$ million seconds in a year)

Exercise 7b (page 48)

1	1 000 000	2	59 244
3	721 568 397	4	2 312 400
5	8 000 000	6	3 000 000 000
7	9215	8	14 682 053
9	108 412	10	12 345
11	100 000 000	12	987 654

Exercise 7c (page 48)

1. (a) \$2 000 000 (b) 150 000 000 km
   (c) 3 000 000 000 (d) 5 500 000
   (e) \$2 100 000 000 (f) 4 200 000 litres
   (g) 400 000 000 (h) \$1 250 000
   (i) 700 000 tonnes (j) \$750 000
   (k) 450 000 (l) \$580 000 000
2. (a) 8 million tonnes
   (b) \$6 million
   (c) 2 billion
   (d) \$3.7 billion
   (e) \$7.4 million
   (f) \$$1\frac{3}{4}$ million
   (g) 0.2 million litres
   (h) $\frac{1}{2}$ billion or 500 million
   (i) 0.3 million tonnes
   (j) $\frac{1}{4}$ million
   (k) 0.98 million barrels
   (l) 0.49 billion or 490 million

Exercise 7d (page 49)

1	9×10^6	2	4×10^3
3	4×10^9	4	6×10^2
5	3×10^5	6	6×10^4
7	5×10^9	8	7×10^1
9	2×10^7	10	8.9×10^4
11	7.2×10^8	12	2.3×10^6
13	5.5×10^1	14	1.7×10^5
15	5.4×10^3	16	2.5×10^7
17	6.3×10^3	18	9.4×10^9
19	4.1×10^5	20	8.5×10^7
21	9.5×10^2	22	3.6×10^3
23	3.6×10^2	24	3.6×10^1

Exercise 7e (page 49)

1	60 000	2	8000
3	900 000	4	400
5	70	6	300
7	50 000 000	8	2 000 000
9	600 000 000	10	63 000
11	8400	12	980 000
13	72	14	360
15	440	16	51 000 000
17	2 500 000	18	670 000 000
19	3700	20	59 000 000
21	85 000	22	3400
23	340	24	34

Exercise 7f (page 50)

1	0.06	2	0.004
3	0.9	4	0.000 008
5	0.0004	6	0.000 06
7	0.003	8	0.000 09
9	0.0007	10	0.16
11	0.034	12	0.0026
13	0.000 28	14	0.084
15	0.0756	16	2.7
17	0.65	16	0.402
19	0.2	20	0.24
21	0.7	22	0.0062
23	0.0402	24	0.03
25	0.72	26	0.072

Exercise 7g (page 51)

1. 5×10^{-3}
2. 8×10^{-2}
3. 6×10^{-4}
4. 4×10^{-6}
5. 2×10^{-5}
6. 9×10^{-7}
7. 3×10^{-1}
8. 3×10^{-3}
9. 3×10^{-5}
10. 3.8×10^{-2}
11. 6.2×10^{-3}
12. 7.1×10^{-1}
13. 8.8×10^{-4}
14. 2.6×10^{-5}
15. 5.5×10^{-6}
16. 9.1×10^{-7}
17. 6.7×10^{-8}
18. 1.5×10^{-4}
19. 1.5×10^{-3}
20. 1.5×10^{-2}

Exercise 7h (page 51)

1. 0.0002
2. 0.000 006
3. 0.005
4. 0.04
5. 0.7
6. 0.000 03
7. 0.006
8. 0.000 09
9. 0.000 000 2
10. 0.000 28
11. 0.000 008 3
12. 0.0051
13. 0.045
14. 0.79
15. 0.000 033
16. 0.0062
17. 0.000 094
18. 0.000 26
19. 0.18
20. 0.0088

Exercise 7i (page 52)

1. 9.6×10^{3}
2. 7.5×10^{8}
3. 8.37×10^{9}
4. 5.2×10^{4}
5. 1.5×10^{2}
6. 2.86×10^{4}
7. 1.67×10^{6}
8. 1×10^{7}
9. 1.101×10^{4}
10. 6×10^{3}
11. 5×10^{8}
12. 5×10^{4}
13. 7.6×10^{-2}
14. 6.4×10^{-4}
15. 4.4×10^{-3}
16. 2.89×10^{-5}
17. 2.35×10^{4}
18. 8.367×10^{6}
19. 7.4×10^{-2}
20. 7.337×10^{3}
21. 6.25×10^{5}
22. 5.49×10^{-4}
23. $1.004\,8 \times 10^{0}$
24. 2.3×10^{-4}

Exercise 7j (page 52)

1. 6×10^{11}
2. 2×10^{4}
3. 8×10^{-7}
4. 3×10^{-6}
5. 4×10^{8}
6. 6×10^{3}
7. 2.8×10^{3}
8. 4×10^{6}
9. 4.5×10^{-2}
10. 6×10^{-15}
11. 3.6×10^{-5}
12. 2×10^{4}
13. 4.35×10^{5}
14. 1.2×10^{1}
15. 2.5×10^{0}
16. 2.47×10^{-1}
17. 7×10^{0}
18. 2.31×10^{-9}
19. 1.2×10^{-3}
20. 6.25×10^{1}

Exercise 7k (page 53)

1. 9.2×10^{9}
2. 3.91×10^{5} km²
3. (a) 1 km² = 100 ha
   (b) 3.91×10^{7} ha
4. 2.54×10^{-5} km
5. 3×10^{7} cm³
6. 39 kg
7. 3.6×10^{3} seconds
8. 1.08×10^{9} km
9. 1100 m
10. 2.19 billion
11. 2.8×10^{4} km
12. $1.987\,8 \times 10^{4}$ m
13. 1.1×10^{7} m³
14. 1.12×10^{-4} m
15. (a) 150
    (b) 1×10^{-4} m

Practice Exercise P7.1 (page 54)

1. (a) 7×10^{3} (b) 4.51×10^{9}
   (c) 3×10^{4} (d) 7.5×10^{8}
   (e) 5.07×10^{5}
2. (a) 5.243×10^{1}, 2.1045×10^{1}, 2.435×10^{0}, 4.325×10^{2}, $2.543\,271 \times 10^{6}$
   (b) 3.14×10^{-2}, 4.13×10^{-3}, 1.34×10^{-1} 1.7504×10^{1}, 5.043×10^{-6}
   (c) 3.6×10^{1}, 3×10^{-1}, 3×10^{-3}, $4.350\,564\,3 \times 10^{7}$, 6.3×10^{-4}, 1.05×10^{-7}, 7.32×10^{3}
3. (a) 370 000 (b) 13 000 000
   (c) 9 000 000 000 000 (d) 207 000
   (e) 906 000 000
4. (a) 524, 141 500, 2430, 43 250 000
   (b) 0.0314, 0.000 041 3, 0.003 124 0.000 001 054
   (c) 400, 0.000 000 26, 0.001 07, 555 000 000
5. (a) 5.6×10^{9} (b) 8.46×10^{7}
   (c) 1.97×10^{7} (d) 7.4×10^{4}
   (e) 5.002×10^{11} (f) 2×10^{9}
6. (a) 1.87×10^{5} (b) 7.535×10^{6}
   (c) 2.312×10^{16} (d) 3×10^{2}

Practice Exercise P7.2 (page 55)

1. (a) 9.2×10^{31} (b) 5.4×10^{68} kg
2. (a) 1.72×10^{8} litres (b) 9.5×10^{7} litres

Practice Exercise P7.3 (page 55)

answers will vary

Exercise 8a (page 56)

1 (a) $\notin$ (b) $\in$ (c) $\notin$ (d) $\notin$

2 (a) 5 (b) 10 (c) 3 (d) 12
(e) 2 (f) 12

3 {even primes greater than 2},
{positive numbers less than 0},
{lines of symmetry in a scalene triangle}
There are many more.

4 {prime numbers}, {negative numbers},
{lines of symmetry in a circle}
There are many more.

5 (a) $\varnothing$, {3}, {4}, {5}, {3, 4}, {3, 5}, {4, 5}, {3, 4, 5}
(b) $\varnothing$, {x}, {y}, {x, y}
(c) $\varnothing$, {0}, {2}, {0, 2}
(d) $\varnothing$, {f}, {o}, {u}, {r}, {f, o}, {f, u},
{f, r}, {o, u}, {o, r}, {u, r}, {f, o, u},
{f, o, r}, {o, u, r}, {f, u, r}, {f, o, u, r}

6 (a) {2, 6} $\subset$ {factors of 18}
(b) {trees} $\not\subset$ {metal objects}
(c) {vehicles} $\supset$ {buses}
(d) {men} $\subset$ {humans}

7
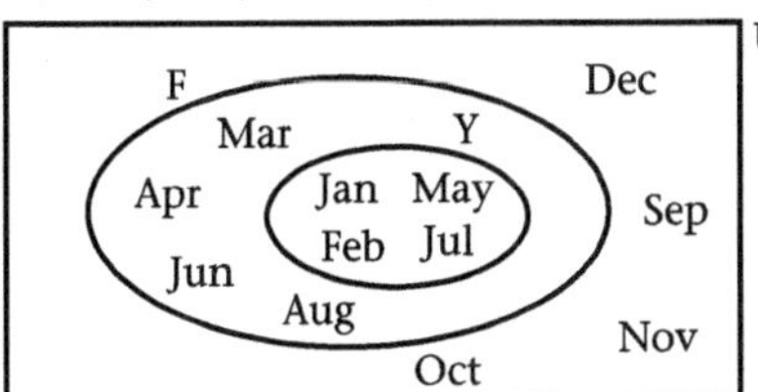

Exercise 8b (page 57)

1 $A \cup B = \{2, 3, 4, 5, 6, 8, 9\}$
$A \cap B = \{3, 5\}$

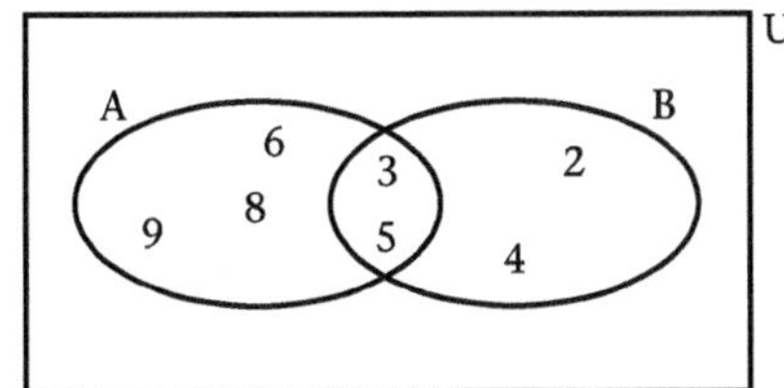

2 C ∪ D = {grapefruit, orange, pear, apple, pawpaw}
C ∩ D = {grapefruit, pear}

3 {Jan, Feb, Mar, Apr, May, Jun}, $\varnothing$

4 {St Georges}

5 (a) {2, 3, 4, 6, 7, 8, 10}
(b) {3, 7}
(c) {3, 4, 6, 7}
(d) {1, 2, 3, ..., 10}

6 (a) and (d) only

7 (f) S

8 (a) {1, 2, 3, 4, 5, 6, 7, 9} (b) {1, 3, 5}
(c) {1, 2, 3, ..., 10} (d) {1, 3, 5, 7, 9}

9 (a) {Kevin, Brian, Ron, Bob, Ned, Rex, Sam, Tom}
(b) {Kevin, Brian, Ron, Rex, Richard}
(c) {Bob, Ned, Sam, Tom, Rex, Ron, Richard}
(d) {Bob, Frank, Kevin, Ned, Andy, Rex, Ron, Sam, Richard, Brian, Tom}, i.e. U
(e) as for (d)
(f) as for (d)
(g) {Ron} (h) {Rex, Ron} (i) {Ron}
(j) {Kevin, Ron, Brian}, i.e. A
(k) {Bob, Ned, Rex, Ron, Sam, Tom}, i.e. B
(l) {Rex, Ron, Richard}, i.e. R

10 (a) {a, b, c, d, e}
(b) {a, b, c, d, f}
(c) {b, c, d, e, f}
(d) {c, d}
(e) {b, d} (f) {d} (g) X
(h) Y (i) Z (j) U
(k) U (l) {d} (m) {d}
(n) {b, c, d} (o) {b, c, d, f} (p) {a, b, c, d}
(q) {b, d} (r) {b, c, d} (s) {a, b, c, d}

Exercise 8c (page 59)

1 (a) {4, 6, 8} (b) {1, 2, 3}
(c) {1, 3, 6, 8} (d) {2, 4}
(e) {1, 2, 3, 4, 8} (f) {8}
(g) U (h) $\varnothing$
(i) {4} (j) {3, 4, 6, 8}

2 (a) {b, d} (b) {a, c, d}
(c) {b, d, e} (d) {a, c, d, e}
(e) {b} (f) {a, c}
(g) U (h) $\varnothing$
(i) {d} (j) {a, b, c, d}
(k) {a, b, c, d} (l) {d}

3 $(A \cup B)' = A' \cap B'$

4 $A' \cup B' = (A \cap B)'$

5 $A' = \{3, 4, 6, 8, 9, 10\}$,
$B' = \{2, 4, 5, 8, 9, 10\}$
$(A \cap B)' = \{2, 3, 4, 5, 6, 8, 9, 10\}$
$(A \cup B)' = \{4, 8, 9, 10\}$

6 $A' \cup B' = \{2, 3, 4, 5, 6, 8, 9, 10\} = (A \cap B)'$
$A' \cap B' = \{4, 8, 9, 10\} = (A \cup B)'$

7 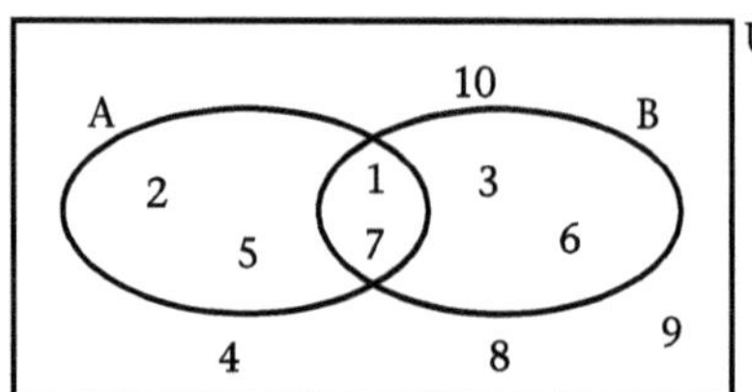

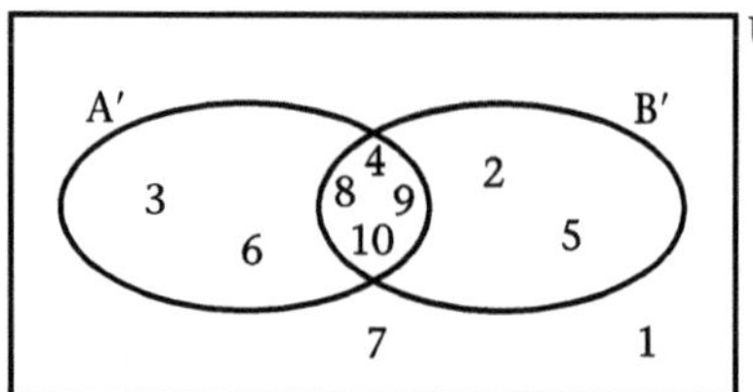

8 (a) S = {Sunday, Tuesday, Wednesday, Thursday, Saturday}
N = {Sunday, Monday, Friday}
S′ = {Monday, Friday}
N′ = {Tuesday, Wednesday, Thursday, Saturday}
(b) $(S \cup N)' = \varnothing$
$(S \cap N)'$ = {Mon, Tues, Wed, Thurs, Fri, Sat}
(c) $S' \cap N' = \varnothing$
$S' \cup N'$ = {Mon, Tues, Wed, Thurs, Fri, Sat}

9 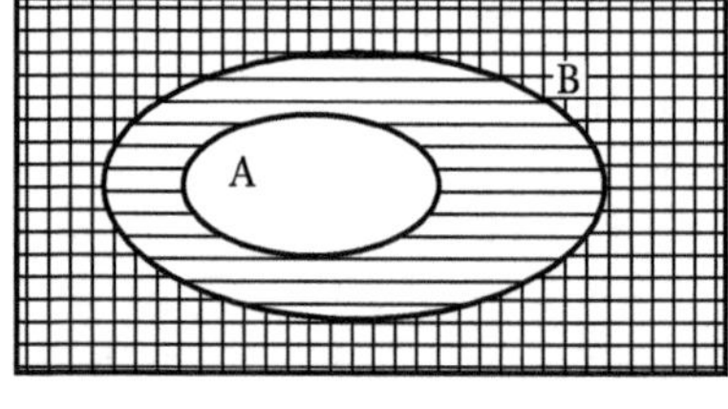

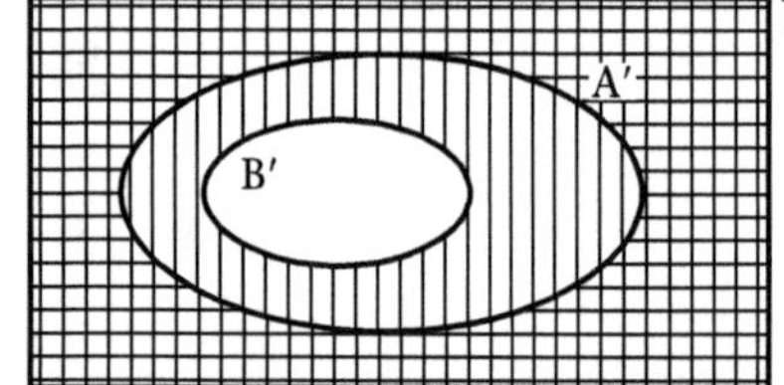

10 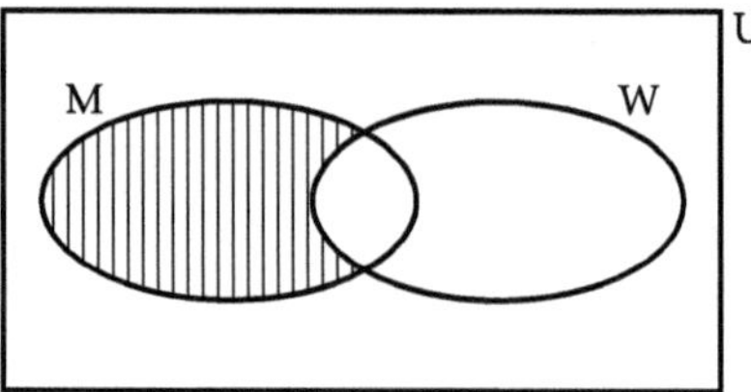

Exercise 8d (page 60)

1 (b) {1, 2, 3, 6} (c) {2, 3, 7}
(d) {9, 10, 11, ...} (e) {−6, −7, −8, ...}
(f) {4} (g) {4, 5}
(h) $\varnothing$

2 (a) {8, 9, 10, ...} (b) {1, 2, 3, ...}
(c) $\varnothing$ (d) {1, 2, 3, ..., 9}
(e) {3} (f) {6}
(g) $\varnothing$ (h) $\varnothing$

3 (a) −1, 4 (b) −8, 9
(c) −4, 0 (d) 2, 9
(e) 5, 5 (f) −7, −2
(g) −7, 0 (h) −5, 1

4 (a) $\{x: -1 < x < 4, x \in Z\}$
(b) $\{x: -2 < x < 2, x \in Z\}$
(c) $\{x: -4 < x < 1, x \in Z\}$
(d) $\{x: -10 < x < 5, x \in Z\}$
(e) $\{x: 8 < x < 11, x \in Z\}$
(f) $\{x: -9 < x < 0, x \in Z\}$

5 (a) {1, 3, 5, 7, 9}
(b) {5, 6, 7, 8, 9, 10}
(c) {7}
(d) {1, 2, 3, 4, 6, 8}

Exercise 8e (page 61)

1 (a) 9 (b) 22
2 $x = 23$, n(U) = 46
3 25
4 61%
5 4
6 9
7 90
8 60%
9 5
10 2

Practice Exercise P8.1 (page 62)

1 (a) (i) {1, 3, 5, 7, 9} (ii) {2, 4, 6, 8}
(iii) {2, 4, 6, 8} (iv) {1, 3, 5, 7, 9}
(b) (i) {odd numbers less than or equal to 9}
(ii) {even numbers less than 10}

2 (a) A = $\{\frac{5}{2}\}$
(b) B = {1, 2, 3, 4, 6, 8, 12, 24}
(c) C = {21, 23, 25, 27, ...}
(d) D = { }
(e) E = {11, 13, 17, 19}

3 (a) (i) A′ = {5, 6, 7, 8}
(ii) $A \cap A'$ = { }
(b) the intersection of a set and its complement is the empty set

4 *answers will vary*

Practice Exercise P8.2 (page 62)

(a)

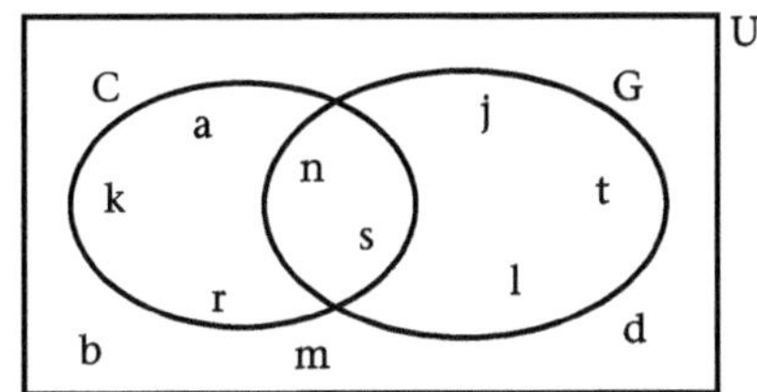

(b) (i) C ∪ G = {a, j, k, n, t, s, l, r}
(ii) C ∩ G = {n, s}

Practice Exercise P8.3 (page 63)

1 (i) A ={l, a, d, e, r} (ii) B = {d, e, a, l, r}
(iii) C = {r, e, t, a} (iv) D = {d, r, e, s}
(v) E = {s, a, d, e, r}

2 (i) n(A) = 5 (ii) n(B) = 5
(iii) n(C) = 4 (iv) n(D) = 4
(v) n(E) = 5

3 (a) Q (b) N (c) N
(d) N (e) V (f) V

Practice Exercise P8.4 (page 63)

1 (a)

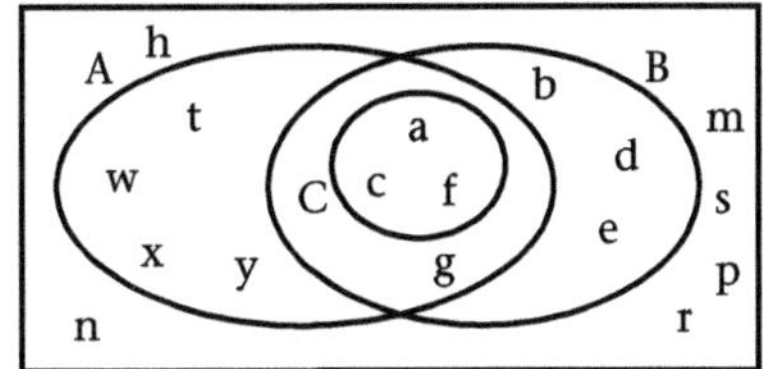

(b) (i) {a, c, f, g}
(ii) {a, c, f}
(iii) {a, c, f}
(iv) {a, b, c, d, e, f, g, t, w, x, y}
(v) (a, c, f, g, t, w, x, y)
(vi) {a, b, c, d, e, f, g}

2 (i) (a) n(P ∩ M) = 3 (b) n(P ∪ M) = 8
(ii) (a) n(X ∩ Y) = 2 (b) n(X ∪ Y) = 5
(iii) (a) n(T ∩ F) = 3 (b) n(T ∪ F) = 6
(iv) (a) n(S ∩ V) = 0 (b) n(S ∪ V) = 7

Practice Exercise P8.5 (page 63)

1

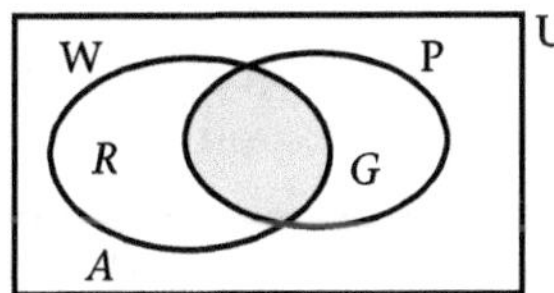

2 (a)

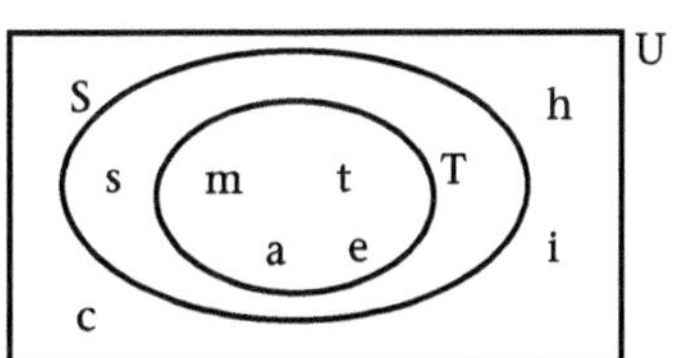

(b) (i) U = {m, a, t , h, e, i, c, s}
(ii) S = {m, a, t , e, s}
(iii) T = {m, a, t , e}
(iv) S ∪ T = {m, a, t , e, s}

Revision exercise 1 (page 64)

1 (a) 41, 48, 55, 62, ...
(b) 26, 33, 41, 50, ...
(c) 48, 96, 192, 384, ...
(d) 30, 42, 56, 72 ...

2 14

3 (a) 3.5×10^6 (b) 5.7×10^3
(c) 2.8×10^1 (d) 4.7×10^{-1}
(e) 8.5×10^{-2} (f) 3×10^{-6}

4 (a) 10^7 (b) $15x^5$ (c) $18a$
(d) 10^4 (e) $2y^2$ (f) $3x^4$

5 $A = 1.11$; $n = -7$

6 (a) 6.76×10^3 (b) 6.2×10^{-3}
(c) 3.2×10^2 (d) 6×10^4

7 6.8×10^7

8 2.84

9 (a) (64 + 59) + 73
(b) $x + 97$
(c) 70 × (80 + 6)

10 2.94

Revision test 1 (page 64)

1 C 2 D 3 B 4 A 5 D

6 (a) $(n + 1)^2 - (n + 1)$
(b) 1 001 000

7 108 × (50 + 3) = 5400 + 324 = 5724

8 (a) 6.6×10^{-3} (b) 6.74×10^4
(c) 1.56×10^5 (d) 1.3×10^{-2}

9 (a) 900 (b) 360 000
(c) 61 000 000 (d) 0.0008
(e) 0.6 (f) 0.0034

10 (a) 661 (b) 1.1×10^{-4}

Revision exercise 2 (page 65)

2. 61 mm
3. 57°
4. 3.5 cm; 7.9 cm^2
5. 720 cm^3, 90
6. 10^8 cm^3
7. 12 000 cm^3
8. 17.75 cm^2
9. 496 cm^3
10. 150 litres

Revision test 2 (page 65)

1. A
2. D
3. D
4. C
5. C
6. 1 m^3
7. 50 cm
8. 6.5 cm
9. 4.6 cm
10. 18.8 cm^2

Revision exercise 3 (page 66)

1. (a) -13 (b) 17 (c) -5 (d) 3
2. (a) -3 (b) 2 (c) -3 (d) 19 (e) -4 (f) -6
3. (c), (a), (d), (b)
4. (a) -48 (b) 6 (c) $-4\frac{2}{3}$ (d) -19 (e) 0. 4 (f) $-\frac{5}{9}$
5. 3 BC
6. (a) $2y - x$ (b) $14x - 30$ (c) $2x + y$ (d) $15y - 6x$
7. (a) $\frac{4x}{m}$ (b) $\frac{10x}{9}$ (c) $\frac{23}{3a}$ (d) $\frac{q-p}{pq}$
8. (a) $ax + ay + bx + by$
   (b) $6pr - 10ps + 3qr - 5qs$
   (c) $2c^2 - 11c + 15$
   (d) $8a^2 - 26ab - 45b^2$
   (e) $b^2 - 10b + 25$
   (f) $4x^2 + 4x + 1$
9. (a) $5(2 + 3b)$ (b) $x(x - a)$ (c) $2a(2b - a)$ (d) $9xy(3x - 4y)$
10. (a) $-1\frac{5}{6}$ (b) -8.2 °C

Revision test 3 (page 66)

1. C
2. D
3. B
4. A
5. D
6. (a) $v^2 - 13v + 36$
   (b) $3b^2 + 14b + 8$
   (c) $10x^2 - 11x - 6$
   (d) $12m^2 - 11mn + 2n^2$
7. (a) $9(2 + c)$
   (b) $\pi r(2h + r)$
   (c) $7x^2y(4y - 3x)$
   (d) $\pi r^2(2r - \frac{1}{3}h)$
8. (a) $\frac{29}{10a}$ (b) $\frac{v+u}{uv}$ (c) $\frac{3a}{10}$ (d) $\frac{19-3y}{28}$
9. (a) $1\frac{1}{4}x$, $\frac{9x}{10}$ (b) $1\frac{1}{4}x - \frac{9x}{10}$ (c) $\frac{7x}{20}$
10. (a) -30 (b) -6.3 (c) 100 (d) -9

Revision exercise 4 (page 67)

1. (a) and (d) are disjoint
   (b) {35, 70, 105, ...}
   (c) {Tobago, Antigua}
2. (a) (i) M ∪ L = {2, 4, 6, 8, 10, 12, 18}
   (ii) M ∩ L = {6, 12}
   (iii) M ∪ U = {2, 4, 6, ..., 20} = U
   (iv) U ∩ M = {6, 12, 18} = M}
   (v) L′ = {14, 16, 18, 20}
   (vi) (M ∪ L)′ = {14, 16, 20}
   (c) (i) 7, (ii) 2, (iii) 4
3.

(a)
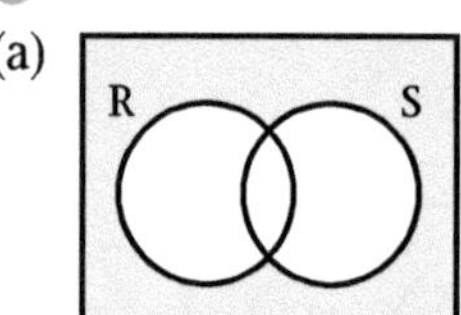

(b)
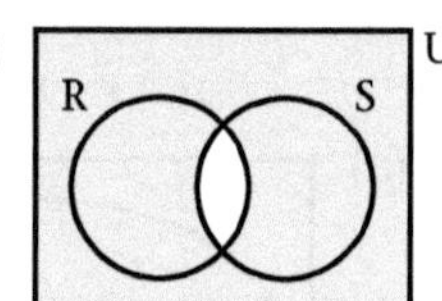

(c)
U
R
S

(d)
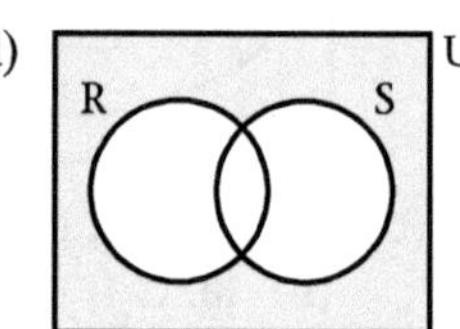

4. (a)

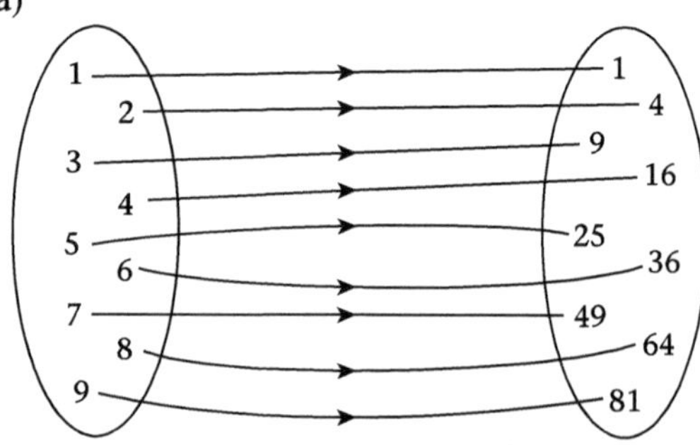

relation: 'is the square root of'
a one-to-one mapping

(b) (1, 1), (2, 4), (3, 9), (4, 16), (5, 25), (6, 36), (7, 49), (8, 64), (9, 81).

6 (a) $\{-4, -3, -2, -1\}$
(b) $\{3\frac{1}{2}\}$
(c) $\varnothing$
(d) $\{-1, 0, 1, 2 \ldots\}$

7 16

9 (a)

No. of weeks	1	2	3	4	5	6
No. of stamps	7	14	21	28	35	42

No. of weeks	7	8	9	10	11	12
No. of stamps	49	56	63	70	77	84

(b) (1, 7), (2, 14), (3, 21), (4, 28), (5, 35), (6, 42), (7, 49), (8, 56), (9, 63), (10, 70), (11, 77), (12, 84)

Revision test 4 (page 68)

1 C 2 C 3 D 4 D 5 B

6

Distance on map (cm)	1	2	3	4	5	6	7	8	9	10
↓	↓	↓	↓	↓	↓	↓	↓	↓	↓	↓
Distance on ground (km)	3	6	9	12	15	18	21	24	27	30

7

No. of pages	1	2	3	4	5	6	7	8	9
↓	↓	↓	↓	↓	↓	↓	↓	↓	↓
No. of words	250	500	750	1000	1250	1500	1750	2000	2250

9

No. of hours	1	2	3	4	5	6	7	8
↓	↓	↓	↓	↓	↓	↓	↓	↓
Overtime pay ($)	25	50	75	100	125	150	175	200

No. of hours	9	10	11	12	13	14	15
↓	↓	↓	↓	↓	↓	↓	↓
Overtime pay ($)	225	250	275	300	325	350	375

10

Excess of speed limit (km)	1	2	3	4	5	6
Fine ($)	25	50	75	100	125	150

Excess of speed limit (km)	7	8	9	10	11	12
Fine ($)	175	200	225	250	275	300

General revision test A (page 69)

1 D 2 C 3 B 4 C 5 B
6 C 7 A 8 C 9 B 10 C

11 (a) 45 (b) 50 12 $7\,\text{cm}^2$

13 approximately 25 times (24.7)

14 (a) (i) $x^2 - 11x + 30$
(ii) $4p^2 - 16pq + 15q^2$
(iii) $m^2 - 16$
(iv) $t^2 - 16t + 64$
(b) (i) $2x(4y + x)$
(ii) $mn(n + m)$

15 (a) $\frac{11a + 7}{6}$ (b) $\frac{2b + 3}{12}$ (c) $\frac{3a + 2}{6}$

16 (b) 5 hours (c) 144 km (d) 240 km

17 (a)

No. of days	1	2	3	4	5	6	7
↓	↓	↓	↓	↓	↓	↓	↓
Litres of water used	250	500	750	1000	1250	1500	1750

(b) 55 litres

18

Radius of circle (cm)	$3\frac{1}{2}$	7	$10\frac{1}{2}$	14	$17\frac{1}{2}$	21
↓	↓	↓	↓	↓	↓	↓
Circumference (cm)	22	44	66	88	110	132

19 (a) 1.35×10^5 (b) 4.2×10^{-2}
(c) 1.53×10^{-4} (d) 6.25×10^6

20 11 boys

Exercise 9a (page 71)

1 R(3), S(−1), T($2\frac{1}{2}$), U($-1\frac{1}{2}$), V(0)

2 B(0.2), C(1.5), D(−0.5), E(−0.9), F(1.8), G(−0.1)

3

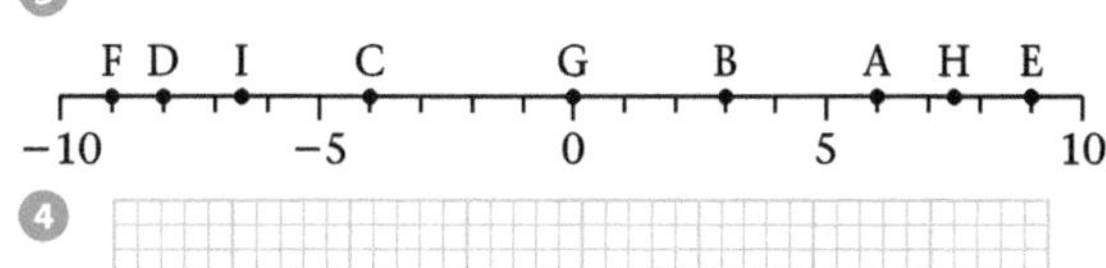

4

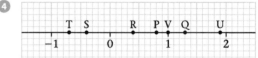

T S R P V Q U
−1 0 1 2

Exercise 9c (page 74)

1 A(1, 1), B(1, 3), C(3, 3), D(2, −1), E(3, −2), F(2,−3), G(−1,−3), H(0,−1), I(−2,0), J(−3, 2)

2 (10, 7), (14, −4), (16, −2), (17, −2), (15, −5), (13, −5), (11, 0), (9, −2), (9, −7), (6, −7), (6, −4), (−2, −4), (−4, −7), (−7, −7), (−6, 5), (−2, 8), (3, 8), (3, 9), (6, 9)

3 (a) C (b) F (c) I
(d) K (e) D (f) A
(g) G (h) J (i) E
(j) L (k) H (l) B

4 T(−2, 1), U(−3, 1), V(0, 4), W(3, 1), X(2, 1), Y(2, −1), Z(−2, −1)

5 (a) (4, 7) (b) (5, 5) (c) (5, 1)
(d) (6, 1) (e) (3, 2) (f) (1, 3)
(g) (2, 2) (h) (2, 7) (i) (4, 4)
(j) (5, 6)
for every point, the x-coordinate = 2

6 (a) (0, 2), (1, 2), (2, 2), (3, 2), (4, 2), (5, 2), (6, 2); for every point, the y-coordinate = 2
(b) (0, 1), (1, 2), (2, 3), (3, 4), (4, 5), (5, 6), (6, 7); for every point, the y-coordinate is 1 more than the x-coordinate

Exercise 9d (page 77)

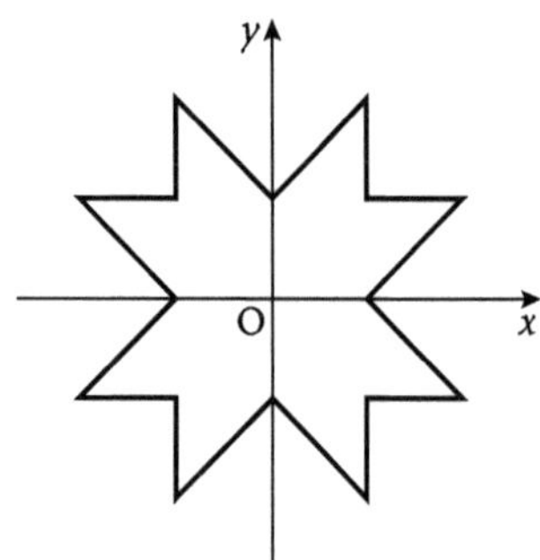

2 the points join to form a star shape (figure above)

3 a horse's head

4 (b) 10 unit^2, 9 unit^2, 6 unit^2, 9 unit^2

5 (a) parallelogram, X(−1, 0)
(b) B, X, D, E lie in a straight line
(c) square, Y($2\frac{1}{2}$, $\frac{1}{2}$)

Exercise 9e (page 78)

1 (a)

1	2	3	4
3	6	9	12

rule: $x \to 3x$

(b)

1	2	3	4	5	6
3	6	9	12	15	18

rule: $x \to 3x$

(c)

1	2	3	4	5	6
5	10	15	20	25	30

rule: $x \to 5x$

2 (a)

no. of hours	1	2	3	4	5	6	7	8
wages ($)	50	100	150	200	250	300	350	400

(b)

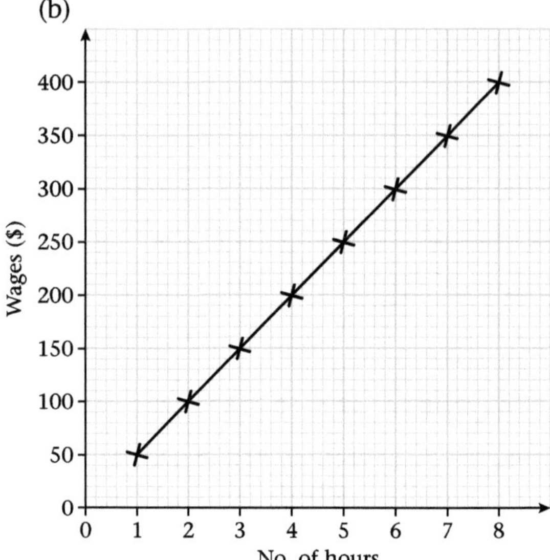

3

no. of sides	4	5	6	7	8	9	10
no. of diagonals	2	5	9	14	20	27	35

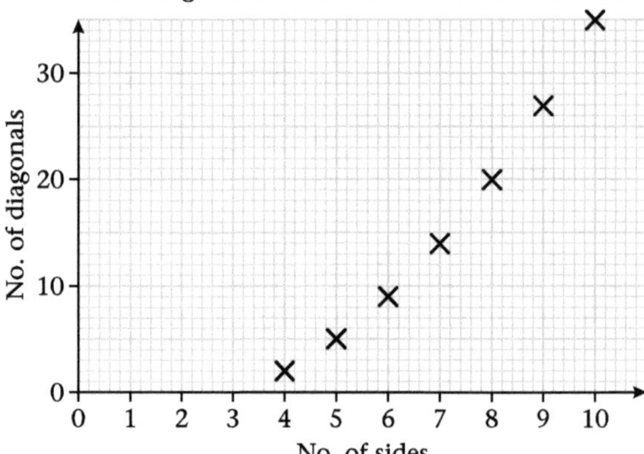

4

no. of points	2	3	4	5	6	7	8
no. of lines	1	3	6	10	15	21	28

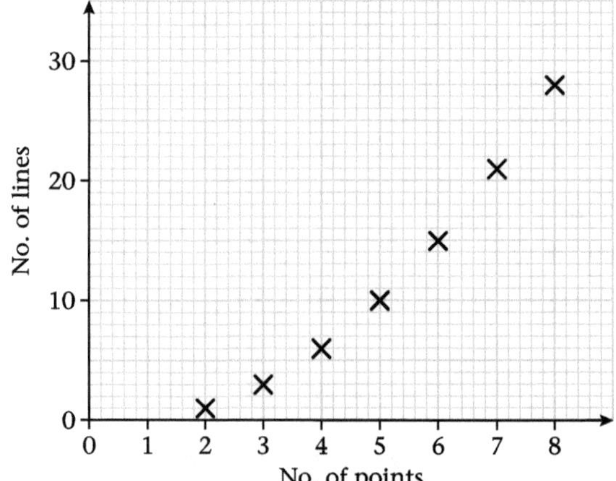

5

x	1	2	3	4	5	6
$3x - 1$	2	5	8	11	14	17

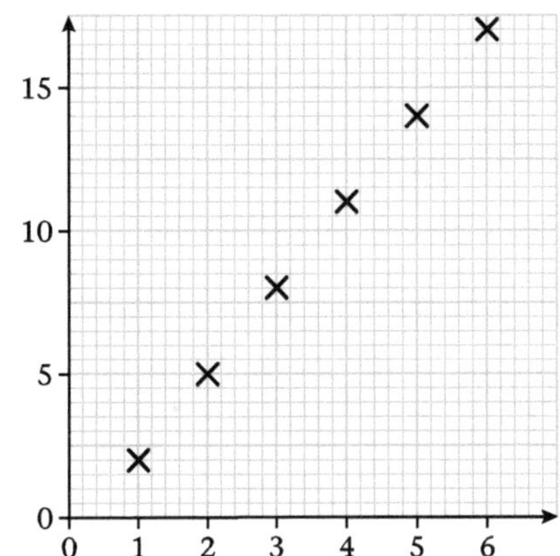

6 (a) (6, 36), (7, 49), (8, 64), (9, 81), (10, 100)
(b) (c) (d) see figure below

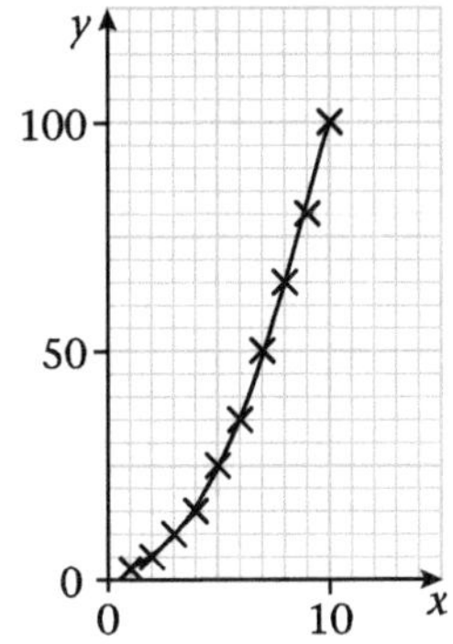

(e) approximate values: 70.5, 42, 4.5, 9.5

Practice Exercise P9.1 (page 80)

1 (a)

x	−2	0	1	3
A	—	—	—	—
B	—	—	—	—
C	−3	−3	−3	−3
D	2.5	2.5	2.5	2.5
E	1	3	4	6
F	2	0	−1	−3
G	−6	0	3	9
H	−5	−3	−2	0

(b) A $x = -3$ for all values of y
B $x = 4.5$ for all values of y
C (−2, −3), (0, −3), (1, −3), (3, −3)
D (−2, 2.5), (0, 2.5), (1, 2.5), (3, 2.5)
E (−2, 1), (0, 3), (1, 4), (3, 6)
F (−2, 2), (0, 0), (1, −1), (3, −3)
G (−2, −6), (0, 0), (1, 3), (3, 9)
H (−2, −5), (0, −3), (1, −2), (3, 0)

(c)

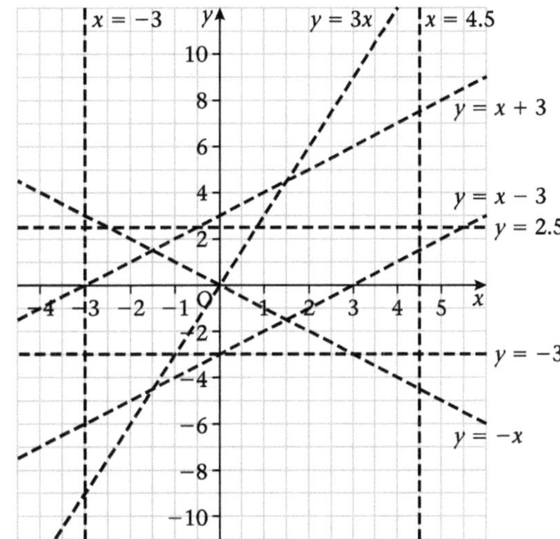

2 (a)

x	−2	0	1	2
1	1	1	1	1
$x + 1$	−1	1	2	3

(c)

x	−2	0	1	2
$2x$	−4	0	2	4
3	3	3	3	3
$2x + 3$	−1	3	5	7

(b) – (c)

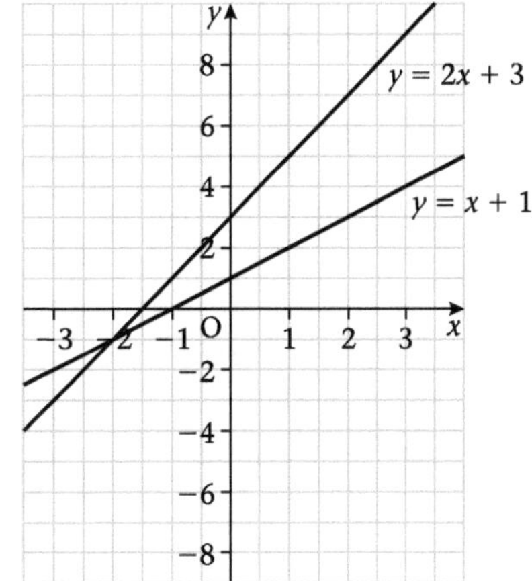

(d) (−2, −1)

❸ (a)–(b)

Pattern #	1	2	3	4	5	6
	.	. .	. . .	. . . .	. . . . .	. . . . . .
	.	. .	. . .	. . . .	. . . . .	. . . . . .
	.	. .	. . .	. . . .	. . . . .	. . . . . .
No. of dots	3	6	9	12	15	18

(c) (1, 3), (2, 6), (3, 9), (4, 12), (5, 15), (6, 18)
(d) multiply by 3
(e)

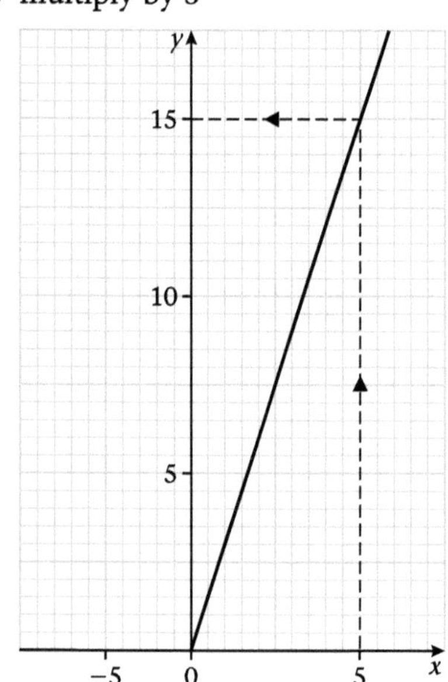

(f) 30

Exercise 10a (page 81)

❶ 7 ❷ 46 ❸ 1020 ❹ 80

Exercise 10b (page 83)

❶ (a) $2x$ (b) $6n$ (c) $6m + 4$
(d) $2y - 5$ (e) $a - 3$ (f) $5d$
(g) $3h$ cents (h) $3t - 7$
(i) $(2k - 9)$ cents
(j) $(2g + 23)$ goals

❷ (a) $4x$ cm (b) $(3a + 4)$ metres
(c) $6c$ cm (d) $8b$ metres
(e) $(2h + 20)$ metres (f) $5t$ cm

Exercise 10c (page 83)

❶ 29 ❷ 13 ❸ 5
❹ 40 ❺ 17 ❻ 6
❼ 7 cents ❽ 5 ❾ 19c, 10c
❿ 56, 79 goals ⓫ 8 m ⓬ 3 m, 6 m
⓭ 15 cm ⓮ 15 m, 5 m ⓯ 3 m
⓰ 3 cm

Exercise 10d (page 84)

❶ $a = 2$ ❷ $a = -3$ ❸ $b = -6$
❹ $x = 1\frac{2}{3}$ ❺ $a = -1$ ❻ $y = \frac{3}{4}$
❼ $n = -\frac{3}{4}$ ❽ $m = \frac{2}{3}$ ❾ $a = 5$
❿ $t = 4$ ⓫ $n = 2$ ⓬ $c = 1$
⓭ $q = -1$ ⓮ $x = 3$ ⓯ $m = -4$
⓰ $x = -\frac{1}{2}$ ⓱ $h = 3$ ⓲ $a = 4\frac{2}{3}$
⓳ $f = 2$ ⓴ $e = 6$ ㉑ $x = 3\frac{1}{8}$
㉒ $x = 1$ ㉓ $x = -12$ ㉔ $n = \frac{2}{3}$

Exercise 10e (page 85)

❶ $x = 4$ ❷ $x = 8$ ❸ $a = 6$
❹ $y = 2$ ❺ $x = -10$ ❻ $x = 3$
❼ $s = 6$ ❽ $b = -9$ ❾ $f = -1$
❿ $x = -3$ ⓫ $a = -14$ ⓬ $b = 2$
⓭ $e = 3$ ⓮ $d = \frac{3}{4}$ ⓯ $x = \frac{3}{4}$
⓰ $x = -2$ ⓱ $y = 5$ ⓲ $y = -6$
⓳ $x = 24$ ⓴ $x = -5$ ㉑ $z = 3$
㉒ $y = 2$ ㉓ $v = -3$ ㉔ $n = 2\frac{1}{2}$

Exercise 10f (page 85)

❶ 9 ❷ 15 ❸ 6
❹ 7, 8 ❺ 13, 15 ❻ $16\frac{1}{2}$ m
❼ 72 kg ❽ 8 ❾ 8
❿ $1.80 ⓫ 150 ⓬ 47 cents

Exercise 10g (page 87)

❶ $x = 15$ ❷ $x = 2\frac{1}{2}$ ❸ $a = 36$
❹ $a = 6$ ❺ $z = 10$ ❻ $x = 2$
❼ $x = 14$ ❽ $a = 19$ ❾ $a = -3$
❿ $y = 19$ ⓫ $n = 5$ ⓬ $a = 3$
⓭ $x = 2$ ⓮ $x = -3$ ⓯ $x = 8$
⓰ $x = 2$ ⓱ $z = \frac{1}{2}$ ⓲ $x = -14$
⓳ $x = 12$ ⓴ $x = 4$ ㉑ $m = 6$
㉒ $x = 1\frac{2}{5}$ ㉓ $x = 7$ ㉔ $t = 3$
㉕ $e = 2$ ㉖ $d = \frac{3}{4}$

Exercise 10h (page 87)

❶ $x = 5$ ❷ $r = 9$ ❸ $m = 4$
❹ $y = 3\frac{1}{2}$ ❺ $s = \frac{2}{5}$ ❻ $n = -\frac{4}{11}$
❼ $t = 3\frac{2}{3}$ ❽ $z = 3\frac{1}{3}$ ❾ $p = 4\frac{1}{2}$
❿ $a = 1\frac{1}{2}$ ⓫ $x = 2\frac{1}{2}$ ⓬ $q = -2\frac{1}{3}$

13 $b = 6$ 14 $y = 5$ 15 $r = 8$
16 $c = \frac{2}{9}$ 17 $d = 15$ 18 $s = -\frac{3}{4}$
19 $z = \frac{2}{15}$ 20 $r = 5\frac{1}{4}$ 21 $t = \frac{2}{3}$
22 $x = 3$ 23 $y = 7\frac{1}{2}$ 24 $d = 6$
25 $f = 3$ 26 $x = 1\frac{1}{2}$

Exercise 10i (page 89)

1 15 2 32 3 7
4 (a) $\frac{15}{n}$ kg (b) 20 fish
5 48 km/h
6 (a) $\frac{1080}{x}$ (b) $\frac{1080}{x}$ $\left(\text{or } \frac{270}{x}\right)$
(c) 54 cents
7 \$240
8 36c, 45c
9 (a) $\frac{1}{2}x$ (b) $\frac{1}{3}x$ (c) \$480
10 (a) $\frac{4d}{5}$ (b) $\frac{3d}{4}$ (c) $d = 30$
11 (a) $\frac{8}{v}$ (b) $\frac{15}{2v}$ (c) 6 km/h
12 (a) $(y - 3)$ years (b) $(y + 4)$ years
(c) 17 years
13 45 km/h
14 (a) $\frac{14.5}{n}$ kg (b) $\frac{21}{2n}$ kg (c) 40 oranges
15 28 fish

Practice Exercise P10.1 (page 90)

1 −2 2 1 3 −5 4 8
5 2 6 −1 7 0 8 −7
9 −3 10 17 11 −40 12 8
13 4 14 2 15 −3 16 3
17 −3 18 $-\frac{1}{2}$

Practice Exercise P10.2 (page 90)

1 $\frac{3}{2}$ 2 −2 3 $\frac{1}{2}$
4 −2 5 3 6 1

Practice Exercise P10.3 (page 91)

1 $n + 4 = 12$ 2 $n - 5 = 13$
3 $3n = 15$ 4 $\frac{n}{7} = 3$
5 $n + 2 = 13$ 6 $n + 9 = 35$
7 $n - 15 = 20$ 8 $n - 14 = 23$
9 $5n = 45$ 10 $10n = 120$
11 $\frac{n}{4} = 12$ 12 $2n + 3 = 9$

Practice Exercise P10.4 (page 91)

1 (a) $4b + 72$
(b) $4b + 72 = 292$
(c) $b = 55$
(d) 55 cents
2 (a) (i) $2C$ (ii) $4C + 150$
(b) $4C + 150 = 3C + 2C$
(c) $C = 150$
(d) (i) 150 ml (ii) 300 ml (iii) 750 ml

Exercise 11a (page 93)

1 (a) 1 : 3 (b) 1 : 2 (c) 3 : 5
2 (a) 9 cm (b) 4 cm (c) 4 cm
(d) 7.3 cm (e) 13 cm (f) 15 cm
(g) 4.5 cm (h) 7.5 cm (i) 15.3 cm
(j) 14.3 cm
3 (a) 60 m (b) 55 m (c) 10 m
(d) 75 m (e) 820 m (f) 18.6 m
(g) 430 km (h) 7.4 km (i) 2.26 m
4 (a) (i) building: 1 cm to 1 m
(ii) window: 1 cm to 20 cm
(iii) roundabout: 1 cm to 10 m
(iv) running track: 1 cm to 20 m
(b) $w = 4$ m, $h = 120$ cm, $d = 36$ m, $t = 10$ m

Exercise 11b (page 96)

1 15 m 2 68 m 5 210 m
6 AC = 12.2 m, XK = 10.8 m, PH = 8.5 m, QY = 7.8 m
7 64 m 8 $121\frac{1}{2}°$ 9 8.9 m
10 PC $\simeq$ 37 m

Exercise 11c (page 97)

1 (a) 3 (b) brown
(c) yellow and green
(d) blue (e) live (L)
(f) 2 (g) 5.1 cm
2 (a) 6 (b) living room
(c) kitchen (d) 3
(e) living room (f) 11
(g) bedroom 3 (h) living room, kitchen
(i) cupboards (j) 8 m, 4 m
(k) $5\frac{1}{2}$ m, 3 m (l) 123 m^2

Exercise 11d (page 98)

1 (a) 9 km (b) 13 km
(c) 13 km (d) 9 km
(e) 18 km (f) 20 km

2 (a) Moortown Road, Portland Road
(b) Queen's Highway, Airport Road
(c) Church Road, Azania Crescent, River Road
(d) Church Road, Azania Crescent
(e) Portland Road, Azania Crescent
(f) Azania Crescent, Portland Road,
(g) Moortown Road, Palm Avenue, Queen's Highway, Airport Road
(h) Azania Crescent then as (g)
(i) Portland Road then as (g)
(j) Moortown Road, Azania Crescent

3 (a) 1500 m (b) 2400 m (c) 2750 m
(d) 750 m (e) 1000 m (f) 1750 m
(g) 2600 m (h) 100 m (i) 150 m
(j) 2100 m

4 (a) (i) 1900 m (ii) 2750 m
(b) (i) 3250 m (ii) 7500 m
(c) (i) 1400 m (ii) 2000 m
(d) (i) 1650 m (ii) 2750 m
(e) (i) 3500 m (ii) 4250 m

Practice Exercise P11.1 (page 100)

1 (a) 6.8 km (b) 21 km (c) 24.8 km
(d) 25.8 km (e) 31.6 km (f) 50.6 km

2 (a) 9.6 cm (b) 14.6 cm (c) 18.24 cm
(d) 23.75 cm (e) 45.04 cm (f) 60 cm

3 (b) 9.7 cm (c) 9.7 m

4

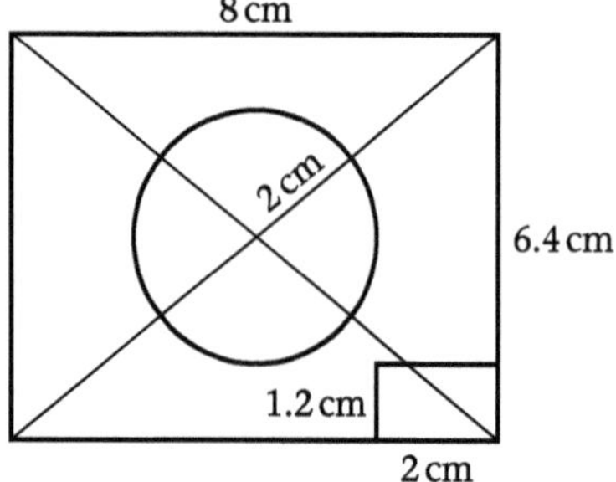

5

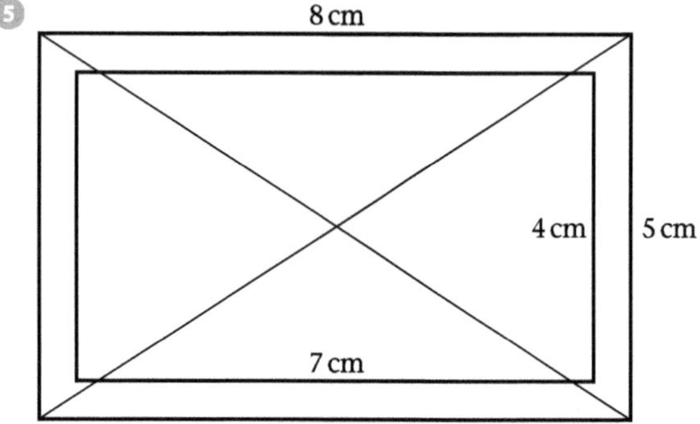

Exercise 12b (page 102)

1 (a) 59°
(b) 122°, 69°
(c) 100°, 90°, 70°
(d) 73°, 97°, 83°, 107°

2 (a) 120° (b) 63° (c) 90° (d) 71° (e) 37°

3 (c) is a trapezium
(d) is a parallelogram

4 (a) $x + 2x + 3x + 4x = 360°$
(b) $x = 36$
(c) 36°, 72°, 108°, 144°
(e) a trapezium

5 (a) $x = 100$; 100°, 100°
(b) $x = 50$; 50°, 100°
(c) $x = 30$; 60°, 90°
(d) $x = 18$; 54°, 162°

Exercise 12c (page 102)

1 3, 4, 5, 6, 7, 8 respectively

2 3, 4, 5, 6, 7, 8 respectively

3 (a) 360° (b) 5
(c) 72° (d) isosceles
(e) $O\hat{A}B = O\hat{B}A = 54°$ (f) 108°
(g) 540°

4 (a) 3 (b) 120°

5 720°

6 (a) 4 (b) 90°
(c) this follows from the sum of the angles of quadrilateral OABC
(d) 1080° (e) 135°

7 (a) pentagon (b) octagon
(c) quadrilateral (d) hexagon
(e) nonagon

8 (c), (d) if done carefully the sum of the angles will be 540°

9 the angles should add up to 720°

10

Polygon	Sum of angles
triangle	180°
quadrilateral	360°
pentagon	540°
hexagon	720°

11 (a) $a : b : c : d = 180 : 360 : 540 : 720$
(b) 1 : 2 : 3 : 4 (c) 900° (i.e. $5 \times 180°$)

Exercise 12d (page 105)

1 (b)

Polygon	Number of sides	Number of triangles	Sum of angles at O	Sum of angles of polygon
quadrilateral	4	4	360°	4 × 180° − 360°
pentagon	5	5	360°	5 × 180° − 360°
hexagon	6	6	360°	6 × 180° − 360°
heptagon	7	7	360°	7 × 180° − 360°
octagon	8	8	360°	8 × 180° − 360°
n-gon	n	n	360°	$n \times 180° - 360°$

(c) $n \times 180° - 360°$

2 (a) 120° (b) 144°

3 5 angles of 108°, 10 angles of 72°, 5 angles of 36°

4 90°, 135°, 135°

5 (a) a rhombus (b) 144°, 144°, 36°, 36°

6 (a) $x = 100$; 100°, 100°, 100°
(b) $x = 80$; 80°, 160°, 160°

7 90° each

8 $x = 60$; largest angle = 150°

9 (a) 20 sides (b) 15 sides

10 (a) 24 (b) 165°

Exercise 12e (page 106)

1 (a) 140° (b) 150° (c) 162°

2 (a) 10 (b) 30 (c) 15

3 (a) $x = 27$ (b) 54°, 81°, 108°, 135°, 162°

4 36°

5 (a) $x = 30$ (b) $x = 100$

Practice Exercise P12.1 (page 107)

1 (a) $9x + 90 = 720$ (b) $x = 70$
(c) 100°, 80°, 70°, 140°, 170°, 160°

2 144°

3 (a) 30° (b) 12

4 8 sides

5 (a) sum of angles = $(n - 2) \times 180°$, where n is the number of sides
(b) $1800 = (n - 2) \times 180$
(c) 12

6 165°

7 67°

8 (a) 1080° (b) 144°

9 (a) 45° (b) 8 sides

Exercise 13a (page 109)

1 (a) \$30, \$17.50, \$45, \$41, \$13, \$35.50
(b) 7 m, 8 m, 4.2 m, 1.8 m, 2.5 m, 9.3 m

2 (b) (i) 37.5 km/h (ii) 5.3 s

3 (a)

Time (min)	0	1	2	3	4	5	6
Distance (m)	0	100	200	300	400	500	600

(c) (i) 570 m (ii) 3.35 min

4 (a)

Length (m)	1	2	3	4	5	6
Cost (\$)	6	12	18	24	30	36

(c) (i) \$22.80 (ii) 2.3 m

5 (a)

Petrol (litres)	0	10	20	30	40	50
Distance (km)	0	70	140	210	280	350

(c) (i) approx. 155 km, (ii) approx. 33 litres

6 (a)

Petrol (litres)	0	10	20	30	40	50	60
Cost (\$)	0	16	32	48	64	80	96

(c) (i) \$35.20 (ii) 37.5 litres

7 (a)

Sugar (kg)	1	2	3	4	5	6
Cost (\$)	3.20	6.40	9.60	12.80	16.00	19.20

(c) (i) \$8.00 (ii) 3.75 kg

8 (a)

Time (h)	0	1	2	3	4	5
Cost (\$)	0	−7.5	−15	−22.5	−30	−37.5

(c) (i) 3.3 h (ii) −11 km

9 (a)

Time (min)	2	4	6	8	10
Distance (km)	3	6	9	12	15

(c) (i) 5.2 km (ii) 6.7 min

10 (a)

Time (min)	$\frac{1}{2}$	1	$1\frac{1}{2}$	2	$2\frac{1}{2}$	3
Distance (km)	9	18	27	36	45	54

(c) (i) 29 km, (ii) 2.2 h

Exercise 13b (page 111)

1 (a)

No. of tickets	0	1	2	3	4	5
Cost ($)	0	8	16	24	32	40

2 (a)

Number of pills	10	20	30	40	50	60
Cost (cents)	80	160	240	320	400	480

(d) $1.36

3 (a)

Sides	3	4	5	6	7
Angle sum	180	360	540	720	900

4 (a)

Tyres	1	2	3	4	5
Cost ($)	240	480	720	960	1200

Exercise 13c (page 113)

1 (b) (i) 1.8°C, (ii) 43 min

2 (b) (i) 182 mm, (ii) 720 g

3 (a)

Week number	0	1	2	3	4	5	6
Mass (kg)	3.4	3.7	4.0	4.3	4.6	4.9	5.2

(c) 37 days (5.3 weeks)

4 (a)

Petrol (kℓ)	1	2	3	4	5
Del. charge ($) Basic charge ($)	180 840	180 1680	180 2520	180 3360	180 4200
Total cost ($)	1020	1860	2700	3540	4380

(c) (i) $3960, (ii) approx. 3.4 kℓ

5 (a)

Area of glass (m^2)	$\frac{1}{2}$	1	$1\frac{1}{2}$	2	$2\frac{1}{2}$
Handling/cutting ($) Basic charge ($)	4 8	4 16	4 24	4 32	4 40
Total cost ($)	12	20	28	36	44

(c) (i) $1\frac{1}{8}$ m², (ii) $22.00

Exercise 13d (page 115)

1 (a) £50 (b) £75 (c) £95 (d) £12.50

2 (a) $55 (b) $90 (c) $115 (d) $52.50

3 (a) 4 min (b) 500 m
(c) (i) 1250 m, (ii) 1750 m
(d) 2100 m (e) $14\frac{1}{2}$ min

4 (a) 50% (b) 40%
(c) 90% (d) 60%
(e) 33% (f) 67%
(g) 17% (h) 83%
(i) 13% (j) 43%
(k) 53% (l) 97%
(m) 25% (n) 75%
(o) 38% (p) 82%

5 (a) 24 (b) 6 (c) 9 (d) 21
(e) 16 (f) 14 (g) 19 (h) 4

6 (a) $2\frac{1}{2}$ h (b) 3.2 h
(c) 80 km/h, 62.5 km/h
(d) $\frac{185 - 60}{3 - 1} = \frac{125}{2} = 62.5$ km/h

7 (a) 45 km
(b) A: 120 km; B: 137.5 km
(c) $37\frac{1}{2}$ km (approx.)
(d) A: 0.6 h; B: 0.8 h

8 (a) 11:00 (b) 12:30
(c) $\frac{1}{4}$ hour (d) 2.2 km
(e) $\frac{1}{2}$ hour (f) 4 km
(g) $\frac{3}{4}$ hour (h) 5 km

9 (a) 12:00 (b) 12:30
(c) 2.6 km (d) 1.6 km

10 (a) 4.4 km/h (b) $5\frac{1}{3}$ km/h
(c) 4.13 km/h (d) 12.4 km/h

Practice Exercise P13.1 (page 118)

1 (a) $y = x + 5$

x	2	0	−4
5	5	5	5
$x + 5$	7	5	1

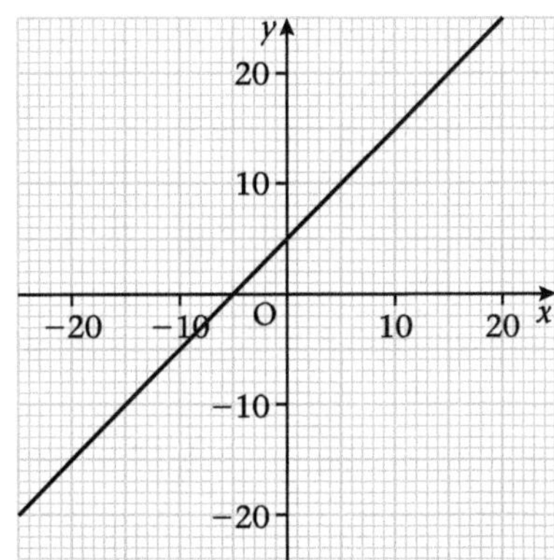

2 (a) $y = 3x - 1$

x	1	0	−2
$3x$	3	0	−6
−1	−1	−1	−1
$3x - 1$	2	−1	−7

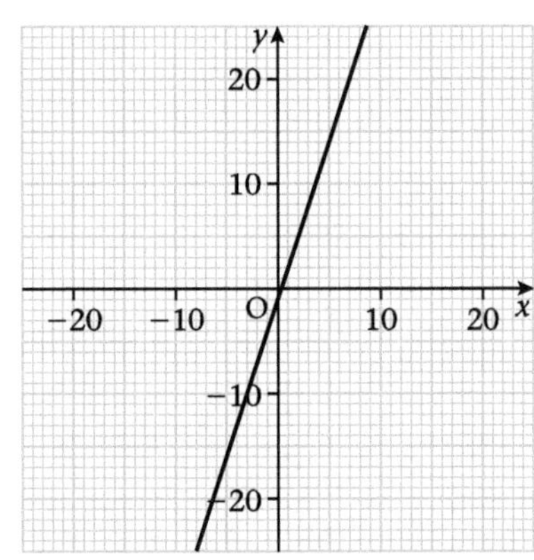

3 $y = x - 4$

x	−5	4	8
−4	−4	−4	−4
$x - 4$	−9	0	4

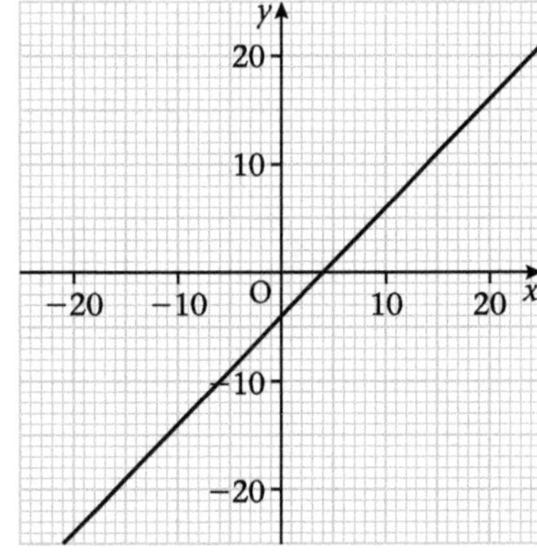

4 $y = 1 - 4x$

x	−2	−1	2
1	1	1	1
$-4x$	8	4	−8
$1 - 4x$	9	5	−7

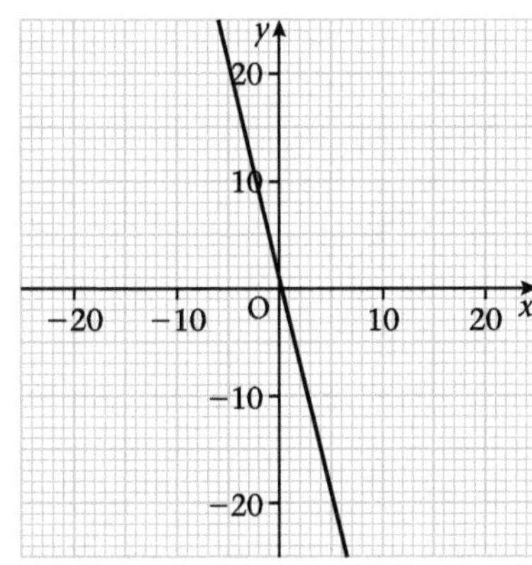

5 $y = x + 4$

x	−7	0	5
4	4	4	4
$x + 4$	−3	4	9

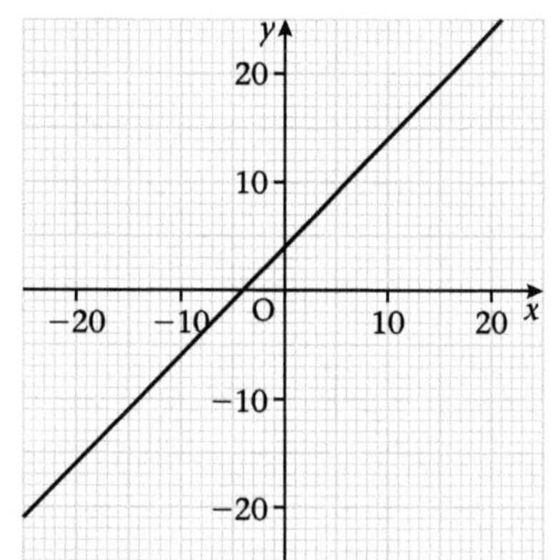

6 $y = -4 - x$

x	−10	0	6
$-x$	10	0	−6
−4	−4	−4	−4
$-x - 4$	6	−4	−10

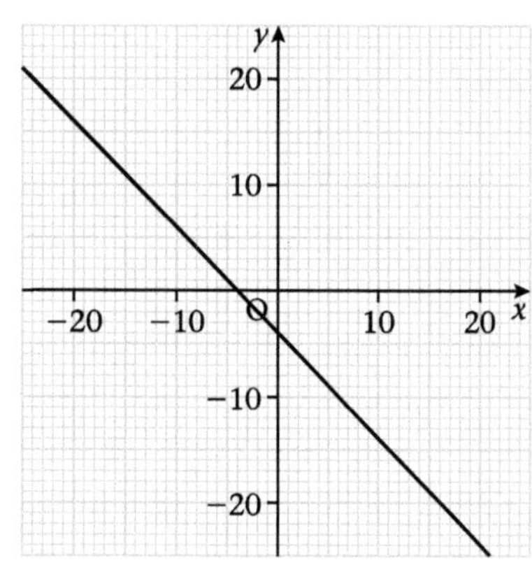

7 $y = 7 - 2x$

x	−1	6	8
7	7	7	7
$-2x$	2	−12	−16
$7 - 2x$	9	−5	−9

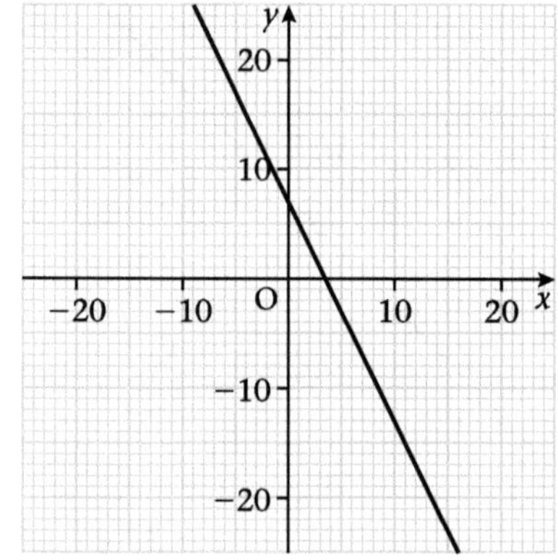

8 $y = 4x - 8$

x	0	2	4
$4x$	0	8	16
−8	−8	−8	−8
$4x - 8$	−8	0	8

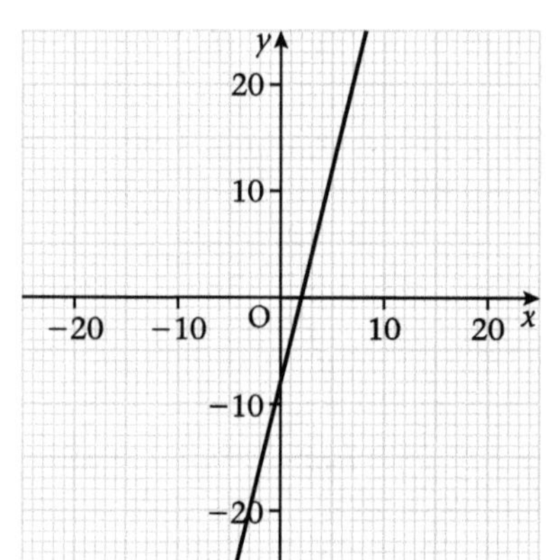

Practice Exercise P13.2 (page 118)

1. $y = 3x - 2,\ -4 < x < 5$

x	-3	0	4
$3x$	-9	0	12
-2	-2	-2	-2
$3x - 2$	-11	-2	10

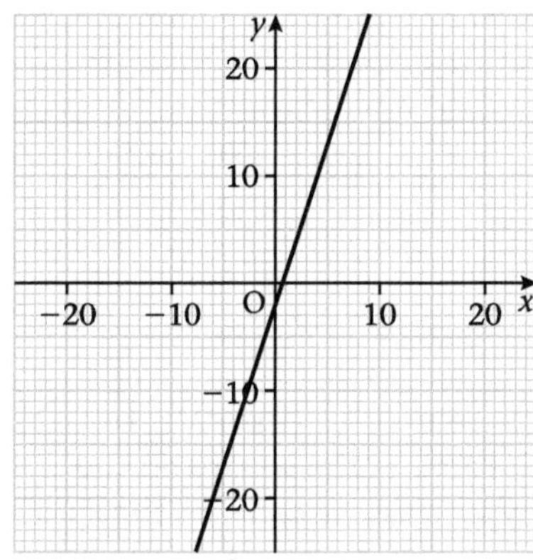

2. $y = -x - 2,\ -5 < x < 4$

x	-5	0	4
$-x$	5	0	-4
-2	-2	-2	-2
$-x - 2$	3	-2	-6

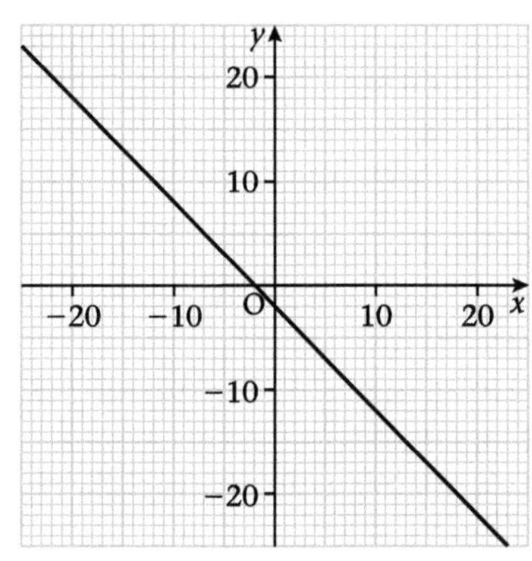

3. $f : x \rightarrow 2x + 5,\ -5 < x < 3$

x	-5	0	3
$2x$	-10	0	6
$+5$	$+5$	$+5$	$+5$
$2x + 5$	-5	5	11

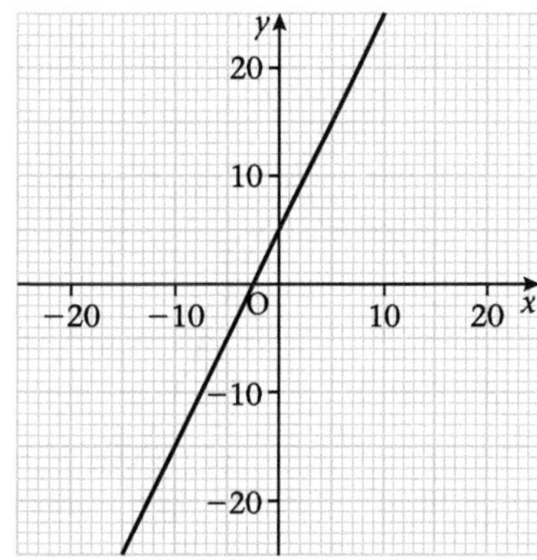

4. $G : x \rightarrow -3x - 1,\ -4 < x < 3$

x	-4	0	3
$-3x$	12	0	-9
-1	-1	-1	-1
$-3x - 2$	11	-1	-10

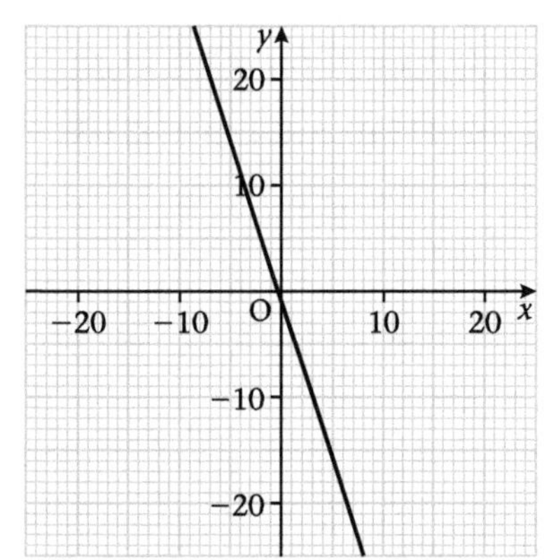

Practice Exercise P13.3 (page 118)

1 (a)

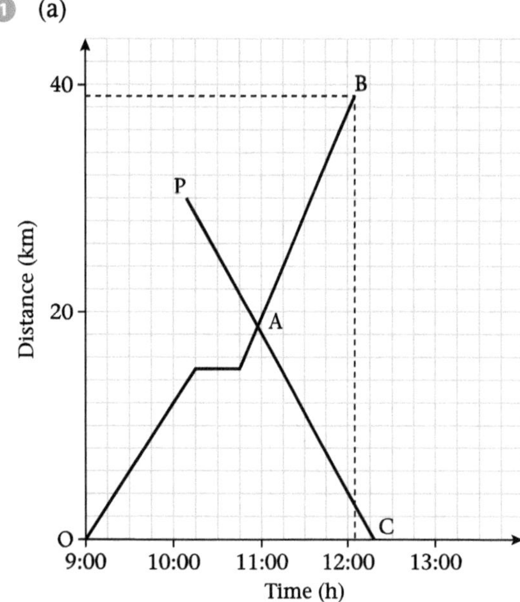

(b) 19 km, 10:57 (c) 12:06, 12:18

2 (a) OA: uniform acceleration for 1 hour from 0 km/h to 10 km/h
AB: constant speed of 10 km/h for 1 hour
BC: uniform acceleration for $\frac{1}{2}$ hour from 10 km/h to 20 km/h
CD: uniform deceleration for $1\frac{1}{2}$ hours from 20 km/h to 0 km/h
(b) (i) 37.5 km (ii) 9.375 km/h

3 (i) 1 cm : TT$10 and 1 cm : Ja$100
(b)

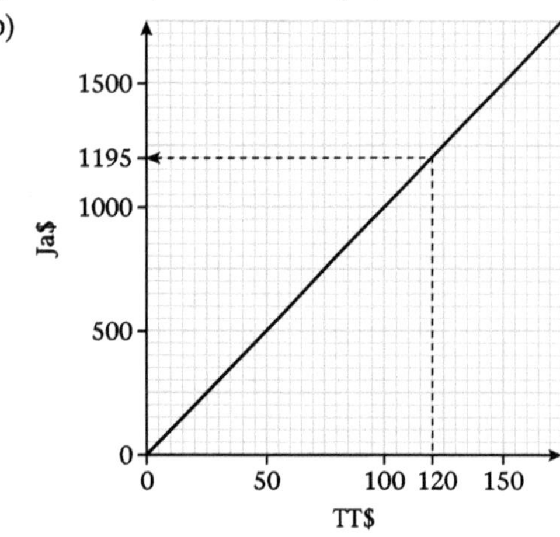

(ii) 1 cm : EC$50 and 1 cm : US$15

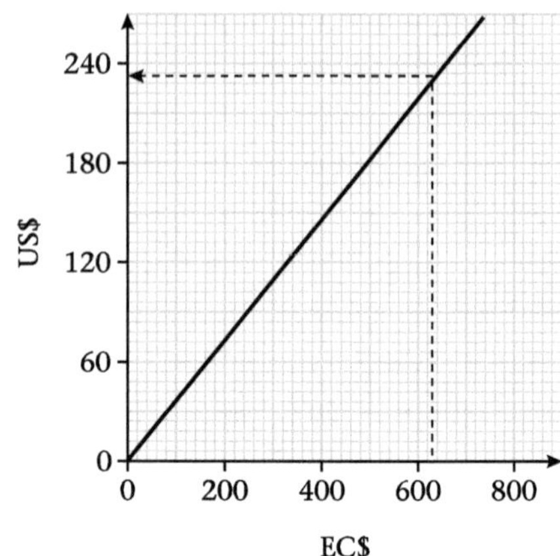

Exercise 14a (page 120)

1 5 cm
2 (a) 25 unit^2 (b) $9 + 16 = 25$ unit^2
(c) the area of the square on the hypotenuse is equal to the sum of the areas of the squares on the other two sides
3 (a) 169 unit^2 (b) $25 + 144 = 169$ unit^2
(c) see 2(c)
4 (b) 10 cm (d) $64\,\text{cm}^2 + 36\,\text{cm}^2 = 100\,\text{cm}^2$
5 (b) 17 cm (d) $64\,\text{cm}^2 + 225\,\text{cm}^2 = 289\,\text{cm}^2$

Exercise 14b (page 122)

1 (a) AC = 10 m (b) AC = 15 cm
(c) AC = 13 m (d) AC = 17 cm
(e) AB = 7 m (f) AB = 15 cm
(g) BC = 60 m
2 841 cm^2
3 (a) 13 (b) 3 (c) 17 (d) 2

Exercise 14c (page 123)

1 (a) (6, 8, 10), (9, 12, 15), (12, 16, 20), (15, 20, 25)
(b) (10, 24, 26), (15, 36, 39), (20, 48, 52), (25, 60, 65)
(c) (14, 48, 50), (21, 72, 75), (28, 96, 100), (70, 240, 250)
(d) (16, 30, 34), (24, 45, 51), (32, 60, 68), (40, 75, 85)
2 (a), (c), (d)
3 $(9, 40, 41) \rightarrow 9^2 = 40 + 41$
$(11, 60, 61) \rightarrow 11^2 = 60 + 61$

4 (12, 35, 37) → $\frac{1}{2}$ of $12^2 = 35 + 37$
(14, 48, 50) → $\frac{1}{2}$ of $14^2 = 48 + 50$

5 (13, 84, 85) (16, 63, 65)
(15, 112, 113) (18, 80, 82)
(17, 144, 145) (20, 99, 101)
(19, 180, 181) (22, 120, 122)
(21, 220, 221) (24, 143, 145)

Exercise 14d (page 124)

1 (a) 1.96 (b) 5.29 (c) 46.2
(d) 51.8 (e) 24.0 (f) 74.0
(g) 31.7 (h) 82.4 (i) 9.92
(j) 3.53 (k) 32.6 (l) 20.6

2 (a) 324 (b) 961 (c) 1020
(d) 225 (e) 841 (f) 1940
(g) 4 970 (h) 424 (i) 3930
(j) 3 580 (k) 6 610 (l) 8260

3 (a) 2.99 (b) 7.90 (c) 6190
(d) 2710 (e) 9310 (f) 2460
(g) 401 000 (h) 648 000 (i) 92 400

4 (a) 16 900 (b) 168 000 (c) 757 000
(d) 254 000 (e) 7 290 000 (f) 69 700 000

5 yes, in general $(N.5)^2 = N \times (N + 1) + 0.25$

Exercise 14e (page 125)

1 (a) 3 (b) 9.49 (c) 1.67 (d) 5.29
(e) 2.17 (f) 6.86 (g) 2.24 (h) 7.10
(i) 6.02 (j) 1.90 (k) 5.07 (l) 1.60

2 (a) 2.65 (b) 8.37 (c) 26.5 (d) 83.7
(e) 1.70 (f) 5.39 (g) 17.0 (h) 53.9
(i) 6.18 (j) 19.5 (k) 61.8 (l) 195
(m) 3.16 (n) 10 (o) 31.6 (p) 100

3 (a) 3.05 (b) 8.84 (c) 21.5
(d) 7.68 (e) 2.41 (f) 70.7
(g) 22.4 (h) 253 (i) 44.5

4 $\sqrt{10} = 3.16$ and $\pi = 3.14$; a difference of 0.02
$\sqrt{10}$ is a good approximation of π

5 (a) $m = 6.32$ (b) $m^2 = 39.9$ (c) $m^2 < 40$;
This is because the tables contain rounded numbers which are not completely accurate.

Exercise 14g (page 125)

1 (a) 2.24 (b) 4.90 (c) 1.97
(d) 16.2 (e) 2.19 (f) 2.24
(g) 4.79 (h) 4.15 (i) 22.4
(j) 0.508 (k) 3.02 (l) 1.93

2 (a) 1.97 (b) 19.7
(c) 388 to 3 s.f (d) $(19.7)^2$

Exercise 14h (page 126)

1 16.1 cm 2 7.3 m
3 2060 m 4 2.5 m
5 570 km 6 10.8 m each
7 6.7 m 8 1.50 m
9 28.3 cm 10 24.8 cm

Practice Exercise P14.1 (page 128)

1 (a), (b), (d), (f) Yes
(c), (e) No

2 $20^2 + 21^2 = 29^2$

3 Triangle ABC is a right-angled triangle; as two angles equal 45°, the third angle must be 90°

4 17.9 cm 5 20.3 cm
6 78 cm 7 1.92 m
8 21.1 cm 9 6.72 m

Exercise 15a (page 129)

2 (a) (i) square, (ii) square, (iii) triangle, (iv) kite, (v) quadrilateral

Exercise 15b (page 130)

1 15.1 (b); 15.2 (a); 15.3 (a), (b), (c)

3 (a) A(1, 6), B(1, 2), C(3, 2)
(b) A(−3, 6), B(−3, 2), C(−1, 2)
Each point moves the same distance to the left.
(c) A(−5, 2), B(−5, −2)

Exercise 15c (page 133)

1 $\overrightarrow{AB} = \begin{pmatrix} 6 \\ 1 \end{pmatrix}$ $\overrightarrow{CD} = \begin{pmatrix} -2 \\ -8 \end{pmatrix}$ $\overrightarrow{EF} = \begin{pmatrix} 0 \\ 3 \end{pmatrix}$
$\overrightarrow{GH} = \begin{pmatrix} -7 \\ 3 \end{pmatrix}$ $\overrightarrow{HI} = \begin{pmatrix} 4 \\ 0 \end{pmatrix}$ $\overrightarrow{IJ} = \begin{pmatrix} -3 \\ 7 \end{pmatrix}$
$\overrightarrow{KL} = \begin{pmatrix} 2 \\ -2 \end{pmatrix}$

2 (a) $\begin{pmatrix} 2 \\ 3 \end{pmatrix}$ (b) $\overrightarrow{AP}$ or $\overrightarrow{BQ}$ or $\overrightarrow{CR} = \begin{pmatrix} 2 \\ 3 \end{pmatrix}$

3 (a) (3, 6), (7, 8), (5, 11)
(b) (4, 2), (8, 4), (6, 7)
(c) (−1, 2), (3, 4), (1, 7)
(d) (−2, 3), (2, 5), (0, 8)
(e) (2, −3), (6, −1), (4, 2)

4 (a) (0, 5), (0, 8), (1, 8), (1, 5)
(b) (4, 3), (4, 6), (5, 6), (5, 3)
(c) (2, 5), (2, 8), (3, 8), (3, 5)

5 (a) $\begin{pmatrix} 1 \\ 4 \end{pmatrix}$ (b) (1, 4) (c) (−2, −2)

6 (a) rhombus
(b) (i) (4, 3), (8, 6), (12, 3), (8, 0)
(ii) (−4, 0), (0, 3), (4, 0), (0, −3)
(iii) (9, 4), (13, 7), (17, 4), (13, 1)

Exercise 15d (page 134)

1 15.1 (b); 15.2 (a), (b); 15.3 (a), (b), (d)
2 (b)

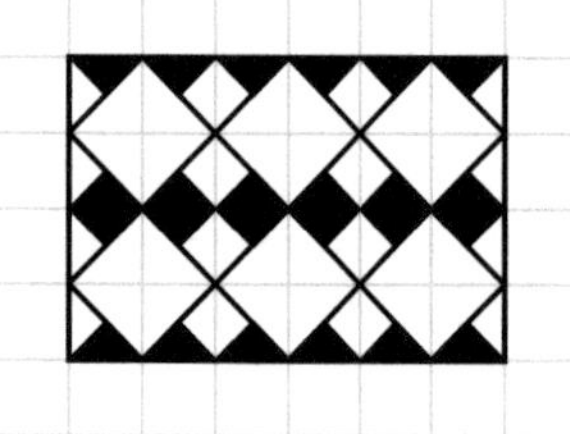

3 (a) and (b)
(i) 1, 1, 5
(ii) 1, 1, 5
(iii) the distances are the same
(iv) 2, 5, 2
(v) 2, 5, 2
(vi) the distances are the same
4 (a) A(1, − 6), B(1, − 2), C(3, − 2)
(b) A(−1, 6), B(−1, 2), C(−3, 2)

Exercise 15e (page 135)

1 (a) $Q\hat{O}Q_1 = P\hat{O}P_1$,
(b) $R\hat{O}R_1 = P\hat{O}P_1 = Q\hat{O}Q_1$
(c) all points are rotated through the same angle
(d) $PR = P_1R_1$ (e) yes
(f) $\triangle PQR$ and $\triangle P_1Q_1R_1$ are congruent
2 (c) $B\hat{O}B_1 = 50° = A\hat{O}A_1$
(d) $C\hat{O}C_1 = B\hat{O}B_1 = A\hat{O}A_1 = 50°$; points rotated through equal angles
(e) triangles ABC and $A_1B_1C_1$ are congruent
3 (a) A(6, −1), B(2, −1), C(2, −3)
(b) A(−1, −6), B(−1, −2), C(−3, −2)
(c) A(−6, 1), B(−2, 1), C(−2, 3)
4 (a) A(3, −2), B(−1, −2), C(−1, −4)
(b) A(−5, −4), B(−5, 0), C(−7, 0)
(c) A(−7, 4), B(−3, 4), C(−3, 6)

Practice Exercise P15.1 (page 136)

1 (a) A = (4, 7), B = (7, 9), C = (8, 4)
(b) A = (−2, 1), B = (1, 3), C = (2, −2)
2 W = (1, −1), X = (−3, 4), Y = (1, 7)
3 W = (3, 3), X = (5, 3), Y = (5, 5), Z = (3, 5)

4

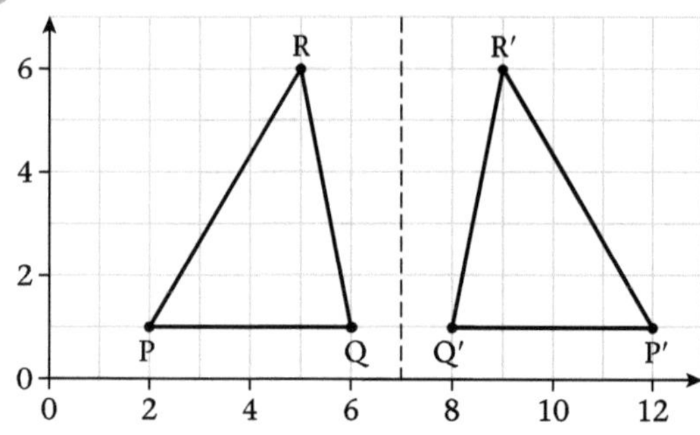

5 (a) P = (1, −2), Q = (2, −5), R = (6, −2)
(b) P = (−2, −1), Q = (−5, −2), R = −2, −6)
6 Yes, $\begin{pmatrix} 4 \\ 3 \end{pmatrix}$

Exercise 16a (page 137)

1 (a) $6 < 11$ (b) $-1 > -5$
(c) $0 > -2.4$ (d) $-3 < +3$
(e) $x > 12$ (f) $y < -2$
(g) $4 > a$ (h) $a < 4$
(i) $15 < b$ (j) $b > 15$
2 (a) true (b) true
(c) false (d) false
(e) false (f) true
(g) false (h) true
3 (a) > (b) < (c) < (d) >
(e) < (f) > (g) < (h) >

Exercise 16b (page 138)

1 (a) $h < 5$ (b) $m < 50$
(c) $x > 5$ (d) $t < 5$
(e) $n < 24$ (f) $m < 20$
(g) $s < 100$ (h) $t > 120$
2 (a) $h < 1.5$ (b) $c < 800$
(c) $b > 12$ (d) $g > 60$
(e) $p > 28$ (f) $m < 55$
(g) $t > 6$
3 $m > 28$ 4 $y > 7$
5 $x > 15$ 6 $x < 60$
7 (a) length > 6 cm
(b) perimeter > 24 cm
8 (a) length < 7 cm
(b) area < 49 cm^2

Exercise 16c (page 139)

1 (a) $a \leqslant 12$ (b) $n \geqslant 5$
(c) $t \leqslant 38$ (d) $s \geqslant 24$
(e) $n < 36$ (f) $v \leqslant 120$

2. (a) $l \leqslant 7$ (b) $s \leqslant 140$
   (c) $h \geqslant 160$ (d) $p \leqslant 5$
   (e) $d \geqslant 6$ (f) $g \geqslant 100$
3. $x < 27, y \geqslant 27, x < y$
4. (i) $a < 42$ (ii) $b \geqslant 42$ (iii) $a < b$
5. (a) circumference $\leqslant 6\pi$
   (b) area $\leqslant 9\pi$

Exercise 16d (page 140)

1. (a) $x < 3$ (b) $x > 2$
   (c) $x \geqslant -2$ (d) $x \leqslant 5$
   (e) $x > -4$ (f) $x \leqslant -4$
2. (a) 1 (b) −2
   (c) −3 (d) 0
   (e) 3 (f) −2
   (g) 4 (h) −1

Exercise 16e (page 141)

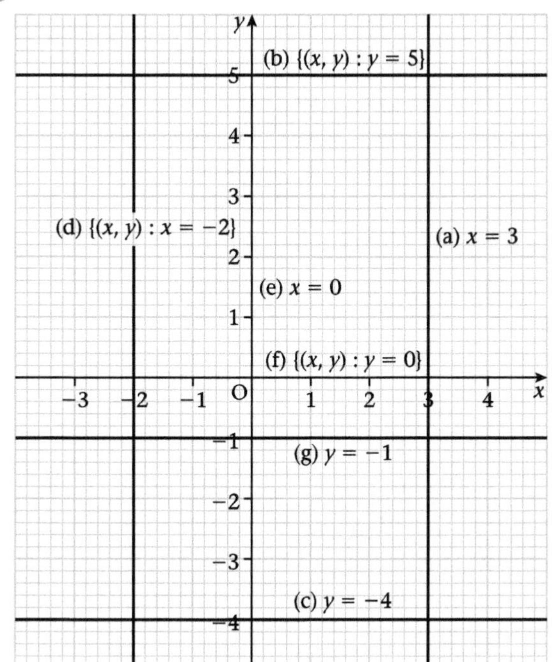

2. (a) $\{(x, y): x \geqslant 1\}$ (b) $\{(x, y): y > -1\}$
   (c) $\{(x, y): x < -2\}$ (d) $\{(x, y): y \leqslant 5\}$

3. (a)

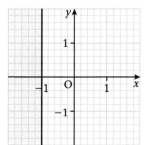

(b)

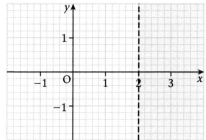

(c)

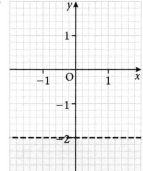

(d)

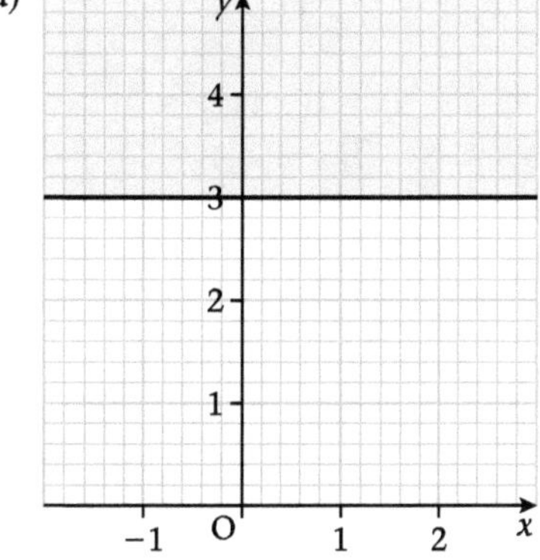

(e)

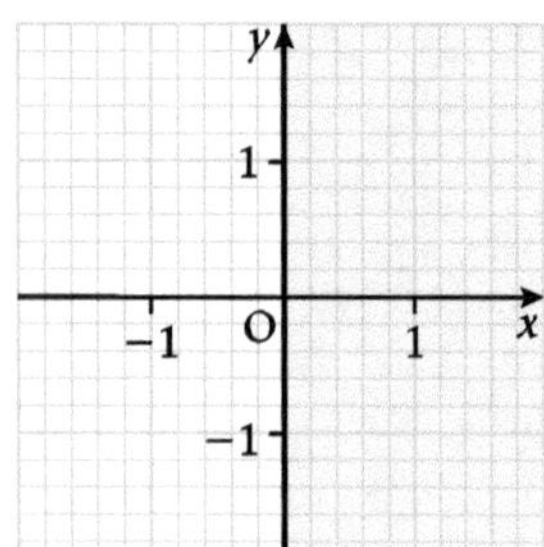

(f)

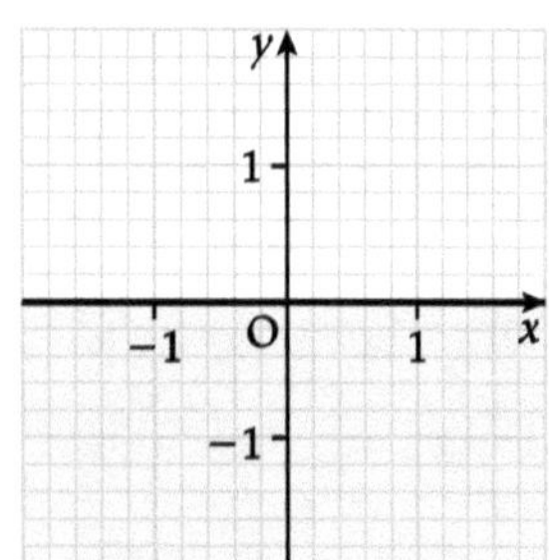

(g)

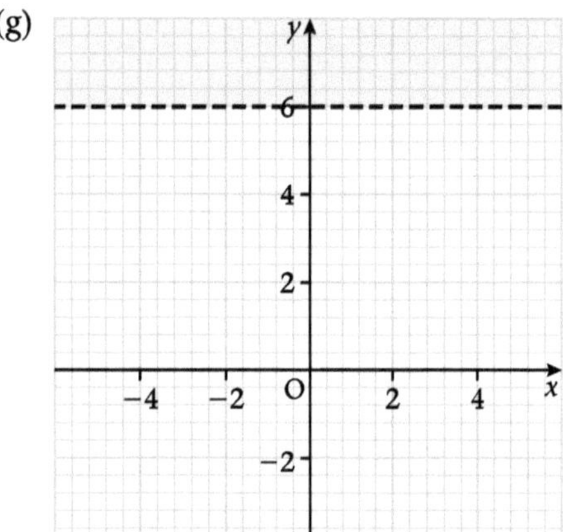

4 (a) $\{(x, y): x < 5\} > \{(x, y): y \leqslant 7\}$
(b) $\{(x, y): x \geqslant -3\} > \{(x, y): y > -4\}$

5 (a)

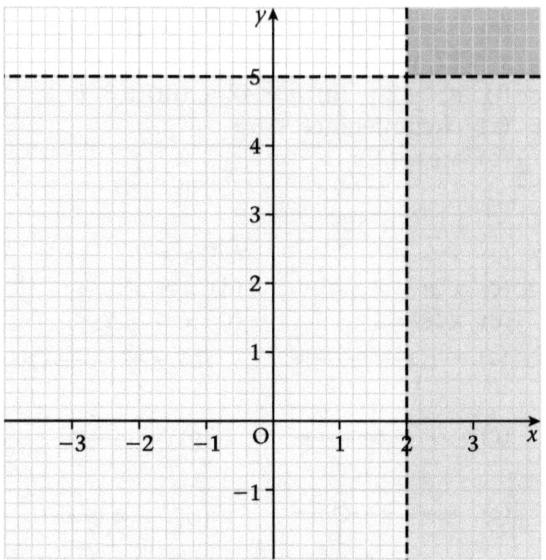

(b)

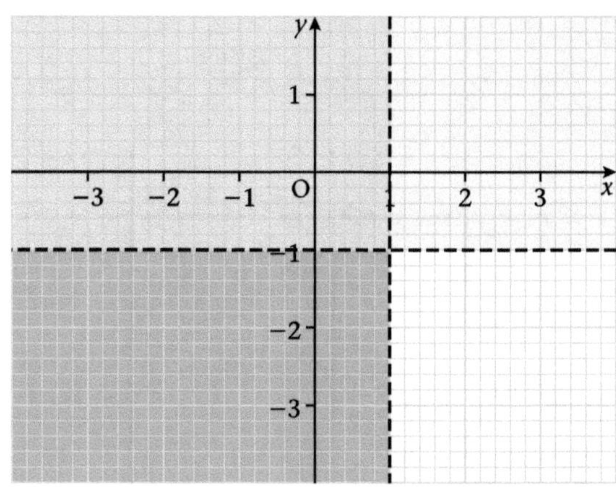

Exercise 16f (page 142)

1 (a) $x < 5$ (b) $x \geqslant 3$
(c) $x < 6$ (d) $x > 5$
(e) $x \leqslant -6$ (f) $x < -5$
(g) $x < 3$ (h) $x \geqslant 9$
(i) $x \leqslant 4$ (j) $x > -2$
(k) $x \geqslant -2\frac{1}{4}$ (l) $x \geqslant 2\frac{2}{3}$
(m) $x < 4$ (n) $x \geqslant 2$
(o) $x < -5$ (p) $x \geqslant -7$
(q) $x > 5\frac{1}{4}$ (r) $x \geqslant -\frac{2}{3}$

2 (a) {5, 6, 7, ...} (b) {2, 1, 0, ...}
(c) {−3, −4, −5, ...} (d) {−3, −2, −1, ...}
(e) {0, −1, −2, ...} (f) {−2, −1, 0, ...}
(g) {5, 4, 3, ...} (h) {4, 5, 6, ...}
(i) {−4, −3, −2, ...} (j) {−4, −5, −6, ...}
(k) {1, 0, −1, ...} (l) {−5, −4, −3, ...}
(m) {−1, −2, −3, ...} (n) {4, 5, 6, ...}
(o) {2, 1, 0, ...} (p) {−4, −3, −2, ...}
(q) {−2, −3, −4, ...} (r) {3, 2, 1, ...}

Exercise 16g (page 143)

1 $x > -4$ 2 $a > 2$
3 $m \geqslant -3$ 4 $d \leqslant -8$
5 $y \geqslant -4$ 6 $z \leqslant 4$
7 $a < 2$ 8 $n \geqslant -1$
9 $r \leqslant -2$ 10 $t \leqslant -1\frac{1}{2}$

Practice Exercise P16.1 (page 143)

1 $8 > 3$ 2 $-8 < -3$
3 $5 \times 3 > 4 + 10$
4 $3 \times 5 + 7 < 3 \times (5 + 7)$
5 $18 \div 3 > 12 \div 4$ 6 $\frac{1}{2} + \frac{3}{4} > \frac{1}{2} \times \frac{3}{4}$
7 $6 \div \frac{2}{3} > 6 \times \frac{2}{3}$ 8 $(-3)^2 > (2)^3$
9 $(5)^2 > \frac{1}{(5)^2}$ 10 $(\frac{2}{3})^2 < \frac{1}{(\frac{2}{3})^2}$

Practice Exercise P16.2 (page 144)

1 $x > 1$ 2 $x \leqslant 3$
3 $-1 \leqslant x < 4$ 4 $-3 \leqslant x < 2$
5 $0 < x \leqslant 6$ 6 $2 < x < 5$

Practice Exercise P16.3 (page 144)

1 $n < 2$ 2 $n \leqslant 2$
3 $n \geqslant 2$ 4 $(n - 5) > 1$
5 $5n \geqslant 15$ 6 $(n \div 7) < 3$
7 $(n + 2) \leqslant 8$ 8 $(n \div 4) > 1$
9 $(n - 5) < 0$ 10 $(n - 4) > 3$
11 $5n \leqslant 15$ 12 $10n < 60$
13 $(2n + 3) \geqslant 5$ 14 $12 < (n + 9) \leqslant 16$
15 $n^2 > 4$

Practice Exercise P16.4 (page 144)

1 (a) {1, 0, −1, −2, −3, ...}
(b) 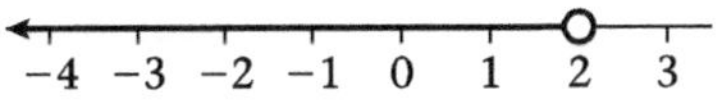 −4 −3 −2 −1 0 1 2 3

2 (a) {2, 1, 0, −1, −2, −3, ...}
(b) 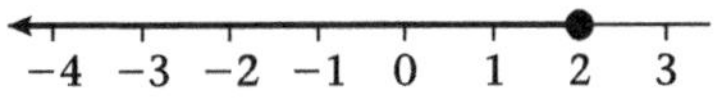 −4 −3 −2 −1 0 1 2 3

3 (a) {2, 3, 4, 5, 6, ...}
(b) 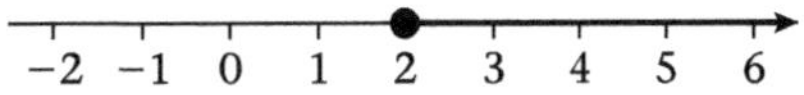 −2 −1 0 1 2 3 4 5 6

4 (a) {7, 8, 9, 10, ...}
(b) 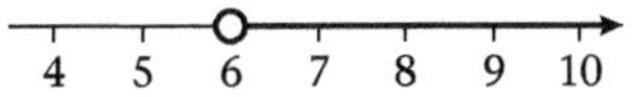 4 5 6 7 8 9 10

5 (a) {3, 4, 5, 6, ...}
(b) 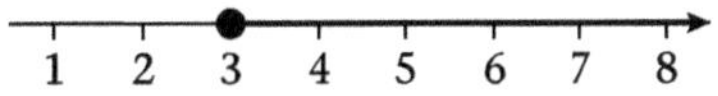 1 2 3 4 5 6 7 8

6 (a) {20, 19, 18, 17, ...}
(b) 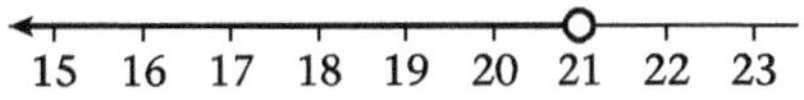 15 16 17 18 19 20 21 22 23

7 (a) {6, 5, 4, 3, 2, ...}
(b) 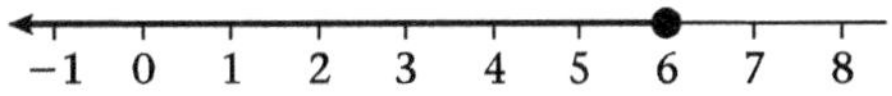 −1 0 1 2 3 4 5 6 7 8

8 (a) {5, 6, 7, 8, 9, ...}
(b) 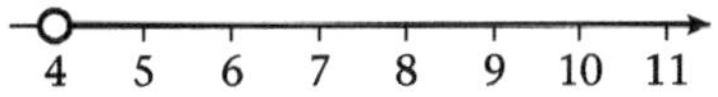 4 5 6 7 8 9 10 11

9 (a) {4, 3, 2, 1, 0, −1, ...}
(b) 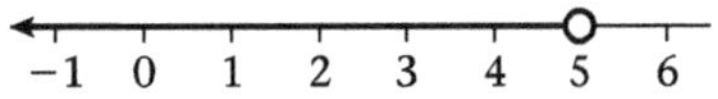 −1 0 1 2 3 4 5 6

10 (a) {8, 9, 10, 11, 12, ...}
(b) 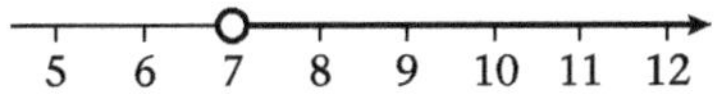 5 6 7 8 9 10 11 12

11 (a) {3, 2, 1, 0, −1, −2, ...}
(b) 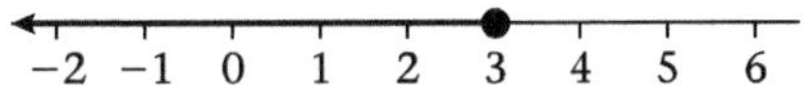 −2 −1 0 1 2 3 4 5 6

12 (a) {5, 4, 3, 2, 1, 0, −1,...}
(b) 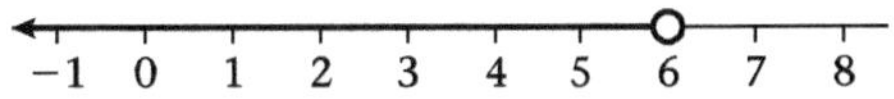 −1 0 1 2 3 4 5 6 7 8

13 (a) {1, 2, 3, 4, ...}
(b) 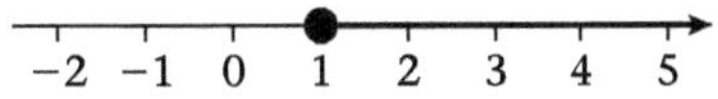 −2 −1 0 1 2 3 4 5

14 (a) {4, 5, 6, 7}
(b) 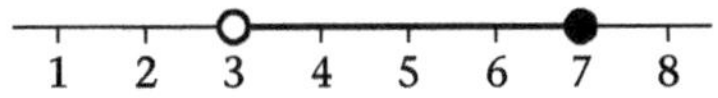 1 2 3 4 5 6 7 8

15 (a) {−1, 0, 1}
(b) 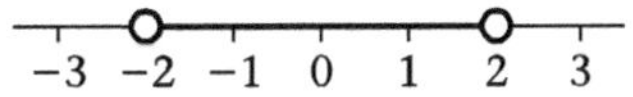 −3 −2 −1 0 1 2 3

Practice Exercise P16.5 (page 144)

1. $\{(x, y) : x \geqslant -1\}$

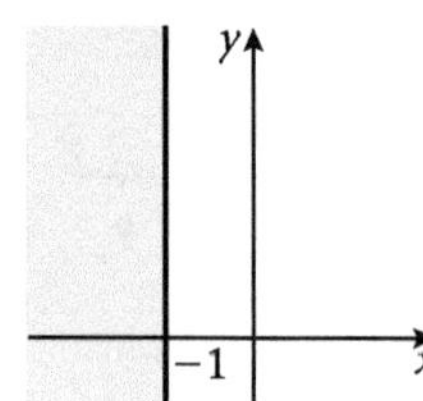

2. $\{(x, y) : y < 1\}$

3. $\{(x, y) : y \geqslant -1\}$

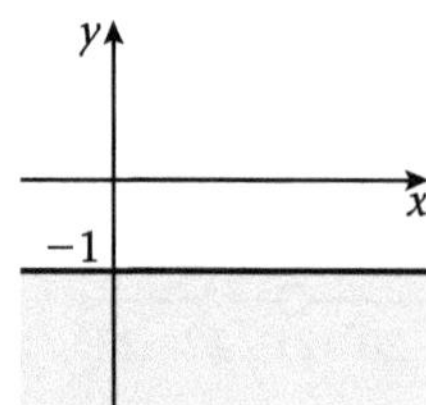

4. $\{(x, y) : x \leqslant -3\}$

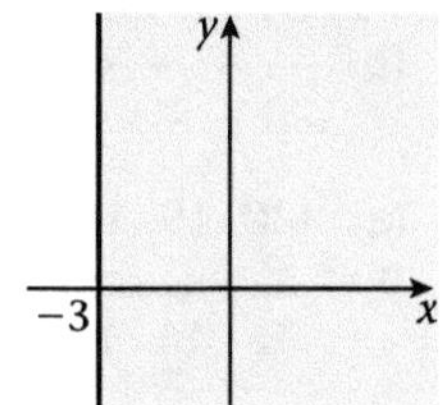

Practice Exercise P16.6 (page 144)

1. $\{(x, y) : x \geqslant 3\}$
2. $\{(x, y) : -3 \leqslant y \leqslant 2\}$
3. $\{(x, y) : x < 5\} \cap \{(x, y) : y < 4\}$
4. $\{(x, y) : -2 < x \leqslant 5\}$

Practice Exercise P16.7 (page 145)

1. $\{(x, y) : x \geqslant -3\} \cap \{(x, y) : x < 2\}$

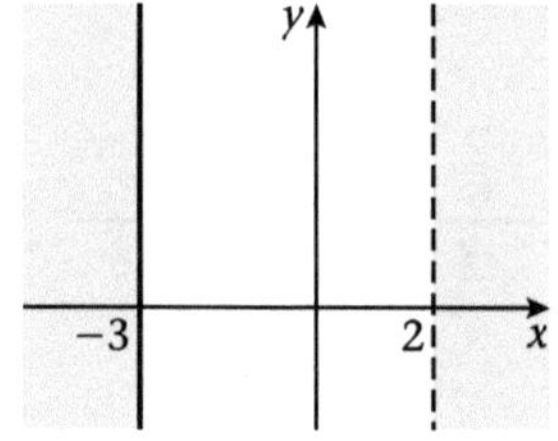

2. $\{(x, y) : y \geqslant -2\} \cap \{(x, y) : y < 2\}$

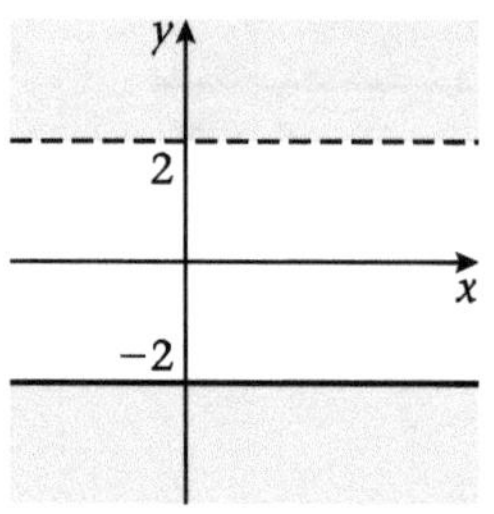

3. $\{(x, y) : x < -2\} \cap \{(x, y) : y \geqslant -1\}$

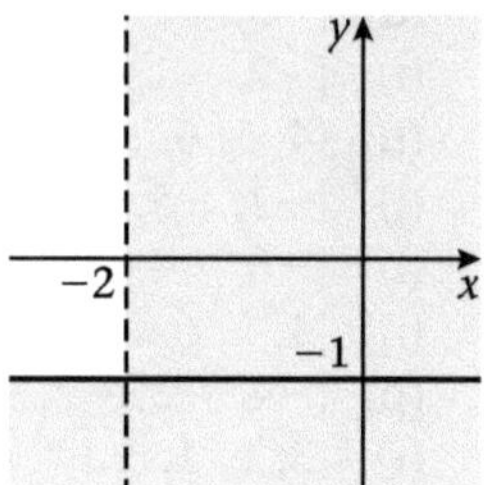

Practice Exercise P16.8 (page 145)

1. $x > -2$
2. $v \leqslant \frac{3}{2}$
3. $r \geqslant 7$
4. $d < 2$
5. $m > -4$
6. $n < 4$
7. $a \leqslant -23$
8. $y > -4$
9. $m \geqslant 6$
10. $a \geqslant -2$

Revision exercise 5 (page 146)

1. G C D E B F H A

   −10 −5 0 5 10

2. P(3, 4), Q(0, 3), R(1, 2), S(3, 0), T(2, −1), U(0, −2), V(−2, −4), W(−4, −1), X(−3, 1), Y(−2, 3), Z(−4, 4)
3. E, F, C, B, G, A, H, D respectively
4. (b) parallelogram (c) $(0, \frac{1}{2})$
5. 1 → 75, 2 → 150, 3 → 225, 4 → 300, 5 → 375, 6 → 450, 7 → 525
6. (a) 11.9 cm (b) 4.17 h (c) 12.5 h
7. (a) 50 km (b) 40 min
   (c) (i) 75 km/h, (ii) 100 km/h
8. (a)

x	0	45	90	135	180
y	180	135	90	45	0

   (c) (i) $y = 140$, (ii) $x = 52$

9. (a)

Value insured (× $1000)	1	2	3	4	5
Standing charge ($)	50	50	50	50	50
Basic rate ($)	60	120	180	240	300
Total cost ($)	110	170	230	290	350

10. (a) $280 (approx.) (b) $155

Revision test 5 (page 147)

1 C 2 A 3 B 4 A 5 C

6 $100 < x \leqslant 250$

7 (a) (2, 1) (b) 90° (c) rhombus

8 (a)

Time (h)	0.25	0.5	0.75	1	1.25	1.5	1.75	2
Distance (km)	1.5	3	4.5	6	7.5	9	10.5	12

9 (a) 6.9 km (b) 1.67 h (1 h 40 min)

10 (a) (i) Bds\$7.40 (ii) EC\$74
(c) (i) EC\$59.20 (ii) EC\$59.40

Revision exercise 6 (page 148)

1 \$70 000

2 (a) (i) $x = -4$ (ii) $n = -4$ (iii) $d = 7$
(iv) $x = 6$
(b) (i) $a = 2$ (ii) $x = 3$
(iii) $x = 9$ (iv) $a = 4\frac{1}{2}$

3 \$44, \$61

4 (a) $m = \frac{5}{6}$ (b) $a = \frac{3}{7}$ (c) $r = 25$
(d) $x = 2\frac{2}{5}$ (e) $m = 5$ (f) $x = 2\frac{1}{2}$
(g) $x = 3\frac{1}{2}$ (h) $a = -3$

5 (a) $2(d + 2d)$ cm *or* $6d$ cm
(b) (i) 14 (ii) 392 cm^2

6 $9x + 13y < 50$

7

(a)

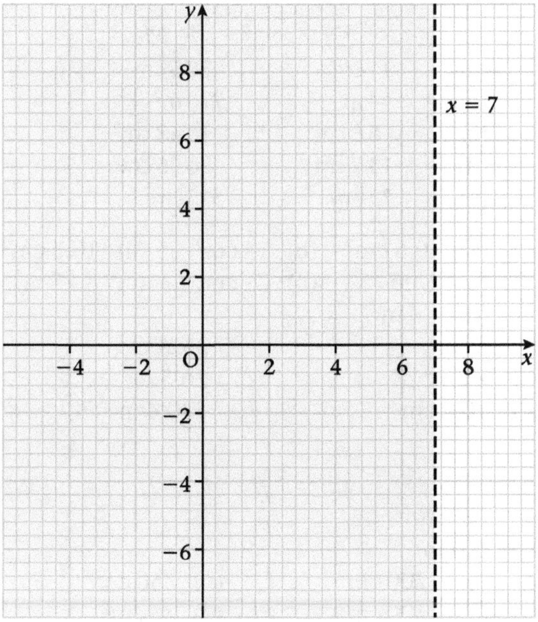

(b)

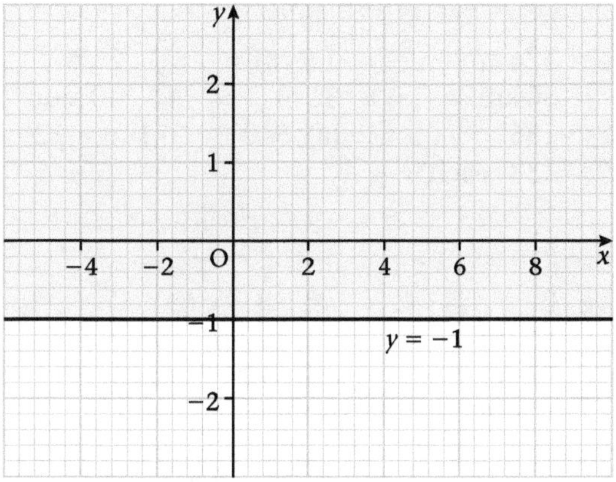

(c)

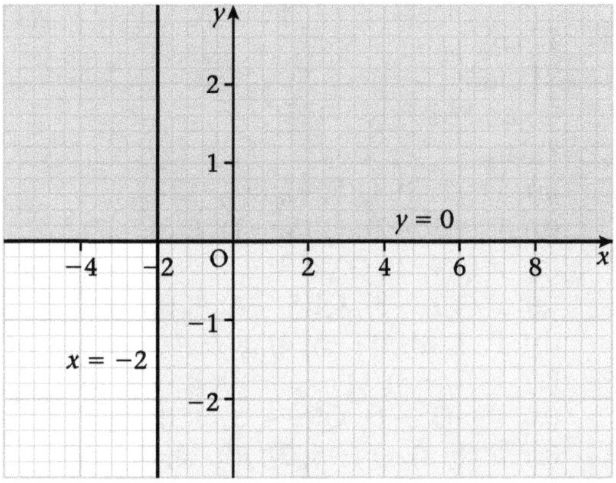

8 (a) $x < 5$ (b) $x < -3$ (c) $x < 2$
(d) $x \leqslant 3$

9 (a) {2, 1, 0, ...} (b) {−6, −5, −4, ...}
(c) {4, 3, 2, ...} (d) {1, 0, −1, −2, ...}

10 (a) $38 - 3n < 20$ (b) 7, 8, 9, 10

Revision test 6 (page 149)

1 C 2 C 3 B 4 A 5 D

6 (a) $x = -6$ (b) $a = 2\frac{1}{3}$ (c) $m = 4$
(d) $y = -2$ (e) $x = \frac{2}{3}$ (f) $a = 28$
(g) $y = 2$ (h) $z = 6$ (i) $x = 4$
(j) $x = 5$

7 $x = 8$

8 (a)

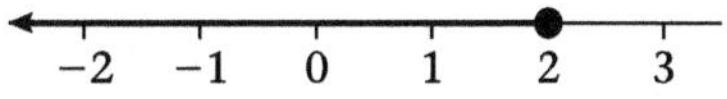

(b)

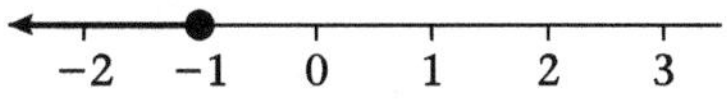

9 (a) {2, 3, 4, ...} (b) {−8, −7, −6, ...}
(c) {7, 8, 9, ...} (d) {−12, −13, −14, ...}

10 $2 \leqslant x \leqslant 11$

Revision exercise 7 (page 149)

1 126 km 2 7.9 cm 3 2 : 5
4 30 km 5 35 m 7 40 sides
8 156° 9 (a) 540° (b) $x = 129$
10 133° each

Revision test 7 (page 150)

1 D 2 C 3 B 4 C 5 A
6 15 cm 7 (a) 4.8 cm (b) 3.5 m
8 72 cm 9 (a) 540° (b) 108°
10 $x = 20$; 80°, 100°, 120°

Revision exercise 8 (page 151)

1 17 cm
2 21 cm
3 (a) and (d)
4 3.3 m
5 (a) P(−1, 3), Q(5, 4), R(4, −2)
(b) P′(−1, −2), Q′(5, −1), R′(4, −7)
(c) P″(−4, 2), Q″(2, 3)
6 (a) P′(−1, −3), Q′(5, −4), R′(4, 2)
(b) P″(1, 3), Q″(−5, 4), R″(−4, −2)
7 (a) P′(1, −3), Q′(−5, −4), R′(−4, 2)
(b) P″(−3, −1), Q″(−4, 5), R″(2, 4)
8 (a) P′(1, 3), Q′(−5, 2), R′(−4, 8)
(b) P″(0, 2), Q″(−1, 8), R″(5, 7)
9 T(−2, 1), U(3, 6), V(−5, 7)
10 (b) P′(−2, 0), Q′(3, −5), R′(4, 3)

Revision test 8 (page 151)

1 B 2 A 3 A 4 C 5 D
6 (a) 23.0 (b) 2300 (c) 230 000
(d) 2.83 (e) 8.94 (f) 92.3
7 (−2, −3)
8 P′(−7, −1), Q′(−6, 1), R′(−3, 1)
9 (a) (−3, 5) (b) (−1, 5) (c) (−4, −4)
10 4.47 m

General revision test B (page 152)

1 B 2 B 3 C 4 A 5 C
6 C 7 A 8 C 9 D 10 C
11 (a) $x = 2$ (b) $c = 8$ (c) $k = 6$
(d) $k = -2\frac{1}{2}$ (e) $x = 4$
12 a hammer
13 (a) $\frac{6}{v}$ (b) $v = 4\frac{1}{2}$
14 $k = 27$
15 (a)

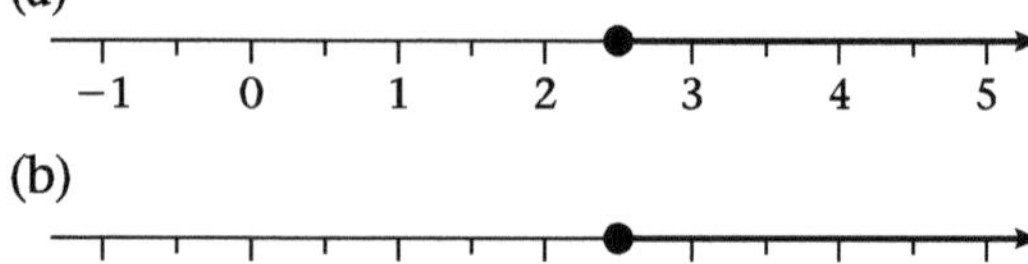

16 (b) AM = 5.4 m
17 (a) \$450 (b) \$34 000 (c) \$250
18 (a) {1, 2, 3, ...} (b) {4, 5, 6, ...}
(c) {2, 1, 0, ...} (d) {6, 5, 4, ...}
19
20 (a) 4.6 cm (b) $\hat{A} = 25°$, $\hat{C} = 65°$

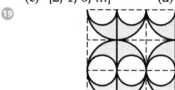

Exercise 17a (page 155)

1 1.26 m² → 1.26 m^2 2 32 m^2 3 84 cm^2
4 10 m^2 5 54 cm^2 6 27 m^2
7 28 m^2 8 28 cm^2 9 28 m^2
10 36 cm^2

Exercise 17b (page 157)

1 (a) 22 m^2 (b) 66 cm^2 (c) 148.5 m^2
(d) 5.28 cm^2
2

	Length of arc	Area of sector
(a)	11 cm	38.5 cm^2
(b)	44 m	770 m^2
(c)	8.8 cm	18.48 cm^2
(d)	13.2 cm	36.96 cm^2
(e)	$73\frac{1}{3}$ m	$513\frac{1}{3}\text{ m}^2$

3 (a) 10.5 cm^2 (b) 14 cm^2 (c) 10.5 cm^2
4 (a) 7 cm (b) 21 cm (c) 5.25 m
(d) 1.4 m
5 5 cm 6 352 cm^2
7 71.5 cm^2 8 16.5 m^2
9 10.4 cm^2 10 530.4 m^2

Exercise 17c (page 158)

1 180 mm
2 30 cm
3 (a) 1200 cm^2 (b) 750 cm^2
(c) 1300 cm^2 (d) 25 cm^2

4 (a) 630 cm^2 (b) 95 cents
5 (a) 300 cm^2 (b) 100 cm^2
(c) 400 cm^2

Practice Exercise P17.1 (page 159)

1 (a) 88.3 cm^2 (b) 188.4 cm^2 (c) 276.7 cm^2
2 (a) 5023 cm^2 (b) 4416 cm^2 (c) 608 cm^2
3 383.28 cm^2
4 Section A = 549.5 cm^2,
Section B = 251.2 cm^2
5 A = 62.8 cm^2, B = 75.36 cm^2, C = 87.92 cm^2
6 (a) 67 824 cm^2 (b) 84 780 cm^2
7 95 m^2

Exercise 18a (page 160)

1 (a) 11 (b) 9 (c) 5 (d) 8
(e) 6 (f) 4 (g) 7 (h) 12
(i) 4 (j) 4.3
2 (a) 4.5 cm (b) \$8.20 (c) 4.8 kg
(d) $3\frac{1}{4}$ (e) 0.63
3 \$12.77 4 3 5 36 6 25°C
7 (a) 15 mm (b) 3 mm
8 27 goals 9 28 years
10 (a) 22 marks (b) 55%
11 (a) 10.7 hours
(b) advertisement (ii) is accurate; advertisement (i) is not accurate since some batteries do not last 10 hours

Exercise 18b (page 161)

1 81 km/h 2 (a) 16 min (b) 15 km/h
3 10 km/h 4 60 km/h
5 \$39.20 6 \$3.20/hour
7 58 8 (a) 60 km/h (b) 45 km/h
9 1.64 m 10 72 km/h

Exercise 18c (page 163)

1

	(a)	(b)	(c)	(d)	(e)
Mode	7	5	12	7	10
Median	9	6	11	7	10
Mean	10	6.5	10	6.3	10.4

2

	(a)	(b)	(c)	(d)	(e)
Mode	4	5	0	7	6
Median	$3\frac{1}{2}$	6	3	7	$4\frac{1}{2}$
Mean	$3\frac{1}{4}$	6	5	8	$4\frac{1}{2}$

3 (a) 17 students (b) grade B (c) grade C

4 (a) size 7 (b) size $7\frac{1}{2}$
5 mode is 1 km; median is 2 km; mean is 2.3 km
6 (a) 15 girls
(b) modal age is 16 years and median age is 16 years
(c) mean is $15\frac{8}{15}$ years

Exercise 18d (page 165)

1 (a) 8, 7, 7 (b) 4, $3\frac{1}{2}$, 3
(c) $4\frac{1}{3}$, $4\frac{1}{2}$, 1 (d) 5, 4, 3
(e) 0.55, 0.35, 0 (f) 156.4, 155.8, 155.8
2 (a) \$123.73 (b) \$123.09
3 83.3 kg
4 24.6 kg
5 23 yr 3 mo
6 59 kg
7 82c/kg
8 32.5 mm
9 \$6000
10 (a) 24 (b) 4 (c) 4 (d) 3.5
11 48 km/h
12 50 km/h
13 61.4
14 (a) Rudy 58, Sonya 67, Tinga 61, Ural 28, Vera 41

Exercise 18e (page 166)

1 mean
2 median
3 mode

Practice Exercise P18.1 (pge 166)

1 (a) 3.7 m, 3.9 m, 4.0 m, 4.0 m, 4.1 m, 4.2 m, 4.3 m, 4.4 m, 4.7 m, 4.8 m, 5.1 m, 5.2 m
(b) 4.25 m
(c) 4.4 m
2 (a) 900 (b) 90 (c) 5
(d) Students' opinions
3 (a) 52 years 6 months
(b) 62 years 6 months
(c) 10 years 5 months
4 (a) Totals are 280, 515, 261, 360, 127
(b) a (c) 1543 (d) 308.6

(e)

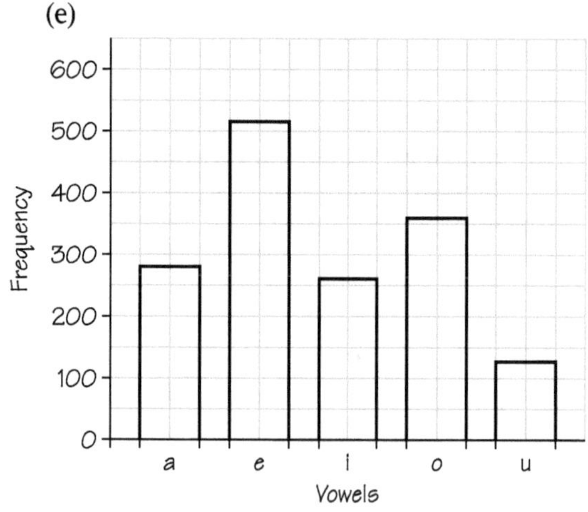

5 (a)

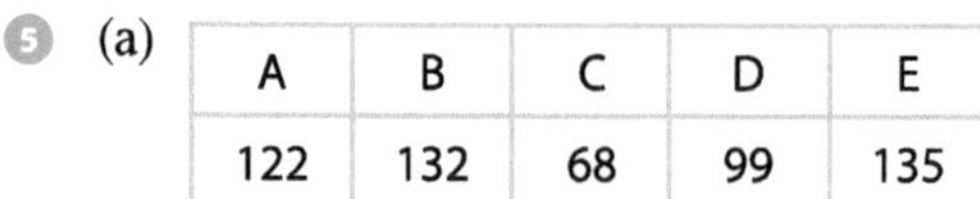

A	B	C	D	E
122	132	68	99	135

(b)

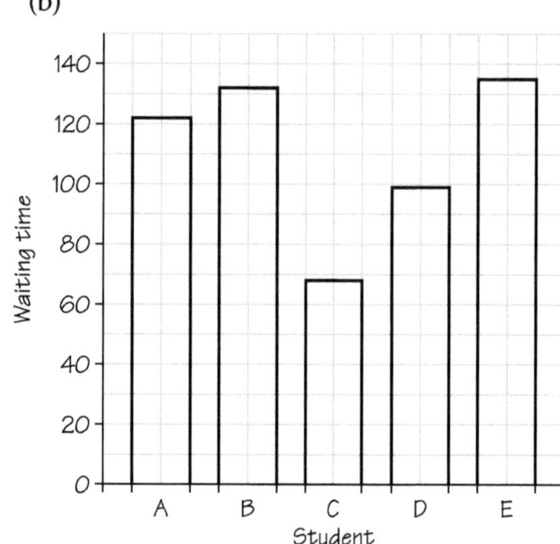

(c) Student E (d) 27 minutes
(e) Student C (f) Friday
(g) Friday

6 (a)

	Test 1	Test 2	Test 3	Total %
Adeff	18	25	27	70
Brandon	21	22	31	74
Jermaine	26	28	38	92
Kyle	22	18	36	76
Rayan	20	24	28	72

(b) 77 (c) 74
(d)

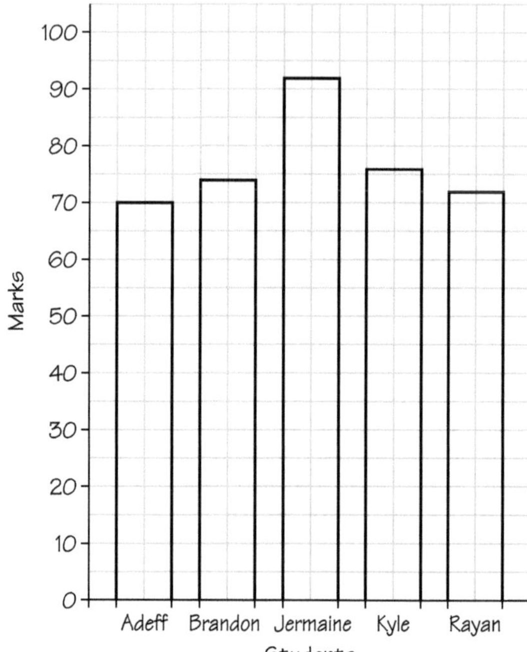

7 Mode = 3, mean = 2.5
8 Sum of ages = 63; mean age of the 10 boys = 8.7 years

Exercise 19a (page 169)

1 tan 30° ≃ 0.58 2 tan 51° ≃ 1.23

Exercise 19b (page 169)

1 0.9 2 1.9 3 0.78
4 2.9 5 1.0 6 0.51
7 4.3 8 0.25 9 0.65

Exercise 19c (page 169)

1 29° 2 16° 3 53°
4 58° 5 70° 6 32°
7 42° 8 48° 9 $35\frac{1}{2}°$

Exercise 19d (page 170)

1 (a) 30° (b) 40° (c) 25° (d) 25°
2 (a) 10° (b) 10°
3 (a) 15° (b) 15°
4 (a) 45° (b) 28°
5 52°

Exercise 19e (page 172)

All answers are corrected to 2 s.f.

1 (a) 6.3 (b) 15 (c) 5.1
2 (a) 12 (b) 8.4 (c) 6.0
3 (a) 11 (b) 11 (c) 5.2
4 1.6 m 5 110 m 6 37 m

Exercise 19f (page 173)

1 0.231 2 2.050 3 0.700
4 1.483 5 3.487 6 28.64
7 0.427 8 0.729 9 1.003
10 0.637 11 0.916 12 0.354
13 1.494 14 2.032 15 0.331

Exercise 19g (page 173)

1 0.525 2 1.13 3 2.40
4 4.96 5 5.05 6 0.486
7 19.7 8 20.4 9 21.2
10 2.5 11 0.821 12 0.211
13 3.06 14 5.91 15 1.47

Exercise 19h (page 174)

1 (a) 11 m (b) 4.9 m (c) 6.7 m
(d) 6.2 m (e) 4.8 m (f) 170 m
2 9.6 m 3 3.25 m 4 2 m
5 3.9 m 6 270 m 7 34 m
8 87 m 9 11 m 10 23.2°

Practice Exercise P19.1 (page 175)

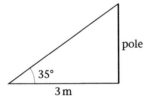

1 3.9 cm
2 (a)

(b) 2 m
3 35 m 4 3.84 m 5 57.1 m
6 (a) 93 m (b) 62.8 m (c) 155.8 m
7 8.24 cm

Exercise 20a (page 177)

1 $2x + 2y$ 2 $35 - 5a$
3 $3n + 27$ 4 $16a - 8b$
5 $-5x - 15y$ 6 $-12p + 4q$
7 $-2m - 2n$ 8 $-3a + 3b$
9 $-4p - 4q$ 10 $-21d + 14$
11 $18k + 27r$ 12 $42s - 6t$
13 $x^2 + 2x$ 14 $y^2 - y$
15 $a^2 + ab$ 16 $3n^2 - 2n$
17 $2ps + 3pt$ 18 $5m - 3mn$
19 $10a^2 - 16ab$ 20 $3x^2 + 27x$
21 $45pr - 40ps$ 22 $-12a^2 + 42ab$
23 $9ab - 12b^2$ 24 $2\pi r^2 + 2\pi rh$

Exercise 20b (page 177)

1 5 2 3 3 mp 4 $5x$
5 $4a$ 6 $13b$ 7 ab 8 $3de$
9 $8p$ 10 $2ax$ 11 3 12 $2a$

Exercise 20c (page 178)

1 $5(a + z)$ 2 $3(2x - 5y)$
3 $mp(7n - 1)$ 4 $5x(y + 3)$
5 $4a(3 + 2a)$ 6 $13b(a - 2)$
7 $ab(b - a)$ 8 $3de(2d - e)$
9 $8p(q + 3p)$ 10 $2ax(5x + 7a)$
11 $3(3xy + 8pq)$ 12 $2a(15d - 14x)$
13 $5m(a - 4b)$ 14 $a^2(5a - 3b)$
15 $\pi r(r + s)$ 16 $d(7d - 1)$
17 $3d(11b - e)$ 18 $3(3pq + 4t)$
19 $b(a - 2)$ 20 $3d(h + 5k)$
21 $x(x + 9y)$ 22 $2a(a + 5)$
23 $a(m + 1)$ 24 $6xy(4x - 1)$

Exercise 20d (page 178)

1 3400 2 122
3 2700 4 6930
5 125 6 44
7 13 400 8 670
9 $3\frac{1}{7}$ 10 530
11 30 12 1400
13 27 000 14 $12\frac{4}{7}$
15 17 400 16 $\pi(R^2 - r^2)$; 176
17 $2\pi r(r + h)$; 660 18 $\pi r^2(h + \frac{1}{3}H)$; 396

Exercise 20e (page 179)

1 $m(3 + u - v)$ 2 $a(2 - 3x - y)$
3 $x(3 - a + b)$ 4 $p(4m - 3n - 5)$
5 $(m + 1)(a + b)$ 6 $(n + 2)(a - b)$
7 $x(a - b + 4c)$ 8 $(a - b)(5x - 2y)$
9 $(5u - v)(3h + 2k)$ 10 $m(u - v + m)$
11 $d(3h + k - 4d)$ 12 $a(5a + b - c)$
13 $x(4x - 3y - 2z)$ 14 $d^2(3d - e + 4f)$
15 $a(4u + v)$ 16 $2a(x - 3y)$
17 $(3u + 2v)(3 - a)$ 18 $(4a - b)(3x + 2y)$

19 $(2a-7b)(h-3k)$ 20 $m(5m-2)$
21 $a^2(2a-3b)$ 22 $4x(x-1)$
23 $(3m-4n)(2d-3e)$
24 $(x-y)(a+2b-3)$
25 $(2m+n)(p+q-r)$
26 $(h+k)(2r-s)$
27 $(u+v)(4x+y)$
28 $(b-c)(2d+3e)$
29 $(a+2b)(a+2b-3)$
30 $(3m-2n)(3m-2n+5p)$
31 $2(2u-3v)(m-3n)$
32 $(x+2y)(a+x+2y)$
33 $(2x+y)(3u-2x-y)$
34 $(f-g)(4e-f+g)$
35 $3(a-3b)(u+2v)$
36 $5(5m+2n)(a+b)$
37 $(x+3y)(m-n+1)$
38 $(2a-3b)(c+d-1)$
39 $(7u-2v)(1+7u-2v)$
40 $(2u-7v)(2u-7v-1)$

Exercise 20f (page 180)

1 $m=1$ 2 $a=5\frac{1}{8}$ 3 $a=5$
4 $x=4$ 5 $h=7$ 6 $x=9$
7 $x=4$ 8 $x=\frac{3}{4}$

Exercise 20g (page 180)

1 5 2 40, 42
3 (a) $\$\frac{67.20}{x}$ (b) 7 watches
4 (a) hours (b) $5\frac{1}{7}$
5 82 books
6 35 mangoes
7 (a) $x+24$ (b) 12

Exercise 20h (page 181)

1 $x=5$ 2 $x=3$ 3 $y=1\frac{1}{2}$
4 $t=3\frac{1}{3}$ 5 $z=-3$ 6 $r=-2$
7 $x=-9$ 8 $k=2\frac{1}{2}$ 9 $a=1$
10 $x=\frac{4}{5}$ 11 $x=2$ 12 $a=10$
13 $y=7$ 14 $b=-9$ 15 $e=-6$
16 $c=12\frac{1}{2}$ 17 $n=4\frac{1}{6}$ 18 $d=-1$
19 $a=\frac{1}{2}$ 20 $x=6$

Practice Exercise P20.1 (page 181)

1 $5(d+2)$ 2 $4m(m+3)$
3 $5(2m+n)$ 4 $x(5y-7z)$
5 $2v(8w+3u)$ 6 $ar(3p-t)$
7 $3d(2+c)$ 8 $v(3w-u)$
9 $d(2d+1)$ 10 $k(h-2k)$
11 $2(x+2y-4w)$ 12 $r^2h(2r^2+3h^2)$

Practice Exercise P20.2 (page 181)

1 $7(w+v)$ 2 $2b(b-1)$ 3 $3(x-2)$
4 $2(b-3)$ 5 $11(m-1)$ 6 $7(n-3)$
7 $7(a+3)$ 8 $2(y-17)$ 9 $7(m-2)$

Practice Exercise P20.3 (page 181)

1 $(7-a)(a+2)$ 2 $2(y+7)$
3 $(1-2y)(2+3y)$ 4 $(4d+3)(2d+1)$
5 $2d(1-3c)(5c+7d)$

Exercise 21a (page 182)

1 \$20 2 \$28 3 \$18
4 \$8 5 \$18 6 \$14
7 \$30 8 \$48 9 \$30
10 \$75 11 \$42 12 \$30
13 \$15 14 \$51 15 \$6.25

Exercise 21b (page 182)

1 \$420 2 \$728 3 \$118
4 \$108 5 \$118 6 \$114
7 \$330 8 \$248 9 \$7230
10 \$675 11 \$392 12 \$280
13 \$1515 14 \$2 601 15 \$256.25

Exercise 21c (page 183)

1 (a) \$7 (b) \$26 (c) \$1090
(d) \$22 000 (e) \$7350 (f) \$1042.75
2 (a) \$440 (b) \$520 (c) \$800
3 (a) \$5 (b) \$1.25 (c) $6\frac{1}{4}$%
4 (a) \$30 200 (b) \$2516.67
5 (a) \$540 (b) \$3540 (c) \$147.50

Exercise 21d (page 184)

1 2 years 2 3 years 3 6 months
4 1 year 5 3 years 6 4%
7 2% 8 4% 9 8%
10 $3\frac{1}{2}$%

Exercise 21e (page 184)

1. $1 750
2. 8%
3. 4 years
4. $ 160 000
5. $300

Exercise 21f (page 186)

1. (a) $30.75 (b) $48.97 (c) $522.24 (d) $658.92
2. (a) $330.75 (b) $848.97 (c) $6922.24 (d) $5908.92
3. (a) $4410 (b) $3573.05
4. (a) $3600 (b) $3820.32
5. $59 045.71
6. $750
7. (a) $10 500 (b) $11 797.80

Exercise 21g (page 187)

1. (a) 30€ (b) TT$126 (c) 2 400 yen (d) G$2000 (e) N1653 (f) Bds$40
2. (a) US$305 (b) US$11.85 (c) US$11.20 (d) US$75 (e) US$450 (f) US$10 000
3. US$0.16
4. US$0.01
5. 9.07€
6. EC$6750
7. US$41.67
8. Bds$1500
9. £138.89
10. (a) NAf94.50 (b) $33.33
11. rises by 83p
12. fallen by 85% of its 1998 value

Practice Exercise P21.1 (page 00)

1. (a) $4000 (b) $4884.08 (c) $884.08
2.

	Principal	Rate of interest	Repayment period	Interest	Amount
Clive	$400	7.2%	2 yrs	$57.60	$457.60
Rose	$2250	7%	8.5 yrs		$3588.75
Evan	$1450	5.8%	6 yrs	$504.60	$1954.60
Reese	$750	6%	3.5 yrs	$157.50	$907.50

Practice Exercise P21.2 (page 188)

1.

Compound interest	At the end of the 1st year		At the end of the 2nd year		At the end of the 3rd year	
	Interest	amt	Interest	amt	Interest	amt
$700 at 5%	$35	$735	$36.75	$771.75	$38.59	$810.34
$450 at 12%	$54	$504	$60.48	$564.48	$67.74	$632.22

2. (a) $389.27; $49.27 (b) $225.74; $35.74 (c) $1825.05; $625.05
3. (a) $71 (b) $1328.23

Practice Exercise P21.3 (page 188)

1. (a) Bds$300 (b) Ja$9238.50 (c) 1711.5 Pesos (d) UK£82.5
2. (a) US$120 (b) TT$59.60
3. (a) (i) 7200 euros (ii) $10 800 (b) $21 600 (c) $30 800 (d) (i) $30 800 (ii) $38 500
4.

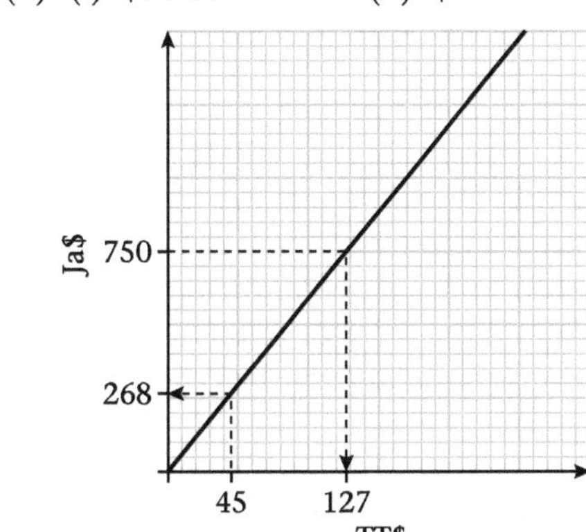

Ja$ (a) 24 (b) 149 (c) 268

5. TT$ (a) 10 (b) 26 (c) 34 (d) 127

6 (a) Bds$ and EC$

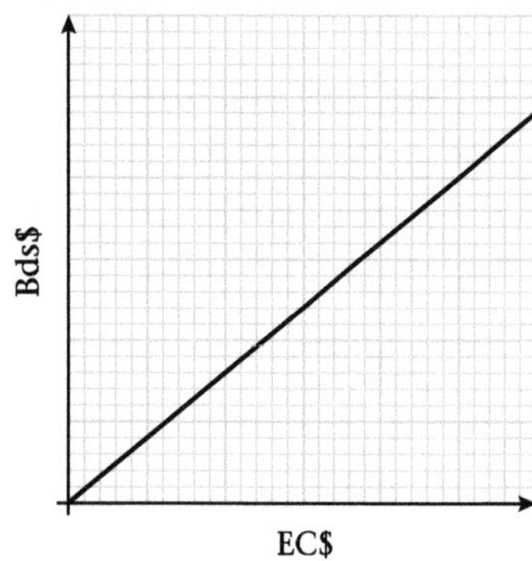

(b) US$ and UK£

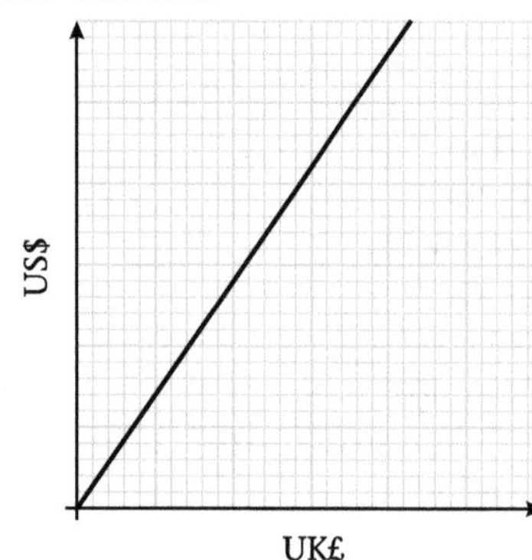

7 $US450 8 EC$780.98 9 Ja$158 090

Exercise 22a (page 191)

2 (a) M is the mid-point of AC
(b) PM is the perpendicular bisector of AC
3 (e) the 3 folds meet at a point
(f) each fold is a perpendicular bisector of one of the sides of △ABC
4 (b) the 3 perpendicular bisectors meet at a point
5 (b) both perpendicular bisectors meet at the centre of the circle
6 (b) a diameter
(c) a square
7 (d) $MN = \frac{1}{2}AC$
8 (c) 7.1 cm

Exercise 22b (page 192)

2 (a) $B\widehat{A}R = C\widehat{A}R$
(b) the bisector of $B\widehat{A}C$
3 (e) the 3 folds meet at a point
4 (d) the 3 bisectors meet at a point
5 (d) in isosceles △XYZ the bisector of $\widehat{Y}$ is the same line as the perpendicular bisector of XZ
6 (e) $\frac{1}{8}$
7 (c) 2.7 cm, 3.3 cm
8 (d) octagon
(e) 57 mm

Exercise 22c (page 194)

3 (b) 117 mm
4 (b) 12.7 cm
5 (b) 7.8 cm
6 (c) AC passes through the centre of the circle, i.e. it is a diameter
7 (b) 5.7 cm
8 (b) 6.9 cm

Exercise 22d (page 196)

4 (d) 5 cm
5 7.9 cm, 13.1 cm
6 (b) 7.3 cm, 9.5 cm
7 (b) 91 mm, 53 mm
8 (b) 7.4 cm

Exercise 22f (page 196)

1 (b) 71 mm
2 (d) 4.7 cm; yes
3 (c) 69 mm
4 (d) yes (e) 2:1
5 (b) yes
6 (c) 8.8 cm
7 (c) each 59 mm
8 (c) 6.6 cm
9 (d) 57 mm
10 (b) 52 mm
11 (b) 5.8 cm
12 (b) 7.9 cm

Practice Exercise P22.1 (page 198)

1 (c) The perpendicular bisectors intersect at a single point
2 5.5 cm
3 Student's drawing
4 Student's drawing
5 A trapezium
6 A trapezium
7 7.5 cm
8 (d) XZ = 8 cm, $Y\widehat{Z}X = 37°$
(e) A right-angled triangle
9 (d) A rhombus
10 (b) The two lines intersect in the middle of the square
(d) The circle touches the square at P, Q, R and S only

Exercise 23a (page 201)

1. (a) 0.34, 0.94
   (b) 0.64, 0.77
   (c) 0.91, 0.42

Exercise 23b (page 202)

1. (a) 5 cm (b) 6.1 m (c) 2.3 km
2. (a) 4.1 m (b) 4.2 km (c) 5.2 cm
3. (a) 4.5 m (b) 1.9 km (c) 3.5 cm
4. 4.3 m 5. 14 cm
6. 9.0 cm 7. 11 cm, 16 cm

Exercise 23c (page 202)

1. 0.829 2. 0.985 3. 0.087
4. 0.755 5. 0.208 6. 0.978
7. 0.276 8. 0.276 9. 0.616
10. 0.616 11. 0.358 12. 0.358
13. 0.688 14. 0.873 15. 0.245
16. 0.942 17. 0.160 18. 0.836
19. 0.710 20. 0.429 21. 0.989
22. 0.862 23. 0.985 24. 0.558
25. 0.578 26. 0.578 27. 0.479
28. 0.479 29. 0.400 30. 0.400

Exercise 23d (page 203)

1. $a = 4.0$ cm $b = 3.0$ cm
   $c = 1.3$ cm $d = 1.5$ cm
   $e = 6.1$ cm $f = 19$ cm
   $g = 5.1$ cm $h = 3.1$ cm
2. AB = 7.3 cm BC = 3.3 cm
3. $x = 3.5$ cm, $y = 13$ cm, $z = 16$ cm
4. BC = 4.2 m, XY = 4.9 cm, PQ = 13 m
5. 3.8 m 6. 45.6° 7. 510 m
8. 4.4 m 9. 6.0 cm

Practice Exercise P23.1 (page 205)

1. 2.2 m
2. 45.3 m
3. 295.0 m
4. AC = 7.0 cm; BC = 13.2 cm
5. (a) The diagonal is 13.0 cm long
   (c) 22.6°
6. Height of triangle = 21.8 cm; $\widehat{QPR} = 24.3°$
7. 0.5 m
8. 48.5°

Revision exercise 9 (page 206)

1. 90 cm² 2. 88 m² 3. 1 628 cm²
4. (a) $7(x - 4)$ (b) $m(5 + 8n)$
   (c) $9b(3a + 4b)$ (d) $7pq(5p - 2q)$
5. (a) $\frac{11k + 15}{12}$ (b) $\frac{17h - 21}{15}$
   (c) $\frac{n + 2}{2b}$ (d) $\frac{1}{3(x + 1)}$
   (e) $\frac{y - 3}{y(y - 1)}$
6. (a) $h(h + 7)$ (b) $m(2m - 1)$
   (c) $(n - 4)(n - 1)$ (d) $8x$
   (e) $(y + 1)(3y + 8)$
7. 700
8. (a) $h = -3$ (b) $x = 1$ (c) $y = \frac{9}{2}$
9. (a) $61\frac{1}{2}$ cm² (b) $115\frac{1}{2}$ cm² (c) 56 cm²
   (d) $38\frac{1}{2}$ cm² (e) 126 cm² (f) $25\frac{3}{4}$ cm²
10. (a) 231 (b) 2566 cm²

Revision test 9 (page 206)

1. C 2. C 3. D 4. B 5. C
6. 176 cm, 2 464 cm² 7. 357 cm²
8. 12.44 cm²
9. (a) $a(5a + b)$ (b) $(a - b)(3x - y)$
10. (a) $\pi r(r + 2h)$ (b) 176

Revision exercise 10 (page 207)

1. 22 years 2. (a) 30 (b) $22\frac{1}{2}$ years
3. 22.63 ($22\frac{19}{30}$) years
4.

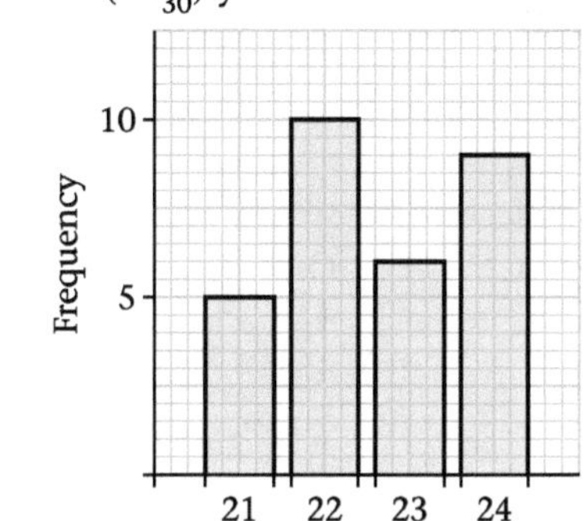

5. 2
6. (a) 23 (b) 22 (c) $22\frac{4}{7}$
7. 6% 8. (a) \$2 475 (b) \$2 613.61
9. (a) US\$28.57 (b) US\$16 666.67
   (c) US\$500 (d) US\$607.14
   (e) US\$360 (f) US\$1825.4
10. \$13.77

Revision test 10 (page 208)

1 A  2 C 3 C 4 D 5 B

6

Profit ($)

500, 400, 300, 200, 100, 0

1 2 3 4 5 6

Week

7 (a) 13 (b) 12.6

8 \$1 546.88 9 \$280 10 \$630.75

Revision exercise 11 (page 208)

1 1.73
2 $x = 39°$, $z = 57\frac{1}{2}°$
3 165 cm
4 parallelogram
5 (b) and (c) 9.4 cm
6 (a) 0.72 (b) 53°
7 (a) 52° (b) 7.09 cm (c) 5.54 cm
8 length 8.66 cm; breadth 5 cm
9 (a) 32° (b) 12.7 cm (c) 25.4 cm
10 11.5 m

Revision test 11 (page 209)

1 C 2 C 3 C 4 B 5 A
6 2.93 cm 7 44°
8 (a) 0.6 (b) $45\frac{1}{2}°$
9 30 m 10 5.9 m

General revision test C (page 209)

1 B 2 C 3 B 4 A 5 C
6 D 7 C 8 C 9 D 10 B
11 7088 cm^2 12 6.5 cm
13 (a) 5 cm (b) 17.6°
15 (a) \$1420
(b) the mode (\$1350) is more representative since 77% of the workforce (i.e. 10 out of 13) receive this wage

16 (a) $x = 3$ (b) $y = 2$ (c) $n = \frac{3}{2}$
(d) $x = 15$
17 (a) \$26 000 (b) \$26 220.50
18 27 m
19 $\widehat{P} = 30°$, PR = 7.3 cm
20 9 m

Practice examination

Paper 1 (page 212)

Section A

1 C 2 B 3 C 4 B 5 C
6 D 7 A 8 B 9 C 10 B
11 B 12 A 13 D 14 D 15 D
16 B 17 C 18 A 19 D 20 D

Section B

21 $\frac{3}{4}, \frac{4}{5}, \frac{17}{20}, \frac{9}{10}$
22 20
23 224_{five}
24 29
25 (a) 0.01 (b) 0.009 2
26 $x + 5$
27 37
28 (a) $5k - 6h$ (b) $x^2 - 8x + 15$
29 (a) 6% (b) \$5618
30 $x = 7$
31 S(0, −3), T(6, −1), U(3, 3)
32 $x = 17$
33 $a = 58$, $b = 122$
34 $\frac{3d + 4}{6}$
35 Bds\$88
36 11
39 \$2 420
40 20 yr, 15 yr

Paper 2 (page 214)

Section A

1 (a) 28 (b) $16\frac{1}{2}$%
2 (a) $\frac{100n}{t}$ (b) $y = -3$
3 (a) 402 (b) eight
4 (a) A(0, 3), B(4, 4), C(5, 0)
(b) (i) C(5, 0) (ii) A(0, 3)
5 1960 m^2 6 818 g
7 3.4 cm

8 (a) {1, 3, 9}
(b) {1, 3, 4, 9, 36}
(c) {2, 6, 12, 18}

9 (a) 35.8 cm (b) 1280 cm^2
No difference

10 (a) $x \geqslant -4$ (b) $\{(x, y): y > 5\}$

11 18°, 81°, 81°

12 (a) 2.4 m (b) 18 m^3

Section B

13 (a) (i) $10 561 (ii) $11 712.40
(b) (i) 2100 (ii) 4.2

14 (b) (i) 9.5 m (ii) 6.7 m (iii) 40°

15 (b) square (c) (i) 4 (ii) 1
(d) A(−2, 0) B(0, 4) C(−4, 6) D(−6, 2)
(e) rotation of 180° about O *or* enlargement of factor −1 with O as centre

16 (a) (i) $h = 2$ (ii) $y = 4$ (b) {2}
(c) (i) $(m - 20)$ g
(ii) $4(m - 20) + m = 330$, $m = 82$

17 (a) (i) 15 (ii) 16 yr (iii) 16 yr
(iv) 15.6 yr
(b) (ii) $1\frac{1}{2}$ km (iii) $7\frac{1}{2}$ km/h

18 (a) $n = 7$ (b) 2 (c) 7 cm
(d) 123 cm^2 (e) 1476 cm^3 (f) 1062 cm^2

Index